D0623740

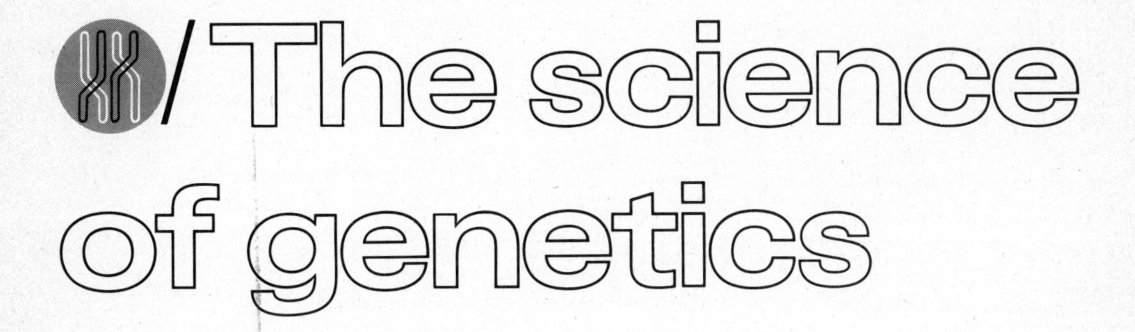

William Hexter / Henry T. Yost, Jr.

Department of Biology, Amherst College
Amherst, Massachusetts

PRENTICE-HALL, INC., ENGLEWOOD CLIFFS, NEW JERSEY

Library of Congress Cataloging in Publication Data

Hexter, William (date)
The science of genetics.

Bibliography: p.
Includes index.
1. Genetics. I. Yost, Henry T. (date)
joint author. II. Title
QH430.H49 575.1 76-2652
ISBN 0-13-794750-X

Figures 17.2, 17.3, 18.3, 18.5, 18.7, 19.7, 20.1, 20.2, 20.3, 20.4, 20.5, 20.7, 21.1, 24.1, 24.8, and Table 20.1 are from H. T. Yost, *Cellular Physiology,* © 1972, by permission of Prentice-Hall, Inc., Englewood Cliffs, New Jersey.

To Rachel and Martha

10 9 8 7 6 5 4 3 2 1

Printed in the United States of America

Prentice-Hall International Inc., *London*
Prentice-Hall of Australia, Pty., Ltd., *Sydney*
Prentice-Hall of Canada, Ltd., *Toronto*
Prentice-Hall of India (Private) Ltd., *New Delhi*
Prentice-Hall of Japan, Inc., *Tokyo*
Prentice-Hall of Southeast Asia, *Singapore*

Contents

Preface

This book is intended to be used as a one-semester introduction to genetics. It is written at a level that should make the material accessible to students with a good high school background in general biology and general chemistry. For students whose high school preparation is inadequate, a good college-level introductory biology course that presents some of the aspects of molecular biology is sufficient.

In writing this book, we have attempted to present the development of genetics as a science. For the most part, we use an historical approach so that the student can become aware of the reasoning that led various scientists to initiate their studies and to show how their methods justify the important conclusions that they drew from their work. To do this, of course, one must be arbitrary. It is not possible to rehearse all of the experimental evidence on which modern genetics is based. We had to select those elements that we felt were most crucial in the development of the structure of genetic theory. That is somewhat easier to do with the material in classical genetics. As one gets closer to the present, it becomes more and more difficult to decide exactly which experiment is the crucial one. In this area we can only hope that we have selected reasonably.

Frequently, we have not reported the first experiment that uncovered an idea but that one which yielded the maximum amount of information from a particular technique. At this level we think that some deviation from historical accuracy is warranted. It is just true that some experiments are better than others, even though they may lead to the same conclusion. To select the poorer example solely for the reason that it was the first report on the subject seems to us to be unsound pedagogically. Our hope is that our readers will see science at its best and will come to understand that, at its best, science is a process of advancing the most reasonable hypothesis consistent with the available data and of then testing that hypothesis through experimental procedure. We hope to persuade the student that science is not just a matter of random observations collected together in some encyclopedic fashion but rather proceeds in a very logical development with experiments that have their origin in previous hypotheses. Our hope, therefore, is that this book will serve a twofold purpose: both to introduce the student to the essential facts of genetics and to instruct the student in the methods of science.

We have consciously deemphasized the introduction of names in the exposition of the material. It is our view that the listing of names tends to interrupt the flow of the text and inevitably leaves the students with extraneous material that they sometimes feel obligated to memorize. We recognize that for a working scientist it is often a great convenience to refer to specific experiments by the names of the authors, but we do not consider this text to be a book for working scientists. It is an introduction to the science of genetics for undergraduates. On the other hand, we recognize the need for the students to be acquainted with the leaders in the field. By including a bibliography at the end of each chapter, we hope to acquaint the students with the reviews and original literature that will help them to make some of the connections between the experimenter and the text. In one way or another, each of the major experiments discussed in a chapter is included in the bibliography at the end of that chapter.

We did not attempt to be comprehensive in making up the bibliography at the end of the chapters; rather, we tried to provide a variety of ways of getting into the literature, ranging from simple and clearly written review articles of a popular nature that should be accessible to anyone in college to some of the more highly specialized material for those students who would like to go beyond the discussion in the chapter itself. We should like to caution the student reading this text that the advanced reviews and papers are often difficult, and we do not think that it is essential for everyone to have mastered that material. It is there for the convenience of the interested few and not to set a standard of achievement for the average student. We hope that by annotating the bibliography we can give direction to the reading, and we advise all students to try to read at least a few of the original papers so that they can develop a sense of the form of good scientific reporting.

At the end of each chapter is included a set of problems. These range from simple definition questions, which serve as a study guide for the chapter, to more difficult problems in which reasoning is required from the material in the text. It is our view that genetics can best be comprehended by working problems. This is particularly true of the classical material, and the student should be aware that solving the problems is an integral part of understanding the text.

Finally, we should like to express our thanks to the people who have helped us in the preparation of this book—in particular to Prof. Carl P. Swanson for his encouragement and critical comment, to those anonymous reviewers who have saved us from error and have provided useful directions toward the organization of the material, and to Martha Yost and Helen Mottla for their assistance in the preparation of the manuscript. Much that is good about this text results from the efforts of these people, and we are indebted to them for their assistance. That which is less than good is our responsibility.

WILLIAM HEXTER
HENRY T. YOST, JR.

"I'll tell you why you're bald and I'm not. It's the genes. I read up on the genes, and, boy, do I know my genes! Take me. I got ***good*** *genes."*

1/Toward an understanding

What do elephants have that no other animals have? Baby elephants.

The answer to this child's riddle is obvious, but when the question is asked, a child more likely than not will give a different answer (trunks is the favorite). Nevertheless, the answer is easily understood by any youngster. He knows that animals have offspring that resemble the parents. Moreover, an observant child also knows that the resemblance is not exact. There is diversity in nature, and although "like begets like," brothers and sisters can be told apart. A child knows that many varieties of dogs share a common ancestry but yet are distinguishable from each other.

Genetics is the science that deals with the underlying causes of these resemblances and differences, and what is common knowledge to any child today was known by man for as long as recorded history. Man with his usual curiosity has moved toward an understanding of why it is that like begets like but not exactly. The early study of genetics was based largely on characters that were easily classified into nonoverlapping groups. But genetics also provides an answer for overlapping groups when the classification may be ambiguous.

An historical inquiry of man's attempt to unravel the laws of genetics begins, as is often the case, with the ancient Greeks. There is also written evidence from the Talmud that nearly two thousand years ago the Hebrews worked out the inheritance of hemophilia, the bleeder's disease. These early beginnings represented shrewd deductions based on pragmatic observation, but they failed to provide any understanding of the laws of heredity. For this reason, they are not regarded as the beginnings of our understanding of heredity. For that our attention turns to the continent of Europe, particularly Germany, in the latter part of the nineteenth century, where many investigators were attempting to unravel the laws of inheritance.

The Idioplasm of Nägeli

One of the most influential of the German scientists in the latter part of the nineteenth century was the Swiss-born botanist Carl Nägeli. Nägeli was interested in the laws of inheritance, and in 1884 he proposed a theory that had great influence on other workers in the field. He conceived of the organism as composed fundamentally of two living substances or plasms, one constituting the main bulk of the protoplasm as a vegetative or nutritive *trophoplasm*, in which is carried on the main operations of nutrition and metabolism. The other, present in much smaller quantity, is a generative *idioplasm* that plays a leading role in reproduction and development and constitutes the physical basis of heredity. Nägeli made no attempt to identify the idioplasm with any morphological component of the cell. Other experimenters, following closely on Nägeli, were to provide that answer.

Fertilization in the Sea Urchin

In 1875 Oskar Hertwig was watching fertilization in the sea urchin and noticed that only one sperm entered an egg. This was the first precise description of fertilization, and it is interesting to note that it was only one hundred years ago that this fundamental fact was established. Hertwig made some crosses between two different species of sea urchins, Species A and Species B. In the first cross he used eggs from Species A females and fertilized them with sperm from Species B males. In a second experiment Hertwig took eggs from Species B females and fertilized them with sperm from Species A males. Hertwig noticed that the two hybrids obtained in these separate experiments were similar in appearance. He also noticed the difference in the amount of cytoplasm involved in the two crosses. He knew that the spermatozoon contained almost no cytoplasm, in contrast with the egg, which contained a large amount of cytoplasm.

Hertwig saw that Cross 1 involved cytoplasm and nucleus from Species A but only nucleus from Species B. In Cross 2 there were cytoplasm and nucleus from Species B but only nucleus from Species A. Since the results from the two crosses were the same, Hertwig reasoned that the cytoplasm was relatively unimportant in determining the appearance of offspring, and he postulated that Nägeli's idioplasm was located in the nucleus. At about the same time, the German botanist Strasburger performed similar experiments with plants and obtained results comparable to those of Hertwig. He also concluded that the idioplasm resided in the nucleus of the cell.

In 1884 Van Beneden was studying fertilization of the egg in the parasitic roundworm Ascaris, and he was particularly interested in what happened to the egg and sperm nuclei immediately after fertilization occurred. Van Beneden noticed that the chromosomes of the offspring were derived in equal numbers from the nuclei of the two conjugating germ cells and therefore were derived equally from the two parents. Although this conclusion by itself is not conclusive in localizing the idioplasm, it provided the stimulus for several investigators to create theories of inheritance based on the contribution and importance of the nucleus.

Boveri's Experiments in Sea Urchins

Although the experiments of O. Hertwig, Strasburger, and Van Beneden were sufficient to answer the question of where to find the idioplasm, additional evidence was provided by a series of elegant experiments by Theodor Boveri.

Boveri was a brilliant experimenter. His experiments in the laboratory might have appeared crude or pointless at the time, but they seemed to work for Boveri, perhaps because *he* knew exactly what he was doing or, more precisely, he knew what questions to ask. It has been said that if you ask the cell the right question, you will get the right answer. Boveri was a master of this technique.

Enucleated Eggs

Boveri noticed that if he took sea urchin eggs and shook them up in a test tube, he was able to fragment the eggs. Some of the fragments contained no nuclei and are referred to as *enucleated fragments.* Boveri then reasoned that if he obtained these fragments from one species of sea urchin and fertilized them by sperm from a different species, he might be able to obtain a hybrid whose characteristics would reveal the relative importance of nucleus and cytoplasm. Ordinarily, larvae obtained from such interspecific crosses would show morphological characteristics intermediate between the two species. Boveri performed this experiment and in a few cases obtained hybrids that showed larval features characteristic of the paternal species only. Many years later in 1918, in a posthumous paper, Boveri points out that certain experimental errors were possible in the foregoing work. It was difficult to be certain that the egg fragment had no egg nuclear material remaining even though it may not be visible. Thus the demonstration of the importance of the nucleus over the cytoplasm from this experiment is suggestive but not conclusive.

Monaster Eggs

During cell division in animal cells, many structures are visible in the dividing cell. Most prominent is a spindle that has starlike structures called *asters* at each end. Normally there are two asters when a cell divides, one at each pole in the cell. (Additional details of cell division will be found in Chapter 7). In a second experiment Boveri, by various devices both mechanical and chemical, was able to interfere with the formation and function of one of the two asters. By doing this, Boveri kept the cell from dividing, even though the chromosomes had already divided. The result was an egg with twice the normal chromosome number. Boveri knew from earlier studies that there is a close correlation between the size of the nucleus and the number of chromosomes it contains. In the egg just described, referred to as a *monaster egg*, the nuclear size is approximately twice that found in a normal egg because it has twice the number of chromosomes.

The hybrid obtained from fertilizing a normal egg from one species by sperm from a different species shows, as expected, larval characteristics of both species. A hybrid obtained by fertilizing a monaster from one species by sperm from a different species showed more characteristics of the maternal species. Since the cytoplasmic contribution to the two eggs was nearly identical, the conclusion is inescapable that the different nuclear contribution was responsible for the different larval results.

Giant Eggs

Boveri had observed that *giant eggs* sometimes arise spontaneously in the sea urchin. These eggs contain twice as much nuclear material and also twice as much cytoplasm as a normal egg. When Boveri fertilized these giant eggs from one species by sperm from a different species, he noted that the hybrid was *matroclinous*—that is, resembled more the mother. As in the monaster experiment, the nucleus in giant eggs had twice the chromosomal complement of normal eggs, so a matroclinous hybrid was not unexpected. The hybrid of the giant egg experiment was, however, identical to that of the monaster experiment even though the giant egg had approximately twice as much cytoplasm. This observation reinforced in Boveri's mind the idea that it was the nucleus that plays the more important role in heredity.

Fragmented Eggs

In another experiment involving interspecific crosses and artificially fragmented eggs, Boveri fertilized a fragment that had a nucleus but that had lost some of its cytoplasm. He reasoned that if the cytoplasm were important, the hybrid resulting from the *fragmented egg* fertilization would be different from the normal fertilization, probably resembling more the paternal species. But, in fact, the hybrid obtained from this experiment was identical to that obtained from normal egg fertilization. The loss of egg cytoplasm apparently made no difference in the appearance of the hybrid.

Convincing as these experiments may seem, some skeptics felt that Boveri was relying on unusual interspecific crosses involving bizarre eggs to discriminate between nuclear and cytoplasmic contributions. He showed, however, that it is possible to distinguish between the roles of the nucleus and of the cytoplasm in a single larva.

Delayed Nuclear Fusion

Boveri had demonstrated that it was possible to paralyze the sperm temporarily by treating it with various chemicals. When such a sperm enters the egg, the resulting sequence of events may be quite different from normal fertilization. The egg nucleus divides normally, but the sperm nucleus does not, nor does it fuse with the egg nucleus. The result is a two-cell stage in which each of the two cells has a daughter nucleus resulting from the division of the female nucleus but only one of the two cells has a male nucleus. By carefully controlling the original treatment of the sperm, the paralysis will last only to this two-cell stage. Belatedly the male and female pronuclei may now fuse, giving a two-cell stage, one-half of which has a normal fusion nucleus and the other half only a female pronucleus. Boveri observed that such a two-cell stage could continue development and form a larva.

The larva obtained from this and other experiments showed that half the larva exhibited typical hybrid characteristics but that the other half possessed only maternal features. This latter half also had nuclei one-half the size of the hybrid nuclei, an observation consistent with the notion that the maternal half of the larva has only half the chromosome number of the hybrid half.

The results of all these experiments compel us to conclude that the nucleus is responsible for heredity and that the cytoplasm plays little if any role. The idioplasm of Nägeli is located in the nucleus!

REFERENCES

VOELLER, B. R. 1968. *The Chromosome Theory of Inheritance.* Appleton-Century-Crofts, New York. (A good collection of papers, including the important features of the investigations of Hertwig, Strasburger, and Van Beneden. In English.)

WILSON, E. B. 1925. *The Cell in Development and Heredity.* 3rd ed. Macmillan, New York. (Still a classic after 50 years. For any historical information concerning the relationship of genetics and cytology, this is the book to consult.)

QUESTIONS AND PROBLEMS

1-1. Define the following terms:

idioplasm	monaster egg
trophoplasm	giant egg

1-2. What evidence suggests that the nucleus, rather than the cytoplasm, is the site of the hereditary components?

1-3. If the cytoplasm were responsible for heredity, would a child tend to look more like his father or his mother? Why?

1-4. Is an understanding of the exact nature of fertilization important for an understanding of heredity? Why or why not?

1-5. What does the study of interspecific hybrids in the sea urchin contribute to the location of the idioplasm?

1-6. On the line scale below, indicate the position of the developed phenotypes of the following items.

(a) Egg fragments without nucleus of species I × sperm of II
(b) Normal egg of II × sperm of I
(c) Giant egg of II × sperm of I
(d) Nucleated fragment of II × sperm of I

Species I — I♀ × II♂ — Species II

2/The monohybrid cross

Gregor Mendel

The significant contributions of the German school to the understanding of fertilization and chromosome behavior discussed in Chapter 1 were unknown in the mid-nineteenth century, a fact that makes what follows an amazing example of abstract scientific reasoning. For it was at that time that a monk from Brünn, Austria (now Brno, Czechoslovakia), Gregor Mendel, was performing the experiments that rightfully have made him the father of genetics. Born Johann Mendel in 1822, son of poor parents who worked as farmers, he came to the monastery at Brünn and was ordained a priest in 1848. In 1851 he was sent by his order to study natural science at the University of Vienna. University records showed Mendel to be an average student, but several teachers commented on his clear and logical mind. He returned to Brünn as a teacher of science in 1853, and in 1856 he began to collect and observe the numerous varieties of the garden pea that seed salesmen offered for sale. These varieties differed in seed, pod, flower, and other characteristics, and they seemed to Mendel to provide suitable material for answering a simple but important question that no botanist up to then had even formulated clearly, much less obtained an answer. After observation and experiments carried out in the monastery gardens for seven years, he obtained the answer he had sought and presented the results of his hybridization experiments, together with the generalizations that we now know as Mendel's laws, at two meetings of the Natural History Society of Brünn in the spring of 1865. The results and the theory were printed in the Annual Proceedings of the Society, which appeared and were distributed to libraries in Europe and America in 1866. It is safe to say that no one who heard Mendel's paper or who read it in the nineteenth century appreciated its significance, for it lay neglected until 1900 when it was discovered independently by Correns in Germany and de Vries in Holland.

Meanwhile Mendel turned to experiments with other plants and with bees and also made meteorological observations, but gradually he became more and more concerned with the administrative work of the monastery, of which he had become Abbot in 1868. The last years of his life were embittered by struggles on behalf of the monastery against the power of the state and, one may imagine, by the frustration of the scientific mind, which had been unable to convince or even interest his contemporaries. He died in 1884, long before his scientific work was recognized.

Why Mendel Succeeded

Mendel was not alone in trying to discover the laws of heredity, and it is interesting to ask why he succeeded, clearly before his time, when so many others had failed. Several aspects of his technique led to his success. First, where previous investigators had often made observations on the plant or animal as a whole organism, Mendel confined his attention to a single character at a time. Secondly, he counted all the progeny that resulted from the cross, thus reducing the phenomenon of inheritance to a measurable quantitative basis. And, thirdly, he kept accurate pedigree records of the members of successive generations so that he could trace the ancestry of any given plant back to the beginning of his experiments.

Mendel's Experimental Organism

Mendel was also perceptive in his choice of material. The garden pea, *Pisum sativum*, proved to be very satisfactory material for experiments on hybridization. Because of the floral structure, contamination by foreign pollen, either wind blown or insect borne, is rare. Self-fertilization is the ordinary occurrence, but artificial cross-fertilization, although laborious, was simple to perform. Mendel had only to open a flower bud and remove the stamens before any pollen had been shed in order to prevent self-pollination. He could then dust the stigma with foreign pollen to produce the fertilization he desired. In many instances, Mendel desired self-fertilization; and since this process is normal for Pisum, the crosses were effected most efficiently. Mendel grew some plants in a greenhouse as a control against cross pollination by insects, and he concluded that contamination was minimal and had no influence on the overall result.

Before analyzing his experiments in detail, it is worth noting that many people consider Mendel a lucky scientist. His detractors claim that he was a bumbling amateur who was blessed with good fortune. Although it may be true that some serendipity was involved in his experiments, a careful reading of his paper makes it clear that he knew what he was doing. Indeed, some think that he knew before he performed the experiments what the answer would be and that he carried them out so as to verify his preexisting prejudice. In any case, Mendel's work stands as a paradigm of scientific method.

Mendel's Experiments and Results

As noted, Mendel obtained seeds from the many seed salesmen who traveled from town to town to sell their products. Clearly indicated on these packets of seeds, as a rule, were their various characteristics—color of the flower, form of the seed, height of the plant, and so on. But Mendel was unwilling to accept even these clear statements of the nature of the product he bought. Every variety of seed used in his experiments was grown for a generation or two to verify that the characteristics were as advertised—that the strains were *true-breeding*. He did so with over 30 different varieties of seeds and for various reasons reduced this number to seven different varieties or characters with which to do his experiments.

Mendel first crossed two varieties of pea plants that differed in the color of the flower. One had white flowers and the second violet-red flowers (designated red). He used as parents, what is now known as the *P generation*, females from the red-flowered variety and males from the white-flowered strain. Mendel also made a *reciprocal cross* in all his experiments, again indicating his great care. In this cross the characters are the same but the sexes of the parents are reversed. Thus it means that he also crossed a female from the white-flowered variety by pollen from the red-flowered variety. Mendel then observed the flower color of the hybrid generation, referred to as the first filial generation, or F_1, and noticed that all the flowers were uniformly red and indistinguishable from the red parent. He also noted the same result in the reciprocal cross. In all Mendel's experiments no difference was noted in the reciprocal cross, indicating to Mendel that in his experiments the original sex of the parents played no role in determining the final results. In subsequent examples the sex of the two parents will not be given unless it makes a difference in the result.

The 3 : 1 Ratio

Mendel allowed the F_1 plants to fertilize themselves to produce a second filial, or F_2, generation. He observed in the F_2 many red-flowered plants and also some white-flowered plants. Because Mendel had the insight to count the number of plants, he noted that approximately three-quarters of the F_2 plants had red flowers and one-quarter had white flowers.

In a second experiment Mendel used two varieties of peas in which the form of the ripe seed differed. In one variety the seed was round; in the second variety the seed was angular or wrinkled, much like a raisin. For this experiment, Mendel crossed the round-seed variety by the angular-seed variety. The seeds of the F_1 were all round. He allowed the F_1 plants to self-fertilize and then noted in the F_2 that, of the 7324 seeds formed, 5474 were round and 1850 were angular—a ratio of 2.96 to 1. It is interesting to look at the results of individual plants in this experiment. (See Table 2-1.)

Table 2-1. Distribution of Seed Shapes by Individual Plants

	Number of Seeds Formed		
PLANT NUMBER	ROUND SEEDS	ANGULAR SEEDS	TOTAL
1	45	12	57
2	27	8	35
3	24	7	31
4	19	10	29
5	32	11	43
6	26	6	32
7	88	24	112
8	22	10	32
9	28	6	34
10	25	7	32
	336	101	437

AFTER MENDEL.

The data from these ten plants illustrate statistical variation, a fact that may or may not have impressed Mendel. Notice, for example, that the ratio varies from almost 5 to 1 (plant 9) to less than 2 to 1 (plant 4) but that the total, 336 to 101, roughly approximates 3 to 1. In fact, an analysis of all Mendel's data by the British biometrician R. A. Fisher indicates that Mendel's results were "too good," thus raising some interesting questions as to why this is the case. The reader may wish to speculate on some of the possible ways that an experimenter could alter his data either consciously or unconsciously so that the final results fit expectation too well.

In a third experiment involving a single characteristic, Mendel used two varieties that differed in the color of the cotyledon, the part of the plant in the seed that provides nutrition to the embryo before it is able to photosynthesize for itself. One strain had yellow cotyledons and the second green. Mendel crossed the yellow variety by the green variety and observed that the F_1 all had yellow cotyledons. He selfed the F_1 plants and observed in the F_2 that three-fourths (6022) of the plants had yellow cotyledons and one-fourth (2001) were green. Mendel then selfed the individual F_2 plants to raise an F_3. All the green F_2 plants, when self-fertilized, gave F_3 plants that had only green cotyledons. However, the yellow F_2 plants sorted themselves into two groups when self-fertilized. One-third gave F_3 plants with only yellow cotyledons, but two-thirds gave progeny of which three-fourths had yellow cotyledons and one-fourth had green cotyledons.

Mendel's Interpretation—Law of Segregation

Mendel concluded that the F_1 hybrid contained a factor from *each* parent. He designated **A** for yellow cotyledon and **a** for green cotyledon. When the F_1 plant forms gametes, these two factors separate from each other. This phenomenon of separation has become known as *Mendel's First Law*, the *Law of Segregation.* That is, there are factors that affect development of color in the cotyledon, and these factors retain their individuality from generation to generation and do not become contaminated when mixed in a hybrid. Such a formulation has also been referred to as the purity of gametes. Figure 2-1 illustrates Mendel's interpretation of the experiment involving cotyledon color. The hereditary elements are carried in pairs; and when gametes are produced, each gamete carries only one element.

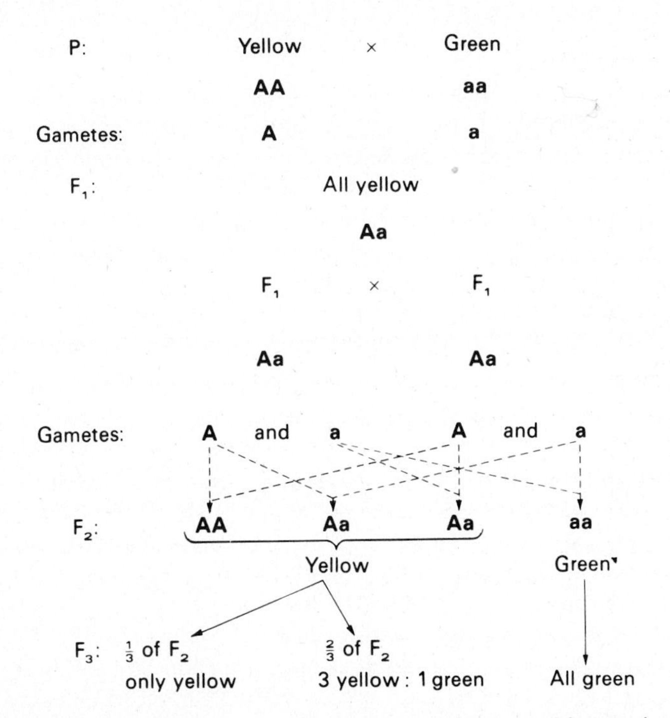

Figure 2-1. Diagrammatic representation of Mendel's Law of Segregation.

In the three examples of Mendel's *monohybrid cross*, a cross in which only one character is followed, the F_1 always had the same appearance as one of the two parents. Mendel defined this result as *dominance.* When the appearance of the F_1 is similar to the appearance of one of the parents, we say that the factor responsible for that character is *dominant*, whereas the factor for the appearance of the other parent, which is not expressed in the F_1, is *recessive*. Thus red flower is dominant over white, round seed is dominant over angular, and yellow cotyledon is dominant over green.

In 1900, when Mendel's paper was discovered by de Vries and Correns, the scientific world was ready for the full impact of his results. One particular individual who was to play an influential role in promoting Mendelism was the English zoologist William Bateson. Bateson was traveling from Cambridge to London to give a lecture when he read Mendel's work for the first time, immediately understood its significance, and changed his lecture to incorporate Mendel's experiments. In science, when an important breakthrough seems to have been made, many scientists will attempt to repeat the work, often using other organisms, to see whether or not the results and interpretations are generally applicable. This situation occurred early in the twentieth century when other breeders attempted to verify Mendel's laws.

Intermediate Inheritance

The first example to be described involved *Antirrhinum majus*, the snapdragon. Typical of many early Mendelian experiments, this one involved the color of the flowers. Two varieties were available, one of which had red flowers, the other white. The F_1 that resulted from the cross of these two plants had neither red nor white flowers but pink-flowered snapdragons. This result was not unusual. As a matter of fact, many flower crosses give this intermediate appearance in the F_1, and it was such results that led to the many theories of blending inheritance that were prevalent in the nineteenth century and earlier. But now, with Mendel's experiments as a model, experimenters self-fertilized the F_1 pink plants and the F_2 showed red-flowered, pink-flowered, and white-flowered plants. Using Mendel's techniques of counting the progeny, they noted that these three colors occurred in a particular ratio. In the F_2 there were 54 red-flowered, 122 pink-flowered, and 58 white-flowered plants, a ratio of 1 : 2 : 1. Recall that this ratio was typical in Mendel's results. The 3 : 1 he observed was due to dominance, but in the snapdragon cross no dominance is present. *Intermediate* inheritance is the designation given to crosses where the F_1 resembles neither parent but is intermediate in appearance.

A second example can be shown in chickens. There are many varieties of chickens with respect to feather pigmentation. In Bateson's laboratory black-feathered chickens were crossed by white-feathered chickens of the Andalusian strain. The F_1 had pigmentation about intermediate between the parental black and the parental white, giving a slate-blue appearance. When F_1 blue was crossed by F_1 blue, the F_2 showed one black to two blue to one white, indicating a typical Mendelian monohybrid ratio. Carrying the experiment one generation further, if the black F_2s were mated by black F_2s, all the F_3 were black. If the white F_2s were mated by white F_2s, all the F_3 were white; but if the F_2 blue was crossed with another F_2 blue, the results in the F_3 were black, blue, and white in a ratio of 1 : 2 : 1.

These F_1 fowl, known as Andalusian blues, often caught the fancy of the chicken breeder, and he wished to obtain a pure-breeding strain of blues; but whenever he crossed them, he was frustrated by the constant segregation of blacks and whites. The reader might see if he can design a method of obtaining a pure-breeding strain of Andalusian blues. All the necessary information may not be available now, but in reading further in this book, you will acquire sufficient facts for the design of a genetic scheme for maintaining pure-breeding Andalusian blues. It should be noted that this is a theoretical problem. As yet no pure-breeding Andalusian blue strains exist.

Codominance

Later in the twentieth century experiments with the ladybird beetle, *Harmonia axyridis*, demonstrated another way factors can behave in development in addition to dominance and intermediate inheritance. Many known varieties differ in the pigmentation pattern found on the elytra, or wing covers. Some have black dots on a yellow background, whereas others have broad black bands on a yellow backcover. When these varieties were crossed, the F_1 had patterns that showed the characteristics of *both* parental strains (see Figure 2-2).

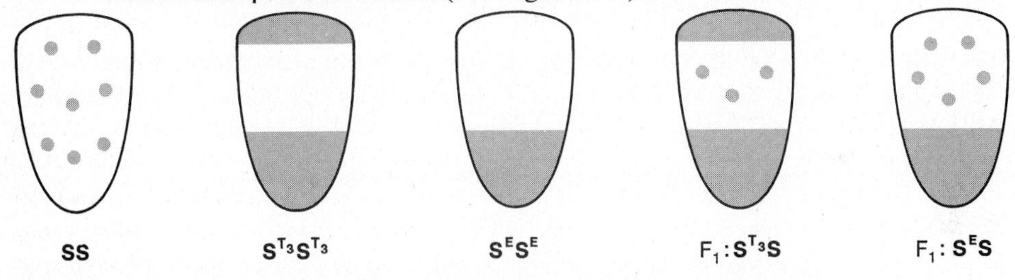

Figure 2-2. Diagram of left elytra of three homozygous strains of *Harmonia axyridis* and the F_1 formed from crossing the strain SS by the strains $S^{T_2}S^{T_3}$ and S^ES^E. The F_1 shows the banded pattern of one parent *and* the dotted pattern of the other where the absence of bands permits the dots to show. (After Tan. 1946. *Genetics*, **31**.)

Neither pattern was dominant, for the F_1 did *not* resemble either parent. Nor can the F_1 be described as intermediate in the same sense that a pink F_1 snapdragon was intermediate between the red-flowered and white-flowered parents. This example in *Harmonia* demonstrates *codominance* because the characteristics of both parents occur simultaneously in the F_1. Many of the blood groups in man show codominant inheritance.

It was not known in the early years of the twentieth century why some crosses showed dominance, others codominance, and still others intermediate inheritance. Undoubtedly the way the factors act during development of the organism was involved, and the mode of this action will become a paramount question some years later. But at this point in history interest centered mainly on the *transmission* of characters, and in all the examples stated the transmission is exactly the same. What is known as Mendelian monohybrid inheritance gives a 1 : 2 : 1 genetic ratio in the F_2.

Segregation of the Gametes

One of the striking features of Mendel's First Law is his hypothesis that an F_1 produces two kinds of gametes in equal frequency (segregation of the gametes). Mendel deduced it from the F_2 results that he had obtained. Is it possible to test this hypothesis directly? Can a population of gametes be examined to see whether approximately one-half are of one kind and one-half the other kind? This experiment was tried in 1921 in the rice plant.

Certain rice varieties known as *starchy* have starch grains that, when stained with iodine, color intense blue, whereas another variety known as *glutinous* lacks starch and stains red or iodine color. It was learned that this difference between starchy and glutinous is due to a single factor, with the factor for starchy dominant. A cross between starchy and glutinous gives the Mendelian result expected from a single factor cross. The interesting question is: Does the F_1 plant resulting from the cross of starchy by glutinous produce, as Mendel proposed, two kinds of gametes in equal frequency? This question can be tested by staining with iodine the pollen grains produced by an F_1 plant. The prediction based on Mendelian principles is that half the pollen grains will stain blue and half red. When this experiment was performed, about 52 % of the pollen grains stained blue and 48 % stained red. (See Figure 2-3.) The rice experiments provide interesting and independent "proof" of Mendel's First Law. The reader might keep in mind, as other genetic principles are demonstrated, how scientists "prove" hypotheses and what the value is of repeating experiments and of designing new experiments to test scientific hypotheses.

Genetic Terminology

It is now appropriate to introduce terminology that more adequately describes the principles of this new science of heredity. It has perhaps been apparent that the language used up to now has been cumbersome. The new terms will simplify communication.

To replace Mendel's "factor," the Danish botanist Johannsen introduced the word *gene* in 1909. At the same time he coined the term *genotype*, which refers to the genetic constitution or hereditary makeup of an organism. **AA**, **Aa**, **aa** are all examples of genotypes. Johannsen also introduced the term *phenotype*, which refers to the appearance of the organism. Red flower, green cotyledon, angular seed are all phenotypes. William Bateson in 1902 proposed the term *homozygous*, which refers to the situation in which both genes in an individual are the same, **AA**, **SS**, **aa**. The term *heterozygous*, from *hetero*, meaning different, refers to the condition of the

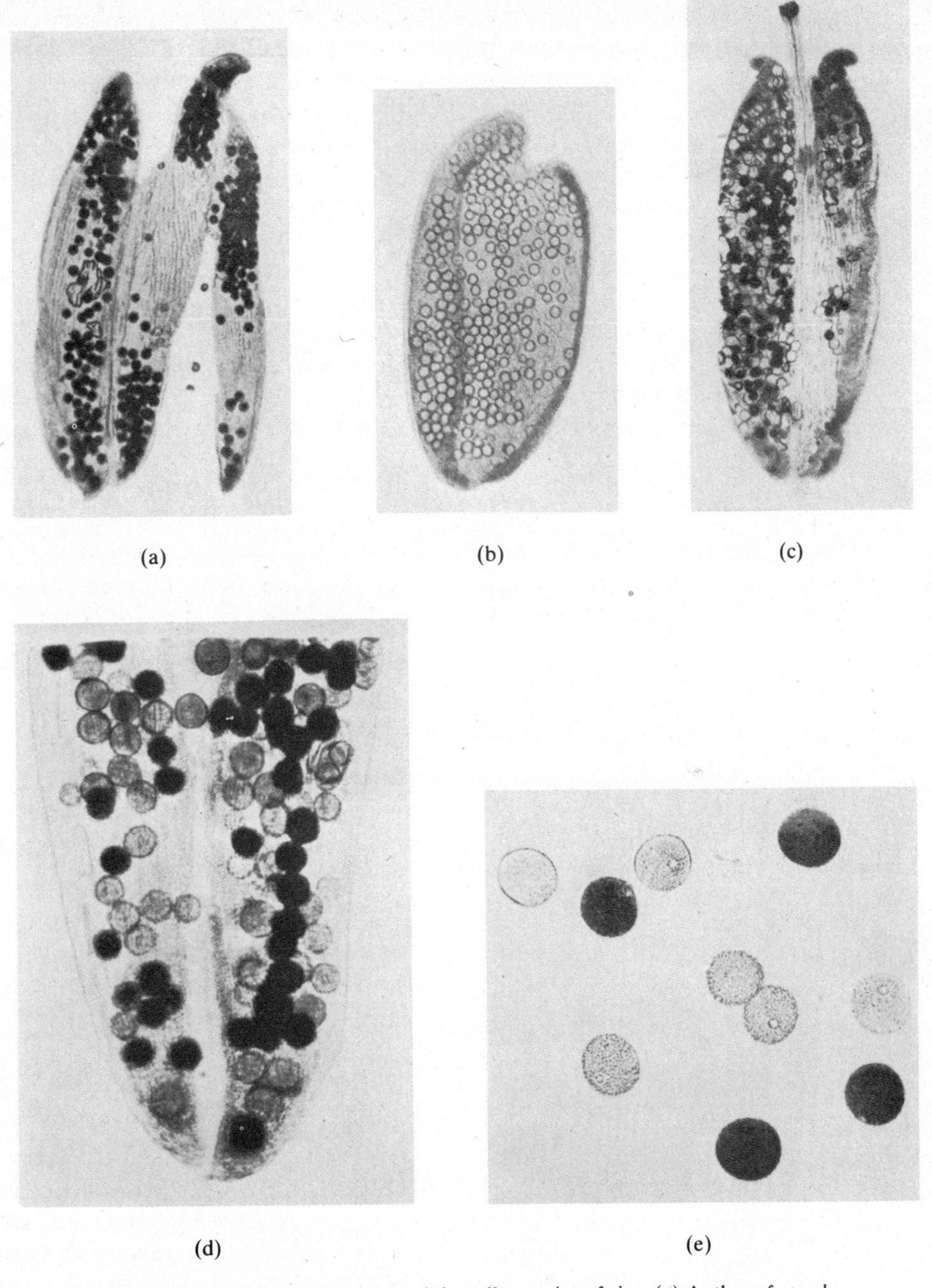

Figure 2-3. Iodine reaction of starch in pollen grains of rice. (a) Anther of *starchy* type, pollen all dark. (b) Anther of *glutinous* type, pollen all light. (c) Anther of F_1, pollen mixture of dark and light. (d) Part of F_1 anther more highly magnified. (e) Free pollen of F_2 showing two types. (After Parnell, F. R. 1921. *J. Genetics* **11**.)

genes being unlike in one individual, **Aa**, **Ss**. Johannsen also introduced the term *allele* (originally allelomorph), from the Greek, meaning "other." An allele is one of the two or more forms of a gene. For example, **A** and **a** are both alleles of the **A** gene. The term allele is sometimes confusing. More will be said about alleles later, when more precise and useful definitions will be given. Finally, the term *genetics* was proposed by Bateson in 1906 to describe the new science of heredity and variation that was rapidly developing.

REFERENCES

BATESON, W., and R. C. PUNNETT. 1905. Experimental studies in the physiology of heredity. *Rep. Evol. Comm. Roy. Soc.* **2**:99–131. (One of several important reports issued by Bateson's laboratory giving the details of his early genetic experiments. This one describes the Andalusian chicken experiment.)

DUNN, L. C. 1965. *A Short History of Genetics.* 261 pp. McGraw-Hill, New York. (A good account of the history of genetics.)

MENDEL, G. 1866. Versuche über Pflanzen-Hybriden. *Verhandlungen des naturforschenden Vereines in Brünn,* **4**(1865)(Abb.):3–47. (The classic. Worth looking at if you can read German.)

RUGER, R., A. MICHAELIS, and M. M. GREEN. 1968. *A Glossary of Genetics and Cytogenetics.* 507 pp. Springer-Verlag, New York. (The best and most comprehensive dictionary of genetic terms.)

STERN, C., and E. SHERWOOD (Eds.). 1966. *The Origin of Genetics. A Mendel Source Book.* 179 pp. W. H. Freeman and Co., San Francisco. (Mendel's paper in translation and his letters to Nägeli. Also the papers of Correns, de Vries, and Fisher, plus some other interesting articles. A valuable book for an understanding of Mendel and his place in history.)

STURTEVANT, A. H. 1965. *A History of Genetics.* 165 pp. Harper and Row, New York. (Like Dunn, a good account of the history of genetics.)

QUESTIONS AND PROBLEMS

2-1. Define the following terms:

dominant
recessive
intermediate
gene
allele
genotype
phenotype
homozygous
heterozygous
true-breeding
codominance
reciprocal cross
P generation
F_1
F_2
monohybrid

2-2. Why was the pea plant a good organism for Mendel to choose for his experiments?

2-3. Why was Mendel successful where others before him were not?

2-4. What is meant by the Law of Segregation?

2-5. In the fruit fly Drosophila, a female with sepia eyes was crossed to a male with red eyes. All the F_1 had red eyes. In the F_2 there were 224 red-eyed flies and 72 sepia-eyed flies.

(a) How does sepia eye appear to be inherited?
(b) What are the genotypes of the parents?
(c) What gametes are produced by the parents?
(d) What is the genotype of the F_1?
(e) What gametes do the F_1 produce?
(f) What is the genotypic ratio in the F_2?

2-6. If an F_2 sepia-eyed female is mated with an F_1 (male from the above cross), what would be the phenotypic ratio of the progeny? What would be the genotypic ratio of the progeny?

2-7. When a white guinea pig is crossed to a yellow guinea pig, all the progeny are cream-colored. From a cross of two such F_1s, repeatedly made, the F_2 showed 32 white, 66 cream, and 30 yellow guinea pigs. How is this trait inherited? Give the genotypes of the parents, F_1 and F_2 progeny.

2-8. In chickens, pea comb is dominant to single comb. If a poultry breeder wanted to be sure that all his pea-combed chickens are true-breeders for the comb characteristic, how would he do it?

2-9. Some people are able to taste the chemical phenylthiocarbamide (PTC), whereas others cannot. From marriages of two tasters, the children are both tasters and nontasters. From the marriages of tasters to nontasters, the children are both tasters and nontasters. From the marriages of two nontasters, all the children are nontasters. What type of inheritance is involved?

2-10. If it were possible to do experimental matings with humans, what crosses would you perform to test your inheritance hypotheses in Problem 2-9? Indicate why you would make the crosses and what your hypothesis predicts for the results of the crosses.

2-11. A guinea pig breeder starts with three guinea pigs, a male with a smooth coat and two females, each with rough coats. When he mated the male to female 1, all the progeny had rough coats. When he mated the male to female 2, half the progeny had rough coats and half had smooth coats. What type of inheritance is involved?

2-12. Give the genotypes of the male, female 1, and female 2.

2-13. The breeder would like to establish true-breeding strains with smooth coats and true-breeding strains with rough coats. How can he do so?

2-14. In the plant the four-o-clock, there are varieties with white flowers and other varieties with red flowers. From a cross of these two varieties, all the F_1 had pink flowers. In the F_2 there were 142 pinks, 68 whites, and 73 reds. How is flower color inherited?

2-15. Give the genotypes and phenotypes for the progeny from the following crosses in the four-o-clock.

(a) white × white
(b) white × pink
(c) red × red
(d) red × pink
(e) pink × pink
(f) red × white

2-16. In corn, starchy pollen grains stain blue with iodine, whereas "waxy" pollen grains stain red. If starchy is crossed to waxy and an F_2 is raised, what would an analysis of the pollen grains produced by the F_2 plants reveal?

3/The dihybrid cross

Mendel not only performed a monohybrid cross involving pea plants that differed by a single pair of genes, but he also asked what would happen if he crossed varieties of peas that differed from each other by two pairs of genes, designated a *dihybrid* cross.

9 : 3 : 3 : 1

For this experiment, Mendel crossed as one parent a plant that had round seed and yellow cotyledon by another that had angular seed and green cotyledon. He verified that these plants were pure-breeding parental strains, and he did reciprocal crosses. Again, it seemed to make no difference which variety he used as the female parent and which as the male parent. Mendel predicted that the F_1 from this cross would be round and yellow, for in the earlier monohybrid cross round proved to be dominant to angular and yellow proved to be dominant to green. There was no reason to suppose that this situation would change. In the F_2 of the earlier experiments he had obtained three round to one angular in one experiment and three yellow to one green in the other. But he was not sure a priori what the result would be in the F_2 when he tested both characters at the same time. Would there be a tendency for the genes to stick together or not? When Mendel selfed the round yellow F_1 he obtained the following F_2 phenotypes: round yellow, round green, angular yellow, and angular green. This kind of result, prior to Mendel, might well have defied analysis; but because Mendel counted the progeny, he was able to reduce the results to a specific ratio. It turned out that $\frac{9}{16}$ of the F_2 were round yellow, $\frac{3}{16}$ were round green, $\frac{3}{16}$ were angular yellow, and $\frac{1}{16}$ were angular green. Mendel pondered why the 9 : 3 : 3 : 1 ratio rather than some other ratio and proposed the following explanation. Again, for simplicity, let us turn to the F_1 that resulted from the cross of the round yellow by the angular green parent. Since Mendel deduced from the monohybrid experiments that genes occur in pairs, it is probable, he assumed, that the round yellow parent was homozygous, **AABB**, and that the angular green parent was homozygous **aabb**. Furthermore, it seemed reasonable to suppose that the F_1 was a double heterozygote **AaBb**; and since **A** and **B** are dominant, as shown in the monohybrid experiments, such an F_1 would be round yellow. But now the critical question was: What kind of gametes will the round yellow F_1 plant produce? Here Mendel predicted that the two pairs of genes would behave independently of each other. He argued that the **A** pair of genes should form two kinds of gametes, **A** and **a**, in equal frequency. He also stated that a similar argument held for the **B** pair of genes and that they should form two kinds of gametes, **B** and **b**, also in equal frequency.

Independent Assortment

A fundamental law of probability states that the probability of two or more independent events occurring simultaneously is the product of their separate probabilities. It therefore follows that if the probability of an **A** gamete being formed is one-half and the probability of a **B** gamete being produced is one-half, then the probability of a gamete having both **A** and **B** simultaneously is one-half times one-half, or one-fourth. A similar argument would follow for the other combinations. Mendel proposed that such an F_1 doubly heterozygous plant would produce four kinds of gametes, **AB**, **Ab**, **aB**, **ab**, in equal frequency. Mendel further reasoned that since sex had no apparent influence on these crosses, what was true for the production of female gametes produced by the F_1 would also be true for the male gametes produced by the F_1. Finally, he stated that if fertilization occurred randomly, he would obtain the results actually observed in the F_2. This situation can be shown schematically, as was often done by early geneticists, by making a checkerboard, or Punnett square (after the British geneticist R. C. Punnett), in which the gametes produced by the male are arranged along the top of the checkerboard and those produced by the female are along the side. A study of Figure 3-1 will verify the 9 : 3 : 3 : 1 ratio he observed.

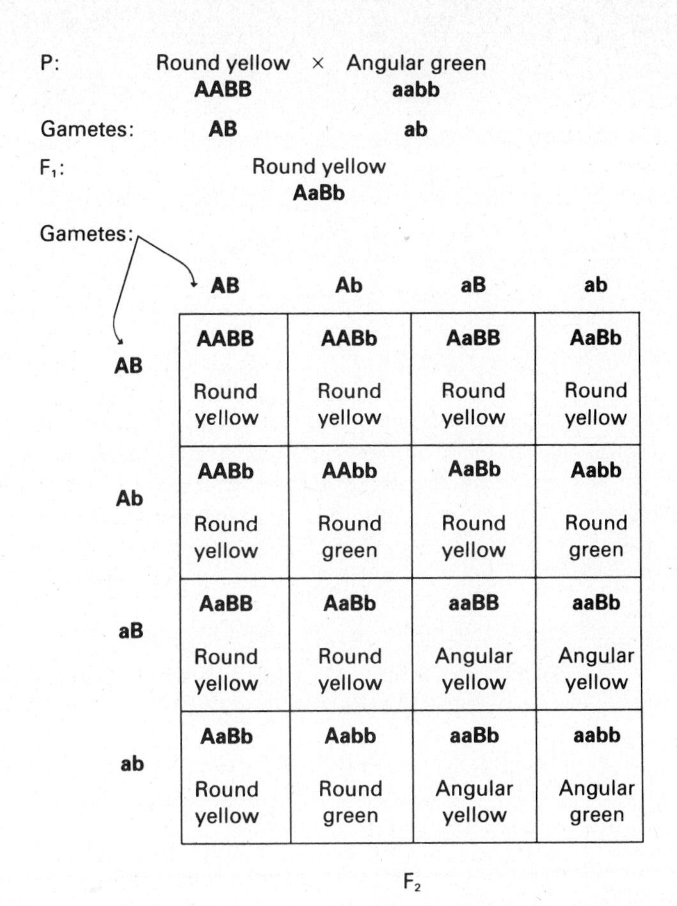

Figure 3-1. Checkerboard showing F_2 distribution of genotypes and phenotypes from a dihybrid cross.

The Test Cross

Although the F_2 results convinced Mendel that his hypothesis of the independent behavior of two pairs of genes (Mendel's Second Law, the Law of Independent Assortment) was correct, he wished, like any good scientist, to test his hypothesis with another experiment. He reasoned that he could test his theory by crossing the F_1 plant, which, according to the hypothesis, is supposedly the double heterozygote, by a plant that is completely recessive, that is angular green. The cross of an organism of unknown genotype by the completely recessive organism is called a *test cross.* Mendel's reasoning was that since the recessive parent will introduce no dominant genes to mask the phenotype, the progeny from such a test cross will be a direct measure of the gametes produced by the supposed heterozygous F_1 parent. And if Mendel's hypothesis was correct, he could predict, in advance of the experiment, that the progeny of the test cross would be of four kinds: round yellow, round green, angular yellow, and angular green, in a ratio of 1 : 1 : 1 : 1. (See Figure 3-2.)

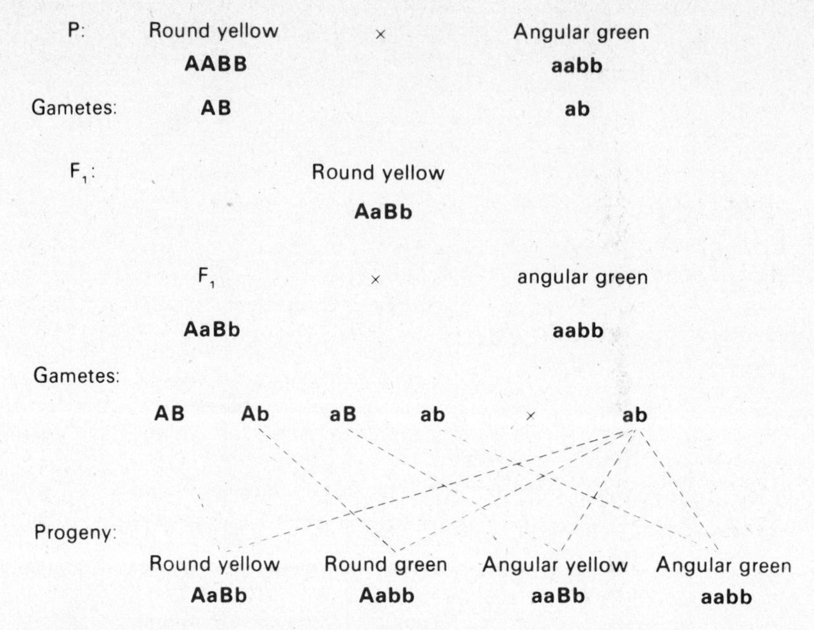

Figure 3-2. Mendel's testcross verifying that an F_1 double heterozygote produces four kinds of gametes in equal frequency.

After performing this experiment, Mendel obtained, in the test-cross progeny, 55 round yellow plants, 51 round green plants, 49 angular yellow plants, and 53 angular green plants. It takes little statistical sophistication to see that this is a very good approximation of the predicted 1 : 1 : 1 : 1 ratio, giving an independent verification of Mendel's Second Law.

Mendel's experiments, now recognized as a classical example of scientific method, showed, in contrast to earlier investigators, that inheritance is particulate. A certain gene, designated **A**, in development is responsible for yellow pigmentation in the cotyledon, and a certain allele, **a**, is responsible for green cotyledon. In an **Aa** plant, during gamete production, these two genes segregate from each other—Mendel's First Law, the Law of Segregation—and neither mix nor join. In a plant that is heterozygous for two pairs of genes, **AaBb**, the segregation of the **A** pair of genes is independent of the segregation of the **B** pair of genes—Mendel's Second Law, the Law of Independent Assortment. By these two laws, Mendel showed that inheritance did not involve blending or mixing but rather was due to discrete, integral entities called genes.

APPENDIX

STATISTICAL ANALYSIS: CHI SQUARE

In the discussion of Mendel's experiments it was mentioned that the results fit well the expectation predicted by the hypothesis. Mendel predicted that in the test cross the progeny would occur in a ratio of 1 : 1 : 1 : 1. The actual results were 55, 51, 49, and 53. Few would doubt that the results did not agree with the expectation.

It is understood that, in large samples, the obtained results will not be exactly what the hypothesis predicts. If a coin is flipped one hundred times, it is unlikely that the actual result will be exactly the 50 heads and 50 tails predicted. If the result is 46 heads and 54 tails, close to expectation, few would question the result. In many experiments intuition might be used to decide whether or not the data are close to expectation. However, intuition is not sufficiently objective and different observers will have different intuitions.

Chi Square

In order to avoid the questionable use of intuition in deciding whether the experimental results agree with the predicted results, statistical tests are used. For many genetics experiments, the statistical test known as chi square (Greek letter χ) is the most appropriate. Stated another way, it is understood that, in an experiment, the actual results are unlikely to fit the theoretical expectation *exactly*. A deviation of some size will exist. Is such a deviation a chance event? Is the deviation due to *errors of random sampling*? Or is the deviation the result of an incorrect prediction? A chi-square analysis will help in answering these questions.

The formula for χ^2 is

$$\chi^2 = \Sigma \frac{(\text{observed} - \text{expected})^2}{\text{expected}}$$

The observed minus the expected represents the deviation, and chi square is the sum of all such deviations squared, divided by the expected.

An example described in Chapter 2 will show how chi square is applied. In the result of Mendel's experiment involving the shape of the seed, out of 7324 F_2 seeds, 5474 were round and 1850 were angular. Based on a hypothesis that predicts 3 round to 1 angular, the theoretical expectation is 5493 round and 1831 angular. Is the deviation Mendel obtained due to errors of random sampling?

SHAPE OF SEED	Number of Seeds OBSERVED	EXPECTED
Round	5474	5493
Angular	1850	1831
	7324	7324

$$\chi^2 = \frac{(5474-5493)^2}{5493} + \frac{(1850-1831)^2}{1831} = 0.26$$

What is the probability that this chi-square value has occurred due to chance? To answer this question, a table of chi square is needed. (See Table 3-1.) The chi-square table contains three parts. Across the top are probabilities, *p*. On the left are degrees of freedom. The remainder of the table represents calculations of chi square. To determine whether the probability that the value 0.26 is random, examine the chi-square values in the table for one degree of freedom. The chi square obtained lies between 0.15 and 0.45. This corresponds to a probability value that lies between 0.70 and 0.50. Therefore there is a probability of between 0.50 and 0.70 of observing a deviation as large or larger than the one Mendel observed in his experiment.

Table 3-1. Chi-Square Table for Mendel's Experiment[a]

DEGREES OF FREEDOM	Probability, p: 0.70	0.50	0.30	0.05	0.01
1	0.15	0.45	1.07	3.84	6.63
2	0.71	1.39	2.41	5.99	9.21
3	1.42	2.37	3.66	7.81	11.34
4	2.19	3.36	4.88	9.49	13.28
5	3.00	4.35	6.06	11.07	15.09
6	3.83	5.35	7.23	12.59	16.82

FISHER. 1972.

[a]Values of chi square are given in body of table.

Degrees of Freedom

Why is there one degree of freedom in this example? Generally there is one less degree of freedom than the number of classes in the example. In the Mendel experiment there were two classes, round and angular. The degree of freedom, therefore, is $2-1=1$ degree of freedom.

The experiment with snapdragons (page 12) can be used as a second example to illustrate chi square. A cross of red by white snapdragons gave 54 red-flowered,

122 pink-flowered, and 58 white-flowered snapdragons in F_2. Based on a hypothesis of 1 : 2 : 1 and a total of 234 plants, we would expect 58.5 red, 117 pink, and 58.5 white-flowered snapdragons. Chi square is calculated by summing all the deviations squared and dividing by the expected.

COLOR OF PLANT	Number of Plants	
	OBSERVED	EXPECTED
Red	54	58.5
Pink	122	117
White	58	58.5
	234	234

$$\chi^2 = \frac{4.5^2}{58.5} + \frac{5^2}{117} + \frac{0.5^2}{58.5} = 0.56$$

Because there are three classes—red, pink, and white—there are 2 degrees of freedom. An examination of the chi-square table shows that 0.56, or 2 degrees of freedom, lies to the left of 0.70. Thus the conclusion is that the probability of such a deviation occurring because of chance alone is greater than 0.70.

Levels of Significance

How should these probability values be interpreted? If the probability is greater than 0.05, the deviation is *not significant.* That is, if the chi square gives a *p* value larger than 1 in 20, it is concluded that the deviation is due to chance alone. If the *p* value is less than 0.05, then it is said that the deviation is *significant.* In other words, if the probability is less than 1 in 20, we conclude that the deviation cannot be due to chance alone. With a *p* value less than 0.05, the hypothesis is carefully examined and usually modified or discarded. Statisticians speak of *levels of significance,* and normally the 5 % level of significance is used. However, in certain cases and with certain investigators, different levels of significance may be adopted. The most commonly used other level of significance is 0.01. If the 1 % level of significance is adopted, then *p* values greater than 0.01 are said to be *not significant,* or due to errors of random sampling, and *p* values less than 0.01 are said to be *significant* and the hypothesis is in doubt.

Errors of the First and Second Kind

In choosing a level of significance, whether 0.05 or 0.01, certain errors may be involved.. There is an *error of the first kind,* which is the rejection of a true hypothesis. If the 5 % level of significance is adopted, an error of the first kind will be made five times in a hundred. That is, occasionally a significant deviation will occur because of chance alone. If the 5 % level of significance was adopted, the hypothesis would be discarded and so a true hypothesis would be rejected at times. It is possible, of course, to obtain improbable results due to chance. If a coin is flipped ten times, the probability of getting all ten heads is $(\frac{1}{2})^{10}$, 1 in 1024. Although this is an improbable event, if each of 1000 people flipped a coin ten times, we would expect, due to chance alone, one of them to obtain ten heads. In other words, improbable events can occur and sometimes an error of the first kind will be made. In order to minimize such errors, lower levels of significance can be adopted. If the 1 % level of significance is chosen, an error of the first kind will be made only one time in 100. Why not choose low levels of significance so as to reduce errors of the first kind? Because there are *errors of the second kind.* An error of the second kind is the acceptance of a false hypothesis. Although it is not possible to quantify how often such errors are made, generally the lower the level of significance, the more likely an error of the second kind will be made. Therefore if very low levels of significance are selected, 0.01 or 0.001, there is a corresponding increase in the likelihood of errors of the second kind. If a very low level of significance is adopted, the probability increases that more than one hypothesis will fit the actual data, and the investigator may choose the wrong one.

There is a dilemma! The higher the level of significance, the more errors of the first kind will be made. The lower the level of significance, the more errors of the second kind will be made. Consequently, many statisticians have accepted 5 % as being as reasonable resolution of this dilemma.

Null Hypothesis

How are experiments analyzed statistically when there is no a priori hypothesis? Such experiments are common in biology. A typical example occurs when an investigator is testing the effect of two different treatments, such as high or low temperature. No theoretical hypothesis predicts exactly what will happen under conditions of two different temperatures. If different results are obtained, is the difference due to the different temperatures? In order to use chi square, an expected must be calculated; and if there is no hypothesis, how is the calculation made? In this case, we make up an hypothesis, called a *null hypothesis.* The null hypothesis states that no difference results from the two treatments. If there is no difference, then the distribution of data will be the same proportionally in the two temperature samples.

An example will illustrate this type of experiment and the statistical analysis used. In Chapter 10 a phenomenon known as crossing over is described. For the moment, let us merely state that, in certain experiments involving the vinegar fly Drosophila, the progeny can be classified according to whether a crossover occurred or not. The question was asked: Will the frequency of crossover events be affected by the temperature at which the flies are raised? Table 3-2 gives the results of an experiment designed to test this question. Chi square will be used to analyze these data statistically. The null hypothesis is formulated and states that there is no difference in the frequency of crossing over in flies raised at 22° compared to flies raised at 31°. If there is no difference, then the number of crossovers and noncross-

Table 3-2. The Effect of Temperature on Crossing Over in Drosophila

HATCHING TEMPERATURE (°C)	*Number of Flies* CROSSOVERS	NON-CROSSOVERS	TOTAL
22	967	2655	3622
31	1282	2265	3547
	2249	4920	7169

AFTER PLOUGH. 1917. *J. EXP. ZOOL.*, **24**.

overs should be proportional to each other for the two temperatures at which the flies were raised.

TEMPERATURE (°C)	*Number of Flies* CROSSOVERS	NON-CROSSOVERS	TOTAL
22	967 (X)	2655	3622
31	1282 (Y)	2265	3547
	2249	4920	7169

In calculating the expected number of crossovers at 22° (X) based on the null hypothesis, the following proportion is used:

$$\frac{X}{2249}=\frac{3622}{7169}$$

Solving for X,

$$X=\frac{3622}{7169}\times 2249=1136$$

If there is no difference in the two temperature treatments, then 1136 crossover flies at 22°C would be expected. Similarly, the expected can be calculated for crossovers at 31°.

$$\frac{Y}{2249}=\frac{3547}{7169}$$

$$=\frac{3547}{7169}\times 2249=1113$$

When all the expecteds have been calculated, a 2×2 *contingency* table results with the expected values in parenthesis.

	Number of Flies		
TEMPERATURE (°C)	CROSSOVERS	NON-CROSSOVERS	TOTAL
22	967 (1136)	2655 (2486)	3622
31	1282 (1113)	2265 (2434)	3547
	2249	4920	7169

Given an observed and an expected, chi square can be used with the following result.

$$\chi^2 = \frac{(967-1136)^2}{1136} + \frac{(1282-1113)^2}{1113} + \frac{(2655-2486)^2}{2486} + \frac{(2265-2434)^2}{2434}$$

$$= 74.0$$

A chi square of 74.0 for one degree of freedom gives a p less than 0.05. It is concluded that the deviation is significant, and the hypothesis should be discarded. What is the hypothesis? There is no difference in crossing-over frequency in flies raised at 22°C compared to flies raised at 31°C. Therefore the conclusion is that the temperature at which flies are raised affects the frequency of crossing over.

Why is there only one degree of freedom? Once the expected, X, was calculated and fitted into the 2×2 contingency table, there was no more freedom to fit any of the other expected values, since the totals must add up. In other words, once 1116 was calculated as the expected for crossovers at 22°, then the expected for crossovers at 31° must be 1113 so that the total crossover expected equals the total observed, 2249. Also, the noncrossovers expected for 22° must be $3622-1136$, or 2486. Only the freedom to calculate one expected is available. After that calculation, all other expected values can be obtained by subtraction. Thus there is only one degree of freedom. A formula for calculating degrees of freedom in contingency tables is $(n-1)(r-1)$, where n is equal to the number of horizontal rows and r is equal to the number of vertical columns. In the crossover experiment $n = 2$ (crossovers and noncrossovers) and $r = 2$ (22° and 31°). The degree of freedom is $(2-1)(2-1) = 1$.

In addition to testing the effect of temperature on crossing over, a second experiment tested the effect of humidity on crossing over. Flies were raised under wet and dry conditions with the results shown in Table 3-3. Because there is no a priori expectation, the null hypothesis is formulated that there is no difference in the frequency of crossing over between flies raised under wet conditions and flies raised under dry conditions. If the null hypothesis is correct, then the distribution of

Table 3-3. The Effect of Humidity on Crossing Over in Drosophila

CULTURE CONDITION	Number of Flies CROSSOVERS	NON-CROSSOVERS	TOTAL
Wet	472	1218	1690
Dry	508	1424	1932
	980	2642	3622

AFTER PLOUGH. 1917. *J. EXP. ZOOL.*, **24**.

crossovers should be proportional to each other for the two culture conditions, wet and dry.

CULTURE CONDITION	Number of Flies CROSSOVERS	NON-CROSSOVERS	TOTAL
Wet	472 (*X*)	1218	1690
Dry	508	1424	1932
	980	2642	3622

$$\frac{X}{980}=\frac{1690}{3622}$$

$$=\frac{1690}{3622}\times 980=457$$

CULTURE CONDITION	Number of Flies CROSSOVERS	NON-CROSSOVERS	TOTAL
Wet	472 (457)	1218 (1233)	1690
Dry	508 (523)	1424 (1409)	1932
	980	1424	3622

$$\chi^2=\frac{(472-457)^2}{457}+\frac{(508-523)^2}{523}+\frac{(1218-1233)^2}{1233}+\frac{(1424-1409)^2}{1409}$$

$$=1.26$$

With one degree of freedom, p is greater than 0.05. Thus the deviation is not significant, and we accept as reasonable the hypothesis, which was that there is no difference in the frequency of crossing over between flies raised in wet or dry cultures. In other words, the moisture of the culture does not affect the frequency of crossing over.

Statistical analysis enables the investigator to determine whether or not the observed results are a good fit to the expected results. For most genetics experiments, chi square is a satisfactory statistical technique. There are, of course, more sophisticated methods of statistical analysis.

REFERENCES

FISHER, R. A. 1972. *Statistical Methods for Research Workers.* 14th ed. Hafner Press, NYC. (Table 3-1 is taken from Table III in this volume.)

STERN, C., and E. SHERWOOD (Eds.). 1966. *The Origin of Genetics. A Mendel Source Book.* 179 pp. W. H. Freeman and Co., San Francisco. (Mendel's paper and other important items on Mendel and his place in history.)

(*Most of the references for Chapter 2 also apply to Chapter 3.*)

QUESTIONS AND PROBLEMS

3-1. Define the following terms:

independent assortment
dihybrid
test cross
null hypothesis
levels of significance
error of the first kind
error of the second kind

3-2. What was the significance of Mendel's experiments?

3-3. In the garden pea, yellow cotyledon is dominant to green and tall plants are dominant to short. A pure-breeding yellow short plant is crossed to a pure-breeding green tall. Give the

(a) genotypes of the parents.
(b) gametes produced by the parents.
(c) genotype and phenotype of the F_1.
(d) gametes produced by the F_1.
(e) genotypes and phenotypes in the F_2 (from a cross of $F_1 \times F_1$).

3-4. In the F_2 there are a number of yellow tall plants. What are the possible genotypes of these plants?

3-5. How could you determine the genotype for *any* specific F_2 yellow tall plant?

3-6. In the snapdragon, a cross of red × white flowers gives a pink-flowered F_1 and a 1 : 2 : 1 ratio of red : pink : white in the F_2. (See Chapter 2.) Also, tall plants are dominant to short plants. A red-flowered tall snapdragon is crossed to a white short one. Give the

(a) genotype and phenotype of the F_1.
(b) genotypes and phenotypes and their ratios for the F_2.

3-7. In Drosophila, a red-eyed short-winged fly crossed to a sepia-eyed long-winged fly gave an F_1 in which all the flies had red eyes and long wings. In the F_2, there were 345 red long, 118 red short, 122 sepia long, and 38 sepia short. How are these two factors inherited?

3-8. What is the *expected* F_2 ratio in the preceding cross? Do a chi-square analysis to test the goodness of fit.

3-9. In the F_2 of a cross of round yellow × angular green, Mendel obtained 315 round yellow, 101 angular yellow, 108 round green, and 32 angular green. Do a chi-square analysis to test these data for goodness of fit.

3.10. In hamsters, black fur is caused by the dominant gene B and brown fur is due to the recessive allele b. Short hair, A, is dominant to long hair, a. As a breeder, you would like a homozygous strain of long-haired black hamsters. If you started with a true-breeding, short-haired black male and a long-haired brown female, what breeding procedure would you follow to obtain the desired strain?

4 / Gene interaction

For the science of genetics, the early years of the twentieth century were used primarily to verify Mendel's results. One of the centers of this activity was in England in the laboratory of William Bateson, the principal advocate of Mendelism. Bateson extended Mendelian techniques to many different organisms. Many of the experiments dealt with the inheritance of comb shape in chickens. (See Figure 4-1.) In one experiment one parental strain of chickens had rose comb and the other parental strain had a single comb. The F_1 were all rose combed, and three-fourths of the F_2 chickens had rose combs and one-fourth had single combs. Consequently, Bateson made the Mendelian interpretation that the inheritance of comb with respect to rose and single was due to a single pair of genes.

In a second experiment, in the early 1900s, Bateson crossed pea-combed chickens by single-combed chickens. The F_1 all had pea combs. In the F_2, three-fourths of the chickens had pea combs and one-fourth had single combs. Bateson concluded that pea and single differed from each other by a single pair of genes. The interesting question is: Is it the same pair of genes? The simplest interpretation of the preceding experiments is that rose and single differ by one pair of genes and that pea and single differ by one pair of genes. To test whether it is the same pair, Bateson crossed rose-combed chickens by pea-combed chickens. All the F_1 had walnut-shaped combs, a result that, prior to Mendel, might have staggered the imagination. But Bateson, taking advantage of Medelian techniques, crossed the walnut-combed F_1s to obtain an F_2. After counting the progeny, he observed that $\frac{9}{16}$ of the chickens had walnut combs, $\frac{3}{16}$ had rose combs, $\frac{3}{16}$ had pea combs, and $\frac{1}{16}$ had single combs. The 9 : 3 : 3 : 1 ratio suggested immediately that two pairs of genes were involved, and therefore "no" is the answer to the question of whether rose by single and pea by single differ in the same pair of genes.

Figure 4-2 illustrates an alternative method to the checkerboard for diagramming genetic experiments. It is particularly useful when dealing with dominant-recessive gene pairs and when only F_2 phenotypes are desired. It is based on the fact that a 3 : 1 phenotypic ratio is obtained in the F_2 for one pair of genes and that independent assortment is the rule for two or more pairs of genes.

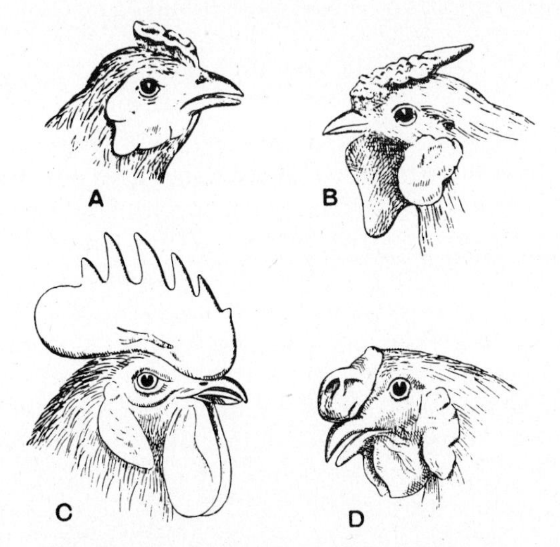

Figure 4-1. Fowls' combs. (From Punnett. 1911. *Mendelism*, Macmillan Co., N.Y.)

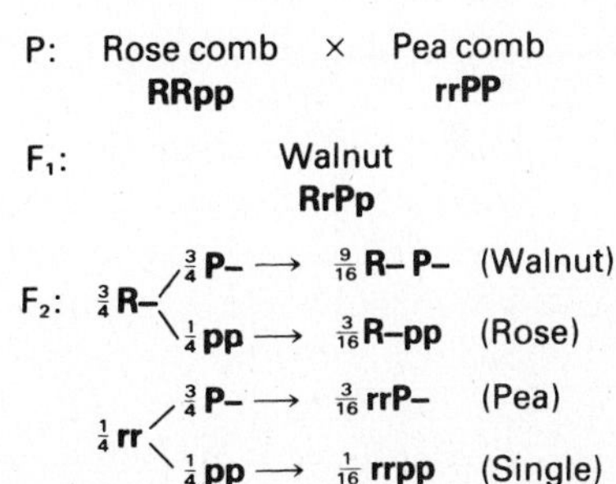

Figure 4-2. An analysis of comb shape in chickens. See text for further description of the crosses and for an explanation of the symbols.

Because only F_2 *phenotypes* are of interest, it is not necessary to distinguish between homozygotes (**RR**) and heterozygotes (**Rr**). This fact is symbolized by a dash (**R**–) to indicate the dominant (**R**) phenotype regardless of whether the organism is **RR** or **Rr**. Thus when it is indicated in Figure 4-2 that $\frac{9}{16}$ of the F_2 are **R–P–**, it means that $\frac{9}{16}$ of the F_2 have at least one **R** and one **P** gene. Since Bateson showed, in the earlier crosses of rose by single and pea by a single, that rose and pea each were dominant to single, it follows that, phenotypically, **RR** resembles **Rr** and **PP** resembles **Pp**.

Mendel showed that a single pair of genes results in a 3 : 1 ratio in F_2, which is symbolized in Figure 4-2 by indicating that three-fourths of the F_2 will be **R–** and one-fourth will be **rr**. Mendel further showed that two pairs of genes assort independently, and this independent assortment is symbolized by listing the **P** phenotypes with each of the **R** phenotypic possibilities. The F_2 probabilities are determined by using the simple law of probability ($\frac{3}{4} \times \frac{3}{4} = \frac{9}{16}$).

The reader should find this method of solving genetic experiments simpler than the checkerboard and less prone to mechanical errors. Also, it is a diagrammatic reminder of segregation and independent assortment.

The chicken comb experiments illustrate that what seems to be complicated phenotypically may be rather simple genetically. The clue to understanding is the 9 : 3 : 3 : 1 ratio, which suggests two pairs of genes. What is happening developmentally is a more fascinating question. The reason why one gene that is responsible for rose comb and a second gene that is responsible for pea comb should together in the same organism give a walnut comb is not known. Nor is it known why the absence of either dominant gene should result in still a fourth phenotype—namely, single comb.

9 : 7 Complementary Factors

Another experiment done in Bateson's laboratory involved the sweet pea, *Lathyrus odoratus.* Many of the early genetic experiments were done with flower colors, as were some of Mendel's. Bateson had seed packets for several different flower colors. He crossed as parents a colored-flowered sweet pea by a white and obtained all colored flowers in the F_1. The F_1 were selfed and the F_2 had three colored to one white. Bateson concluded that white and colored differed by a single pair of genes. In a second experiment colored-flowered plants crossed to another white variety, designated white-2, resulted in all colored F_1. In the F_2, there were three colored to one white. Again the question is: Are white-1 and white-2 the same? To find out, Bateson crossed white-1 by white-2 and observed in the F_1, somewhat to his surprise, colored flowers. He selfed the F_1 and in the F_2 he counted 382 colored plants and 269 white plants. Bateson said that this ratio "indicates some complication." He concluded ultimately that it was a ratio of 9 : 7. The fact that the ratio could be interpreted in this way suggested to Bateson that the most probable hypothesis involved two pairs of genes, and he only needed to imagine a developmental scheme that would result in the phenotypes he observed. His final interpretation was that coloration in the flowers occurred only when the two dominant genes were present together.

The experiments are diagrammed in Figure 4-3. The white-1 parent is designated **aaBB** and is white because it does not have the **A** gene. The white-2 parent is **AAbb**, being white because it does not have the **B** gene. The F_1 would be **AaBb** and, according to the hypothesis, is colored. Selfing the F_1 gives $\frac{9}{16}$ F_2 plants

P: White-1 × White-2
aaBB **AAbb**

F_1: Colored
AaBb

F_2: $\frac{3}{4}$ **A–** → $\frac{3}{4}$ **B–** → $\frac{9}{16}$ **A– B–** Colored
$\frac{3}{4}$ **A–** → $\frac{1}{4}$ **bb** → $\frac{3}{16}$ **A– bb**
$\frac{1}{4}$ **aa** → $\frac{3}{4}$ **B–** → $\frac{3}{16}$ **aaB–** White
$\frac{1}{4}$ **aa** → $\frac{1}{4}$ **bb** → $\frac{1}{16}$ **aabb**

Figure 4-3. The cross of white flowered plants in the sweet pea, *Lathyrus odoratus* resulting in a 9 : 7 ratio in the F_2.

that have at least one **A** and one **B** and should be colored, $\frac{3}{16}$ plants that have one **A** and **bb**, which should be white, $\frac{3}{16}$ that have **aa** and one **B**, white, and $\frac{1}{16}$ the double recessive, **aabb**, also white.

The hypothesis can be tested by selfing the F_2 colored plants. If correct, the prediction is that $\frac{1}{9}$ of the progeny of selfed F_2 colored plants should be only colored, $\frac{4}{9}$ of the progeny should be 3 colored to 1 white, and $\frac{4}{9}$ should be 9 colored to 7 white. The reader should verify this prediction.

It is important to understand the distinction between the *transmission* of genes and their *action* in the development of organisms. It is the transmission of genes that demanded the most attention in the early twentieth century. Gene action forms an important part of the history of genetics later in the century. The early geneticists had the imagination, however, to propose many interesting schemes of gene action, and their speculations, although often unsupported by evidence, indicate how gene action was regarded in those early years. For example, suppose that the **A** gene in the sweet pea example is responsible for a colorless precursor and that the **B** gene is responsible for an enzyme that is needed to convert the colorless precursor to a colored pigment. In the absence of either **A** or **B**, the necessary ingredients for colored pigment will not be present. Or a second possibility is that there is a colorless precursor that is responsible for colored pigment. One of the genes, **B**, is responsible for the enzyme necessary to convert the precursor to the colored pigment. Enzymes often function within very limited environmental ranges; it might be supposed that the other gene, **A**, is responsible for the pH of the cellular environment such that the **A** allele gives a pH that allows the enzyme to function, thereby converting the colorless precursor to the colored pigment, whereas the recessive allele **a**, is responsible for a second pH in which the enzyme does not function at all. (See Figure 4-4.) The reader should be able to think of many other possible ways in which these genes might act in development. The point here is not to propose what is the correct scheme but to begin to think, in a general way, about what genes may be doing.

Genes that act together to produce an effect that neither can produce separately are called *complementary* genes. In the sweet pea, **A** and **B** are complementary genes because together they result in colored flowers.

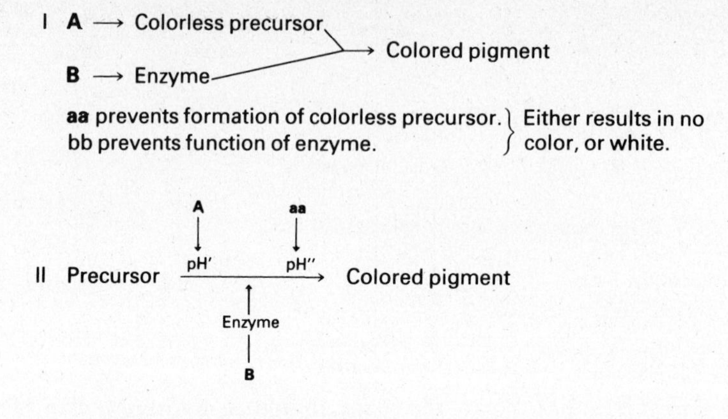

aa results in pH″ in which enzyme is nonfunctional.
bb prevents function of enzyme.
} Either results in no color, or white.

Figure 4-4. A hypothetical scheme to imagine the action of genes in determining color in sweet pea. I and II are variations on a model of enzyme action. Other possibilities exist.

13 : 3

In an experiment that crossed two white chickens, a white Leghorn by a white Wyandotte, all the F_1 were white. This might seem an uninteresting and unproductive experiment, but when F_1s were mated, the F_2 gave both white and *colored* chickens in a ratio of 13 : 3. Bateson was significantly influenced by the ratio of sixteenths to propose two pairs of genes as an explanation for the results. He needed only to postulate a scheme of gene interaction to explain the observed phenotypic ratio. His hypothesis stated that there were two pairs of genes, **I,i**, in which **I** is an inhibitor of pigmentation, and **C,c**, where the homozygous recessive, **cc**, results in an albino or white chicken. Thus the only genotype that results in pigmentation is **iiC–**.

Given these assumptions, the original cross can be explained in the following way: the parental genotypes would be **IICC** for white-1, **iicc** for white-2. The first variety is white because of the color inhibitor gene **I**. The second variety is white because of the lack of the pigmentation gene, **C**. The F_1 is **IiCc**, which is white due to the dominant inhibitor gene. In the F_2, three-fourths of the chickens will receive the **I** gene and therefore will be white regardless of the segregation of the **C** gene. The remaining one-fourth will be **ii**. Of these, one-fourth will be **cc**. Thus $\frac{1}{4} \times \frac{1}{4} = \frac{1}{16}$ will be white due to the genotype **iicc**, which, when added to the $\frac{12}{16}$ white due to the presence of the **I** gene, gives a total in the F_2 of $\frac{13}{16}$ white. The remaining $\frac{3}{16}$ will be **iiC–**, colored. (See Figure 4-5.)

The reader is again reminded that the transmission of genes is exactly the same as Mendel proposed. Only the way the genes interact during development is different, and it is the variety of interactions that gives results that at first appear

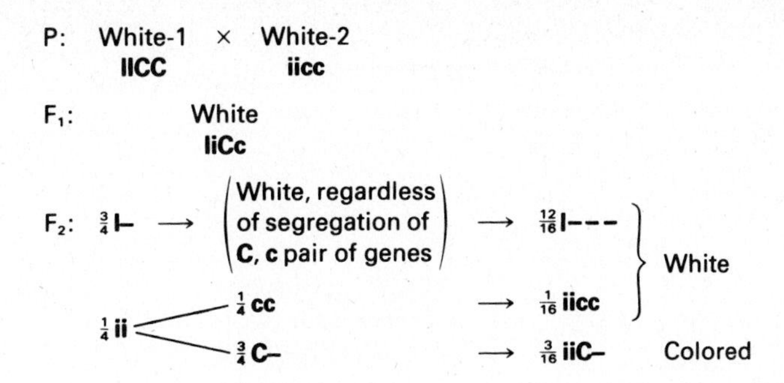

Figure 4-5. The cross of a white Leghorn by a white Wyandotte chicken resulting in a 13 : 3 ratio in the F_2.

difficult to interpret. Another example of gene interaction can be found in mammals. In many mammals, pigmentation of the coat in nature is gray or *agouti*. A wild rabbit seen in the backyard will almost certainly be agouti. Also known to rabbit breeders are albino rabbits that are all white with pink eyes and a variety of colored rabbits, such as black, brown, and cinnamon.

9 : 3 : 4 Epistasis

When a black rabbit was crossed by an albino, the F_1 were all black. Mating the F_1 resulted in an F_2 of 3 black to 1 albino. The conclusion from this cross is that coat-color inheritance in these two varieties of rabbits is due to a single pair of genes. In a second experiment the black rabbit was crossed to another albino strain, designated albino-2. In this case, the F_1 were agouti and the F_2 were $\frac{9}{16}$ agouti, $\frac{3}{16}$ black, and $\frac{4}{16}$ albino. A ratio expressed in sixteenths immediately suggests a hypothesis involving two pairs of genes.

Assume one pair of genes, **C**,**c**, which determines whether there is any pigmentation at all. A rabbit with **cc** will be albino. If **C** is present, the rabbit will be pigmented and the color will depend on other genes. To explain the experiments, a second pair of genes, **A** and **a**, is postulated in which **A** results in agouti and **a** in black. In the cross of black by albino-2, it is postulated that the black rabbit is **aaCC** and the albino-2 rabbit is **AAcc**. The F_1 is the double heterozygote, **AaCc**. It is pigmented due to **C**, and the color is agouti due to the presence of **A**. In the F_2, $\frac{1}{4}$ of the rabbits are **cc** and are albino regardless of the **A** pair of genes. Thus $\frac{4}{16}$ of the rabbits will be albino. Of the remaining $\frac{3}{4}$ that are pigmented, $\frac{3}{4}$ of them, or $\frac{9}{16}$, are **A–** and agouti, and $\frac{1}{4}$, or $\frac{3}{16}$, are **aa** and black. (See Figure 4-6.)

The first cross of black by albino-1 can be explained by assuming that both strains of rabbits are **aa** but that the black rabbit is **CC** and the albino rabbit is **cc**. Since in effect, then, these strains differ by only one pair of genes, we expect the cross to behave as a simple Mendelian monohybrid.

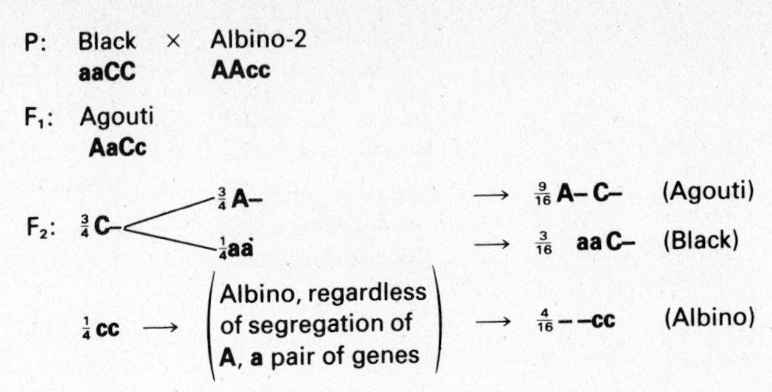

Figure 4-6. The cross of a black rabbit by an albino rabbit resulting in a 9 : 3 : 4 ratio in the F_2.

Genetic analysis has revealed a number of independent genes that determine coat color in rabbits. Table 4-1 gives a few examples. Similar examples of coat-determining genes can be demonstrated in the guinea pig and mouse.

Bateson in 1907 introduced the term *epistasis* to describe the phenomenon in which one pair of genes masks the expression of one or more pairs of nonallelic genes. In the rabbit example the gene **c** is epistatic to the other color genes because if the rabbit is **cc**, it is albino regardless of what the other genotypes are. The **cc** genotype covers up or is epistatic to all others. Stated another way, the color genes are *hypostatic* to the **c** gene. The expression of the color genes is masked if **cc** is present.

Table 4-1. Some of the Possible Combinations of Coat Color Genes in the Rabbit

Genotype	Phenotype	
1. CCAABBDDee	agouti (wild type)	if genotype were cc all rabbits would be albino
2. CCaaBBDDee	black	
3. CCAAbbDDee	cinnamon	
4. CCAABBddee	dilute wild	
5. CCaaBBddee	blue	
6. CCAABBDDEE	steel gray	

15 : 1 Duplicate Factors

When word of Mendel's achievements reached the Scandinavian countries, it was not surprising that plant breeders would turn their experimental attention to the genetics of grains, which constitute an important part of the Scandinavian economy. The

Swedish geneticist Nilsson-Ehle investigated seed color in oats. He had oats with black seeds and a second variety in which the seeds were white. Nilsson-Ehle crossed the black strain by the white strain and observed in the F_1 oat plants in which all the seeds were black. The F_1 were selfed and the F_2 had black seeds and white seeds in a ratio of 15 : 1. Nilsson-Ehle recognized that this ratio is easily explained with a two-gene hypothesis. He postulated that two pairs of genes were involved in the color of the seeds, **A**,**A′** and **B**,**B′**. The only genotype that would give white seeds would be **A′A′B′B′**. Given those assumptions, the experiment is diagrammed as follows: the black parent is **AABB**, the white parent **A′A′B′B′**. The F_1 is a double heterozygote and is black due to the presence of at least one **A** or one **B** gene. In the F_2, only $\frac{1}{16}$ of the plants will have all the "prime" genes, **A′A′B′B′**. All other F_2 plants

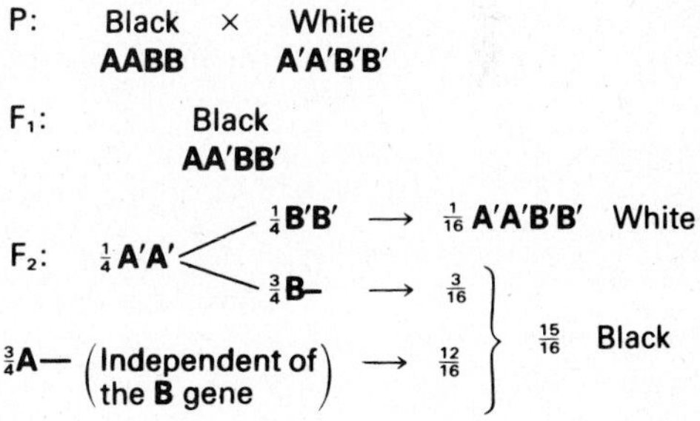

P: Black × White
AABB **A′A′B′B′**

F_1: Black
AA′BB′

F_2: $\frac{1}{4}$**A′A′** < $\frac{1}{4}$**B′B′** → $\frac{1}{16}$ **A′A′B′B′** White
$\frac{3}{4}$**B–** → $\frac{3}{16}$
$\frac{3}{4}$**A–** (Independent of the **B** gene) → $\frac{12}{16}$
} $\frac{15}{16}$ Black

Figure 4-7. The cross by Nilsson-Ehle of black seeded oats by white seeded oats resulting in a 15 : 1 ratio in the F_2.

have at least one **A** or one **B**, resulting overall in $\frac{15}{16}$ black and $\frac{1}{16}$ white (see Figure 4-7). Note that because there is no true dominant-recessive relationship, the symbolism of capital and small letters was discarded in favor of the use of all capitals with primes to distinguish between alleles. Genes that interact such that any one of several independent genes produces the same or a similar effect are designated *duplicate genes.* In the oats experiment gene **A** duplicates in its developmental effect very nearly the developmental effect of gene **B**.

REFERENCES

BATESON, W., and E. R. SAUNDERS. 1902. *Reports to the Evolution Committee of the Royal Society*, **I**:1–160.

BATESON, W., E. R. SAUNDERS, and R. C. PUNNETT. 1905. *Reports to the Evolution Committee of the Royal Society*, **II**:1–131.

(*Both reports contain the results of the many experiments performed in Bateson's laboratory that demonstrated Mendelian principles in other organisms and revealed several examples of gene interactions.*)

QUESTIONS AND PROBLEMS

4-1. Define the following terms:

complementary factors
epistasis
hypostasis
duplicate factors

4-2. For the following crosses in poultry, give the phenotypes and genotypes for the parents and the progeny. (Assume that the parents are homozygous unless there is evidence to the contrary.)

(a) rose comb × single
(b) pea comb × single
(c) F_1 from (a) × F_1 from (b)
(d) rose × pea
(e) F_1 from (d) by F_1 from (d)

4-3. How many different genotypes are possible for the walnut phenotype?

4-4. Design an experiment to show how you could tell the genotype of a walnut-combed rooster.

4-5. In corn, the aleurone is that layer of cells below the seed coat. Three genes are necessary for color of the aleurone, A, C, and R. A plant homozygous for any one of the recessive alleles, a, c, r, will have aleurone that is white. Give the genotypes and phenotypes of the progeny from the following crosses.

(a) AACCRR × aaccrr
(b) From (a), $F_1 \times F_1$
(c) AaCcRr × aaccrr

4-6. A cross of a white aleurone plant by a colored gave the following ratios in the progeny. Indicate the genotypes of the parents.

(a) All colored
(b) 1 colored : 1 white
(c) 1 colored : 3 white
(d) 1 colored : 7 white

4-7. From a cross of two white plants, an F_1 was obtained that had colored aleurone. What is the genotype of the parents if the F_2 gives a ratio of

(a) 9 colored : 7 white?
(b) 27 colored : 37 white?

4-8. In onions, a white strain can be due either to the dominant allele, I, or to the recessive allele, c. A yellow strain is due to the genotype iiC– only. In the following problems, give the genotypes of the parents and of the F_1.

(a) white × yellow
    F_1 yellow
    F_2 3 yellow : 1 white
(b) white × yellow
    F_1 white
    F_2 3 white : 1 yellow
(c) white × white
    F_1 white
    F_2 13 white : 3 yellow

4-9. In corn, there is a fourth gene, P, which in the presence of the color genes, A, C, and R (Problem 4-5), makes for purple aleurone. The allele, p, makes for red aleurone. The P–p gene pair has no effect if the plant is homozygous recessive for any of the color genes. In the following problems, assume three pairs of genes, A–a, R–r, P–p, and find the genotypes of the parents.

(a) A red plant by a red plant produces $\frac{9}{16}$ red : $\frac{7}{16}$ white.
(b) A red plant by a purple plant produces $\frac{9}{32}$ purple : $\frac{9}{32}$ red : $\frac{7}{16}$ white.
(c) A purple plant by a white plant produces $\frac{1}{8}$ purple : $\frac{1}{8}$ red : $\frac{3}{4}$ white.
(d) A white plant by a white plant produces $\frac{3}{16}$ purple : $\frac{1}{16}$ red : $\frac{12}{16}$ white.

4-10. In squashes, there is a gene, A, for white fruit color. The recessive allele, a, gives green fruit. The allele, A, is epistatic to the gene, B, for yellow fruit so that A–B– and A–bb squashes are white. However, aaB– squashes are yellow and aa bb are green. In the following crosses, give the genotypes of the parents.

(a) A yellow by a yellow produces $\frac{3}{4}$ yellow, $\frac{1}{4}$ green.
(b) A yellow by a white produces $\frac{1}{2}$ white, $\frac{3}{8}$ yellow, $\frac{1}{8}$ green.
(c) A white by a white produces $\frac{3}{4}$ white, $\frac{3}{16}$ yellow, $\frac{1}{16}$ green.
(d) A white by a white produces $\frac{3}{4}$ white, $\frac{1}{8}$ yellow, $\frac{1}{8}$ green.

4-11. In oats, when a green plant was self-fertilized, it produced 199 green seedlings and 14 white seedlings. Explain this result.

4-12. Assume that the dominant genes A, B, and C are complementary factors resulting in a certain phenotype P. Homozygosity for the recessive alleles of any one or all the above genes does not lead to P.

(a) In a cross of AABBCC × aabbcc, what will be the phenotype of the F_1?
(b) What proportion of the F_2 will have the phenotype P?
(c) What proportion of the progeny will be P if the F_1 is crossed to aabbcc?

4-13. Assume three pairs of genes acting in a duplicate fashion such that the presence of any one dominant gives phenotype Z. If AABBCC is crossed by aabbcc, what proportion of the progeny will be Z if the

(a) F_1 is selfed?
(b) F_1 is crossed to aabbcc?

5/Quantitative inheritance

Earlier an experiment by Nilsson-Ehle was described that involved duplicate genes in oats. In a similar experiment in 1909 on grain color in wheat, he observed, in the F_2, from a cross of red-grained wheat by white-grained wheat that about $\frac{63}{64}$ of the plants had red grains and only $\frac{1}{64}$ were white. The result suggested that there were three independently segregating pairs of genes. The primary interest in this experiment is not the analysis of a genetic cross involving three pairs of genes. If a two-gene pair experiment is understood, then a three-gene pair experiment is no more difficult. The principles are the same. But what makes this example of grain color in wheat interesting are Nilsson-Ehle's analysis and interpretation.

He saw that it was somewhat misleading to describe all $\frac{63}{64}$ F_2 plants as red-grained because the intensity of the red pigment in the various F_2 plants was not identical. Some of the F_2 grains were as dark as the original red parent. Others were intermediate, resembling, in fact, the F_1. Still other classes of F_2 grains ranged

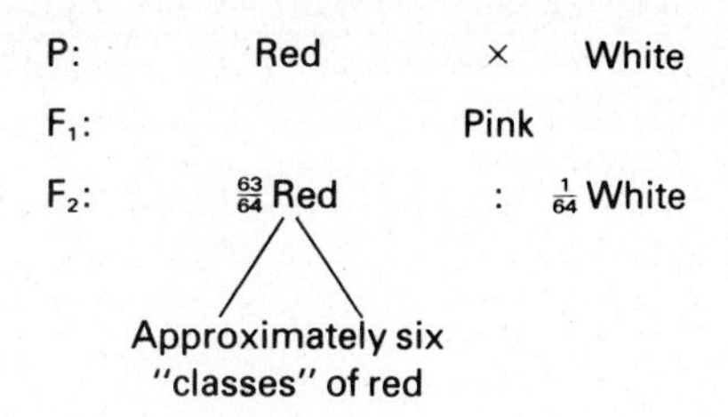

Figure 5-1. Cross by Nilsson-Ehle involving grain color in wheat resulting in an F_2 ratio of 63 : 1.

between the F_1 and the red parent, on the one hand, and the F_1 and the white parent on the other. After examining the F_2 segregation more carefully, Nilsson-Ehle was able to distinguish about six phenotypic classes of red (see Figure 5-1). He made two assumptions about the inheritance of grain color in wheat. First, as already mentioned, three pairs of genes are involved. Secondly, these genes are behaving as duplicate genes. Each is contributing to pigmentation approximately equally. With these assumptions, it follows that the original red parent is $\mathbf{A}^1\mathbf{A}^1\mathbf{B}^1\mathbf{B}^1\mathbf{C}^1\mathbf{C}^1$ and the original white parent is $\mathbf{A}^2\mathbf{A}^2\mathbf{B}^2\mathbf{B}^2\mathbf{C}^2\mathbf{C}^2$. The F_1 would then be expected to be the triple heterozygote and pink, according to the assumptions. An interesting question is: What kind of gametes would such an F_1 plant produce? Remembering Mendel's First Law and Second Law, it is easy to show that the probability of any given gamete getting an $\mathbf{A}^1$ gene is one-half, of getting a $\mathbf{B}^1$ gene is one-half, and of getting a $\mathbf{C}^1$ gene is one-half. Using the law of probability, it follows that the probability of getting all three genes, $\mathbf{A}^1$,$\mathbf{B}^1$, and $\mathbf{C}^1$, in one gamete is one-half × one-half × one-half, or one-eighth. If this is true of one parent, and since sex does not seem to be involved in this cross, it should be also true of the other parent. Therefore the probability of getting in the F_2 a plant whose genotype is $\mathbf{A}^1\mathbf{A}^1\mathbf{B}^1\mathbf{B}^1\mathbf{C}^1\mathbf{C}^1$ is $\frac{1}{8} \times \frac{1}{8}$, or $\frac{1}{64}$. That is, $\frac{1}{64}$ of the F_2 should be as red as the original red parent. A similar argument would show that $\frac{1}{64}$ of the F_2 should be $\mathbf{A}^2\mathbf{A}^2\mathbf{B}^2\mathbf{B}^2\mathbf{C}^2\mathbf{C}^2$ and thus white as the original white parent.

It was mentioned that, in the F_2, about $\frac{63}{64}$ of the red plants arranged themselves into six classes representing various shades of red. How can this situation be explained? The original assumption was that each "1" gene contributed equally to pigmentation. The more "1" genes the plant has, the more red it is. If, in the F_2, a plant gets all the "1" genes, $\mathbf{A}^1\mathbf{A}^1\mathbf{B}^1\mathbf{B}^1\mathbf{C}^1\mathbf{C}^1$, it will be as red as the red parent. As already shown, the probability of such a plant is $\frac{1}{64}$. How many of the F_2 plants will have five "1" genes and be not quite as red as the red parent? Figure 5-2 shows the answer.

Three F_2 genotypes have five "1" genes and one "2" gene. The probability for each genotype is calculated by finding the probability for each pair of genes separately. To determine the probability of getting $\mathbf{A}^1\mathbf{A}^2\mathbf{B}^1\mathbf{B}^1\mathbf{C}^1\mathbf{C}^1$ in the F_2, take one pair of genes and determine what the probability is of obtaining the F_2 genotype for that pair only. For example, the probability of getting $\mathbf{A}^1\mathbf{A}^2$ in the F_2 after selfing an $\mathbf{A}^1\mathbf{A}^2$ F_1 is $\frac{2}{4}$ (remember the 1 : 2 : 1 ratio?). In a similar way, the probability of getting $\mathbf{B}^1\mathbf{B}^1$ in F_2 after selfing a $\mathbf{B}^1\mathbf{B}^2$ F_1 is $\frac{1}{4}$. The same argument holds for $\mathbf{C}^1\mathbf{C}^1$. Therefore the probability of obtaining $\mathbf{A}^1\mathbf{A}^2\mathbf{B}^1\mathbf{B}^1\mathbf{C}^1\mathbf{C}^1$ is $\frac{2}{4} \times \frac{1}{4} \times \frac{1}{4} = \frac{2}{64}$. By using

PROBABILITY	GENOTYPE	PROBABILITY	PHENOTYPE	NUMBER OF "1" GENES
$\frac{1}{64}$	$A^1A^1B^1B^1C^1C^1$	$\frac{1}{64}$	Red	6
$\frac{2}{64}$	$A^1A^2B^1B^1C^1C^1$			
$\frac{2}{64}$	$A^1A^1B^1B^2C^1C^1$	$\frac{6}{64}$	Very slightly less red	5
$\frac{2}{64}$	$A^1A^1B^1B^1C^1C^2$			
$\frac{1}{64}$	$A^2A^2B^1B^1C^1C^1$			
$\frac{1}{64}$	$A^1A^1B^2B^2C^1C^1$			
$\frac{1}{64}$	$A^1A^1B^1B^1C^2C^2$			
$\frac{4}{64}$	$A^1A^2B^1B^2C^1C^1$	$\frac{15}{64}$	Slightly less red	4
$\frac{4}{64}$	$A^1A^2B^1B^1C^1C^2$			
$\frac{4}{64}$	$A^1A^1B^1B^2C^1C^2$			

Figure 5-2. A partial list of genotypes in F_2 from selfing an F_1 that $\mathbf{A^1A^2B^1B^2C^1C^2}$. The assumption is that each "1" gene adds to the pigmentation and that all "1" genes are equal in their effect. Figures 5-2 and 5-3 show the result of such an F_2 phenotypic segregation without listing all the genotypes.

this method, it is reasonably simple to obtain all the F_2 genotypes and their probabilities without actually using a checkerboard. Figure 5-2 is the beginning of such an attempt, and the reader is urged to finish the analysis for all genotypes. Figures 5-3 and 5-4 show, in different ways, the distribution of phenotypic classes in the F_2.

NUMBER OF "1" GENES	PROPORTION OF F_2
6	$\frac{1}{64}$
5	$\frac{6}{64}$
4	$\frac{15}{64}$
3	$\frac{20}{64}$
2	$\frac{15}{64}$
1	$\frac{6}{64}$
0	$\frac{1}{64}$

Figures 5-3. The proportion of the F_2 plants that will have all six "1" genes, five "1" genes, etc.

An examination of these figures shows that most of the individuals are of intermediate pigmentation. Only about $\frac{1}{64}$ resemble one original parent, and $\frac{1}{64}$ resemble the other original parent. If the probable assumption is made that some variability of pigmentation exists within any one genetic class because of environmental causes, then an almost complete gradation of color from darkest red through lightest pink to white is obtained. Nilsson-Ehle has thus shown by this example that so-called blending inheritance can be explained by Mendelism, if enough genes are

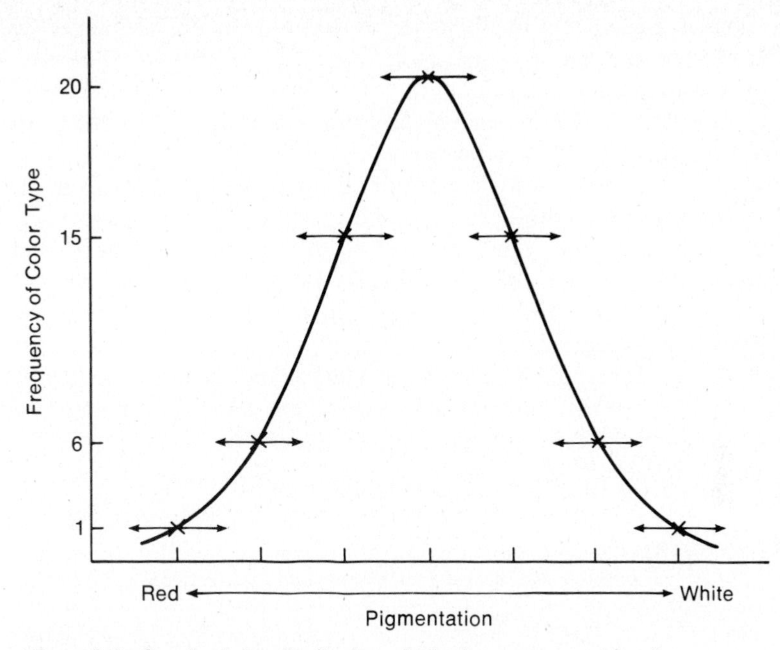

Figure 5-4. Graph of the distribution of F_2 phenotypes resulting from a cross involving grain color in wheat. The assumptions are three pair of duplicate genes affecting pigmentation. The horizontal arrows suggest that for each genotype environmental factors might also affect final grain pigmentation.

assumed. In Mendel's experiments and in later ones, traits were analyzed that were little affected by the environment and that could be accurately described and classified into nonoverlapping phenotypes. However, not all individual differences can be readily classified in this way. Such variations as height, weight, and mental ability in man, plus traits that are important economically in domestic plants and animals, such as yield of fruit or seeds and egg, meat, and milk production, usually fall not into a few discontinuous phenotypic classes but into many small continuous ones.

Multiple Factors

At first these traits seemed to defy analysis in terms of Mendelian genetics, but Nilsson-Ehle and others showed that if enough genes are assumed, the inheritance of these traits can be understood. Such traits have been called *quantitative traits*, and their inheritance is known as *quantitative inheritance*. It is an interesting historical note that this important experiment of Nilsson-Ehle, as well as similar ones by other geneticists, in the early 1900s resolved one of the great debates raging at the time between the Mendelists, on the one hand, largely led by Bateson, and the biometricians on the other. The latter, disciples of the great British scientist Sir Francis

Galton, argued that there must be two kinds of inheritance. Galton had been interested for many years, even prior to the discovery of Mendel's work, in quantitative traits, such as mental ability in man. Because no simple genetic analysis was evident, Galton was skeptical of Mendelism as such. Nilsson-Ehle, to a large extent, dissipated this controversy.

It has been shown that if three pairs of genes are involved, the probability of recovering in the F_2 the genotype of one of the original parents is $(\frac{1}{4})^3$, or $\frac{1}{64}$. Assume that five gene pairs affect the size of an individual. In the F_2, the probability of recovering one individual genetically identical to one parent would be $(\frac{1}{4})^5$, or $\frac{1}{1024}$. And if ten pairs of genes are involved, given the same assumptions, the probability of recovering in the F_2 a genotype identical to one parent would be $(\frac{1}{4})^{10}$, or $\frac{1}{1,048,576}$. Ten pairs of genes is not a large number to assume for the inheritance of quantitative traits. Independent assortment and recombination generate a very large number of different genotypes from a relatively small number of heterozygous gene pairs. (See Table 5-1.)

Table 5-1. The Number of F_2 Genotypes and Phenotypes Expected Due to the Segregation and Independent Assortment of Several Pairs of Genes[a]

				Number of Phenotypes	
NUMBER OF GENE PAIRS	NUMBER OF HOMOZYGOTES	NUMBER OF HETEROZYGOTES	NUMBER OF GENOTYPES	DOMINANT	INTERMED
1	2	1	3	2	3
2	4	5	9	4	9
3	8	19	27	8	27
⋮					
n	2^n	3^n-2^n	3^n	2^n	3^n

[a]Number of gene pairs indicates by how many pairs of genes the original parents differ (for example, AABB × aabb differ by two pairs of genes). The other columns refer to the number of *different* possibilities expected in the F_2. With an original difference of three pairs of genes, you would expect 3^3, or 27, F_2 genotypes.

It has been shown that the inheritance of traits that must be measured (that is, the inheritance of quantitative traits) can be explained by assuming the action of many pairs of genes. This kind of inheritance was originally known as *multiple factor inheritance*, using the Mendelian term for gene. Geneticists now prefer the term *polygenic inheritance*, which refers to a gene system in which many genes have a slight and essentially equal effect on the final phenotype.

REFERENCES JOHANNSEN, W. 1909. *Elemente der exacten Erblichkritslehre.* 515 pp. Jena, Germany.

NILSSON-EHLE, H. 1909. Kreuzungsuntersuchungen an Hafer und Weizen. *Acta Univ. Lund. Ser.* 2, **5** (No. 2), 1–122.

(Two of the influential papers establishing the idea of polygenic inheritance as an explanation for the heredity of quantitative traits.)

QUESTIONS AND PROBLEMS

5-1. Define the following terms:
trihybrid
quantitative trait
polygenic inheritance

5-2. If a cross were made between two pure-breeding parental lines, is it possible that any of the F_2 progeny could be more extreme in phenotype than either grandparent? Explain why or why not.

5-3. It is assumed that, in hamsters, size is determined by genes having an equal and additive effect. From a total of 4025 F_2 progeny resulting from crosses between large and small strains, 17 were as large as the average of the large parent variety and 16 were as small as the average of the small parent variety. How many pairs of genes are involved?

5-4. Assume that the difference in height between two human populations, one averaging 5 feet 0 inches in height and the other averaging 6 feet 0 inches in height, depends on three independent genes (AABBCC = 6 feet, aabbcc = 5 feet). Assume that there is no dominance and that gene A adds 3 inches, gene B adds 2 inches, and gene C adds 1 inch to the height. What is the expected ratio for height among the progeny of the following crosses?

(a) AABBCC × aabbcc
(b) AaBbCC × aabbcc
(c) AaBbCc × aabbcc
(d) AaBbcc × aabbC
(e) AaBbCc × AaBbCc

5-5. Assume that in peaches a plant of the genotype aabbcc produces peaches with a base weight of 4 ounces. Every time a capital-letter gene is substituted, $\frac{1}{2}$ ounce is added to the weight of the peach; so that AABBCC peaches weigh 7 ounces. Plant 1, when self-fertilized, produces F_1 plants with only 5-ounce peaches. Plant 2, when self-fertilized, produces peaches ranging in weight from 4 to 6 ounces. When plant 1 is crossed to plant 2, the F_1 range in weight from $4\frac{1}{2}$ ounces to $5\frac{1}{2}$ ounces and in later generations (after self-fertilization) from 4 to 7 ounces. Give the genotypes of plants 1 and 2.

5-6. A strain of cucumbers is aabbcc and is 150 centimeters tall. Assume that the addition of each and any dominant gene (A, B, or C) adds 5 centimeters to the height, up to a height of 180 centimeters.

(a) List the genotypes of all plants 155 centimeters high.
(b) List the genotypes of all plants 175 centimeters high.

(c) If any plant from (a) is crossed by any plant from (b), what is the tallest plant possible from such a cross?
(d) What is the shortest plant possible from such a cross?

5-7. Two homozygous strains of corn are hybridized. They are distinguished by five different pairs of genes, all of which assort independently and produce independent phenotypic effect (no interactions are observed).
(a) In the F_2, what is the number of possible genotypes?
(b) What is the number of homozygous genotypes?
(c) If all gene pairs act in dominant-recessive fashion, what is the number of F_2 phenotypes?
(d) What proportion of the F_2 is homozygous for all dominants?
(e) What proportion of the F_2 is homozygous for all recessives?
(f) What proportion of the F_2 shows all dominant phenotypes?
(g) If the heterozygotes of all five gene pairs are intermediate in phenotype, how many different F_2 phenotypes would be produced?
(h) If the heterozygotes in three gene pairs are intermediate in phenotype and in the other two gene pairs are dominant, what would be the number of all F_2 phenotypes?

5-8. The capacity or incapacity of mice to permit growth of a tumor transplanted to them is genetically determined. A susceptible strain was crossed to a resistant strain. All F_1 animals were susceptible. Among several hundred F_2 animals, none was susceptible. How can these results be interpreted in a *general* way?

6/Multiple alleles and lethal genes

Multiple Alleles

In the discussion of coat color in mammals in Chapter 4, it was mentioned that the wild-type rabbit had a gray coat, designated agouti, and that another familiar variety of rabbit was the albino, which had white fur and pink eyes. When a cross was made of agouti by albino, the F_1 were agouti and the F_2 were 3 agouti and 1 albino. Another variety of rabbit known as the Himalayan also has pink eyes and the fur is white except for the feet, tail, ears, and tip of the nose, which are black. The F_1 from a cross of agouti by Himalayan are all agouti and the F_2 are 3 agouti to 1 Himalayan. From both experiments, it is concluded that there is a single gene difference between agouti and albino, on the one hand, and between agouti and Himalayan on the other. Is it the same pair of genes? An obvious way to obtain the answer is to cross a Himalayan by an albino. Before doing the experiment, it is useful to see what predictions are possible, based on two mutually exclusive assumptions, a single pair of genes or two or more pairs of genes. One of the important features of the scientific method is the design of experiments that provide the data for an unambiguous choice between two or more hypotheses.

It can be argued that if Himalayan and albino are due to changes in *two* pairs of genes, then each type should carry homozygously the other gene and a reversion to wild-type in F_1 should occur. For example, if Himalayan is due to **bb** and albino is due to **aa**, then Himalayan should also be **AA** or it would not give a monohybrid ratio when crossed with agouti. For similar reasons, albino should be **BB**. The assumption must be that agouti, under a two-gene hypothesis, is **AABB**. Thus a cross of Himalayan by albino gives **AaBb**, and under the simplest assumptions it should be agouti. That is, a two-gene hypothesis predicts agouti in the F_1. The two-gene hypothesis also predicts that **A–B–** rabbits occur in the F_2 and they should be agouti (see Figure 6-1).

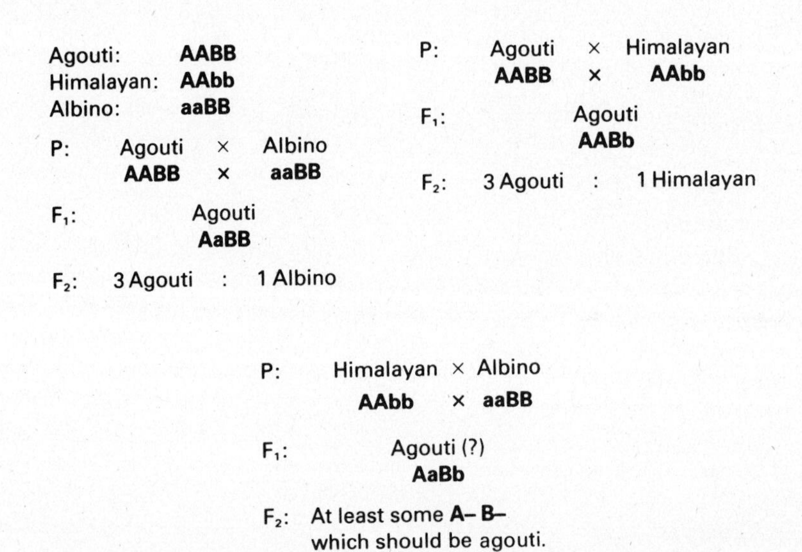

Figure 6-1. Crosses involving agouti, albino, and Himalayan rabbits assuming two pair of genes involved. If this hypothesis is correct, the F_1 is expected to be agouti and certainly some agouti are expected in the F_2.

In the actual cross of a Himalayan rabbit by an albino rabbit, the F_1 were all Himalayan! This result indicates that a two-gene hypothesis is wrong. Selfing the F_1 gave 3 Himalayan to 1 albino in the F_2, a typical monohybrid ratio. This result strongly suggests the hypothesis that Himalayan and albino differ by a single pair of genes.

This example, besides showing how scientists can distinguish between different hypotheses, introduces the concept of *multiple alleles*. It is clear from the foregoing experiments that agouti is dominant to both albino and Himalayan, and so **C** is used to designate the agouti allele. To be consistent in the symbolism, **c** is used to designate the albino allele and the cross of agouti by albino is symbolized **CC** × **cc**. In the cross of agouti by Himalayan, **C** still designates agouti. The symbol for Himalayan should be chosen to indicate that it is recessive to agouti and allelic to albino. To distinguish it from the albino allele already symbolized **c**, the Himalayan allele is designated $\mathbf{c}^h$. A cross of agouti by Himalayan is **CC** by $\mathbf{c}^h\mathbf{c}^h$, resulting in an F_2 of 3 agouti to 1 Himalayan. It follows that a cross of Himalayan by albino is $\mathbf{c}^h\mathbf{c}^h$ by **cc**, giving an F_1 of $\mathbf{c}^h\mathbf{c}$ and an F_2 of 3 Himalayan and 1 albino. Another variety of rabbit, known as chinchilla, when crossed with agouti gives an F_2 ratio of 3 agouti to 1 chinchilla and when crossed with albino gives a ratio of 3 chinchilla to 1 albino. It is now recognized that the chinchilla rabbit represents still another allele in the series designated $\mathbf{c}^{ch}$.

Definition of Allelism

Two mutant genes are defined as alleles if they produce a mutant phenotype when crossed and if they never occur together in the same gamete. In the rabbit example the cross of a Himalayan by albino gave Himalayan, a mutant phenotype. Therefore this cross satisfied the first criterion for allelism. An analysis of the second part of the definition—that two true alleles never occur together in the same gamete—is presented in Chapter 16.

Geneticists have discovered many cases of multiple alleles in various organisms, and it is no longer possible to think in the simple Mendelian terms of only two alternatives for any one gene. In theory, a large number of alternate forms of a given gene may exist. The antigens for several human blood groups have been shown to be controlled by a multiple allelic series, often involving many alleles.

Lethal Genes

Early in the twentieth century the French geneticist Cuénot, while examining the inheritance of coat color in the mouse, experimented with a mutant variety that had a yellow coat. Cuénot noticed that the yellow variety never bred true, and so he concluded that all yellow mice were hybrids. Table 6-1 contains results of experiments on the yellow variety that were obtained by many experimenters over a number of years.

Notice in Table 6-1 that the cross of yellow by nonyellow gave a genetic ratio of 1 : 1 and that the cross of yellow by yellow resulted in a ratio of 2 : 1. This latter ratio is interesting and has not been encountered before. It was also observed that in the cross of yellow by yellow the litter size was smaller, by about one-fourth than the litter size of yellow × nonyellow. These facts could be reconciled by Cuénot if he assumed monohybrid inheritance and that, of the three F_2 classes

Table 6-1. Segregation of the Yellow Allele in the Mouse[a]

COAT COLOR OF PARENTS	*Number of Offspring* YELLOW	NONYELLOW
Yellow × nonyellow	2378	2398
Yellow × yellow	2386	1235

FROM GRÜNEBERG, H. 1952.

[a]Data are the sum of the results of several experiments performed by different authors.

postulated—homozygous yellow, heterozygous yellow, and homozygous nonyellow—only the last two survived. This hypothesis can be symbolized by letting **Y** represent yellow and **y** nonyellow. Since Cuénot noted that the yellow mouse never bred true, he assumed that the yellow mouse was **Yy**. The nonyellow mouse is assumed to be **yy**. A cross of **Yy** × **yy** results in a 1 : 1 ratio of yellow : nonyellow. A cross of two yellows, **Yy** × **Yy**, should give a 1 : 2 : 1 ratio; but if it is assumed that the genotype **YY** is lethal, then a 2 : 1 ratio of yellow to nonyellow and a decrease in litter size by one-fourth are expected.

It would be interesting to see whether the embryonic deaths could be confirmed by sacrificing **Yy** females that had been impregnated by **Yy** males. This experiment might be inconclusive, for if death occurs very early in development, as is often the case, actual observation of aborted embryos may not be possible. However, in the yellow mouse, it was shown that death occurs quite late in development, and so the hypothesis was further substantiated by the observation of aborted embryos.

One final question might be asked. Is the gene **Y** classified as a dominant or a recessive? Clearly, it takes only one **Y** to produce a mutant coat color, but two **Y** genes are required for genetic death. It is left to the reader to ponder this problem.

REFERENCES

GRÜNEBERG, H. 1952. *The Genetics of the Mouse.* 2nd ed., 650 pp. Martinus Nijhoff, The Hague. (Contains a good account of the lethal gene, yellow, in the mouse.)

STURTEVANT, A. H. 1913. The Himalayan rabbit case, with some considerations on multiple allelomorphs. *Am. Nat.* **47**:234–239. (Reprinted in L. Levine, *Papers on Genetics.* C. V. Mosby Co., St. Louis, 1971.) (A short but straightforward explanation of crosses in rabbits interpreted as a multiple allelic series.)

QUESTIONS AND PROBLEMS

6-1. Define the following terms:
multiple alleles
lethal gene

6-2. Assume an allelic series of four recessive genes, a^1, a^2, a^3, and a^4, none of which is dominant to the others, the heterozygote always being phenotypically different from any homozygote or from any other heterozygote. Assuming all possible genetic combinations, how many different phenotypes are possible from this allelic series?

6-3. In rabbits, C = agouti, c^{ch} = chinchilla, c^h = Himalayan, c = albino. The agouti C is dominant to the other three alleles, c is recessive to all three other alleles, and chinchilla is dominant to Himalayan. What will be the phenotypes of progeny from the following crosses?

(a) CC × cc
(b) Cc × Cc
(c) Cc^{ch} × Cc
(d) Cc^h × cc
(e) $c^{ch}c$ × c^hc
(f) $c^{ch}c^h$ × c^hc

6-4. When an agouti rabbit was crossed to another agouti rabbit, the following offspring were produced. Give the genotypes of the parents.

(a) 3 agouti : 1 albino
(b) 3 agouti : 1 chinchilla

6-5. For the following crosses in rabbits, give the genotypes of the parents:

(a) A chinchilla by an albino produces $\frac{1}{2}$ chinchilla and $\frac{1}{2}$ Himalayan.
(b) A chinchilla by chinchilla produces $\frac{3}{4}$ chinchilla and $\frac{1}{4}$ albino.
(c) A chinchilla by Himalayan produces $\frac{1}{2}$ chinchilla, $\frac{1}{4}$ Himalayan, and $\frac{1}{4}$ albino.

6-6. In humans, the three alleles I^A, I^B, and I^O contribute to the A–B–O blood group in the following way. The I^O allele is recessive to I^A and I^B. The I^A and I^B alleles act in an additive fashion such that the genotype I^AI^B produces both antigens, with the result that such an individual belongs to blood group AB. Give all possible genotypes for the following phenotypes.

(a) Blood group A
(b) Blood group B
(c) Blood group O
(d) Blood group AB

6-7. If an AB marries an AB, what are the genotypes and phenotypes of possible children?

6-8. Can the following children be produced? Explain your answer.

(a) An O child from the marriage of two A individuals.
(b) An O child from the marriage of AB by O.
(c) An O child from the marriage of AB by A.
(d) An O child from the marriage of an A by B.
(e) An AB child from the marriage of an A by O.
(f) An A child from the marriage of an AB by B.

6-9. In Drosophila, crosses of flies with Curly wings by other Curly-winged flies always gives $\frac{2}{3}$ Curly to $\frac{1}{3}$ normal. How would you explain this result?

6-10. If your hypothesis in Problem 6-9 is correct, what would be expected from the cross of Curly by normal?

6-11. Crosses of "Dexter" cattle by "Kerry" cattle produce equal numbers of Kerry and Dexter. Crosses of Kerry by Kerry produce only Kerry. Crosses of Dexter by Dexter produce $\frac{1}{4}$ Kerry, $\frac{1}{2}$ Dexter, and $\frac{1}{4}$ still-born calves. Give the genotypes of the parents and offspring for each of the three crosses.

6-12. In a *Drosophila subobscura*, there are four strains differing in eye color: wild-type, pink, and two strains that show the same orange phenotype, orange-1 and orange-2. The following matings were performed with the results shown:

wild-type × orange-1 (gives wild-type)
wild-type × orange-2 (gives wild-type)
orange-1 × orange-2 (gives wild-type)
orange-2 × pink (gives orange-2)

What would be the results of the following matings?

(a) wild-type × pink
(b) orange-1 × pink

6-13. What are the relationships of the genes involved?

7/Cell division

In the interval separating Mendel's work in the 1860s and the rediscovery of his laws in 1900, a remarkable amount of work was done, largely in Germany, that has provided us with a basis for understanding the process of cell division. Although Hooke in England and Malpighi in Italy had made observations of cells in both plants and animals in the seventeenth century, it was not until 1838–1839 that Schleiden and Schwann formalized the theory that the cell is the basic unit of structure in all organisms. Furthermore, although cell division had been observed in the living cells of both plants and animals early in the nineteenth century, the recognition that the process of cell division was universal and was the basis for the continuity of living systems was not accepted prior to 1858, when Virchow, on the basis of his studies of abnormal cells, put forward the famous dictum that all cells arise from preexisting cells. (It was not until 1875 that Pasteur was able to extend this dictum to the microorganisms to show that all forms of life are basically the same.) Thus the work of Mendel is essentially contemporaneous with the recognition of the cell as a unit of inheritance, but it precedes any understanding of the *mechanism* of cell division, a mechanism that was elucidated only in the last quarter of the nineteenth century.

The description of cell division that follows is given in terms of our present understanding of the process. No attempt has been made to explain how the various statements made were shown to be correct. Furthermore, the description is as simple as possible, and the terminology has been kept to a minimum. None of this is meant to imply that the extensive description of cell division, which is the work of the cytologist and the cytogeneticist, is unimportant. Rather, as geneticists, there are only certain aspects of cell division on which we care to focus. Consequently, it is not necessary to name every organelle and discuss the many unusual events. We are only concerned at this time with the possible genetic effects of cell division.

Mitosis

Figure 7-1(a) presents a diagram of a typical animal cell (that is, a cell that has never existed anywhere except in the mind of the artist). The cell contains a rather prominent nucleus, which is separated from the rest of the material of the cell (the cytoplasm) by a nuclear membrane. In an unstained cell, the nucleus is essentially clear; in a stained cell, however, it can be seen to have a granular appearance, the stain being taken up specifically by what appear as small granules distributed randomly through the nucleus. Generally, within the nucleus, one can see one or more spherical bodies that appear to be homogeneous, whether stained or unstained. These bodies, called the *nucleoli* (sing. *nucleolus*), differ in composition from the rest of the nucleus, since they do not stain in the same way. They are structures of great importance to the cell's metabolism; but since they are not directly pertinent to cell division, they will be omitted from this discussion (see Chapter 21). It is enough to say that they are present during interphase, the period of time when the cell is not dividing (that is, the period during which the cell performs most of its important functions), but they disappear during division and are reformed at the next interphase.

We now know that the granular appearance of the nucleus is caused by the differential staining of certain regions of the chromosome. In fact, during interphase the chromosomes are very long and thin, so thin that, for the most part, they are beyond the powers of resolution of the light microscope. In certain regions, however, the chromosomes appear to be compacted, and here the staining is accentuated so that we are able to see them as granules within a nucleus. It is important to understand from the very beginning that, in this sense, the granules are an artifact and that the chromosomes are continuous structures at all stages of the cell cycle even though they are not always visible as such. The chromosomes in interphase are much the same as a bowl of spaghetti. It is not easy to make out the beginning or end of individual strands of spaghetti in the bowl, for they are much too long and thin and greatly intertwined.

Prophase

When the cell is ready to divide, the nucleus swells and the chromosomes become more directly visible. This process occurs because the chromosomes are shortening by coiling up in much the way a stretched screen door spring will shorten and thicken when the tension is reduced. The coiling proceeds continuously until the chromosomes become the very short, thick rods that are characteristic of the next stage of division, metaphase.

Just outside the nuclear membrane in Figure 7-1(a), two small dots can be seen. These dots represent bodies called *centrioles* that are prominent characteristics of most animal cells and that are found in all cells of both plants and animals having motile sperm. These bodies are absent in the cells of higher plants. However, the process of division is essentially the same in all cases, and thus we can presume that in the higher plants the role played by the centriole is taken over by some other less easily visualized center of organization within the cytoplasm. For this reason, a discussion of cell division given in terms of the animal cell is perfectly satisfactory for all cells, the absence of the centrioles in higher plants being inconsequential from our point of view.

During prophase the centrioles begin to separate and migrate along the edge of the nuclear membrane (Figure 7-1b). Simultaneously with this migration, the cytoplasm becomes fibrous in the region immediately surrounding each centriole, giving rise to a "starlike" body called the *aster*. This separation continues throughout the prophase until the two centrioles and their surrounding astral material lie directly opposite one another on the two sides of the nucleus, which by this time is generally greatly distended (Figure 7-1c).

In late prophase the nuclear membrane breaks down, and a fibrous structure is organized between the two centrioles, which because of its shape (Figure 7-1d) is given the name *mitotic spindle*. The spindle is the apparatus on which the separation of the chromosomes will be effected. At some time during the formation of the spindle, the chromosomes become attached to it through a region of their structure known as the *centromere* (Figure 7-2). The centromere is a small but complex region of the chromosome, containing two small bodies, the *kinetochores*, the structures to which the fibers of the spindle actually attach. Since there are two kinetochores, on opposite sides of the chromosome, one can be attached to a fiber extending from it to one pole of the spindle while the other is attached in the same manner to the opposite pole. The contraction of these fibers will pull the two halves of the chromosome toward the different poles when the division process is completed.

It is already possible to see, by late prophase, that the chromosome is mostly a double structure (Figure 7-2). Using the centromere as a marker, the chromosome can be described as having two "arms," one on the right and one on the left. The placement of the centromere varies from chromosome to chromosome. In certain cases, it may be directly in the middle and in others displaced very far to one end or the other. However, the length of the two arms in each chromosome appears to be an invariant feature of chromosome structure, even if one arm is very small. In addition to the two arms of the chromosome, Figure 7-2 shows that, except that in the centromere region, the chromosome is "split" lengthwise into two

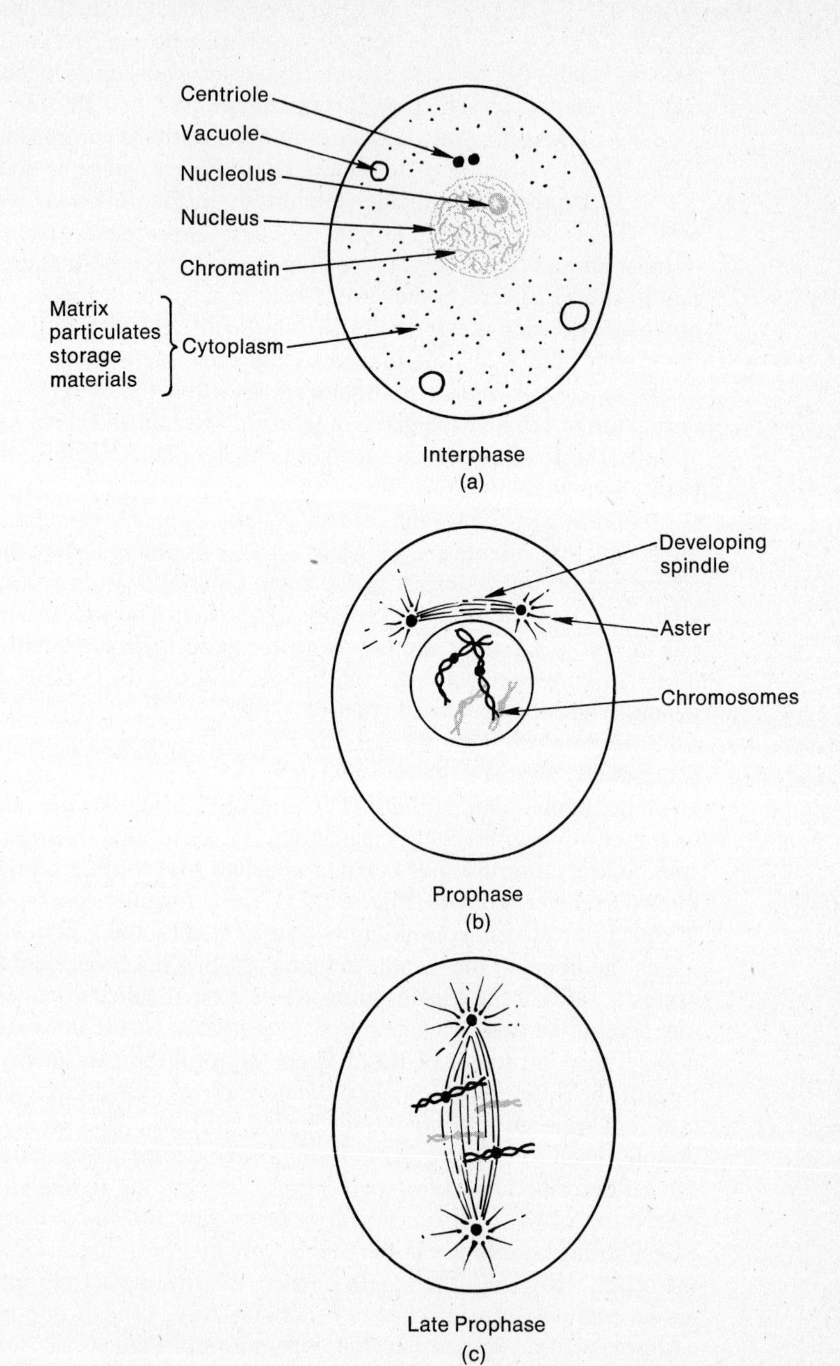

Interphase
(a)

Prophase
(b)

Late Prophase
(c)

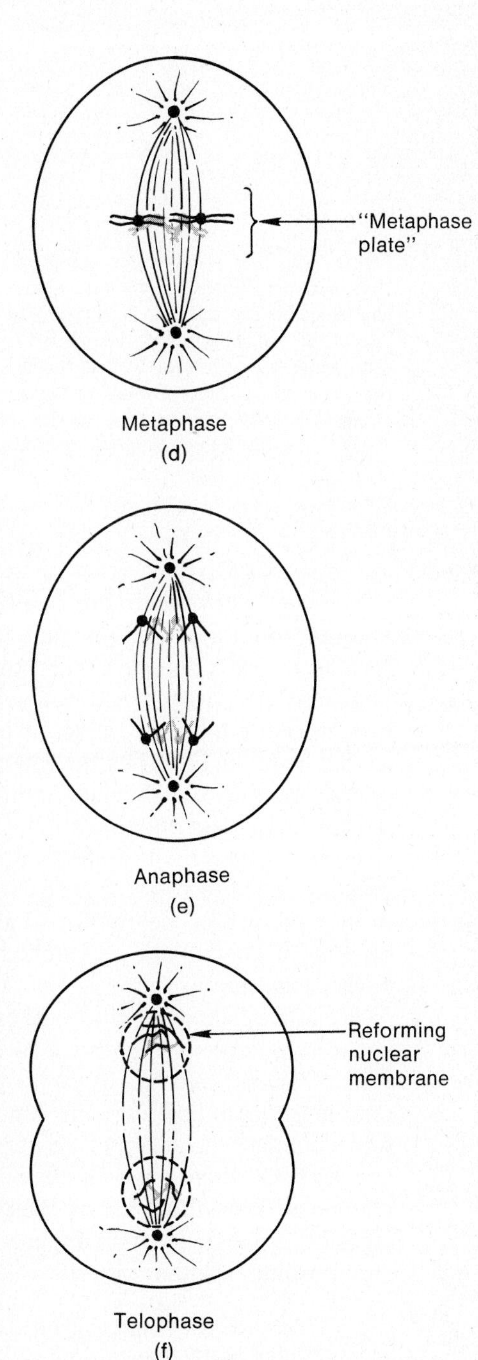

Figure 7-1. Mitosis. (a) The interphase cell. The nucleus shows little structure except for the pronounced nucleolus. The stained material that makes up a network in the nucleus is called the chromatin. The diagram is somewhat exaggerated, since the chromatin is often more diffuse. The chromatin is composed of regions of the interphase chromosomes that stain relatively darkly. (b) Prophase. Only the nucleus and spindle elements are diagrammed. The nucleolus has disappeared. The nucleus enlarges and the chromosomes begin to coil up into compact, darkly staining bodies. (See Figure 7-2.) The centrioles have separated and begin to move apart. Fibrous protein structures develop between and around them (the spindle and the asters, respectively). (c) Late prophase. The nuclear membrane has disappeared and the chromosomes have attached to the spindle fibers by their centromeres. (d) Metaphase. The tension on the spindle fibers is equal in both directions and the chromosomes are brought to an equilibrium position at the midline of the spindle (the metaphase plate). (e) Anaphase. The centromere divides and the two daughter chromatids separate, each going to a different pole of the spindle. (f) Telophase. The cytoplasm begins to divide. The two sets of chromosomes have been pulled clear of the plane of division and now begin to lose their tightly coiled structure. A new nuclear membrane begins to form. Ultimately, two identical cells will be formed.

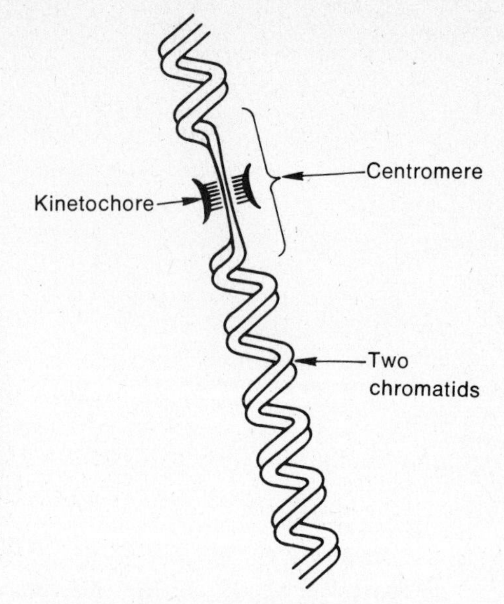

Figure 7-2. Late prophase chromosome. The chromosome is shown as a double structure that has begun to coil up very regularly. The two "arms" of the chromosome are joined by the centromere (kinetochore). Each arm consists of two chromatids. The centromere is the point of attachment of the chromosome to the spindle fibers. It divides longitudinally at the onset of anaphase, permitting the two identical chromatids to separate and move to the opposite poles.

chromatids. (Thus there are two right arms and two left arms.) These are the two "halves" of the chromosome that will separate during division, and it is important to recognize that these two chromatids are *exact* copies of one another. At the beginning of the cell cycle, there is only one copy. Then during interphase, by a process that is not yet completely understood, another exact duplicate is organized, so that by prophase, when the cell is ready to divide, each chromosome is composed of two identical units* (right and left as drawn in Figure 7-2). In the centromere region the process of duplication is not complete. Here the chromosome is a single structure. Consequently, the two chromatids cannot separate until the duplication process is completed in the centromere. Completion of duplication occurs at the end of metaphase (see below), at which time the two chromatids become completely separate, each with its own kinetochore and each attached by fibers to one of the two poles of the spindle.

Since each chromatid is an exact copy of the other, it is not correct to call them half chromosomes. Each is a fully functional chromosome in its own right and will proceed to act as such in each of the two cells produced by the division. This terminology can lead to some confusion. When is a chromatid a chromosome and when is it half of one? For clarity, we will use the term chromatid only when it is essential to discuss the duplicate nature of the late interphase, prophase, or metaphase chromosome. In these cases, chromatid refers to each of the two identical copies shown in Figure 7-2. In all other cases, we will use the word chromosome, whether or not it is a double structure.

*The same thing happens to the centriole. Each daughter cell receives one, which is replicated during interphase.

Metaphase

In time the fibers connecting the kinetochores to the poles of the spindle shorten. Since each can exert an equal force, the chromosomes line up in the middle of the spindle and are held there for a brief period prior to the separation (Figure 7-1e). The alignment of the chromosomes in the middle of the spindle is frequently given the special name *metaphase plate.*

At this point the centromeres divide, and now nothing is holding the two chromatids of each chromosome together. Consequently, the spindle fiber can shorten still further and thus separate the two chromatids of each chromosome completely.

Anaphase

The contraction of the fibers uniting the chromosomes with the poles of the spindle continues until the two chromosomes are completely clear of one another. Then the remainder of the spindle fiber material elongates and thus pulls the two groups of chromosomes farther apart from one another. This process is very rapid. The motion of the chromosomes through the cytoplasm results in a characteristic V- or J-shape with the centromere preceding and the arms trailing behind.

Telophase

In telophase many things happen at once. The chromosomes have reached the poles of the spindle and begin to elongate. The nuclear membrane is reconstituted, and the division of the cytoplasm occurs through an infolding at the midline of the cell in animals (Figure 7-1f) or the growth of a new cell wall in plants. Clearly, it is important that these various events be timed accurately. Should the division of the cytoplasm occur before the complete separation of the chromosomes, some loss of chromosomal material might occur. It is an interesting although poorly understood aspect of cell division that coordination of the various stages is sufficiently exact to ensure that each nucleus of each daughter cell has exactly equal amounts of the chromosomal material and at least roughly equal amounts of the cytoplasmic material in most cases.

The two processes are not always linked, however. In certain rare cases, cytoplasmic division (*cytokinesis*) may occur without nuclear division (*karyokinesis*). In other cases, and far more frequently, nuclear divisions occur without division of the cytoplasm (in the development of the gametophytes of higher plants, in the early stages of insect development, etc.). Consequently, we must consider the two processes as distinct but normally correlated.

When the division process is complete, the nuclei are reestablished in the interphase condition through an uncoiling of the compacted chromosomes, and each cell is now in a position to develop according to its own plan, including the resynthesis of a new exact copy of itself in preparation for the next round of division.

General Comments

We might well ask why the chromosome bothers to go through the cycle of coiling and uncoiling that is characteristic of the division process. In one sense, it is obvious that if the chromosome is not compacted at the time of division, the pinching-off of the two daughter cells might cut off the ends of the chromosomes. To understand the magnitude of the problem, consider the case of man. About 6 feet of fully extended chromatin is divided among the 46 chromosomes, giving 1.5 inches for the length of an "average" chromosome. All this material must be separated into daughter cells whose diameter is only about 0.001 inch! That is a neat trick unless an efficient mechanism exists to compact the chromatin to a very small percentage of its extended length.

Clearly, a compact structure is desirable to effect the complete separation of the chromosomes in mitosis. Why, then, do the chromosomes extend so greatly during interphase? We have only the answer that apparently they cannot function unless they are in the extended condition (Chapters 22 and 25). That very fact should give us pause. Considering the amount of material present, it seems unlikely that all the chromatin could be fully extended. In fact, it is probable that some is but that much, possibly even most, is not. The speckled appearance of the nucleus results from the staining of the compacted regions of the chromatin, the most extended parts connecting the compacted regions remaining invisible. Even so, the amount of material contained in the nucleus is great indeed and that should give more meaning to our comparison of the interphase nucleus to a bowl of spaghetti.

Whatever functions the chromosome performs apparently are carried out during interphase, and we must assume that essentially the chromosomal material is

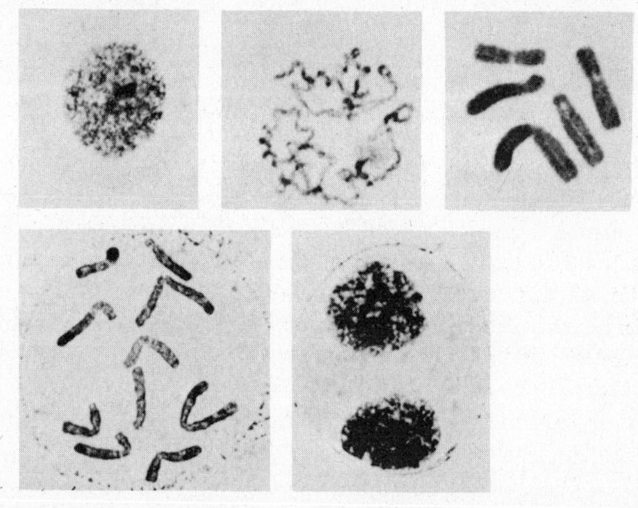

Figure 7-3. Mitosis. The pictures show five stages of mitosis in the microsopore of *Irillium erectum.* The cells are stained to show the chromosomes very distinctly; the spindle cannot be seen. From top left to lower right the five stages are interphase, prophase, metaphase, anaphase and telophase. Note the distinctly doubled structure of the tightly coiled metaphase chromosomes, particularly clear in the lower right chromosomes of the metaphase cell. The centromere does not stain and can be seen either as a constriction (bottom chromosome) or as a clear area joining the two arms of the metaphase chromosomes (particularly evident in the two upper chromosomes). (Photographs courtesy of Dr. A. H. Sparrow.)

inactive from late prophase through telophase. In this regard, the time table of the cell cycle (interphase → interphase) is instructive. Human cells grown in tissue culture take on the order of 18 hours to complete the entire cycle, but they spend only 45 minutes in the process from prophase through telophase. Mitosis is a very brief event in the life of a cell.

A quick review of Figure 7-1 and a comparison to Figure 7-3 should make it evident that the consequence of mitosis is the production of two daughter cells, each of which has an identical chromosomal constitution, and each of which is identical in its chromosomal constitution to the mother cell from which it was derived. To the geneticist, this means that if the chromosomes play any role in inheritance, a subject covered in the next chapter, we can say that the consequence of mitosis is genetic identity, and all the elaborate apparatus of cell division is apparently present to ensure this identity. Any biological process will occasionally be faulty. Mitosis is one of the most regular of all cellular processes, and the number of failures observed under normal conditions is exceedingly rare. This fact suggests that the equal division of the chromosomes is an extremely important process.

Meiosis

In 1884 it was demonstrated that although each parent contributes an equal number of chromosomes in fertilization, the chromosome number of the offspring is the same as each of the parents. This situation can only be explained if each parent contributes one-half its normal chromosomal constitution. (If each parent contributed a full set, the chromosome number would double at each generation.) This observation led to the postulate that the formation of germ cells (sperms and eggs), or as they are more usually called by geneticists, gametes, must go on by a special kind of division, a *reduction division*. (Obviously mitosis is completely unsatisfactory for this process, since the daughter cells produced by it always have exactly the same chromosome number.) It is interesting that this postulate was made without any evidence to support it. However, the logical necessity turned out to be a fact. It was soon demonstrated that a special kind of cell division, which we now call *meiosis*, accounts for the production of gametes.

The nomenclature surrounding meiosis is even more extensive than that for mitosis, and a number of books have been devoted entirely to this form of cell division. However, for our purposes, we can afford to be brief. We merely wish to look at meiosis from the standpoint of the geneticist, capitalizing on the essential similarities between meiosis and mitosis and emphasizing those few but all-important differences in the two processes.

Since the function of meiosis is the production of gametes, it is a division process that is restricted to those parts of an organism normally producing gametes, the testis or the ovary of an animal (or the equivalent parts in plants). Despite the localization of this process, however, its essential similarity to mitosis suggests that it is a derivative process. The various stages of the division cycle are similar (we use the same names); the chromosomes undergo the same behavior, coiling up to become very compact at metaphase, and so on. Furthermore, the mechanism of division is exactly the same in both cases. A spindle forms; the chromosomes attach to it by fibers running from the poles to the kinetochores, and so forth. The basic division process is the same, but there are some marked differences.

One major difference is that meiosis consists of two divisions, separated by a very brief interphase; in fact, in some cases, there is no interphase at all, the chromosomes proceeding directly from a telophase condition to a new prophase. However, since there are two parts to the cycle, it is convenient to speak of them as Meiosis I and Meiosis II, terminology used throughout this text.

Meiosis I Prophase

The most striking difference between mitosis and meiosis is the extended prophase of Meiosis I. In certain organisms, meiotic prophase may last hours, even days, rather than minutes as it does in mitosis. The changes that go on in the chromosomes during this period are both exceedingly slow and markedly different from those that occur in mitosis. For this reason, it is possible to divide the prophase stage into a number of different parts (see Figure 7-4). Each of the stages has been given a name that corresponds to the condition of the chromosome (leptotene = thin threads, pachytene = thick threads, and so forth). These names are used in the figure and in the following discussion for ease of identification, but since most have little meaning for anyone other than a specialist, it is hardly worth memorizing them. Furthermore, it is most important to recognize that meiotic prophase is continuous, as is any division process.

The chromosomes first appear as very thin threads. In certain cases, it is possible to see that they are already doubled, but frequently they do not appear to be. We know by other evidence, however, that they do have a double structure at this point. It is only the fact that they are extended for a longer period of time than they would be in a mitotic prophase that reminds us that they look single. Throughout prophase the chromosomes become thicker and shorter, but something very special happens. As they shorten and thicken, they also synapse—chromosomes that look alike appear to stick to one another and gradually align so that they lie exactly next to one another as pairs. Consequently, in mid prophase the homologous (structurally similar) chromosomes (*homologs*) have synapsed to form a structure composed of four chromatids, two in each chromosome. Although each set of two chromatids is still joined by its centromere, the four units become intimately associated in many regions. This quadapartite structure is frequently referred to as a *tetrad*, a term that will be useful later on.

We do not understand how the synapsis of homologous chromosomes occurs. However, studies with the electron microscope have revealed a unique structure in the regions of tight pairing. This structure, called the *synaptinemal complex*, suggests that the two chromosomes become intimately entwined with one another and may be united by a set of small fibers (of unknown composition) running between the two strands. Unfortunately, too little is known about this structure at the present time to say much more. Significantly, however, the synaptinemal complex is found only in cells undergoing meiosis and only during the time of pairing of the homologs. In those cells in which it is possible to prevent the pairing from occurring experimentally, the synaptinemal complex is also missing. It thus appears that the pairing of the homologs involves a structural union of the two that may be very important in holding the associated chromosomes together during the subsequent stages of the division process. We will discuss the possible role of the synaptinemal complex in more detail when considering the molecular structure of the chromosome (Chapter 22).

Let us briefly review the events occurring in the prophase of Meiosis I. Leptotene is similar to the earliest prophase stages of mitosis except that the chromosomes are very long and thin. In general, the chromosomes have a beaded appearance at this time, apparently the result of regions of tight coiling separated by extended regions of the chromosome. The "beads" are called *chromomeres*, and, interestingly, they always appear in precisely the same relationship to one another in any one organism. The fact that the distribution of these regions along the chromosome is not random suggested early to cytologists that the chromosome might have a linear organization. At zygotene, the chromosomes become visibly synapsed. It is probable that some association has existed prior to this stage, but it is impossible to see it with the light microscope. Once the synapsis has started, it proceeds along the length of the chromosome until complete, a stage referred to as pachytene. Shortly after, the chromosomes appear to move apart. The double structure of the paired chromosomes is now quite evident, and so this stage is called diplotene. The regional separation of the paired homologs seem to be the result of a relaxation of the pairing forces, resulting from, or correlated with, a local dissolution of the synaptinemal complex.

It is in diplotene that we can see most clearly the four chromatids of the tetrad. Where joined to one another, these chromatids give the appearance of a crosslike structure. Consequently, these regions have been given the special name *chiasma* (plural, *chiasmata*). The chiasmata represent regions of close pairing of chromosomes at the very least. They may also represent regions where physical exchange of parts of chromosomes occurs. Figure 7-5 shows homologs joined by chiasmata in the diplotene stage.

By diplotene, the coiling of the chromosomes has become more pronounced. The coiling continues the process of shortening the chromosome and thus forces the pairs farther apart from one another. At the stage called diakinesis, the contraction of the chromosome forces the chiasmata to move laterally, so that in many cases the paired homologs are joined only at the ends, the chiasmata having been "terminalized."

In large chromosomes with many chiasmata, the terminalization is never complete, so that extended regions of the chromosome may appear joined even into metaphase (Figure 7-6 on p. 69). However, it is rare that the terminalization ever results in the complete separation of the paired homologs. They remain joined together throughout prophase and into metaphase.

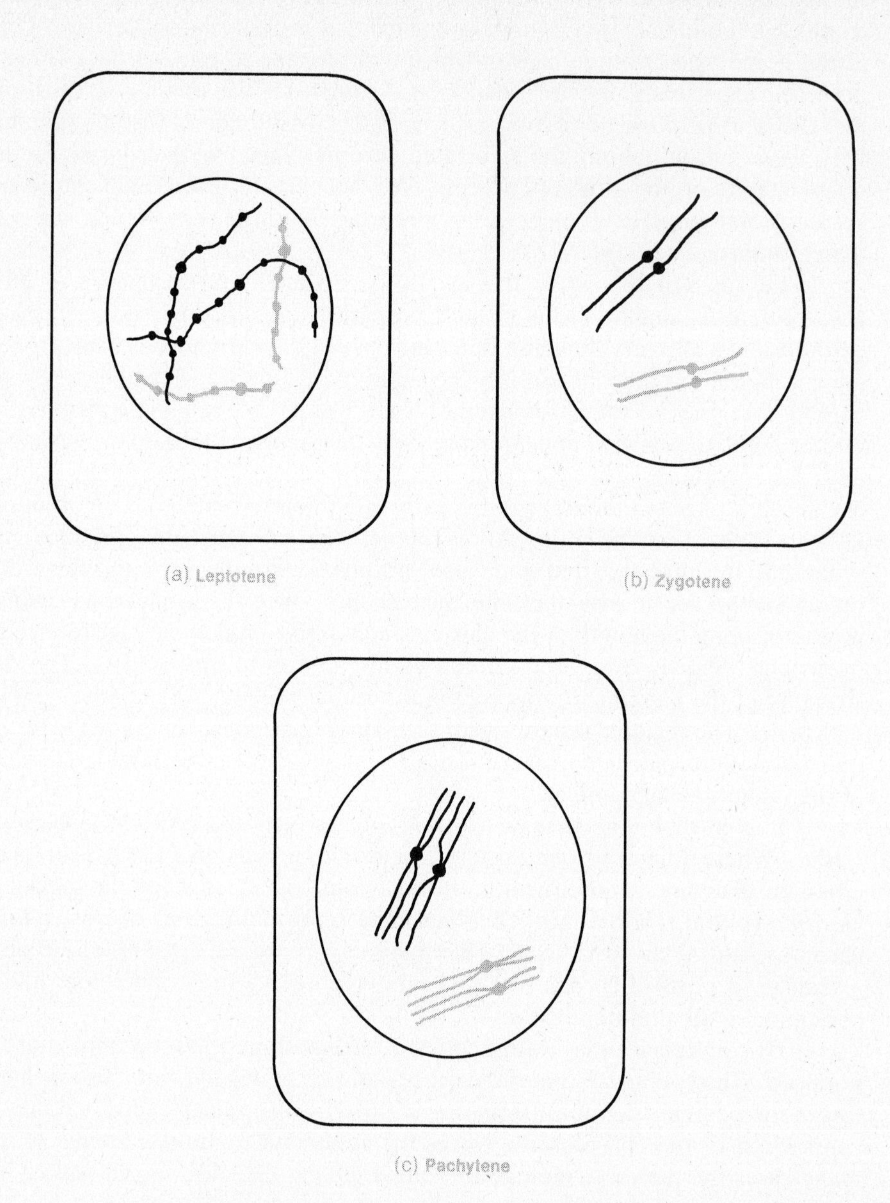

(a) Leptotene

(b) Zygotene

(c) Pachytene

It is interesting that the process of condensation of the chromosomes by coiling in meiosis finally reduces the size of the chromosomes dramatically; by the time they are ready to orient on the spindle, they are very condensed bodies, generally much shorter than the same chromosomes in mitotic division. The usefulness of this supercontraction remains obscure.

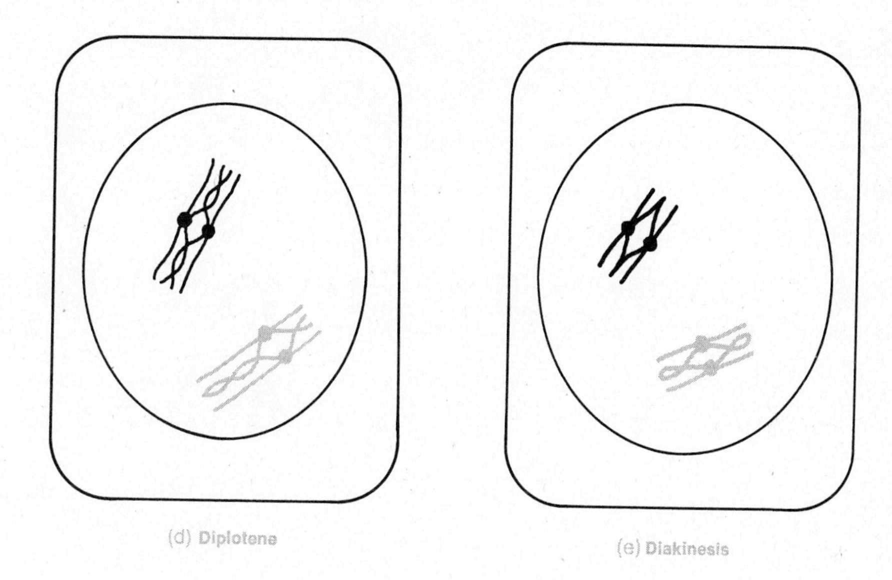

Figure 7-4. Meiosis I prophase. The first meiotic division is unique in that it has a very long and complex prophase. (a) Initially the chromosomes appear as long, thin, "beaded" structures. The small beads are called chromomeres. They are relatively tightly coiled regions of the chromosome. At this stage the chromosomes appear to be singly stranded units; however, we know from other evidence that the replication of the chromosomes has already occurred (in the preceding interphase). Apparently the two chromatids of the chromosome are very closely appressed and cannot be seen as distinctly separate structures in the highly extended condition characteristic of this stage. (b) Soon the extended chromosomes begin to pair in a two-by-two fashion. Structurally similar chromosomes always pair with one another. (c) The pairing continues until it is complete. By this time the double structure of the two chromosomes has become visible, probably as a result of the thickening of the chromosome caused by its progressive coiling. (d) The shortening of the chromosomes forces them apart in many regions, and it can be seen that they are joined only in certain places. The regions of contact are called chiasmata. (e) At the end of prophase, the chromosomes are tightly coiled. Many of the chiasmata have been pushed to the ends of the paired structure (terminalized), and in many cases the pairs appear to be joined only at the ends.

Meiosis I Division

During the course of meiotic prophase a spindle has been formed in the usual manner, and the paired homologs are now connected to the spindle in much the same manner as they were in mitosis. However, in this case, the *pair* of chromosomes moves as a single unit, behaving as if it were one chromosome. The centromere of one chromosome of the

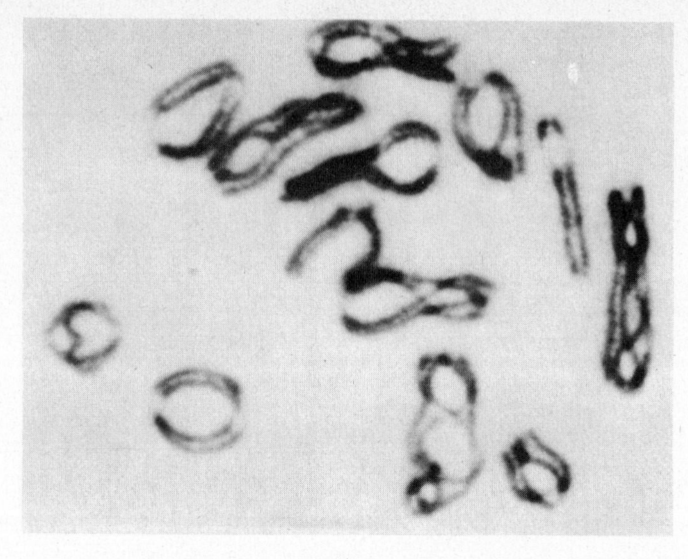

(a)

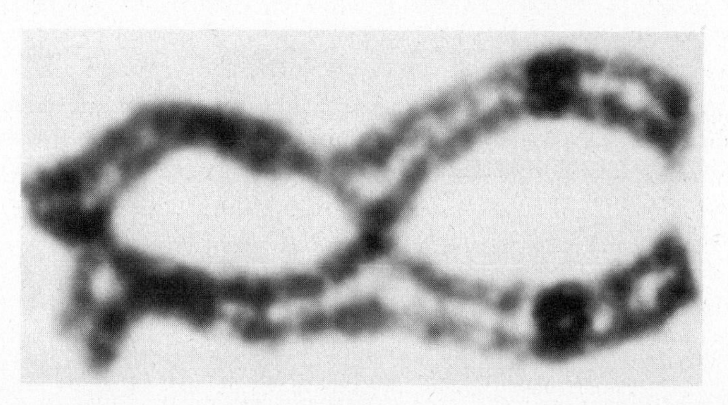

(b)

Figure 7-5. Chiasmata. (a) The photograph shows the diplotene stage of the first meiotic prophase in the salamander *Oedipina uniformis.* Note the numerous places at which the chromosomes are joined, apparently by the crossing of the various chromatids. These are the chiasmata. It is important to note that all four chromatids of any pair of chromosomes can be involved, unlike the rather simplified diagram in Figure 7-4. This is particularly evident in the pair at the far right and the large pair at the bottom of the figure. However, whenever one chiasma has formed at a particular locus between two chromatids, another one cannot form at the same locus between the other two chromatids. Apparently structural interference prevents this from occurring. (b) An enlargement of one pair of chromosomes. Note that two chiasmata have formed in the long arm of the pair. Note also that the two are not in the same chromatids. The chromatids are twisted around one another to permit the chiasmata to lie in the same plane. The twisting is seen most readily near the right chiasma. (Photographs courtesy of Dr. J. Kezer.)

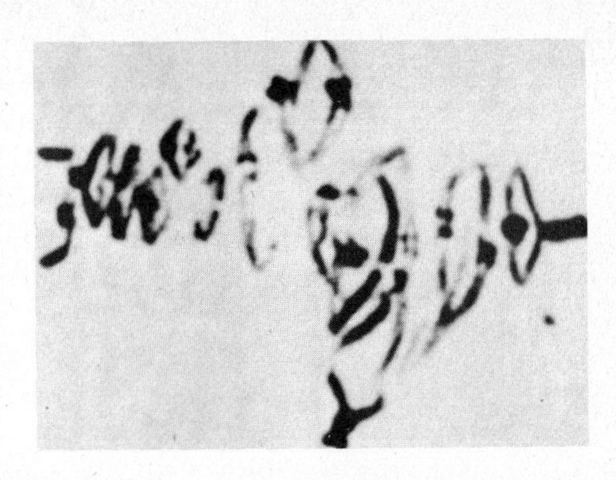

Figure 7-6. Meiosis I metaphase. The picture is of the first meiotic division in the spermatocyte of *Amphiuma*. The tension of the spindle fibers has pulled the paired homologs apart except where they are tightly joined by the chiasmata, which have largely been terminalized. Unlike mitosis, the paired structure insures that whole chromosomes (each having two chromatids) will separate and move to the opposite poles, thus reducing the chromosome number by one-half. (Photograph from Connelly, G. M., and A. H. Sparrow. 1965. *J. Heredity* **56**:91.)

pair is attached by a fiber to one pole of the spindle, and the centromere of the other chromosome of the pair is attached to the opposite pole. Consequently, when the fibers shorten, they will separate the two from one another. When this division is observed in vivo, the chiasmata sometimes appear to be points of tight union, since the chromosome may be considerably stretched before these points of pairing separate. In other cases, the paired homologs separate quite easily. The difference in behavior is unexplained. Whether the synaptinemal complex disintegrates at this time and permits the separation or is destroyed by the mechanical pulling of the fiber is uncertain, although the former seems more reasonable. Nevertheless, the final separation of the paired homologs can sometimes require considerable force.

The motion of the separating chromosomes is precisely the same as in any anaphase, the exception being that in each chromosome we can see two chromatids still joined by their centromere moving toward the poles. A division of the cytoplasm is effected, the chromosomes pass through telophase and into a brief interphase prior to the beginning of the second division.

In one sense, the first meiotic division halves the chromosome number of the individual, and for this reason it is frequently called a reduction division. However, since each chromosome is still composed of two chromatids, each of which, as we have seen, is a completely functional chromosome, it is not strictly correct that the reduction has occurred, and a second division is required to complete the process.

Meiosis II

In some cases, there may be no interphase at all, the cells going directly from telophase to prophase; but in the majority a brief interphase separates the first and second meiotic division. In all cases, however, there is no synthesis of new chromosomal material. The chromosomes are already duplicated (remember that whole *chromosomes* separate at the first meiotic division and each is composed of *two* chromatids). Consequently, the onset of the second meiotic division is rapid.

Unlike Meiosis I, the meiotic prophase of the second division is brief. There are no longer any homologous chromosomes, so that pairing could not occur in any case. It is only necessary for the chromosomes to coil up and orient on the spindle for the second division to be effected. Thus, in this *mechanical* sense, the second division of meiosis is essentially identical to a mitotic division. The centromere divides and the two chromatids of each chromosome are separated into the daughter cells; each receives one chromatid of each of the different kinds of chromosomes.

The end result of meiosis is the production of four cells, each of which has one-half the normal chromosome number of the parent. These cells now develop into the gametes that will effect fertilization and by that union reestablish the normal chromosomal number of the individual. For this reason, it is convenient to have terms that can be used to describe the chromosome number. We consider the normal condition of most cells of higher organisms to be *diploid*—that is, having two sets of each kind of chromosome (the two homologs that pair in prophase I). Each of the cells produced by meiosis have one-half this number and are called *haploid*. For purposes of abbreviation, the haploid number is generally written as n and the diploid number as $2n$. In the human, $2n = 46$, $n = 23$; in the onion, $2n = 24$, $n = 12$. Higher levels of ploidy would be $3n$, $4n$, and so forth. Gametes are characteristically haploid. The cells of the body (somatic cells) are generally diploid. The union of two haploid gametes produces a diploid zygote, which will develop into a normal, diploid adult.

General Comments

The diagram in Figure 7-7 shows an overview of the entire meiotic process. Here we have shaded the chromosomes differently to make them easy to follow. Obviously one cannot do so in real life. However, because the chromosomes are frequently morphologically distinct (relative position of centromere, differences in length, position of the chromomeres, and so forth), we can distinguish the different chromosomes quite well. What we can see in this way is made clear in the diagram by the different shades. Notice that in Figure 7-7(a) the two black chromosomes went to one pole as a result of the first metaphase, and the two gray ones went to the other pole. However, in theory, another arrangement is possible (Figure 7-7b). By following this alternate metaphase from the first division through the completion of meiosis, we end up once again with four cells, each of which has half the chromosome number. However, it is important to note that the constitution of the chromosomes (assuming the differences in color have meaning) is quite different.

In speaking now of the genetic consequences of meiosis, we can see that there are several important implications. First, the process reduces the chromosome number by one-half, but this reduction is not haphazard. As a consequence of the pairing of homologs, each daughter cell gets one member of each pair of homologous chromosomes. Secondly, if, as we have assumed, the two homologs that are shaded differently are in fact at least somewhat different, and if it is true that the two arrangements in metaphase occur at random, we would predict that the meiotic process would result in the production of different combinations of chromosomes within the various gametes. Unlike mitosis, in which all resulting cells are identical, meiosis can give nonidentical products. Once again we must emphasize that this result occurs if our assumption of differences in the chromosomes is correct. It is this assumption that is discussed in the next chapter.

Gametogenesis

Meiosis is required for the sexual process in which the gametes of two different individuals unite to form a third individual. Consequently, we cannot end a chapter on cell division without some comment on the mechanism of production of the gametes themselves.

The production of sperm in animals is a direct result of meiosis. The four products of the division at Meiosis II differentiate into motile sperm. In this process, a tail is formed, which gradually elongates into a very long, thin, flagellar structure. Simultaneously with tail formation, droplets of cytoplasm are extruded until eventually the functional sperm is little more than a nucleus containing tightly compacted chromosomes, some mitochondria,* and the tail to propel the sperm in its long swim in search of an egg. In its final form, the functional sperm consists of four major parts: an acrosome, which is a small body containing enzymes that can digest intercellular substance through which the sperm may need to pass in order to get to the egg cell; the nucleus, which contains the highly compacted, inactive chromosomes; the midpiece, which contains the mitochondria that provide the energy to keep the sperm swimming; and the tail.

The production of the egg is rather different from that of the sperm. In animals, the egg generally has a large cytoplasmic mass, which contains the stored nutrients for the development of the zygote. The sperm contributes nothing but a nucleus. To achieve the maximum size of the egg, the process of meiosis is modified in such a way that it produces a single functional gamete and three, nonfunctional, tiny cells, which are discarded. This is achieved by having the division process occur on the surface of the egg, so that at the end of Meiosis I one very small cell (polar body) is produced and one very large cell. In Meiosis II, the first polar body may or may not divide; but the large presumptive egg cell undergoes a similar superficial division, producing a small secondary polar body and the final, very large haploid egg cell. The polar bodies are discarded, and a single, large egg cell is produced. Thus, in the male, four motile sperm are produced for each meiotic division, but in the female only a single egg is produced as a result of meiosis.

Plants

In most species, gametes are produced from a haploid plant ($1n$), the gametophyte. Consequently, they are produced by a mitotic rather than a meiotic mechanism. On the other hand, once these gametes have fused, they produce a diploid ($2n$) sporophyte, which is generally very different from the gametophyte that produced the gametes. Meiosis occurs in the sporophyte and gives rise to haploid spores, which, in turn, germinate to give rise to gametophytes. This difference in the life cycle of animals and plants means that in plants meiosis really produces spores rather than gametes. Nevertheless, in the higher plants at least, the gametophyte is of short duration and so tiny (generally an

*The mitochondria are small, numerous cell organelles that carry out a great deal of the intermediary metabolism. They supply the energy for most cell processes and, as such, have been called "the powerhouse of the cell."

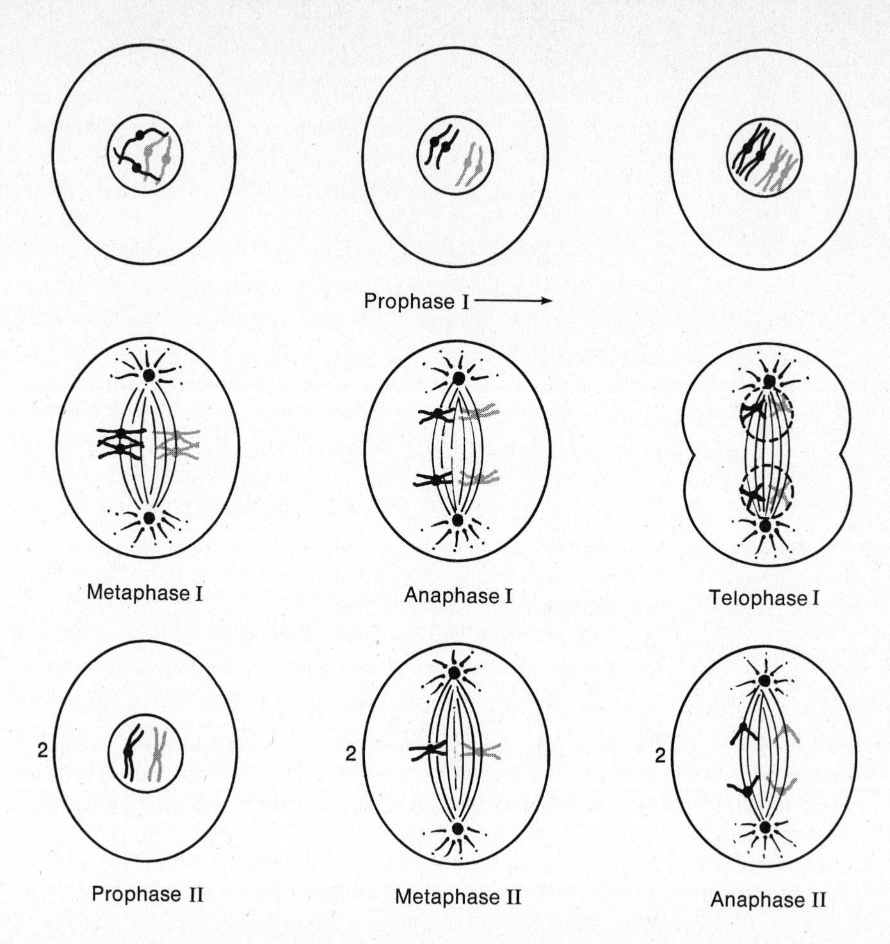

internally localized single cell) that the distinction is relatively unimportant from the point of view of the geneticist. However, to be strictly correct, it should be understood that the pollen of the higher plants is not the sperm but a small spore (microspore) that will, on germination, give rise to the tiny gametophyte (the pollen tube) that grows out from the pollen grain. This will produce the sperm, by mitosis, that will then fertilize the egg. Similarly, in the female line, the principal product of meiosis is a large spore (megaspore) that immediately undergoes a series of mitotic divisions to give rise, finally, to a single egg cell. What follows is a brief account of this process as it is found in corn (maize), a plant of great interest to geneticists and one with which it is hoped the student is familiar, at least in general terms.

The pollen is produced in the tassel, a structure consisting of many individual male (staminate) flowers. The reproductive organs of these flowers, the anthers, contain cells, each of which undergoes meiosis to produce four apparently identical microspores. Each microspore undergoes a mitotic division of the nucleus without an accompanying division of the cytoplasm. The two nuclei so produced differentiate, so that they are morphologically distinct although chromosomally identical.

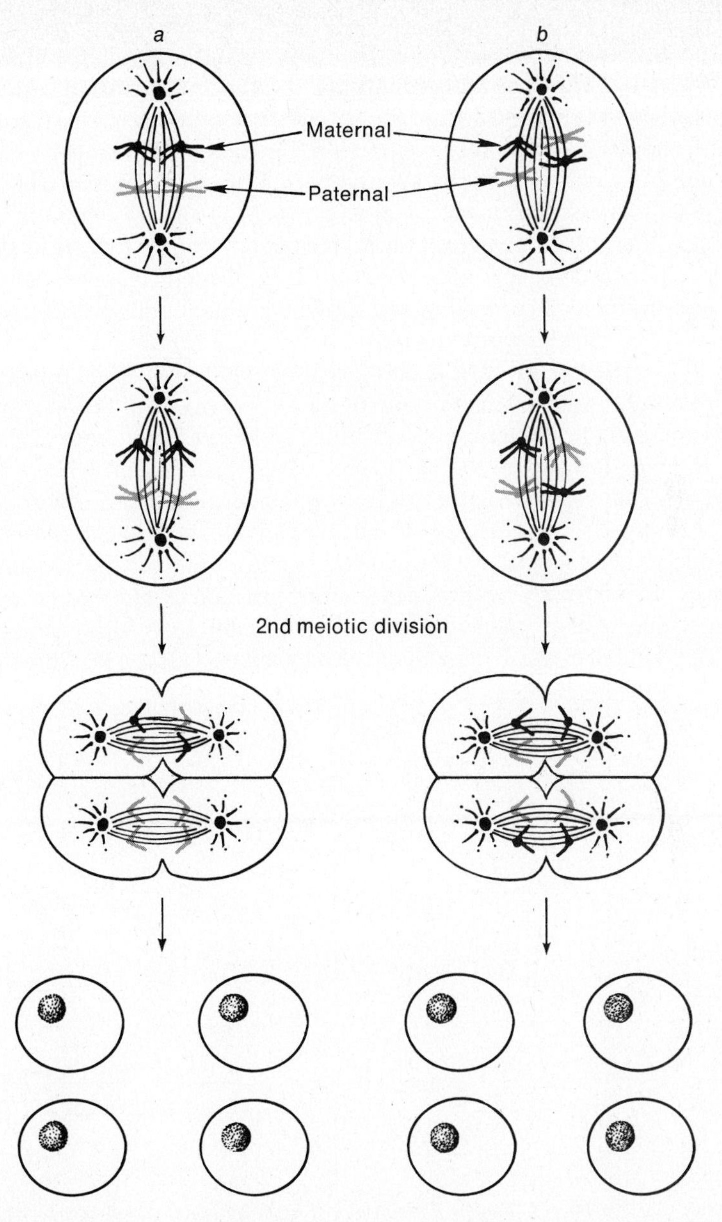

Figure 7-7. Meiosis. The diagram presents an overview of the entire meiotic process. It is prepared to demonstrate the major consequence of meiosis: the chromosome number is reduced by one-half but the *quality* of the chromosomal constitution of each daughter cell is unchanged. Each cell produced by meiosis has one of each *kind* of chromosome. However, the distribution of maternal and paternal chromosomes may be quite different (compare *a* and *b*). At the end of meiosis four haploid cells are produced. These may be equal in size and all survive to function or they may be unequal in size and only one survive.

Each binucleate microspore then develops a heavy cell wall that will prevent dehydration when the anthers rupture and the pollen is shed. At this point the pollen is ready to be wind-borne to the female flowers of other corn plants, where it is caught by the "silks," the receptive portion (pistil) of each female flower found on the corn cob. The pollen penetrates the silk, forming a long, thin, unicellular pollen tube (the male gametophyte) that extends down to the ovary where the egg cell is embedded. During the growth of the pollen tube one of the two nuclei divides mitotically to produce two sperms. The other nucleus of the original pollen grain is apparently concerned with controlling the growth and metabolism of the pollen tube and plays no direct role in reproduction.

The ear of corn is composed of many individual female flowers, and each "kernel" is the product of one of these. Within each flower a single cell undergoes meiosis to produce four spores, three of which degenerate while the remaining one enlarges to become the megaspore. It is this cell that will eventually produce the egg. The nucleus of the megaspore undergoes three divisions to produce eight seemingly similar nuclei, which float in a common cytoplasm. A distribution of nuclei then takes place: three move to one end to become the antipodals, which play no further role in reproduction; two remain in the middle of the megaspore to become "polar" nuclei, which will eventually fuse with each other; the remaining three move to the other end of the megaspore. Of the last three nuclei, two become

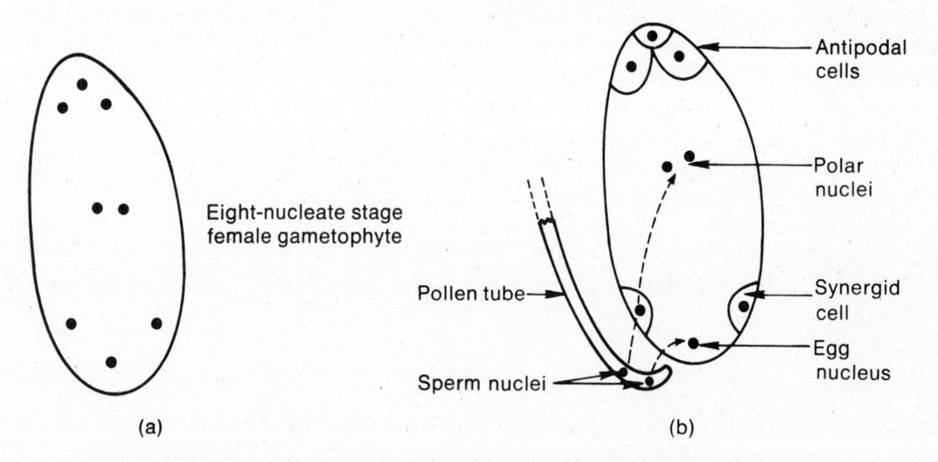

Figure 7-8. Female Gametophyte, eight-nucleate Stage. In many higher plants, meiosis produces four megaspores, one of which survives. The surviving cell enlarges greatly and undergoes three mitotic divisions of the nucleus without division of the cytoplasm. The result is a large cell containing eight haploid, identical nuclei (a). Cytoplasmic division removes five nuclei, the three antipodals and two synergids (b). The two polar nuclei unite with one of the two sperm nuclei to give rise to the triploid endosperm nucleus. The egg nucleus unites with the other sperm nucleus to give rise to the diploid zygote nucleus. Subsequent division of the cytoplasm separates the two regions of the egg.

synergids and are separated off by cell walls, and the last one becomes the egg nucleus. The megaspore is now ready for fertilization to take place (Figure 7-8).

The pollen tube penetrates the tissues surrounding the megaspore and discharges its two sperm into the interior of it. One sperm fuses with the egg nucleus to form the diploid ($2n$) zygote nucleus. The other fuses with the two polar nuclei to form a triploid ($3n$) fusion nucleus. Unequal cytoplasmic division now separates the megaspore into two cells, the smaller of the two being the zygote and the larger one containing the fusion nucleus. Repeated mitotic divisions of the zygote will give rise to the embryo, while comparable divisions of the fusion cell will produce a nutritive tissue, the *endosperm*, that will nourish the young embryo when the seed (kernel) is planted. The endosperm is therefore equivalent to the yolk of an animal egg, which nourishes the embryo during the early stages of its development. By far the greater part of the corn kernel that we eat is the endosperm.

The preceding discussion of gametogenesis in corn serves two purposes. It is frequently useful to use characters of the endosperm in plants; and in order to understand such crosses, it is essential to recognize that the endosperm is triploid. Furthermore, this brief review of the gametogenesis of higher plants should serve to make clear that the function of meiosis is the same in all cases whether or not it directly produces sperm or eggs. The variations in reproductive mechanisms exhibited by organisms all have a common goal: union of the right sperm with the right egg in an often hostile environment. No matter how tortuous the route, however, the ultimate function of meiosis is to produce haploid gametes that can unite with other haploid gametes to produce a diploid organism.

REFERENCES

SCHRADER, F. 1944. *Mitosis.* Columbia University Press, New York. (A classic review of the early literature on the mechanism of cell division. Now a little out of date, but it presents the problem clearly and fully.)

SWANSON, C. P. 1957. *Cytology and Cytogenetics.* Chapter 3. Cell division and syngamy. Prentice-Hall, Englewood Cliffs, N.J. (A far more extensive review of the material treated in this chapter.)

VOELLER, B. R. 1968. *The Chromosome Theory of Inheritance.* Appleton-Century-Crofts, New York. (A collection of classic papers, frequently only excerpts, all in English. It includes papers on the discovery of fertilization, the significance of meiosis, and so forth. The papers by Weismann, Fleming, Hertwig, Roux, Strasburger, and Van Beneden are particularly pertinent to the material discussed in this chapter.)

WILSON, E. B. 1925. *The Cell in Development and Heredity.* 3rd ed. Chapter 2, Cell division, and Chapter 6, Maturation and reduction: Meiosis. The Macmillan Co., New York. (The classic text in cytology. Although out of date and far too comprehensive for the average undergraduate, it is well worth examining for the historical material.)

YOST, H. T. 1972. *Cellular Physiology*. Chapter 15, pp. 704–738. Cell division. Prentice-Hall, Englewood Cliffs, N.J. (An account of the mechanism of cell division in molecular terms, for the advanced student.)

QUESTIONS AND PROBLEMS

7-1. Define the following terms:

chromosome	chromatid
chromatin	chromomere
centromere	centriole
kinetochore	spindle
fiber	cytokinesis

7-2. What is accomplished by mitosis?

7-3. The replication of the chromosomes is not finally completed until metaphase has been reached. For what reason is one region of the chromosome replicated so late?

7-4. When does replication of the majority of the chromosome occur? How many chromatids does an unreplicated chromosome contain? How many does a replicated one have?

7-5. Define the following terms:

synapsis	homolog
chiasma	diploid
haploid	tetrad
reduction division	

7-6. What is the importance of meiosis?

7-7. In what ways are mitosis and meiosis similar? In what important ways do they differ?

7-8. Tradescantia has a diploid ($2n$) chromosome number of 12. What is the chromosome number of a root tip cell? Of a pollen tube nucleus? Of a zygote nucleus? Of an endosperm nucleus?

7-9. If an animal has six chromosome pairs, what proportion of its gametes will receive the following distribution of centromeres?

(a) All from the father
(b) All from the mother
(c) All from the father or all from the mother
(d) All from the father and all from the mother

7-10. What happens in the meiosis of an organism that has an odd number of chromosomes?

7-11. If the diploid chromosome number were 17, what would be the haploid number of the four sperm produced from meiosis? In the egg?

7-12. A cross of a horse by an ass produces a mule. The mule is almost always sterile, a condition that usually results from the crosses of two different species. Suggest a reason.

7-13. Triploid ($3n$) organisms generally have abnormal meiosis. Why? Is their mitosis normal? Why?

7-14. Give a reason why many organisms have adopted the practice of producing four functional sperm but only one functional egg.

8/The chromosome theory of inheritance

The work of Gregor Mendel presented certain formal regularities. He showed that if experimental crosses are made in a particular way, certain predictable results occur. In a monohybrid genetic cross, it is predicted that in the F_2 there will be a 1 : 2 : 1 genetic ratio. Similarly, in a dihybrid genetic cross, another predictable genetic ratio will be observed in the F_2. Therefore it can be stated that formal regularities will occur again and again as a result of breeding experiments.

The question now foremost in the minds of the early geneticists was: What is the *causal mechanism* responsible for the Mendelian formal regularities?

One of the earliest theories directed toward this question was the Roux-Weismann theory of the 1880s. Wilhelm Roux had argued that the division of the chromosomes by what appeared to be a longitudinal splitting must imply that they contain many different "qualities" that are arranged in a linear way on the chromosomes. Splitting of the chomosomes in mitosis must be for the purpose of dividing these qualities equally among the daughter cells. Although the argument is teleological, in essence it is correct.

The German zoologist and theoretician August Weismann, proceeding from Roux's ideas, reinforced the concept of a linearly arranged series of "qualities" on the chromosomes that were segregated equally by mitosis. Moreover, as pointed out in Chapter 7, Weismann realized that there must be a form of nuclear division different from mitosis or otherwise the fusion of an egg and sperm that were mitotic in origin would result in a zygote with a chromosome number twice that characteristic of the species. Weismann proposed that there must be a division (meiosis) leading to the formation of gametes that reduced the chromosome number by one-half such that fertilization restored the normal number.

Sutton-Boveri Theory

But it was an American graduate student, William Sutton, at the Department of Zoology, Columbia University, and, to a somewhat lesser extent, the famous German zoologist Theodor Boveri who in 1902 made the correct correlation between Mendel's factors and chromosomes in what has become known as the Sutton–Boveri theory.

A quick review of Mendel's First Law, the Law of Segregation, will recall that when an **AA** parent is crossed by an **aa** parent, the gametes formed by each parent are **A** and **a**; the F_1 is **Aa**; and the gametes formed by such an F_1 are half **A** and half **a**. Sutton and Boveri recognized the parallel between Mendel's genes and the behavior of chromosomes during meiosis and therefore postulated that **A** was on one chromosome and **a** on the other (see Figure 8-1).

Does the Sutton–Boveri theory explain Mendel's Second Law, the Law of Independent Assortment? Recall that a cross between **AABB** and **aabb** produced **AB** and **ab** gametes, respectively, and that fertilization produced the F_1 double heterozygote, **AaBb**. Further recall that the gametes produced by such an F_1 were **AB**, **Ab**, **aB**, and **ab** in equal frequency.

Prerequisites to Sutton–Boveri Theory

Sutton and Boveri assumed that two pairs of factors lie on two pairs of chromosomes, and independent assortment can be explained if we assume that the chromosomes also assort at random (see Figure 8-2). With these assumptions, Sutton and Boveri provided a causal explanation for the formal regularities of Mendelism—namely, that the genes lie on the chromosomes. Although an interesting and attractive hypothesis, it lacks that essential ingredient necessary for scientific hypotheses—proof. Experimental evidence must be sought to provide the justification for what has come to be called the *Chromosome Theory of Inheritance.* But before this theory can be directly tested, certain prerequisites of the theory must first be verified. There are four such prerequisites: (1) the individuality or continuity of chromosomes; (2) pairing between homologous chromosomes; (3) a qualitative

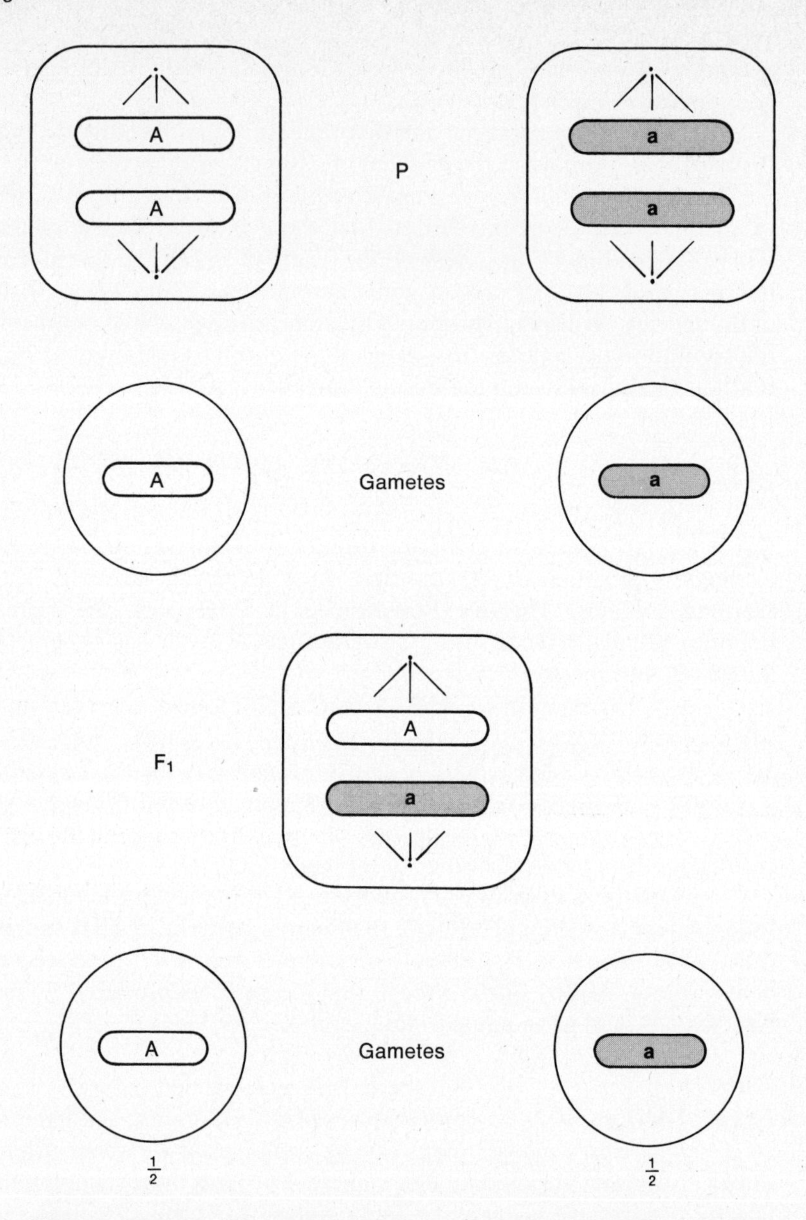

Figure 8-1. Diagram illustrating Sutton–Boveri hypothesis of genes on chromosomes to explain Mendel's Law of Segregation. For simplicity, the chromosomes are drawn as simple, unduplicated rods.

difference between different pairs of chromosomes; and (4) different chromosome pairs recombine independently. These four prerequisites will be discussed one at a time. Each will be defined, and it will be shown why it is essential that they be satisfied before proof for the chromosome theory itself becomes meaningful.

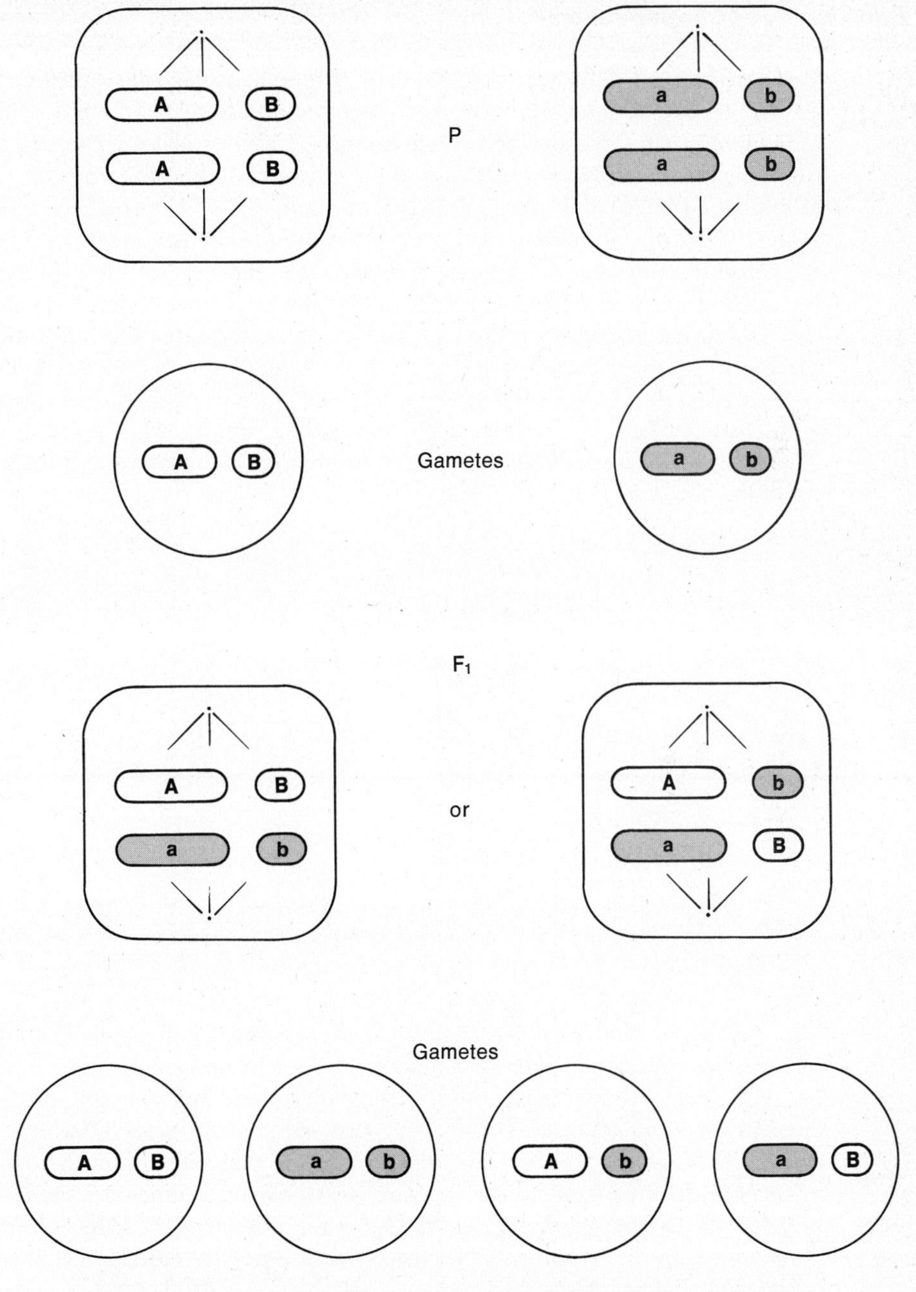

Figure 8-2. Diagram illustrating Sutton–Boveri hypothesis of genes on chromosomes to explain Mendel's Law of Independent Assortment. For simplicity the chromosomes are drawn as simple, unduplicated rods. The two independent pairs of chromosomes are differentiated by size.

Individuality of Chromosomes

There was considerable discussion in the late nineteenth and early twentieth century as to whether the chromosomes were continuous and maintained their individuality. In the discussion of cell division (Chapter 7) it was pointed out that at interphase the chromosome material, or chromatin of the cell, is so long and attenuated that distinct chromosomes are not visible. During cell division the individual chromosomes are observed; but when the cell goes again into interphase, the chromosomes once more become indistinct as separate entities. In the succeeding cell division, the chromosomes are again seen as distinct bodies. The question is: Are the chromosomes seen in the first metaphase the same chromosomes seen at the following metaphase? (See Figure 8-3.) One of the prevalent theories of the time was that the chromosomes were composed of one long thread, or *spireme*, which was then broken up during each succeeding interphase into the individual chromosomes. The spireme theory is contradictory to the concept of the individuality of the chromosomes.

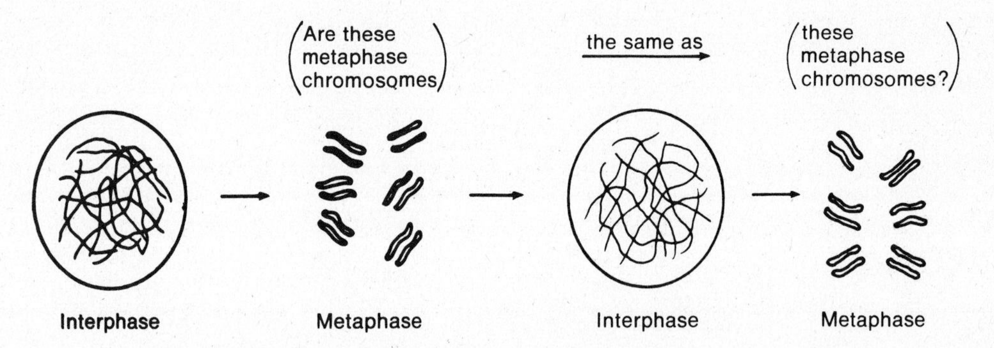

Figure 8-3. The individuality or continuity of the chromosomes. Since the chromosomes are not visible as distinct bodies during interphase the question arises whether the same chromosomes are seen after interphase as are seen before.

The German zoologist Rabl noted that during cell division the configuration of chromosomes at anaphase resulted in all the chromosomes being oriented in a particular way. Rabl claimed that during interphase he was still able to see the chromosomes as separate bodies, but this cytological observation was open to question. More importantly, however, he showed that when the chromosomes were again visible during the succeeding prophase, all had their apices oriented in the same manner as in the preceding anaphase. (See Figure 8-4). This observation is consistent with the hypothesis that the chromosomes are individual and continuous from one cell division to the next.

Theodor Boveri in 1889, working with the parasitic roundworm *Ascaris megalocephala*, provided even more elegant demonstration of the individuality of the chromosomes. Boveri had observed that the nuclei of Ascaris commonly show a

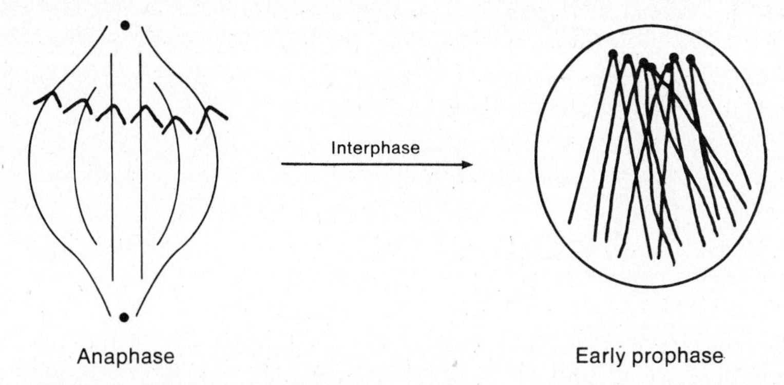

Figure 8-4. Diagram of Rabl's observation of chromosomes during cell division. Rabl claimed that the chromosomes had the same orientation in early prophase that they had in the preceding anaphase. This is interpreted as evidence for the continuity of the chromosomes.

number of finger-shaped lobes or pockets that are formed during telophase by the free ends of the V-shaped chromosomes. These nuclear lobes thus provide landmarks that indicate the position in which the chromosomes have arranged themselves in the resting nucleus. In both a polar and side view of cell division, Boveri noticed that several possible arrangements occurred. Furthermore, he saw that sister cells are usually identical in the number and shape of nuclear pockets formed. Boveri made a final observation while observing cell division in Ascaris, which was that when the chromosomes separated at anaphase, the ends of the chromosomes seemed to stay attached to each other and were the last parts of the chromosome to separate when they advanced toward their respective poles. (See Figure 8-5.)

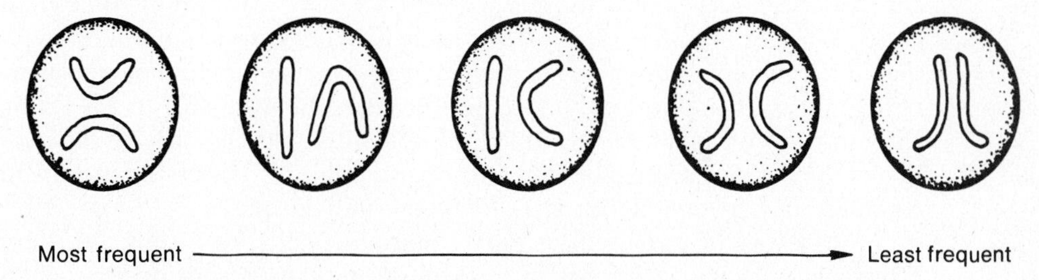

Figure 8-5. Boveri's analysis of the individuality of the chromosomes in Ascaris. Here are some of the several arrangements of the two chromosomes that he observed at metaphase. These are polar views observed by looking at the dividing cell from the top, or pole.

Thus, by observing how the chromosomes enter the telophase nucleus, Boveri was able to draw this conclusion: in the following prophase the chromosomes always

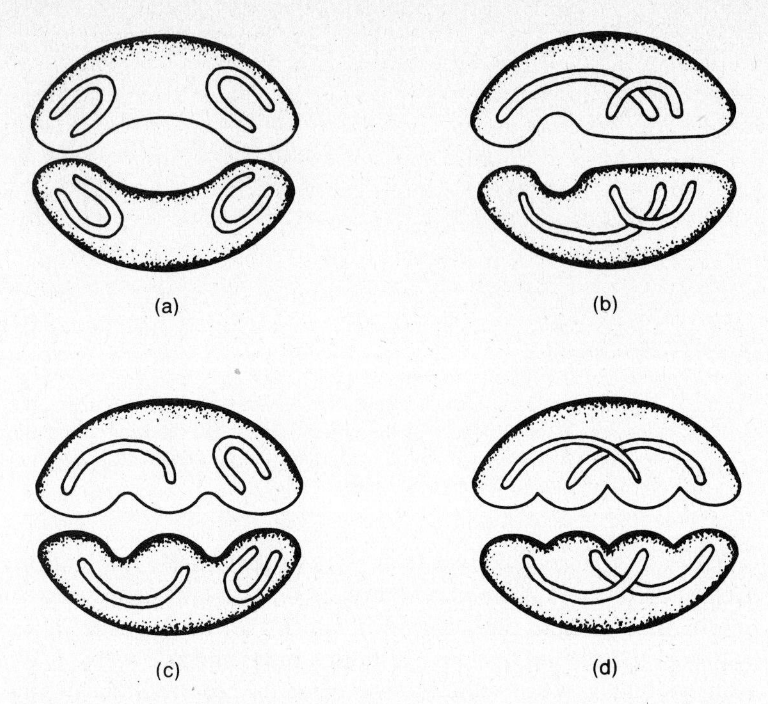

Figure 8-6. Four different results of cell division in Ascaris observed by Boveri. Note that although (a), (b), (c), and (d) differ from each other, in each case the two sister cells are mirror images of each other.

reappear with their free ends lying in lobes identical to those observed in the dividing cell at the preceding telophase. The argument was further strengthened by the fact that although the number and position of the nuclear lobes, and hence the chromosomes, vary widely from cell to cell, they are identical in sister cells. (See Figure 8-6.) From this evidence Boveri concluded that it was not a likely hypothesis that a continuous chromosome thread was broken up during interphase and redistributed as separate chromosomes. It seemed improbable to Boveri that the chromosomes would know just what their configuration was and what the configuration of the sister cell would be. It was more plausible for Boveri to conclude that the chromosomes are continuous and retain their individuality.

Later in the twentieth century confirming evidence of this hypothesis was provided by chance occurrence of the loss, during cell division, of either part or the whole of a chromosome. From such cells where a chromosome was lost or broken, the loss or break is always repeated *exactly* in subsequent cell generations.

Pairing Between Homologous Chromosomes

In many of the first organisms studied the chromosomes were nearly all the same size and shape. For example, the salamander has 24 chromosomes and all are V-shaped and about the same size. Investigators looking at these chromosomes thought they were all alike and thus were uncertain as to whether pairing between homologs actually occurred. It was possible to observe pairing; but because the chromosomes were all identical, it was not certain whether the pairing was random or specific.

Montgomery studied the chromosomes of Hemipterans, or true bugs, and observed size and shape differences between various ones. During meiosis he observed that the largest chromosomes always paired with the largest chromosome, the smallest with the smallest, and so on. Montgomery concluded that pairing between chromosomes was not random but involved two identical homologous chromosomes. This same conclusion was reached by Sutton while studying grasshopper chromosomes and subsequently by many others.

Qualitative Difference Between the Chromosomes

If the Sutton–Boveri theory is correct, then different pairs of chromosomes have different qualities. Stated in modern terms, this means that different chromosomes have different genes. The **A** gene is on one chromosome and the **B** gene is on another, and the assumption is that these genes are different. They are responsible for different developmental processes. If that is true, then it should follow that there is a qualitative difference between the chromosomes in addition to the quantitative difference observed in many organisms that have chromosomes of varying sizes. This prerequisite is not easy to verify, and it took the experimental genius of Theodor Boveri to provide the answer.

Tetraster Eggs

For some time it had been known that when a sea urchin egg is accidentally fertilized by two sperm, the first cleavage figure is multipolar. The egg has four poles established instead of the normal two and at the first division divides into four cells or *blastomeres.* Subsequent divisions usually result in a normal *blastula,* an early developmental stage in the sea urchin. Such blastulae occasionally produce normal pluteus larvae but only in a very small percentage of cases. A wide range of larvae are observed, from nearly normal to very abnormal. Also, in the four quadrants resulting from the first cleavage, four different stages of development might be observed.

Boveri's analysis of this phenomenon was as follows: in normal development where a haploid sperm with 18 chromosomes fertilizes the haploid egg that also has 18 chromosomes, the resulting zygote has 36 chromosomes, the normal diploid number for the sea urchin. The first cleavage results in two cells, each of which has 36 chromosomes. (See Figure 8-7.) In the dispermic fertilization, in addition to the egg nucleus with its 18 chromosomes, there are two sperm, each bringing in 18 chromosomes plus one centriole. In each sperm, the one centriole divides into

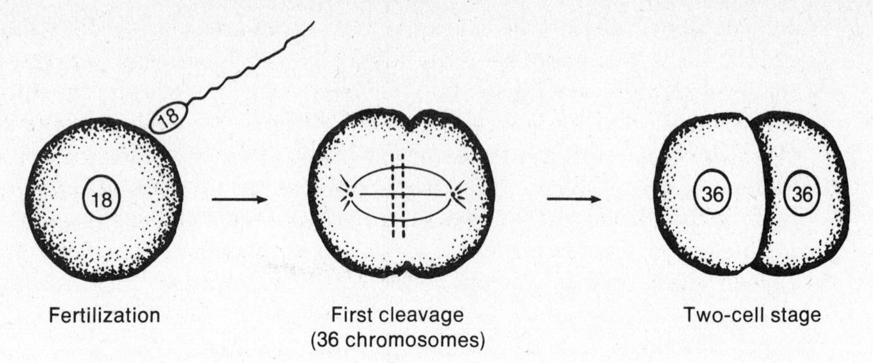

Figure 8-7. Normal fertilization in the sea urchin. The egg and the sperm each have the haploid chromosome number of 18. At the first cell division the 36 chromosomes are divided mitotically to the two daughter cells.

two, giving a total of four centrioles. These four centrioles at the first division set up four poles, and the 54 chromosomes are distributed randomly among the four spindles. The first cleavage results simultaneously in four blastomeres, each of which gives rise to approximately one-fourth of the larva. Such four-spindled eggs resulting from double fertilization are often called *tetraster* eggs. (See Figure 8-8.)

From earlier work with sea urchins, Boveri had learned that there was a close correlation between the number of chromosomes in the nucleus and the size or volume of the nucleus. Therefore, by measuring the volume of the nucleus, he was able to get a good estimate of the number of chromosomes in each nucleus. He tentatively concluded that the difference in chromosome number resulting from the random distribution of the chromosomes was probably responsible for the differences in development in the four quadrants and the overall poor development that he observed in these dispermic fertilizations.

However, Boveri found that sometimes he would observe four blastomeres with approximately the same chromosome number in each and yet he still observed some normal development and some abnormal development. Moreover, as pointed out in Chapter 1, he also knew that parthenogenetic and giant eggs, with 18 and 54 chromosomes, respectively, developed normally. In a further experimental refinement Boveri would place these zygotes at the four-cell stage into calcium-free sea water to separate the blastomeres. He could then trace each blastomere individually in development. He observed that four normal larvae resulted, although each larva contained only 18 chromosomes in the parthenogenetic experiment and 54 chromosomes in the giant egg experiment. His conclusion was that chromosome number was not the essential factor for normal development. Two cells with identical chromosome number do not necessarily develop in a similar fashion.

Boveri now hypothesized that the chromosomes had different qualities. He schematized this in the following way. Each chromosome was designated by a letter:

a, b, c, or d. For simplicity, only four chromosomes are assumed instead of the 18. Boveri postulated that every quadrant must get *at least one chromosome of each kind for normal development.* (See Figure 8-9.)

The example was designed in such a way that the chromosome number in each blastomere was the same in each fertilization, and yet the results were different. Boveri attributed the differences to his hypothesis that the chromosomes are qualitatively different and that, for normal development, there must be at least one of each chromosome pair. How could Boveri prove his hypothesis?

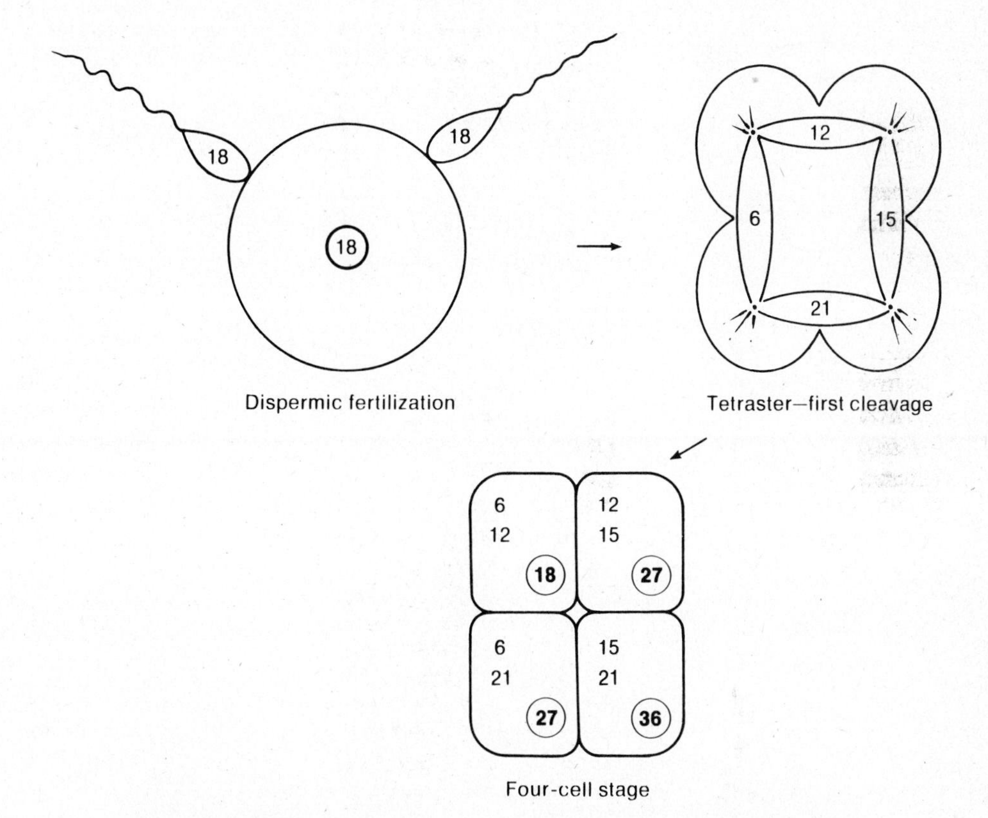

Figure 8-8. Dispermic fertilization in the sea urchin. The double fertilization results in a tetraster egg with four spindles present at the first cleavage. The 54 chromosomes arrange themselves randomly among the four spindles. The chromosomes divide mitotically with half going to one pole and half to the other. For example, the six chromosomes segregate so that six go to the upper left cell and six to the lower left. The 12 chromosomes segregate so that 12 go to the upper left cell and 12 to the upper right, etc. In this manner the four cells end up with the chromosome number indicated in the circle.

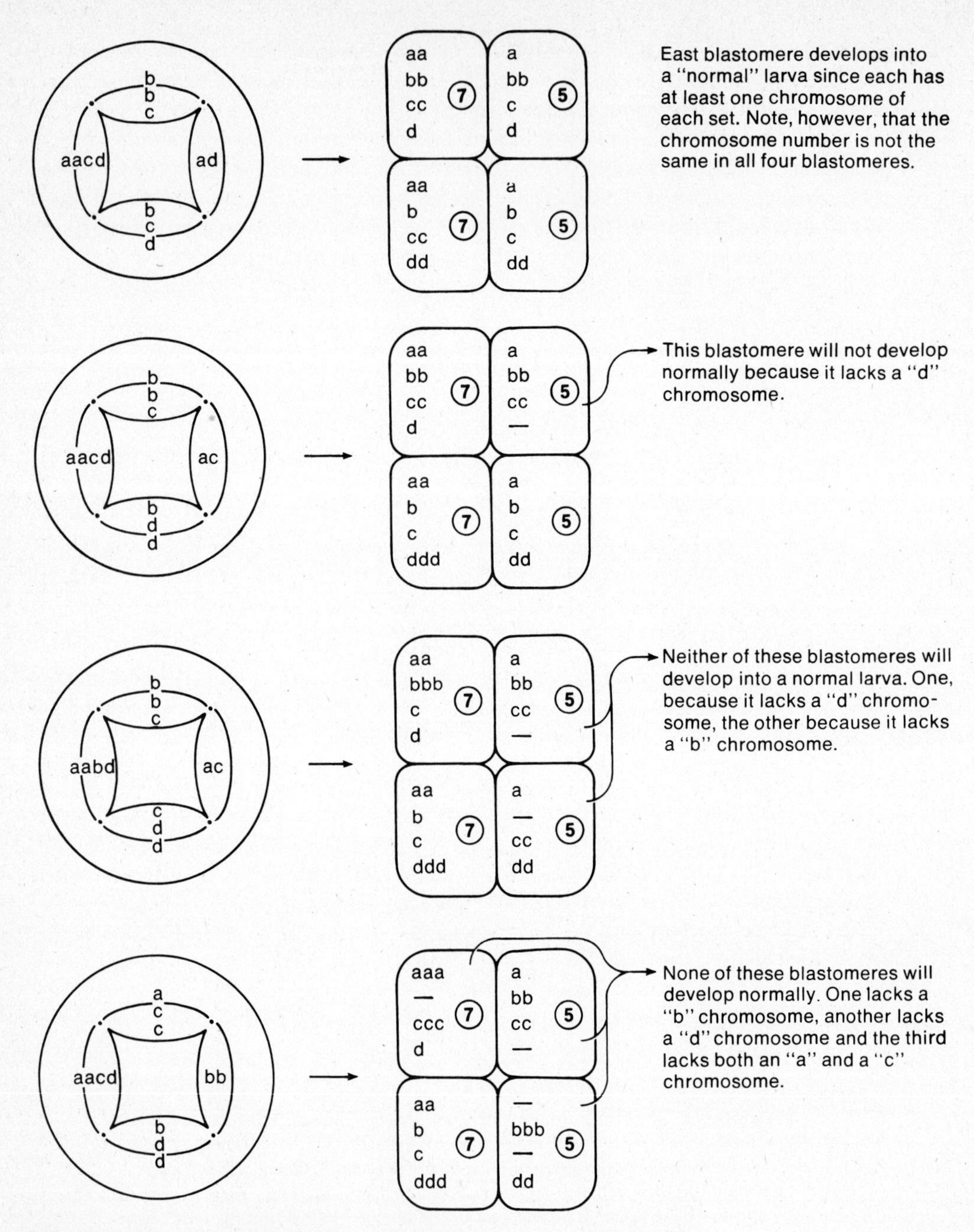

Figure 8-9. Boveri's scheme to illustrate his hypothesis that normal development in the sea urchin requires at least one of each chromosome. Shown is the tetraster resulting from dispermic fertilization and the four blastomeres resulting from the first cleavage.

Triaster Eggs

In addition to tetraster eggs, Boveri could also obtain triaster eggs, which resulted when one of the two centrioles failed to divide after double fertilization. Thus there were three centrioles in the zygote. This result can also be obtained artificially by paralyzing one centriole chemically. (See Figure 8-10.) Boveri reasoned that if his hypothesis was correct, it

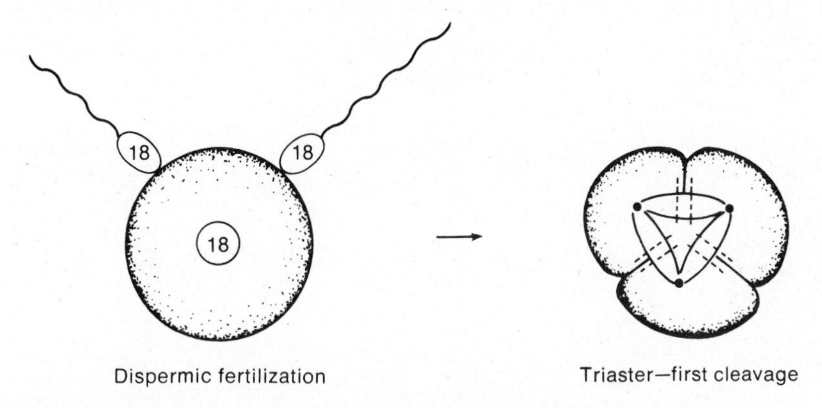

Figure 8-10. Dispermic fertilization in the sea urchin and the failure of one centriole to divide. The three remaining functional centrioles establish three spindles and the 54 chromosomes distribute themselves randomly among the three spindles.

could be assumed that a higher percentage of normal larvae would occur in the triaster experiments than in the tetraster eggs for the simple reason that the 54 chromosomes distributed themselves among fewer spindles in the triaster egg. (See Figure 8-11.) He calculated that the probability of all blastomeres in a triaster egg getting at least one chromosome is 4 %. The calculated probability for all blastomeres in a tetraster egg getting at least one chromosome is 0.0026 %. In actual experiments Boveri observed that 719 triaster eggs gave 58 perfect larvae, or 8.0 %, whereas 1500 tetraster eggs resulted in only two normal larvae (and one of these was questionable), a percentage of 0.13 %. Boveri's prediction of more normal larvae from triasters compared to tetrasters was verified, and the agreement with the calculated probabilities is amazingly close in spite of the many variables involved in sea urchin development.

Boveri wished to make additional comparisons of predictions based on the hypothesis of qualitatively different chromosomes and that normal development demands at least one chromosome of each pair. Given this problem today, a bright college student would probably propose a statistical analysis, using probability theory to predict what results would be expected based on a random distribution of chromosomes. Boveri, however, did not possess the expertise to perform this statistical analysis, but he was able to test it in a pragmatic way.

He made a tray with four quadrants and placed on it three sets of 18 balls, numbered 1 through 18. He shook the tray so that the balls were distributed,

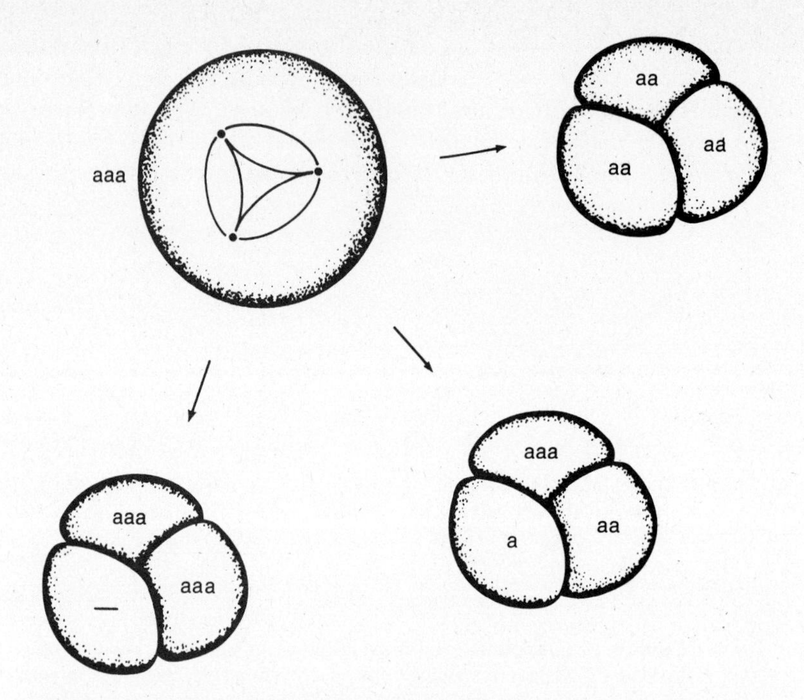

Figure 8-11. A scheme showing the distribution of one chromosome (present three times from two sperm plus the egg) in a triaster egg. The three chromosomes can assort on the three splindles in only three ways (one chromosome on each spindle; two on one spindle and one on another; or all three on one spindle) giving the three possible results shown. Note that of the nine blastomeres only one fails to get an "a" chromosome.

presumably randomly, to the four quadrants. He then observed the results. He repeated the procedure a total of 200 times and tabulated how often none of the quadrants received at least one ball for each of the 18; how often one of the four quadrants received at least one ball for each of the 18; how often two quadrants, three quadrants, or, finally, all four quadrants received at least one ball for each of the 18. This gave Boveri an empirical expectation with which to compare his observational results. He also did a similar experiment with the tray divided into three quadrants to simulate a triaster experiment. (See Figure 8-12.)

Table 8-1 shows the interesting results that Boveri obtained. Notice that the agreement between observed and expected is fairly good, considering the crudity of the empirical experiment. Also notice that there is a higher percentage of normal development in triaster eggs (43 %) than in tetraster eggs (18 %). These results confirm Boveri's hypothesis that normal development is dependent not on the number of chromosomes but on which ones are present. In other words, individual chromosomes must possess different qualities.

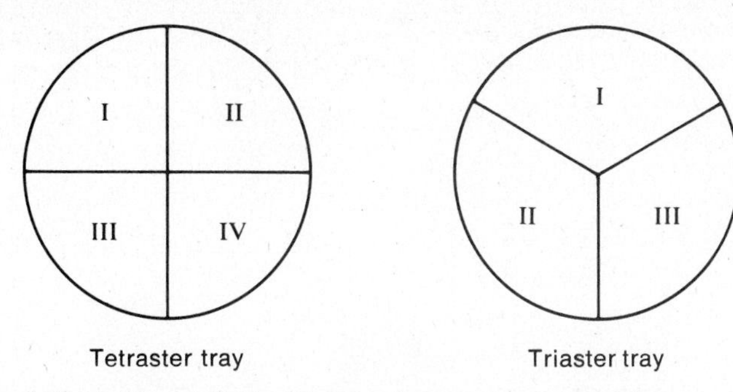

Figure 8-12. A diagram illustrating the trays made by Boveri to test empirically the distribution of chromosomes in dispermic eggs in the sea urchin.

Table 8-1. Results of Boveri's Dispermic Fertilization of Sea Urchin Eggs and Empirical Results of Tray Experiments[a]

Tetraster, 23 Dispermic Eggs			*Triaster, 34 Dispermic Eggs*		
NUMBER OF BLASTOMERES	NORMAL GASTRULA (%)	EMPIRICAL EXPECTATION (%)	NUMBER OF BLASTOMERES	NORMAL GASTRULA (%)	EMPIRICAL EXPECTATION (%)
4	0	0	3	14	11
3	4.5	0	2	23	42
2	4.5	2	1	40	36
1	54.5	34	0	23	11
0	36.5	64			
Total: 18 % normal gastrula			Total: 43 % normal gastrula		

[a]In each experiment, Boveri placed the eggs in calcium-free water following the first cleavage. He then observed the percentage of the blastomeres that developed into normal gastrulae. This percentage was compared to the percentage obtained in his tray and balls experiment. The latter percentage reflects the frequency with which each quadrant in the tray received at least one of each set of balls.

A later experiment confirmed this Boveri hypothesis. It was done in 1922 by an American geneticist, Blakeslee, working with the jimson weed, Datura. Datura has 12 pairs of chromosomes, and it has been possible to obtain plants that contain an extra chromosome, thus having 25 instead of the normally expected 24. Blakeslee collected plants that had an extra chromosome for each of the 12 pairs. Now if the chromosomes were qualitatively similar, all the plants containing an extra chromosome should appear more or less alike because the difference in size among the various pairs of chromosomes in Datura is not great. If, on the other hand, the chromosomes are qualitatively different, then it might be expected that 12 different types of plants would occur. Blakeslee examined one particular character in Datura, the seed capsule, and observed that the capsule phenotype varied in each of the 12 "mutant" strains of Datura. (See Figure 8-13.) From this result, Blakeslee came to

Figure 8-13. Twelve primary extra-chromosomal types. Above, capsule of a normal in Jimson Weed. Below, capsules of the 12 primary $2n + 1$ types each characterized by having a different extra chromosome. Beneath each capsule is given the common name and a numbered model of the extra chromosome which is responsible for the mutant form. The "gene-pills" which are in excess in each of the extra chromosomes influence the development of the capsule in characteristic ways. The whole plant as well as the capsule is affected by the extra chromosomes. (From Blakeslee, A. F. 1934. New Jimson Weeds from Old Chromosomes. *J. Hered.* **25**.)

the inevitable conclusion that the 12 chromosomes differed qualitatively from each other and that each contained different genes affecting the morphology of the capsule.

Different Chromosome Pairs Recombine Independently

An essential prerequisite to the Sutton–Boveri theory is the hypothesis that, at the first division of meiosis, different pairs of chromosomes align themselves on the metaphase plate randomly. If only two pairs of chromosomes are considered, the prediction would be that two different metaphase alignments would occur in equal frequency. (See Figure 8-14.) Unfortunately, however, the chromosomes viewed

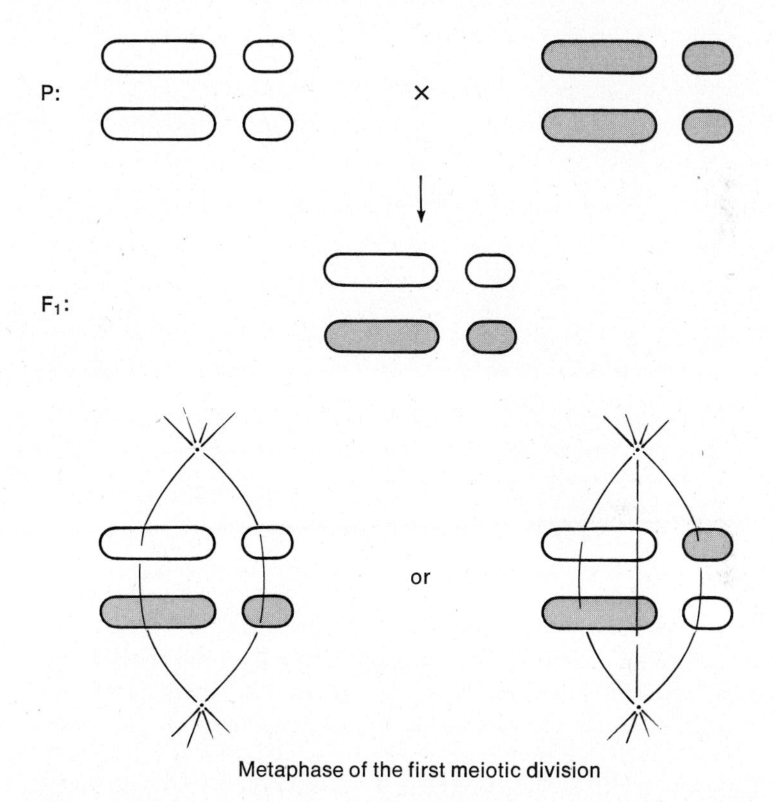

Figure 8-14. Diagram of the postulate that different pairs of chromosomes recombine independently. The hypothesis states that at metaphase of the first meiotic division in the hybrid two alternate arrangements of the chromosomes are possible and moreover that the two possibilities occur with equal frequency. For simplicity the chromosomes are drawn unduplicated.

through the microscope are not different shades, as shown in Figure 8-14, so that routine observation to verify the prediction is not possible. More favorable material must be found to establish the cytological proof of the hypothesis. Such evidence was found in the grasshopper, by Carothers. She observed one pair of chromosomes

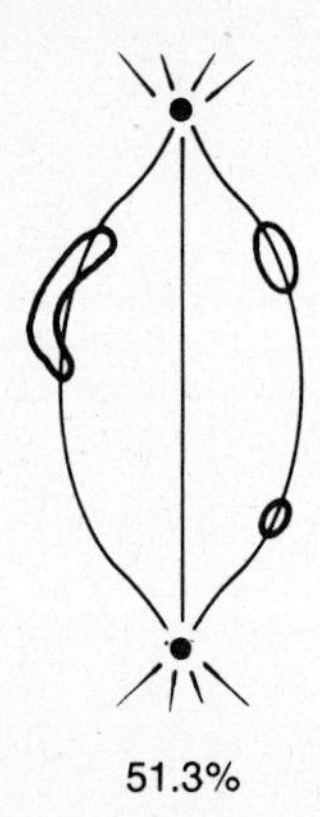

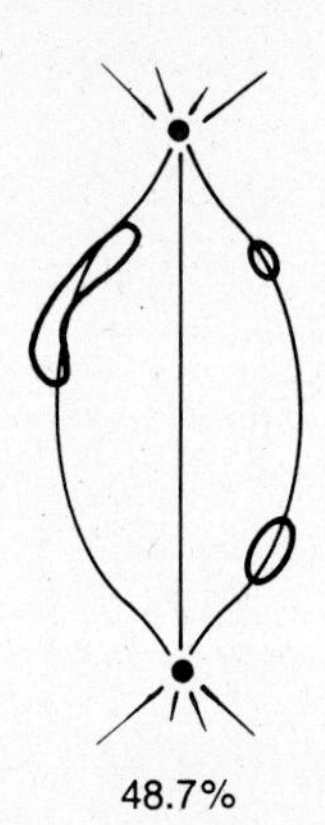

Figure 8-15. Diagram of Carother's observation in the grasshopper. If the hypothesis of independently assorting chromosomes is correct, then the two metaphases diagrammed should occur with equal frequency. Note that in the left metaphase the larger of the two homologues is segregating with the unpaired chromosome whereas in the right metaphase the smaller of the two homologues is segregating with the unpaired chromosome. The observations indicated a nearly 50–50 segregation verifying the hypothesis.

in which one homolog was larger than the other. She also noted that there was a chromosome that had no homolog. If the hypothesis that different chromosome pairs assort independently is true, then Carothers predicted that two meiotic arrangements should be observed in equal frequency. (See Figure 8-15.)

Out of 300 dividing cells observed, Carothers found that the large homolog segregated with the unpaired chromosome in 154 cells, and the smaller chromosome segregated with the unpaired chromosome in 146. It is clear from this evidence that the two pairs of chromosomes are assorting independently.

Thus the four prerequisites for the Chromosome Theory of Inheritance have been satisfied. Verification of the theory itself can now be examined.

REFERENCES

BLAKESLEE, A. F. 1922. Variations in Datura due to changes in chromosome number. *Am. Nat.* **56**:16–41. (Available in *The Chromosome Theory of Inheritance*, edited by B. R. Voeller. Appleton-Century-Crofts, New York, 1968.) (The paper that verifies Boveri's hypothesis of a qualitative difference between different pairs of chromosomes.)

BOVERI, TH. 1902. Über mehrpolige Mitosen als Mittel zur analyse des Zellkerns. *Vech. d. Phys. Med. Ges. Wurzburg, N. F.* **35**:67–90. (Available in English, in *Foundations of Experimental Embryology*, edited by B. H. Willier and J. M. Oppenheimer. Prentice-Hall, Englewood Cliffs, N.J., 1964.) (Boveri's contribution to the Sutton–Boveri theory.)

CAROTHERS, E. E. 1913. The mendelian ratio in relation to Orthopteran chromosomes. *J. Morph.* **24**:487–511. (Excerpts available in *The Chromosome Theory of Inheritance*, edited by B. R. Voeller. Appleton-Century-Crofts, New York, 1968.) (Demonstration that different pairs of chromosomes recombine independently.)

MONTGOMERY, TH. 1901. A study of the chromosome of the germ cells of Metazoa. *Trans. Amer. Phil. Soc.* **20**:154–236. (Excerpts available in *The Chromosome Theory of Inheritance*, edited by B. R. Voeller. Appleton-Century-Crofts, New York, 1968.) (Shows that homologous chromosomes pair with each other.)

SUTTON, W. S. 1902. On the morphology of the chromosome group in *Brachystola magna. Biol. Bull.* **4**:24–39.

SUTTON, W. S. 1903. The chromosomes in heredity. *Biol. Bull.* **4**:231–251. (Available in *Classic Papers in Genetics*, edited by J. A. Peters. Prentice-Hall, Englewood Cliffs, N.J., 1959.) (These two papers are Sutton's brilliant contribution to the chromosome theory of inheritance.)

WILSON, E. B. 1925. *The Cell in Development and Heredity.* 3rd ed., 1232 pp. Macmillan, New York. (For a complete description of all the important contributions to the chromosome theory, this is the book to read.)

QUESTIONS AND PROBLEMS

8-1. Define the following terms:

spireme
tetraster eggs
triaster eggs
blastomere
dispermic fertilization

8-2. What is meant by the following and why are they important?

(a) Chromosome Theory of Inheritance
(b) Continuity of the chromosomes
(c) Pairing of homologous chromosomes
(d) Qualitative difference between different pairs of chromosomes
(e) Different chromosome pairs recombine independently.

8-3. Assume that a dispermic egg receives three chromosomes, a, b, and c, from each gemetic nucleus. Assume further that three centrioles are present at cleavage and that three spindles are formed. If, in such an egg, two spindles receive the chromosomes aab, ab,

(a) what chromosomes are found on the third spindle?
(b) what are the constitutions of the nuclei in the three cells resulting from the cleavage division?
(c) what, according to Boveri, is the developmental fate of the three cells?

8-4. Given the same assumptions as in Problem 8-3, if the two spindles receive the chromosomes abc, abb,

(a) what chromosomes are found on the third spindle?
(b) what is the developmental fate of the three cells?

8-5. If the two spindles receive the chromosomes aaa, ccc,

(a) what chromosomes are found on the third spindle?
(b) what are the constitutions of the nuclei in the three cells resulting from the cleavage division?
(c) what is the developmental fate of the three cells?

8-6. In sea urchins, the chromosome number of the sperm nucleus is 18. In Boveri's experiments with dispermic fertilization, two cases of chromosome distribution were as shown below.

(1)

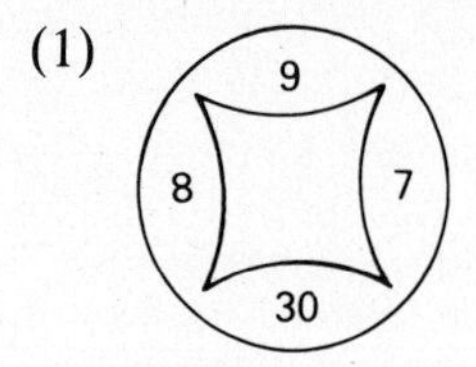

(2)

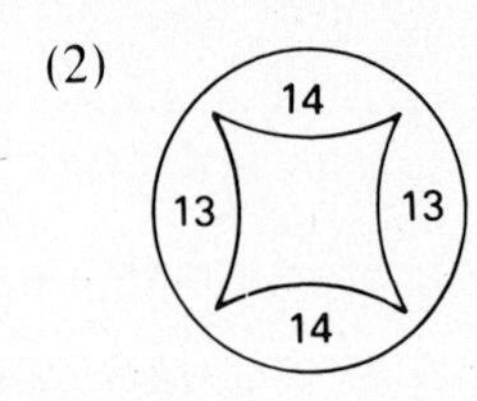

(a) What chromosome numbers will be found in the four nuclei to be formed by each of the eggs? Indicate your answer by drawing a four-cell stage and putting the appropriate numbers in each cell.
(b) After separation from each other, how many of the four cells of each egg are certain to develop successfully? (Barring accidents.) Explain.
(c) After separation from each other, how many of the four cells of each egg are certain *not* to develop normally? Explain.

9/Sex linkage

In the discussion of Mendel's experiments (Chapter 2) it was emphasized that he always made his genetic crosses reciprocally. That is, he switched the sexes in the parental cross to see whether it made any difference which parent brought in which characteristic. It never did! In the twentieth century, with other investigators testing Mendelism in many different organisms, it was not long before exceptions were found. The first example occurred in 1906 and was done by Doncaster and Raynor in moths. A later example (1908) involving chickens will be used to illustrate this exception to Mendelism.

Exceptions to Mendelism

Exceptions to Mendelism in Chickens

The experiment involved a feather pigmentation pattern in the Plymouth Rock strain of chickens. In one variety of chickens, the coloring on the feathers was arranged in such a way that the chicken had a barred appearance as though the black and white pigments alternated with each other. The other variety of chicken was black, or nonbarred. The cross involved a nonbarred female by a barred male. In the F_1, all the chickens were barred. If the cross was genetic and if a single pair of genes was involved, then it might be deduced that barred was dominant to nonbarred. This experiment will be continued and its results explained differently from the way previous genetic crosses have been described. First, instead of crossing the F_1 by F_1, a test cross will be made. The F_1 barred chickens, both female and male, are mated by nonbarred chickens of the appropriate sex. Finally, instead of describing the progeny from the test cross, the genotype of the F_1 chickens will be given as inferred from the test cross progeny.

The results of the cross of barred F_1s by nonbarred chickens suggested that the F_1 male chicken was heterozygous for barred and nonbarred but that the F_1 female chicken contained only the barred allele. When the reciprocal cross was made of a barred female by a nonbarred male, the F_1 females were all nonbarred and the F_1 males were all barred. From a mating of the F_1 chickens separately to a nonbarred chicken of the appropriate sex, it was determined that the F_1 males were heterozygous for barred and nonbarred but that the F_1 females contained only the nonbarred allele. (See Figure 9-1.)

P: Nonbarred ♀ × Barred ♂

F_1: Barred ♀ Barred ♂

Mate F_1 × nonbarred and conclude from progeny obtained that:

F_1♂ is heterozygous for barred and nonbarred.
F_1♀ has only the barred allele.

P: Barred♀ × Nonbarred ♂

F_1: Nonbarred ♀ Barred ♂
Mate F_1 × nonbarred and conclude from progeny obtained that:

F_1♂ is heterozygous for barred and nonbarred.
F_1♀ has only the nonbarred allele.

Figure 9-1. Reciprocal crosses involving barred and nonbarred chickens. Instead of mating $F_1 \times F_1$, a test cross was made of F_1 by nonbarred and from the progeny observed the genotypes of the F_1 chickens were determined.

Note that this striking result differs from simple Mendelism in two obvious ways. First, Mendel always obtained a uniform F_1. In previous crosses, all the F_1s were the same phenotypically. However, in the second chicken cross of barred times nonbarred, the F_1s were not all the same. Secondly, recall that Mendel always got the same result when he made a reciprocal cross. But, clearly, in this experiment that is not true.

A few years later, in the Department of Zoology at Columbia University in New York City, an American zoologist and embryologist, Thomas Hunt Morgan, was pondering the Chromosome Theory of Inheritance. In fact, Morgan had written a paper, published in *Science*, in which he raised serious questions about the chromosome theory. He was, to say the least, a skeptic. But Morgan was also an experimental scientist, and he wondered whether he could provide experimental evidence to verify his skepticism.

He began by searching for an organism in which he might perform some simple genetic experiments. He had heard about the use of a small fly, *Drosophila melanogaster*, often called the fruit fly or vinegar fly. Drosophila was already being used for genetic experiments in 1906 by the American geneticist Castle and his group at Harvard. From Castle, Morgan learned of the advantages of using this organism in experiments. It is small and can be easily raised in the laboratory in bottles and fed on a simple medium of bananas and yeast. From a single pair of flies, Morgan could obtain several hundred progeny, which would make quantification and determination of ratios fairly simple. And, as Morgan was to learn, Drosophila possesses many clear-cut characteristics very similar in kind to those that Mendel had used in his various experiments with peas. And, finally, as Morgan was to learn, Drosophila has only four pairs of chromosomes. The advantage of studying an organism with only a few pairs of chromosomes will be apparent later.

So Morgan began to breed Drosophila in the laboratory with the hope of performing genetic experiments. One day a single white-eyed male appeared. This was in striking contrast to red-eyed Drosophila, which is the normal, or *wild type*, a genetic term synonymous to normal, eye color. Morgan bred this white-eyed male to the red-eyed females he had in the laboratory and after several generations of breeding obtained a pure breeding strain of white-eyed flies. He was now ready to analyze the inheritance of this eye color in Drosophila.

Exception to Mendelism in Drosophila

He crossed a red-eyed female by a white-eyed male and obtained all red-eyed flies in the F_1. So he reasoned that if white eye was inherited and was due to a single pair of genes, then white was recessive to red. In order to keep the experimental plan similar to that already described in chickens, it will be assumed that Morgan made analogous crosses. The F_1s were crossed by white-eyed flies of the appropriate sex, and the progeny revealed that the F_1 female appeared to be heterozygous for red and white but the F_1 male contained only the red allele. Morgan made the reciprocal cross of a white-eyed female by a red-eyed male and obtained in the F_1 red-eyed females and white-eyed males. He again test crossed these individually to white-eyed flies and from the progeny concluded that the F_1 female was heterozygous for red and white and that the F_1 male had only the white allele. (See Figure 9-2.)

As with the chickens, this cross contradicts simple Mendelism on two counts—one, the difference obtained in reciprocal crosses and, two, that in at least part of the cross the F_1s were not all the same.

P: Red-eyed ♀ × White-eyed ♂

F_1: Red-eyed ♀ Red-eyed ♂

Mate F_1 × white-eyed and conclude from progeny obtained that:

F_1♀ is heterozygous for red and white.
F_1♂ has only the red allele.

P: White-eyed ♀ × Red-eyed ♂

F_1: Red-eyed ♀ White-eyed ♂

Mate F_1 × white-eyed and conclude from progeny obtained that:

F_1♀ is heterozygous for red and white.

F_1♂ has only the white allele.

Figure 9-2. Reciprocal crosses involving red- and white-eyed Drosophila. Instead of mating $F_1 \times F_1$ a test cross was made of F_1 by white-eyed and from the progeny observed the genotypes of the F_1 flies were determined.

Sex Chromosomes

Morgan now sought an explanation for this exception to Mendelism and found it through his understanding of the cytology of Drosophila and organisms in general. After reviewing the literature, he found that in 1891 the German zoologist Henking, in looking at chromosomes during spermatogenesis of a Hemipteran, or true bug, noticed that one chromosome had no homolog. Henking deduced that two kinds of sperm were produced in equal frequency, one with this body and one without it. Henking, however, had no clear idea of the nature of this element and for this reason, perhaps, called it an "X" body, using the algebraic symbol for unknown. About a decade later, in 1902, the American cytologist McClung, at the University of Pennsylvania, made the first suggestion that this X body, which McClung called the *accessory chromosome*, was concerned with sex determination. He emphasized the parallel between the two classes of sperm formed because of the presence or absence of the accessory chromosome, and the two sexes. He then suggested the hypothesis that a particular chromosome is the sex determiner, but final clarification came a few years later.

Verification came largely from the Department of Zoology at Columbia University, where E. B. Wilson, one of the great American biologists, and his colleague Nettie Stevens were analyzing gametogenesis in a number of different organisms. They cleared up many of the early discrepancies in the literature when Wilson clearly demonstrated that in many organisms the female had two X chromosomes, to use Henking's nomenclature, and the male only one X chromosome, all other chromosomes being identical in the two sexes. Stevens showed that in other species the male, instead of lacking the second X chromosome, possesses a homolog to the X but that this homolog is usually smaller. Wilson designated it the Y chromosome. Thus when a female produces eggs, each egg contains an X chromosome, whereas in spermatogenesis in the male, half the sperm contain an X chromosome and half contain none, or a Y chromosome, depending on the species involved. (See Figure 9-3.)

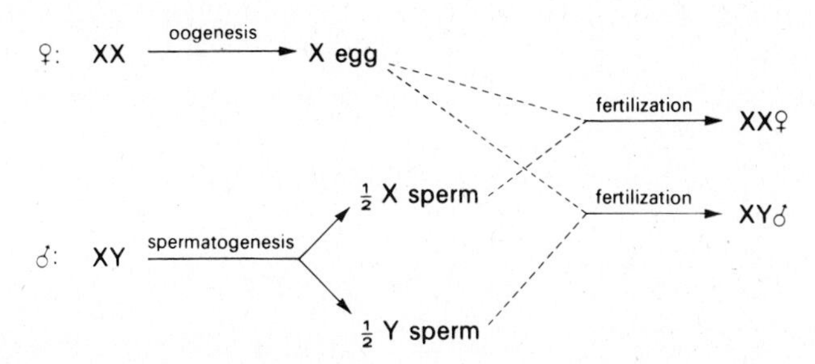

Figure 9-3. Wilson's explanation for the role played by chromosomes in sex determination. The female produces only an X-bearing egg. The male produces two kinds of sperm in equal frequency. One-half are X-bearing and upon fertilization give an XX individual which is female. One-half are Y-bearing (or in some species, have no sex chromosome) and upon fertilization give an XY (or XO) individual which is male.

Morgan, whose office was just down the hall from E. B. Wilson's at Columbia and who conferred with him about scientific matters, was now prepared to make his interpretation of the Drosophila experiments involving eye color. Morgan proposed that the cross could be explained if it is assumed that the alleles for red and white eyes are carried on the X chromosome and if it is further assumed that the Y chromosome contained no genes for eye color. In Figure 9-4 it can be seen that the mating of the F_1 female to a white-eyed male will demonstrate that the F_1 female is heterozygous for the red and white alleles and that the mating of the F_1 male by a white female reveals that the F_1 male has only the red allele.

In the diagram of the reciprocal cross of white-eyed female by red-eyed male (Figure 9-5), again it is shown that the mating of the F_1 female to a white-eyed male reveals that she is heterozygous for the red and white alleles, whereas the mating of the F_1 white-eyed male to a white-eyed female shows that only the white allele is present.

The German geneticist Richard Goldschmidt quickly pointed out that the chicken and moth experiments could be explained in a similar way if, in these organisms, it is the female that has only one X with genes on it and the male has two Xs. Seiler, a student of Goldschmidt, showed cytologically that, in moths, the male does have two X chromosomes and the female only one. Later studies have extended this finding to other moths and butterflies, as well as to birds.

New terminology is needed to describe the experiments and interpretations of Morgan. *Sex linkage* refers to the location of genetic factors on the *sex chromosomes.* The sex chromosomes are those chromosomes, usually symbolized X and Y, that play a role in the determination of sex. All other chromosomes are known as

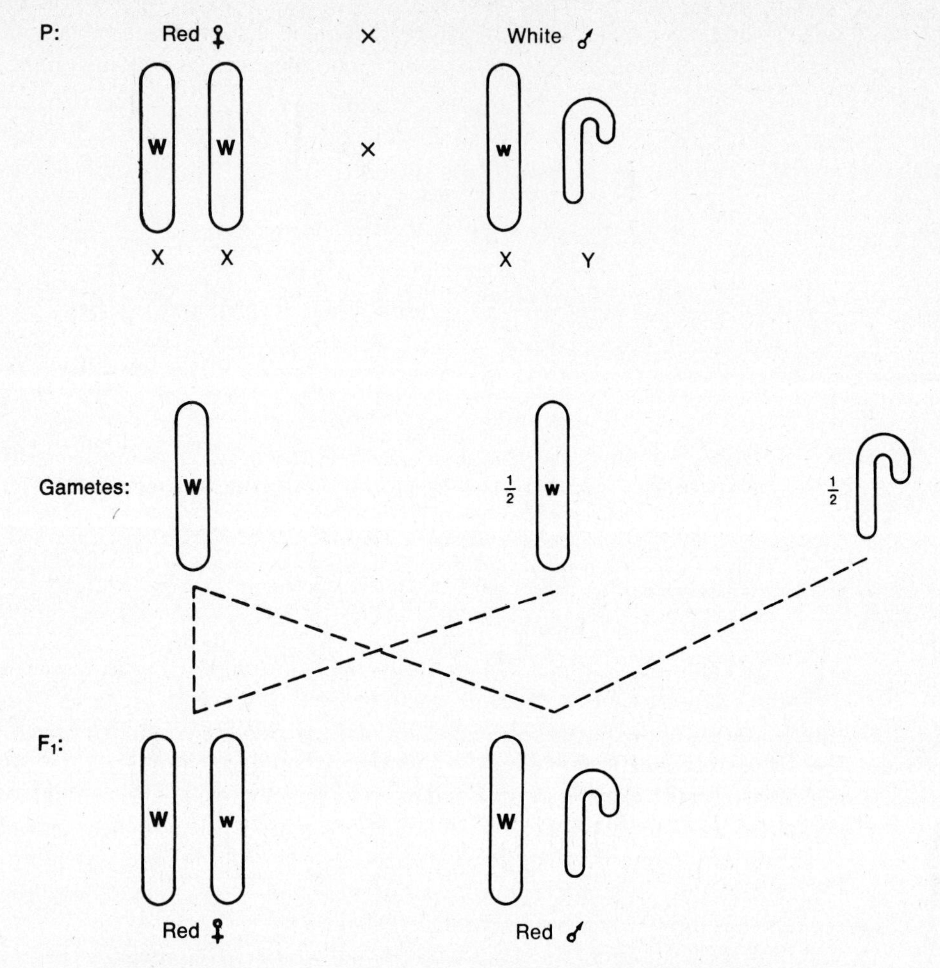

Figure 9-4. Morgan's interpretation of the inheritance of white eye in Drosophila. He postulates that the genes involved in this cross are on the chromosomes that determine sex in Drosophila. **W** = red eye; **w** = white eye.

autosomes. The *heterogametic* sex is the sex that forms two kinds of gametes. In Drosophila, the male that forms both X- and Y-bearing gametes is the heterogametic sex. In humans, the male is also the heterogametic sex. In birds and lepidoptera (moths and butterflies), it is the female that is the heterogametic sex, producing two kinds of ova. In Drosophila and man, the female is the *homogametic* sex, producing only one kind of gamete with respect to the sex chromosomes, whereas, in birds and lepidoptera, the male is the homogametic sex. The terms homozygous and heterozygous cannot be used to refer to that condition postulated by Morgan in

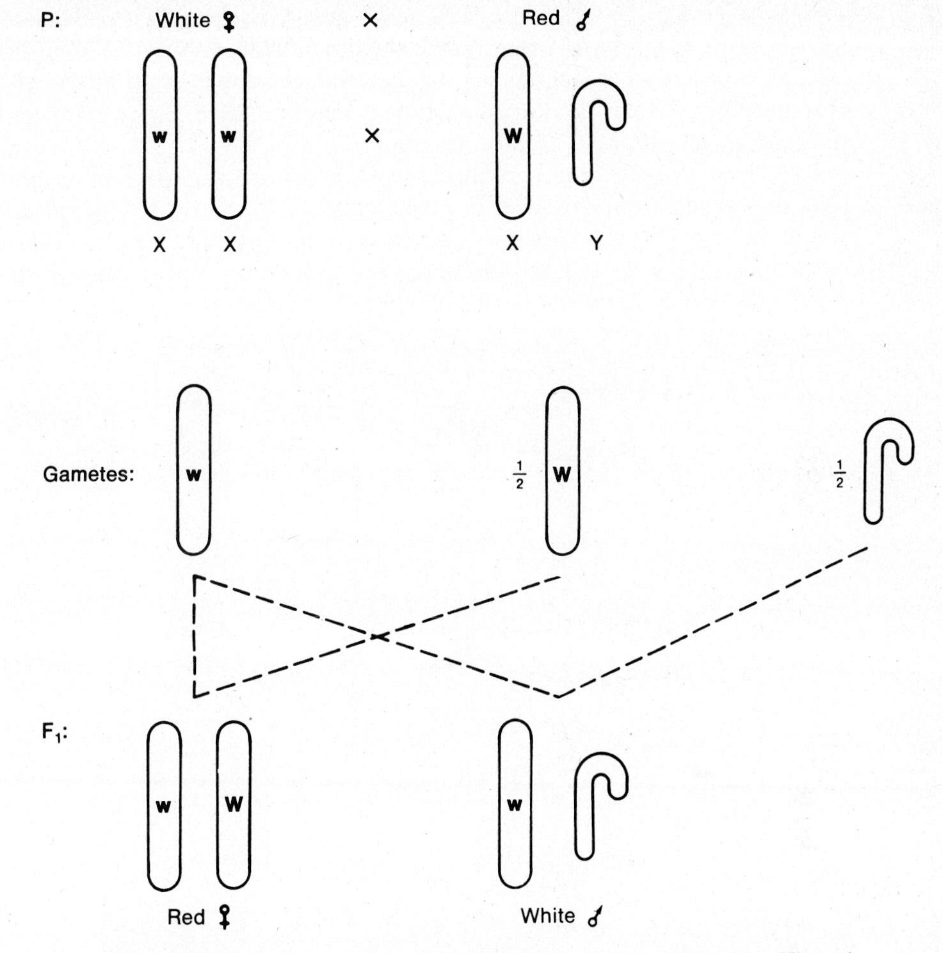

Figure 9-5. The reciprocal cross involving white eye color in Drosophila. **W** = red eye; **w** = white eye.

which only one allele is present, for the Y chromosome contains no genes. The term *hemizygous*, half a zygote, is used to describe the condition in which only one allele is present.

It should be emphasized that Morgan's assumptions for the inheritance of eye color in Drosophila present not only a new and slightly different kind of inheritance but also represent, if true, the first experimental evidence for the chromosome theory! Morgan's explanation is plausible only by the placement of a specific gene on a specific chromosome. It is perhaps ironic that Morgan, originally skeptical of the chromosome theory, now provides the first, formidable evidence for its establishment.

Y Linkage

Usually when a geneticist speaks of sex linkage, he is referring to the situation of genes on the X chromosome, but the word correctly refers to the inheritance of genes on either of the sex chromosomes; and it is interesting to see whether there is any evidence for the presence of genes on the Y chromosome.

The Danish geneticist Winge was investigating a particular morphological pattern of pigmentation in the guppy fish Lebistes. The characteristic pattern, which was found in males, was known as *maculatus (mac)* and consisted primarily of a large black spot on the dorsal fin and a large red spot below and in front of the dorsal fin. Figure 9-6 contains the results of Winge's experiments.

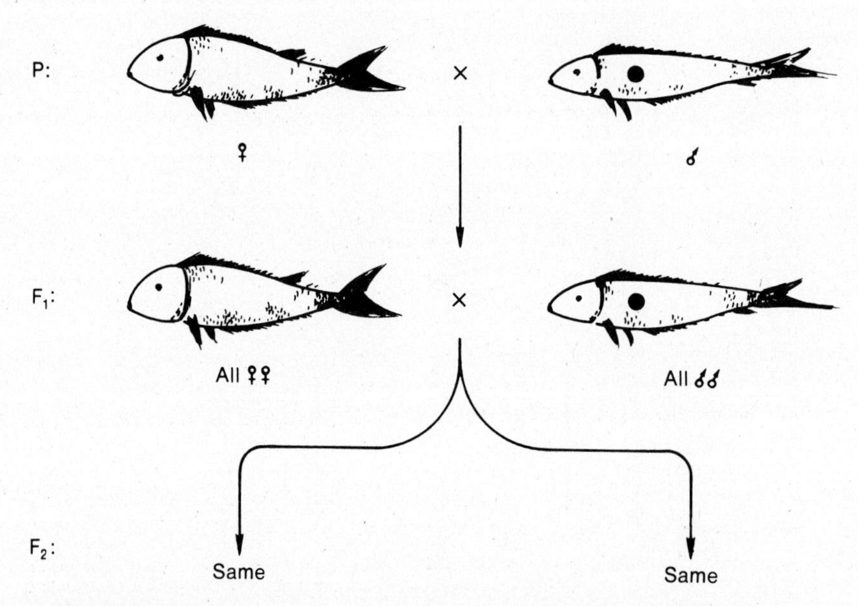

Figure 9-6. Inheritance of the *maculatus* pattern in the guppy, Lebistes. *Maculatus* is characterized by a black spot on the dorsal fin and a red spot below the dorsal fin in the male. When a non-*mac* female is crossed by a *mac* male all the sons are *mac* like the father and all the daughters are non-*mac* like the mother. A cross of $F_1 \times F_1$ repeats the pattern in F_2 of non-*mac* females and *mac* males.

Observe that the trait is always passed from father to son. It never occurs in the daughters and is never passed to sons from their mothers. Winge knew from cytological studies that, in Lebistes, the male is the heterogametic sex, and he postulated that *maculatus* is due to a gene on the Y chromosome. From this particular cross alone, the data do not indicate whether the *mac* gene is dominant with a recessive allele on the X or whether there is no allele to the *mac* gene on the X chromosome.

In Drosophila, Curt Stern discovered a strain of flies in which the bristles of the female were shorter than normal. This mutant was named *bobbed*, after the bobbed hairstyle common in women of the 1920s. When Stern made a cross of a bobbed female by males with wild-type (normal) bristles, the F_1 females were all bobbed and the F_1 males were all wild-type. The cross of F_1 by F_1 produced similar results in the F_2. All the females were bobbed, and all the males were wild-type. This pattern could be continued for many generations. Stern postulated that the mutant gene bobbed was located on the X chromosome with a wild-type allele on the Y. (See Figure 9-7.) If true, this hypothesis contradicts to some extent Morgan's

P:	Bobbed ♀	×	Wild-type ♂
F_1:	All ♀♀ bobbed		All ♂♂ wild-type
F_2:	All ♀♀ bobbed		All ♂♂ wild-type
P:	$X^{bb}X^{bb}$ **Bobbed ♀**	×	$X^{bb}Y^{Bb}$ **Wild-type ♂**
F_1:	$X^{bb}X^{bb}$ **Bobbed ♀**		$X^{bb}Y^{Bb}$ **Wild-type ♂**

Figure 9-7. The cross of a bobbed female by a wild-type male in Drosophila and Stern's explanation for it by assuming a wild-type allele for bobbed on the Y-chromosome. **Bb** = wild type; **bb** = bobbed.

assumption that the Y chromosome contains no genes. However, subsequent genetic analysis indicates that, for most of the genes in Drosophila, such as eye-color genes, wing-mutant genes, and most bristle mutants, no alleles are, in fact, present on the Y chromosome. So, as a generalization, Morgan is correct, with the bobbed gene representing an exception.

Stern further postulated that if his hypothesis was true, it should be possible to obtain a bobbed male sometime in the future. This prediction is based on the assumption that if the normal allele changed to produce a bobbed gene on the X, a similar change should eventually occur with the normal gene on the Y. This prediction was, in fact, verified when a male appeared that was bobbed and that, when crossed with a bobbed female, gave all bobbed progeny.

Sex-linked traits are often the first discovered in any organism and are the most easily analyzed. The reason should be apparent to the student who thinks about it for awhile. In man, there are two very well-known sex-linked traits, and many others have become increasingly well known in recent years. Perhaps the most famous of all sex-linked traits in man is hemophilia, the so-called bleeder's disease. This disease was prevalent in the royal family of Europe in the nineteenth century and is alleged to have originated with Queen Victoria. Hemophilia is inherited as a sex-linked recessive.

A second well-known sex-linked trait in man is color blindness. Although there are several types, the characteristic red-green form of color blindness is inherited as a sex-linked recessive. There are, however, other kinds of color blindness, some of which are autosomally inherited.

Nondisjunction

When a new science undergoes rapid development, as in the case of genetics in the early part of the twentieth century, new journals often arise to receive the burgeoning number of scientific reports. So it was in 1916 that the first edition of *Genetics*, an American journal for the study of inheritance, was published. Now scientific journals are not unlike their commercial counterparts, the popular magazines, in trying to attract as many readers as possible. To do so, they may often attempt to load the first edition with the best papers available as an attraction for prospective subscribers. Such was the case with *Genetics*. The first paper in volume 1, number 1, in *Genetics* was by one of the students of T. H. Morgan, Calvin Bridges, and the title is revealing and significant. Bridges called it "Nondisjunction as Proof of the Chromosome Theory of Heredity." For, as always in science, there were skeptics, among whom was the very influential geneticist William Bateson, who provided so much early impetus to the study of genetics. Bateson was not convinced by Morgan's analysis of sex linkage as a proof for the chromosome theory. Nor was he persuaded by the elegant evidence of Morgan's concerning linkage, which will be discussed in detail in Chapter 10. And so Bridges' title was probably selected with Bateson in mind. The results of this classical experiment and Bridges' analysis will be given in some detail.

The first suggestion of something unusual occurred in Morgan's original cross of red-eyed females by white-eyed males. In addition to the results already described, a few white-eyed males were noted. Bridges' analysis began here and especially in the reciprocal cross of white-eyed females by red-eyed males. (See Figure 9-5.) In addition to the expected progeny of red-eyed females and white-eyed males, an occasional white-eyed daughter was obtained and an occasional red-eyed son occurred. These unusual flies, designated the *primary exceptions*, occurred in a frequency of approximately 1 in 2000 flies. In addition, Bridges discovered that the red-eyed primary exceptional male, which in every phenotypic aspect was a normal male, was always sterile.

When Bridges mated a primary exceptional white-eyed female to normal red-eyed males obtained from stock, more unusual flies were observed in the progeny. Although the great majority of flies were red-eyed females and white-eyed males, some white-eyed females and red-eyed males again occurred but this time in a frequency of about 4 % compared to $\frac{1}{2000}$ noted above. These exceptions were designated *secondary exceptions*, and Bridges observed that the secondary exceptional male was fertile. When he mated the secondary exceptional female, she behaved as the primary exceptional female had, yielding in the offspring about 96 % normal and about 4 % exceptional progeny. (See Figure 9-8.)

In considering an explanation for these unusual results, Bridges reasoned that, in order for a white-eyed female to have a white-eyed daughter, the latter must have (1) two X chromosomes and (2) no red-eyed genes. Bridges further considered what occurred in normal meiosis. The egg has an X chromosome with the white allele on it and can be fertilized by either the X-bearing sperm with the red-eyed allele or the Y-bearing sperm. This process results in the expected progeny, as

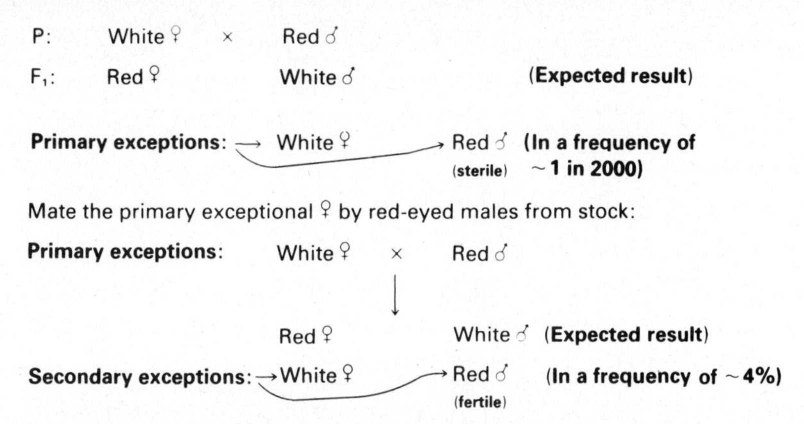

Figure 9-8. The results obtained by Bridges in Drosophila from a mating of white-eyed females by red-eyed males.

Morgan had already pointed out. (See Figure 9-9.) Bridges next postulated that, on rare occasions during meiosis, the two X chromosomes do not disjoin (separate) from each other but stay together, either both ending in the egg or both ending in the polar body. The failure of homologous chromosomes to separate at meiosis is called *nondisjunction*. If each of these rarely produced eggs can be fertilized by either an X-bearing or Y-bearing sperm, four possible progeny could result. (See Figure 9-10.) Further analysis by Bridges of his experiments showed that if a zygote receives no X chromosome at all, but only a Y, it always dies. If a zygote gets three X chromosomes, it nearly always dies. It will be shown in Chapter 27 that some of these flies do, on rare occasions, survive and give us insight into the problem of sex determination. But for now attention will be confined to the two progeny that, for the most part are the only survivors.

A fly that has two X chromosomes, both received from the mother, and a Y chromosome received from the father is a female and is white-eyed due to the presence of the two **w** genes. She is the primary exceptional female. A fly that receives one X chromosome with a **W** gene from the father and no Y chromosome is the primary exceptional male. Subsequent experiments by Bridges and Curt Stern showed that the sterility of the primary exceptional male is due to the absence of the Y chromosome. Later work has shown, beyond any doubt, that the Y chromosome contains many fertility genes necessary for the normal functioning of the sperm. Again, this is a contradiction of Morgan's original assumption that the Y chromosome is genetically inert.

The primary exceptional female thus contains two X chromosomes, each with a **w** allele and the Y chromosome. Cytologists have demonstrated that when three homologous chromosomes synapse during meiosis, they synapse only two at a

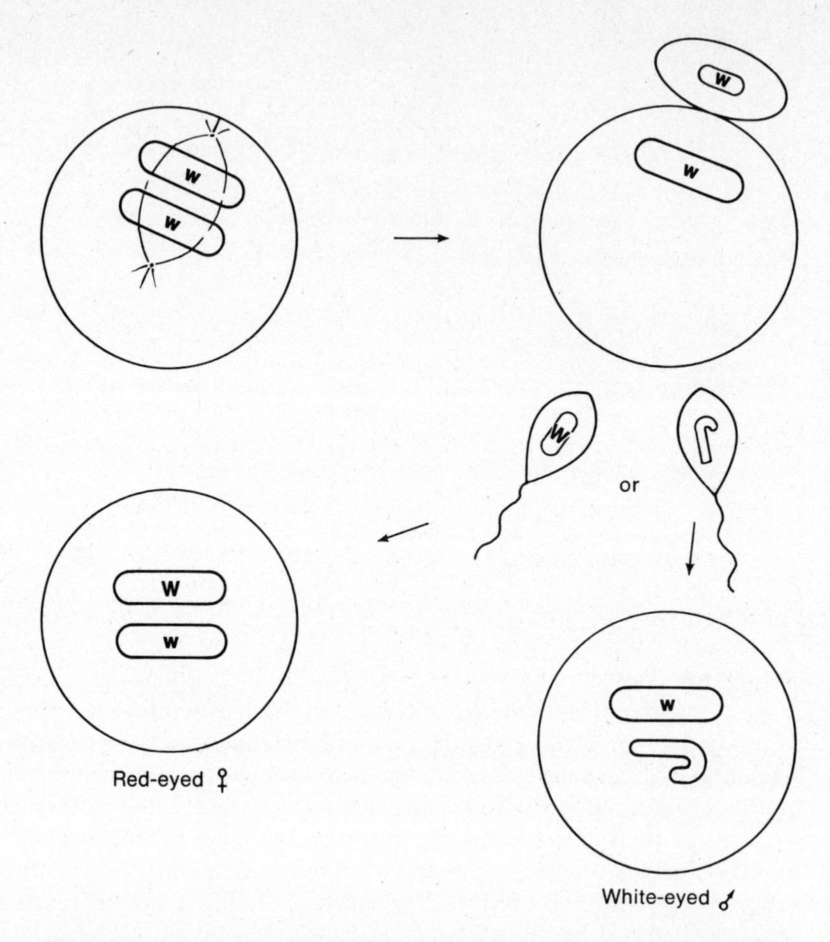

Figure 9-9. A simple view of meiosis to explain the normal results obtained from the cross of a white-eyed female by a red-eyed male. For simplicity the chromosomes are represented as unduplicated and only a single meiotic division is diagrammed. **W** = red eye; **w** = white eye.

time. That is, two of them pair and the third segregates at random. This results in two chromosomes going to one pole and one to the other. Such pairing behavior should apply to the sex chromosomes of Drosophila. Although the Y chromosome is not an exact homolog of the X, it does have partial homology. Bridges deduced from his experiments that the two X chromosomes pair and disjoin about 92 % of the time and stay together about 8 % of the time in meiosis in a primary exceptional female. The results are: 46 % of the eggs will have one X; 46 % will have one X and a Y; 4 % will have two Xs; and 4 % will have a Y. These eggs can each be fertilized by an X-bearing sperm with a **W** gene or Y-bearing sperm. (See Figure 9-11.) Again, it should be noted that zygotes with three X chromosomes usually die as do

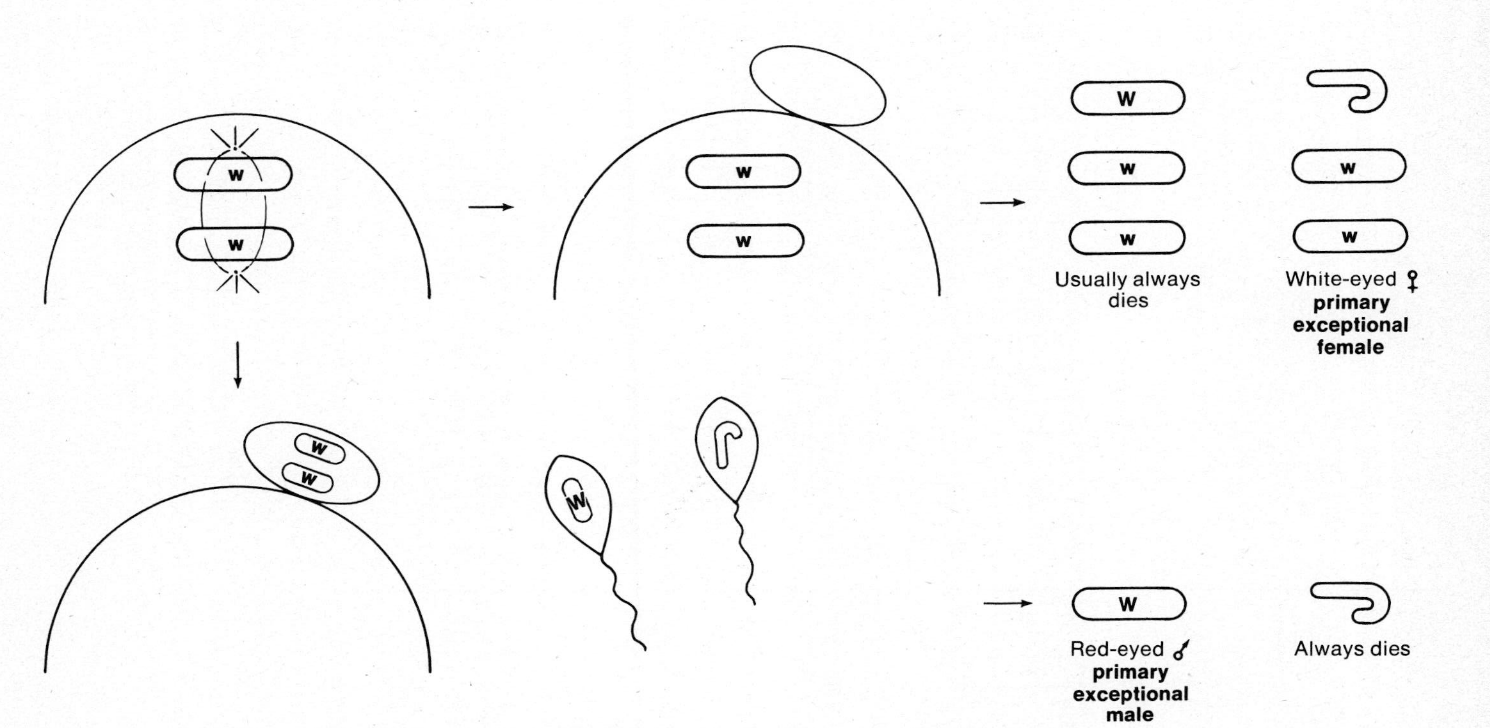

Figure 9-10. Bridges hypothesis to explain the exceptional progeny. Occasionally in meiosis the two X chromosomes fail to disjoin resulting in both X's in the egg and none in the polar body or none in the egg and both in the polar body. Each of these eggs can be fertilized by an X-bearing sperm or a Y-bearing sperm resulting in the four indicated zygotes. Only two of the four possibilities usually survive and they are the primary exceptions. **W** = red eye; **w** = white eye.

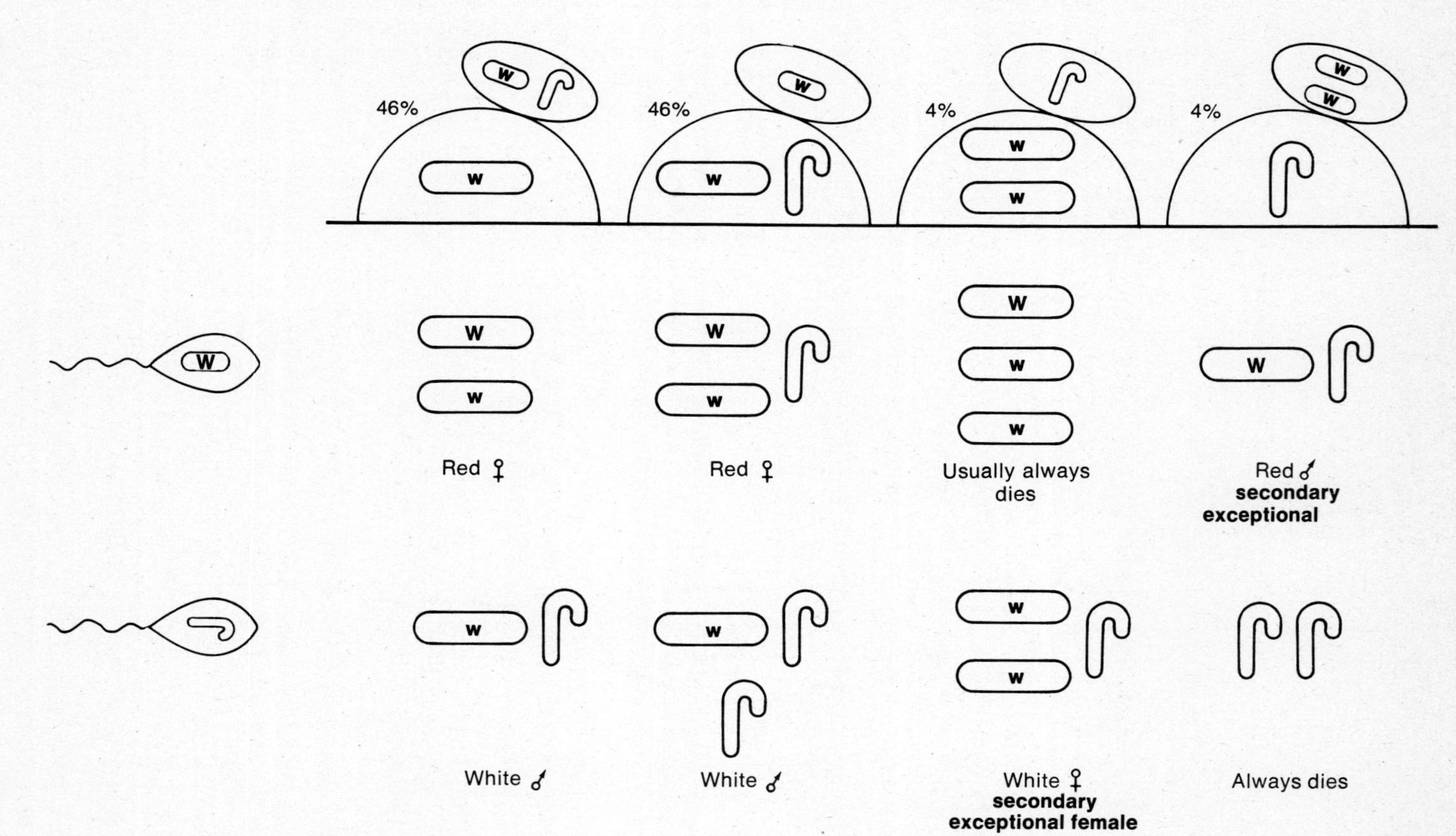

Figure 9-11. Bridges analysis of meiosis in the primary exceptional female. Four kinds of eggs are produced in the indicated frequencies. Resulting fertilization by the two kinds of sperm results in secondary exceptional females and secondary exceptional males (which have a Y chromosome and are fertile) in a frequency of about 4%.

those with no X chromosome. Among the remaining survivors is an XXY white-eyed female and an XY red-eyed male. These two represent the secondary exceptions and occur in a frequency of about 4%. Observe that the secondary exceptional male would be expected to be fertile, because of the presence of a Y chromosome.

Bridges next attempted to verify his hypothesis by direct observation of the chromosomes. The normal chromosome configuration of Drosophila is shown at the top of Figure 9-12. It is interesting to note that peculiar to the Dipterans (two-

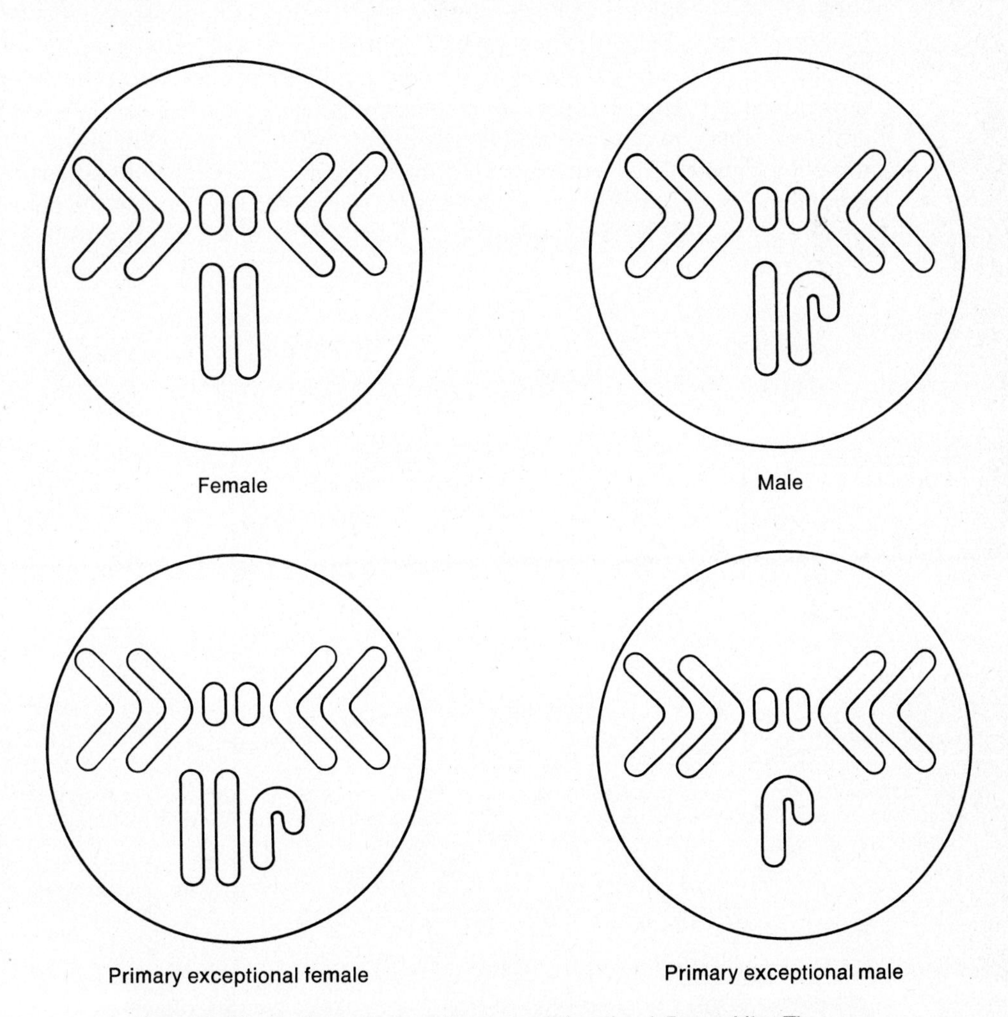

Figure 9–12. Cytology of normal and nondisjunctional Drosophila. The upper figures show the normal chromosomal configuration to consist of three pairs of autosomes (two large pairs of V-shaped chromosomes and a small dotlike pair) plus the pair of sex chromosomes. In the female there are two rodlike X chromosomes and in the male a rod X and a smaller Y chromosome. The lower figures show that the primary exceptional female is XXY and the primary exceptional male is XO.

winged flies), which includes Drosophila, is the characteristic that the chromosomes even pair in somatic cells, giving the diagrammatic picture seen in the figure. Bridges made sections of ovaries and testes and, using good scientific methods, made predictions as to what the exceptions would look like. He predicted that the primary exceptional female and the secondary exceptional female should be XXY and that the primary exceptional male should be XO. The bottom of Figure 9-12 shows the cytology of the primary exceptions. His predictions were verified in every case. However, skeptics again questioned whether the cytological proof that Bridges proposed was sufficient. They pointed out that he was relying on genetically exceptional progeny, and therefore it might not be surprising that such exceptional progeny might give exceptional cytological results. Bridges circumvented this criticism, which he never took seriously in any event, by examining the genetically normal progeny of the primary exceptional female. He predicted that, among the normal red-eyed females, half would have a Y chromosome and, among the normal white-eyed males, half would have two Y chromosomes. (See Figure 9-11.) He

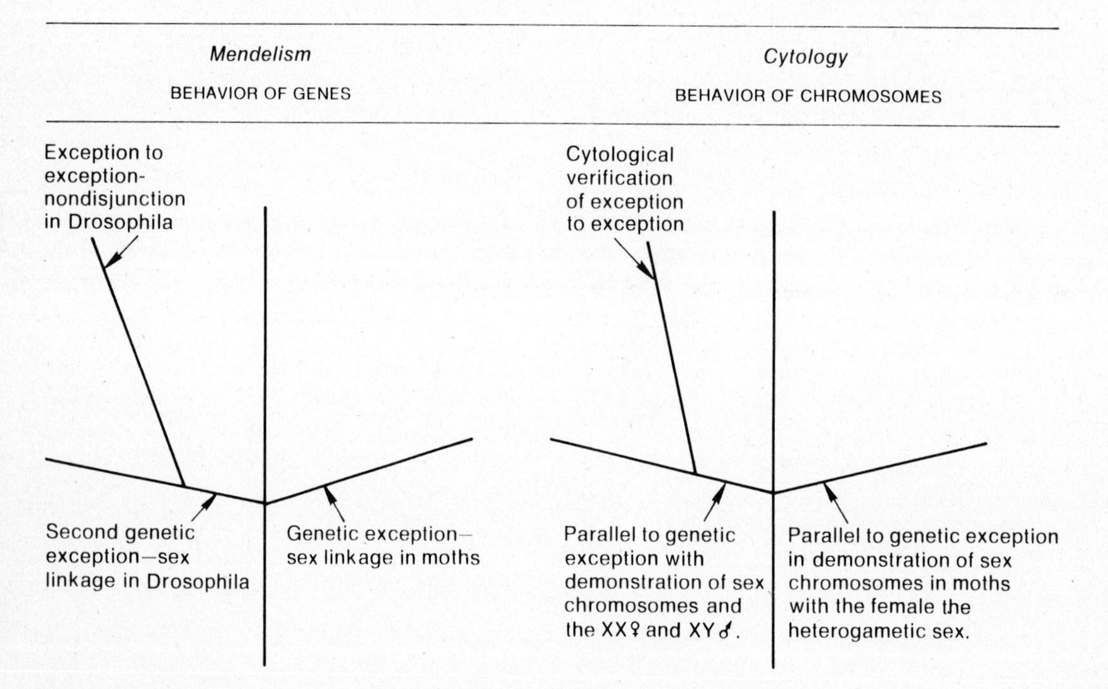

Figure 9-13. A diagram summarizing the persuasiveness of the Chromosome Theory of Inheritance. The two parallel vertical lines symbolize the Sutton–Boveri theory, which pointed out the parallel between chromosome behavior at meiosis and Mendel's laws. The two nearly horizontal lines represent exceptions to Mendelism which have a cytological explanation in the exception of the sex chromosomes. Also depicted is Bridges' exception to the exception which also could be explained by a cytological exception.

tested this prediction cytologically and showed it to be true, leaving little doubt about the correctness of his hypothesis.

It should be noted that the results of Bridges' experiments say something about sex determination in Drosophila. It seems evident that the Y chromosome is playing little role in determining sex, for a fly that is XXY is female. Also, a fly can be XO without a Y and morphologically a normal male, albeit a sterile one. More will be said about sex determination in Chapter 27.

Bridges' masterly analysis of nondisjunction was of crucial importance in the history of genetics, for it proved beyond a reasonable doubt that the sex-linked genes are actually carried on the X chromosome, a fact that had been inferred but not proved by Morgan's observations of normal sex-linked inheritance. (See Figure 9-13.)

REFERENCES

BRIDGES, C. B. 1916. Nondisjunction as proof of the chromosome theory of heredity. *Genetics* **1**:1–52, 107–163. (Excerpt available in *The Chromosome Theory of Inheritance*, edited by B. R. Voeller. Appleton-Century-Crofts, New York, 1968.) (A brilliant paper that convinced nearly everyone that the genes were on the chromosomes.)

MORGAN, T. H. 1910. Sex-limited inheritance in *Drosophila. Science* **32**:120–122. (Available in *Classic Papers in Genetics*, edited by J. A. Peters. Prentice-Hall, Englewood Cliffs, N.J., 1959.) (In spite of out-of-date terminology and an awkward nomenclature, this paper should be read as the first significant publication from the Morgan laboratory of Drosophila genetics.)

WILSON, E. B. 1925. *The Cell in Development and Heredity*. 3rd ed., 1232 pp. Macmillan, New York. (Wilson gives an excellent review of the cytology of the sex chromosomes and places Morgan's work in historical perspective.)

WINGE, O. 1922. One-sided masculine and sex-linked inheritance in *Lebistes reticulatus. J. of Genetics* **12**:145–162. (A good example of Y-linked inheritance.)

QUESTIONS AND PROBLEMS

9-1. Define the following terms:

sex linkage
sex chromosome
autosome
homogametic
heterogametic
hemizygous
primary exception
secondary exception
nondisjunction
wild-type

9-2. Why was Morgan's analysis of the inheritance of white eye in Drosophila important for the chromosome theory?

9-3. Why was Bridges' analysis of nondisjunction in Drosophila important for the chromosome theory?

9-4. In canaries, cinnamon color is dependent on a recessive sex-linked gene (a); green coloring is due to its dominant allele (A). (In birds, the female is the heterogametic sex.) Give the genotypes and phenotypes of the offspring of the following crosses.
(a) Green female × cinnamon male
(b) Green female × heterozygous green (Aa) male
(c) Cinnamon female × heterozygous green male

9-5. How would you arrange the cross in canaries so that the sex of the F_1 can be determined by coloration?

9-6. Assume that the Y chromosome of a bird contains a dominant gene for a special trait. A female and male produce 30 offspring, half males and half females. Which parental and which F_1 birds will show the trait?

9-7. In Drosophila, scalloped wings (sd) is a sex-linked recessive. Also, sepia eyes (se) is an autosomal recessive. Give the genotypes and the phenotypes for the offspring from the following crosses.
(a) sd sd se se female × Sd Se Se male
(b) Sd Sd Se Se female × sd se se male
(c) F_2 from cross of F_1 × F_1 from (a)
(d) F_2 from cross of F_1 × F_1 from (b)

9-8. In Drosophila, from a cross of a normal female × scalloped male, the following offspring were produced:
Females: $\frac{3}{8}$ normal, $\frac{3}{8}$ scalloped, $\frac{1}{8}$ sepia, $\frac{1}{8}$ sepia scalloped
Males: $\frac{3}{8}$ normal, $\frac{3}{8}$ scalloped, $\frac{1}{8}$ sepia, $\frac{1}{8}$ sepia scalloped
What are the genotypes of the parents?

9-9. From a cross of a normal female × normal male, the following offspring were produced:
Females: $\frac{3}{4}$ normal, $\frac{1}{4}$ sepia
Males: $\frac{3}{8}$ normal, $\frac{3}{8}$ scalloped, $\frac{1}{8}$ sepia, $\frac{1}{8}$ scalloped
What are the genotypes of the parents?

9-10. In humans, color blindness is due to a sex-linked recessive gene. Give the genotypes and phenotypes of the progeny from the following matings.
(a) A color-blind woman × a color-blind man
(b) A color-blind woman × a normal man
(c) A normal woman with no history of color blindness in her pedigree × a color-blind man

(d) A normal woman who has a color-blind father × a normal man
(e) A normal woman who has a color-blind father × a color-blind man
(f) A normal woman who has a color-blind mother but no other history of color blindness in her pedigree × a normal man
(g) A normal woman who has a color-blind mother × a normal man who has a color-blind father

9-11. In Drosophila, there is a mutant, Beadex, in which the wing margins are excised. A cross of a Beadex female by a normal male gave all Beadex F_1. In the reciprocal cross of a normal female by a Beadex male, all the F_1 females were Beadex and all the F_1 males were normal. Explain the inheritance of Beadex.

9-12. From repeated crosses of a white-eyed Beadex female by a normal male, the following offspring were produced:

3194 Beadex females — 2 white Beadex females
3129 white Beadex males — 1 normal male.

(a) Account for each of the offspring and give their genotype.
(b) If one of the two white Beadex females is mated to a normal male, what phenotypes of offspring, and in what approximate percentages, are to be expected?
(c) If the one normal male is mated to a white Beadex female, what types of offspring and in what approximate percentages, are to be expected?

9-13. In Drosophila, the mutant Abruptex is a sex-linked dominant. An Abruptex female is mated to a normal male.

(a) What will be the phenotypes of the primary exceptional offspring?
(b) What are the chromosomal constitutions of these exceptions?

10/Linkage and recombination

PART I

In earlier chapters simple cases of Mendelian inheritance were considered in which no more than a single pair of genes for each chromosome pair was followed. With the active study of heredity following the rediscovery of Mendel's laws, it was soon found that the number of characters that individually followed Mendelian principles was greater than the number of chromosome pairs to which the differentiating genes could be assigned. For a few years this apparent discrepancy was advanced as a strong argument against the Chromosome Theory of Heredity. It was pointed out that if the theory were true, exceptions to Mendel's Second Law, the Law of Independent Assortment, should be discovered. Since there were none, the Chromosome Theory of Inheritance was found lacking.

The proponents of the chromosome theory, however, replied that this argument was valid only as long as the principle of independent assortment held. With the ever-increasing number of experiments involving the inheritance of simple characters, it soon became evident that independent assortment was, in fact, far from universal in its application. Ironically, the first exception to Mendel's Second Law was discovered by the greatest skeptic of the chromosome theory, Bateson, in 1905.

No Independent Assortment

Work of Bateson and Punnett

Bateson and his coworker, Punnett, were doing a dihybrid experiment in sweet peas. Two varieties were available, one having purple flowers and long pollen grains and the other having red flowers and round pollen grains. From earlier experiments involving these traits separately, Bateson and Punnett had determined that purple was dominant to red and long was dominant to round. In monohybrid experiments each trait gave a normal 3 : 1 ratio in the F_2.

In a dihybrid experiment Bateson and Punnett crossed a purple-long plant by a red-round plant. The F_1 plants were all purple, long. The F_2 results are shown in Table 10-1. Notice that there is a discrepancy from the expected 9 : 3 : 3 : 1 ratio. Notice also that the parental combinations—namely, purple long and red round—occurred more frequently than expected (6169 instead of 4345) and that the new combinations, purple round and red long, occurred only 783 times instead of the expected 2607.

Table 10-1. Cross 1. Results of a Dihybrid Experiment Involving Flower Color and Pollen Grain Shape in the Sweet Pea[a]

P: Purple long × Red round
PPLL **ppll**

F_1: Purple long
PpLl

F_2:

TYPE OF PLANT	Number of Plants: OBSERVED	Number of Plants: EXPECTED	EXPECTED RATIO
Purple long	4831	3910.5	$\frac{9}{16}$
Purple round	390	1303.5	$\frac{3}{16}$
Red long	393	1303.5	$\frac{3}{16}$
Red round	1338	434.5	$\frac{1}{16}$

DATA FROM PUNNET. 1919. *MENDELISM*, 5TH ED. MACMILLAN, NEW YORK.

[a]P = purple; p = red; L = long; l = round.

Bateson and Punnett thought that the dominant gene for long pollen might have a tendency to stay associated with the dominant gene for purple flower with which it was introduced in one of the parental plants. Similarly, they argued that perhaps the recessive gene for round pollen tends to stay associated with the recessive gene for red flower. To test this hypothesis, Bateson and Punnett performed the cross in another way. In this instance, they crossed a purple-round plant by a red-long plant and again obtained F_1 plants that were all purple long. Selfing the F_1 gave the results seen in Table 10-2. Again notice that there is a discrepancy from the expected 9 : 3 : 3 : 1. The data may appear, on first approximation, to fit expectation, but a statistical analysis would show that the observed data are significantly different from the expected. Moreover, a close examination of the data reveals again that it is the parental types—in this instance, purple round and red long—that are too numerous (192 instead of 157) and the new combinations that are less numerous (227 instead of 262) than expected. Apparently the F_1 plants produce more parental gametes (**PL** and **pl** in the first cross; **Pl** and **pL** in the second cross) instead of the equal frequency of four kinds of gametes that should have resulted if the genes had assorted independently. The second experiment also indicates that there does not seem to be any tendency for specific genes to stick together, as Bateson and Punnett first thought, but only for parental genes to stay associated, whatever the parental arrangement may be.

Table 10-2. Cross 2. Results of a Dihybrid Experiment Involving Flower Color and Pollen Grain Length in the Sweet Pea[a]

P: Purple round **PPll** × Red long **ppLL**

F_1: Purple long **PpLl**

F_2:

TYPE OF PLANT	Number of Plants: OBSERVED	Number of Plants: EXPECTED	EXPECTED RATIO
Purple long	226	235.8	$\frac{9}{16}$
Purple round	95	78.5	$\frac{3}{16}$
Red long	97	78.5	$\frac{3}{16}$
Red round	1	26.2	$\frac{1}{16}$

DATA FROM PUNNETT, 1919.

[a]P = purple; p = red; L = long; l = round.

If the F_1 plants from the two experiments are test crossed to a completely recessive plant, it would be possible to determine exactly what gametes the F_1 plants produce and in what ratio. Table 10-3 shows the results of such test crosses. It is clear that the F_1 plants are not producing four gametes in equal frequency. Mendel's Second Law cannot be applied universally! The genes are not assorting independently!

Table 10-3. Results of Two Different Crosses Involving Flower Color and Pollen Grain Length in the Sweet Pea

	Number of Gametes		
TYPE OF PLANT	CROSS 1	CROSS 2	EXPECTED
Purple long	7	1	1
Purple round	1	7	1
Red long	1	7	1
Red round	7	1	1

Coupling and Repulsion

Bateson and Punnett coined several terms to describe their understanding of the experiments just discussed. They used the word *coupling* to refer to the situation in which one parent introduces both dominant genes and the other parent brings in both recessive genes, as in the case of Cross 1. The term *repulsion* refers to the situation in which one parent brings in one dominant gene and one recessive gene and the other parent brings in the other dominant and the other recessive, as in Cross 2. These terms are presented for their historical significance; they are not widely used today. Bateson and Punnett, in fact, elaborated a rather intricate and almost metaphysical hypothesis to explain their results, in which the terms coupling and repulsion played a prominent part. Even though their theory is no longer considered seriously, the terms coupling and repulsion are still useful in the descriptive sense and will be so used.

Meanwhile, at Columbia University, the Drosophila school was proceeding with its unraveling of the chromosome theory. T. H. Morgan, the leader of the group, began these historical experiments. However, associated with Morgan from the beginning was an amazing group of students, the most prominent of which were C. B. Bridges (already mentioned), A. H. Sturtevant, and H. J. Muller. These four, joined by others later, primarily were to provide that moment in science when a major breakthrough occurs. For this work, which was developing in the 1910s, Morgan received the Nobel Prize in 1933, the first American to do so in medicine and physiology. It is perhaps symbolic that Morgan divided the money from the prize among the children of Bridges and Sturtevant, for the work in this laboratory was truly communal; and although particular names are associated with particular investigations, it is probably true that all the members of the laboratory shared in every discovery.

The number of newly discovered mutants in Drosophila accumulated quickly; and, as mentioned earlier, if the chromosome theory was correct, there would soon be several of these genes on the same chromosome and some genes would not assort independently when tested in dihybrid experiments. Some actual data from Morgan's laboratory analyzed by Bridges show such deviations from independent assortment. (See Table 10-4.)

Table 10-4. The Inheritance of Purple Eye and Vestigial Wing in Drosphila[a]

(1) P: Wild-type × Purple vestigial (coupling)
PrPrVgvg **prprvgvg**

F_1: Wild type: Mate F_1♂ × Purple vestigial ♀
PrprVgvg **prprvgvg**

TYPE OF DROSOPHILA	GENE COMBINATION	NUMBER OF FLIES
Wild-type	**PrprVgvg**	519
Purple vestigial	**prprvgvg**	552
Vestigial	**Prprvgvg**	0
Purple	**prprVgvg**	0
		1071

(2) P: Purple × Vestigial (repulsion)
prprVgVg **PrPrvgvg**

F_1: Wild-type: Mate F_1♂ × Purple vestigial ♀
PrprVgvg × **prprvgvg**

TYPE OF DROSOPHILA	GENE COMBINATION	NUMBER OF FLIES
Wild-type	**PrprVgvg**	0
Purple vestigial	**prprvgvg**	0
Vestigial	**Prprvgvg**	358
Purple	**prprVgvg**	346
		704

(3) P: Wild-type × Purple vestigial (coupling)
PrPrVgVg **prprvgvg**

F_1: Wild-type: Mate F_1 ♀ × Purple vestigial ♂
PrprVgvg **prprvgvg**

TYPE OF DROSOPHILA	GENE COMBINATION	NUMBER OF FLIES
Wild-type	**PrprVgvg**	2876
Purple Vestigial	**prprvgvg**	2433
Vestigial	**Prprvgvg**	270
Purple	**prprVgvg**	270
		5849

Table 10-4. (*Continued*)

(4) P: Purple × Vestigial (repulsion)
prprVgVg **vgvgPrPr**

F_1: Wild-type: Mate F_1 ♀ × Purple vestigial ♂
PrprVgvg **prprvgvg**

TYPE OF DROSOPHILA	GENE COMBINATION	NUMBER OF FLIES
Wild-type	**PrprVgvg**	265
Purple vestigial	**prrprvgvg**	234
Vestigial	**Prprvgvg**	2038
Purple	**prprVgvg**	2203
		4740

DATA FROM MORGAN. 1914. *BIOL. BULL.* **26**:195–204.

[a]Cross is made twice in coupling (1 and 3) and twice in repulsion (2 and 4). In first two crosses (1 and 2), the F_1 male is test-crossed to a double recessive female. In the last two crosses, the F_1 female is test-crossed to a double recessive male. **Pr** = red eye, **pr** = purple eye; **Vg** = long wing, **vg** = vestigial wing.

No Independent Assortment in Drosophila

Bridges was working with two traits in Drosophila: an eye color, purple, recessive to wild-type eye color, and a wing mutant, vestigial, recessive to the wild-type wing. Four crosses were made, two in coupling and two in repulsion; and it is important, in analyzing these data, to pay strict attention to the sex of the F_1 flies that are test crossed.

In the first experiment Bridges crossed wild-type by purple vestigal. In Bateson's terminology, this is a cross in coupling. The F_1 were all wild-type. Bridges made a test cross of the F_1 *male* by the completely recessive purple vestigial female. Note (Table 10-4) that, in this cross, the parental genes stayed together absolutely.

In a second cross, Bridges mated vestigial by purple, a cross in repulsion, and again the F_1s were all wild-type. Once more he test crossed an F_1 *male* by a purple vestigial female. The results (Table 10-4) again show that the parental genes stay associated, and only parental progeny were recovered from this cross.

For his third cross, Bridges mated wild-type by purple vestigial, a cross in coupling again, similar to Cross 1. The F_1 were all wild-type. In this instance, however, he test crossed an F_1 *female* by the completely recessive male. In this experiment (Table 10-4) there is an excess of parental types, indicating that the parental genes tend to stay together, but there are also some new combinations.

Finally, in the fourth experiment, again in repulsion, he crossed vestigial by purple. The F_1 wild-type *female* was test crossed by the completely recessive male. Again, note (Table 10-4) that the parental genes tend to stay together but that some new combinations occur. Such results were becoming commonplace in Morgan's laboratory, where the Drosophila work was making remarkable progress. Morgan, in seeking an interpretation, pondered two specific questions: (1) Why is there no free combination or independent assortment? and (2) Why is there some recombination?

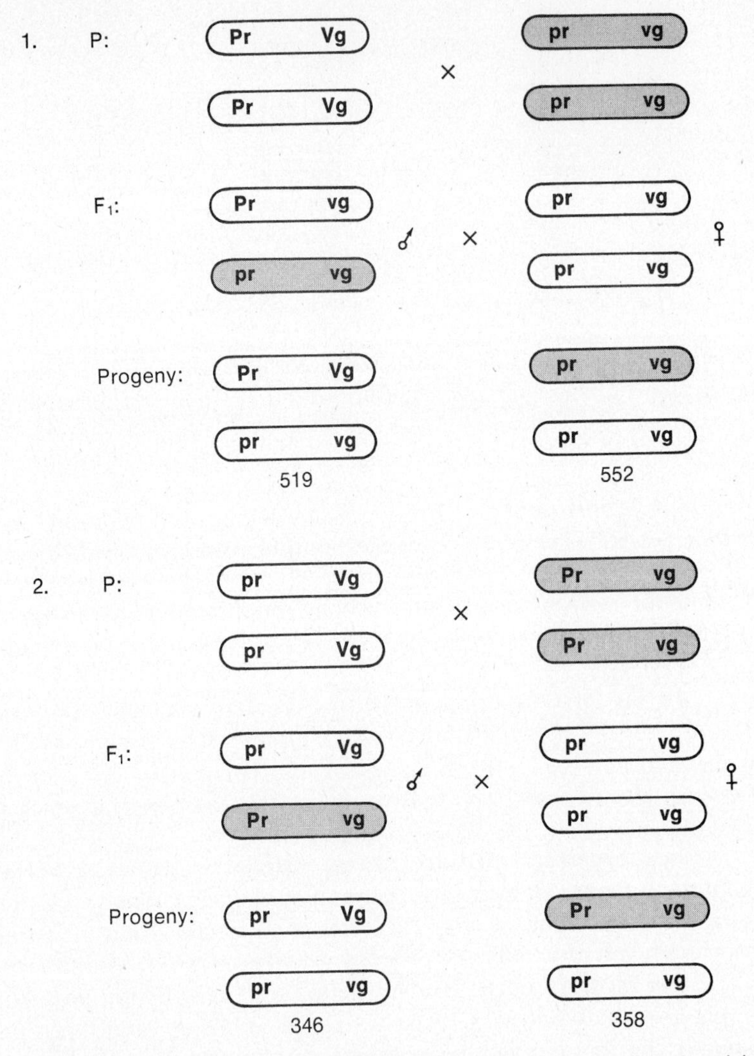

Figure 10-1. Morgan's interpretation of the failure of the genes purple and vestigial in Drosophila to assort independently. He proposes that the two genes are linked on the same chromosome. **Pr** = red eye, **pr** = purple eye, **Vg** = long wing, **vg** = vestigial wing.

Linkage

Morgan saw coupling and repulsion as two aspects of a single phenomenon, which he called *linkage.* He reasoned that linked genes tended to remain in their original combination because they were on the same chromosome. Returning to the first two crosses with purple and vestigial, Morgan's interpretation may be pictured as follows. (See Figure 10-1.) It was apparent to Morgan that if two genes were linked on the same chromosome, they would be inherited as the chromosomes were inherited, giving only two kinds of gametes and only two classes of progeny. More difficult to visualize is how in Crosses 3 and 4 some new combinations were produced.

Crossing Over

Morgan solved this difficulty by supposing that the pair of synapsing chromosomes could break and reunite in a new way. (See Figure 10-2.) If the mechanics of the diagram are taken literally, then it is easy to see that, in addition to the parental gametes produced by the female, there will be nonparental gametes producing the progeny with the new combinations of genes.

Morgan was also aware of supporting cytological evidence for his hypothesis. A few years earlier Janssens had published a paper on meiosis in salamanders. He observed what appeared to be chromosomes in a configuration similar to that postulated by Morgan. Janssens even came to the conclusion that there actually was a break and exchange of chromosome parts, although his reasoning was teleological, and the cytological evidence, while suggestive, was in no way conclusive. Morgan's hypotheses of linkage and breakage and exchange would explain why some genes tend to stay together and not assort independently and also why this tendency to stay together is not absolute but allows new combinations to be formed.

The new concepts introduced by Morgan were accompanied by the addition of several new terms to genetic vocabulary. The term *linkage* has already been mentioned. Genes are said to be linked if they occur on the same chromosome. The cross-link configuration observed by Janssens is called a *chiasma* (Chapter 7). The appearance of chiasmata does not necessarily imply that the chromosomes have broken and reunited in a new fashion. *Recombination* is a genetic phenomenon and refers to a new combination of genes. Finally, Morgan introduced the term *crossing over,* a cytological phenomenon defined as the exchange of corresponding segments of a pair of homologous chromosomes by breakage and reunion following pairing. Crossing over is inferred genetically from the recombination of linked genes and cytologically from the formation of chiasmata between pairs of chromosomes.

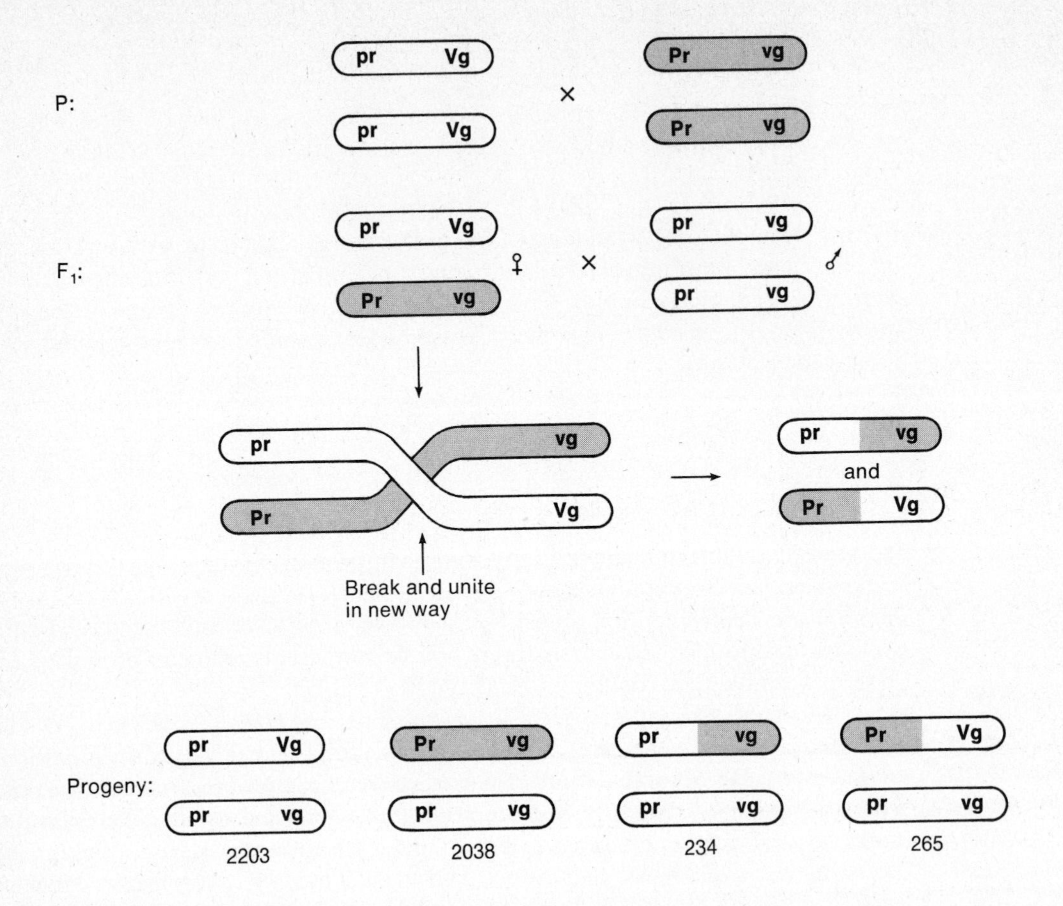

Genetic Nomenclature

Genetics, like other branches of science, has developed its own nomenclature. To accommodate simple Mendelism, capital and small letters are used to indicate a gene pair. But it should be evident that this nomenclature is inadequate to describe linkage. In Table 10-4 the F_1s of Crosses 1 and 2 had the same symbolism and yet gave different results. The new concepts of linkage and crossing over forced the Morgan group to design a new nomenclature, and Bridges was its principal architect.

Bridges proposed a small letter symbol for genes recessive to wild-type. Vestigial, therefore, would be symbolized **vg**. For the wild-type allele, instead of writing **Vg**, use the symbol $\mathbf{vg}^+$. Since purple is also a recessive gene, it is symbolized beginning with a small letter, in this case **pr**, and the wild-type allele for purple would be $\mathbf{pr}^+$.

A mutant gene, which is dominant to wild-type, is symbolized with a capital letter. There is a mutant in Drosophila known as Bar eye, which will be discussed in detail later. It is a dominant mutant in which the eye, instead of being round and

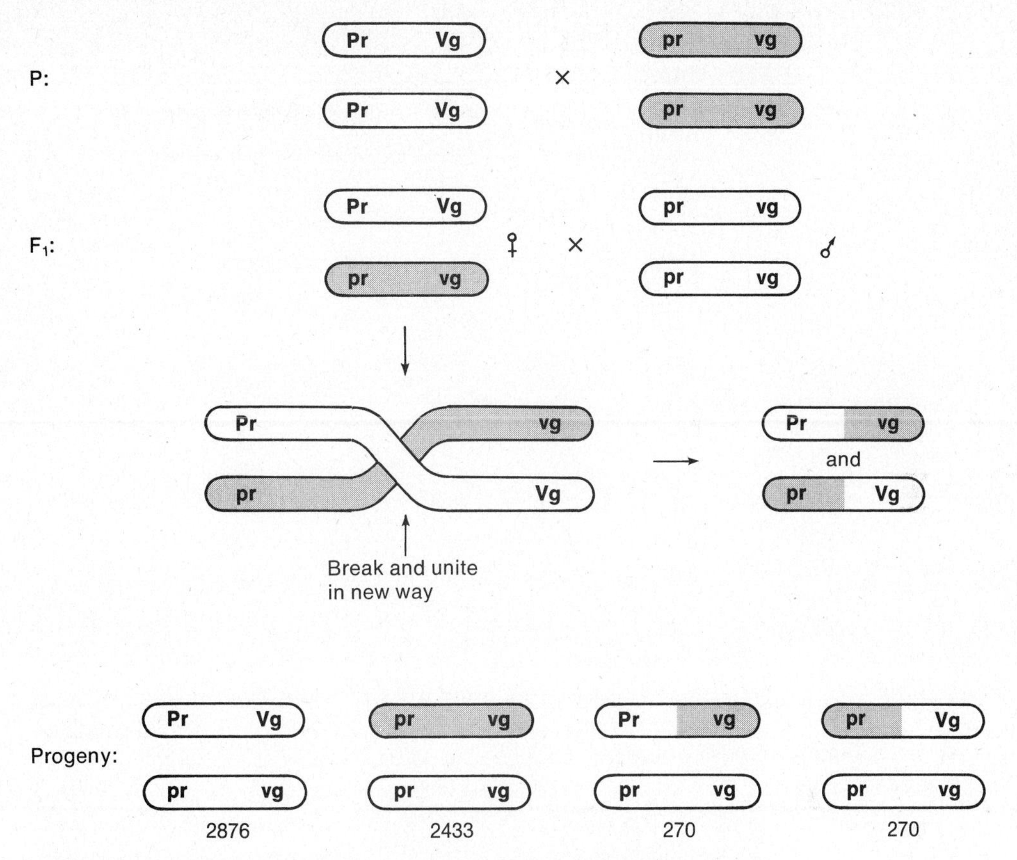

Figure 10-2. Morgan's interpretation of the failure of the genes purple and vestigial to assort independently but yet give new combinations in addition to parental types. He proposes that during meiosis in the female synapsis of homologous chromosomes may result in breakage and reunion. For simplicity the chromosomes are not duplicated. **Pr** = red eye, **pr** = purple eye; **Vg** = long wing, **vg** = vestigial wing.

composed of some 750 facets, is much reduced in size and bar-shaped with approximately 200 facets. Because it is dominant to wild-type, it is symbolized as **B**. To be consistent, the wild-type allele for Bar is $\mathbf{B}^+$. Thus all wild-type alleles are designated by a +.

Linkage means genes on the same chromosome. Homologous chromosomes could be drawn as two single lines with the genes written on the lines. To simplify this, however, instead of drawing two lines for the two homologous chromosomes, a single line is drawn and the genes on the two homologous chromosomes are written in above and below the line. There is one further simplification. When writing the symbol for the wild-type allele, the mutant symbol (for example, **vg**) is omitted; only

pr = Purple eye
pr$^+$ (or simply, +)=Wild-type (red) eye
vg = Vestigial wing
vg$^+$ (or simply, +)=Wild-type (long) wing

Cross 1 (Table 10-4) is now symbolized:

$$P: \frac{+\ \ +}{+\ \ +} \times \frac{pr\ \ vg}{pr\ \ vg}$$

$$F_1: \frac{+\ \ +}{pr\ \ vg}♂ \times \frac{pr\ \ vg}{pr\ \ vg}♀$$

Cross 2 becomes:

$$P: \frac{pr\ \ +}{pr\ \ +} \times \frac{+\ \ vg}{+\ \ vg}$$

$$F_1: \frac{pr\ \ +}{+\ \ vg}♂ \times \frac{pr\ \ vg}{pr\ \ vg}$$

Figure 10-3. Bridges' nomenclature for genetics crosses in Drosophila to indicate linkage arrangements.

the + is written. (See Figure 10-3.) This symbolism proposed by the Drosophilists will be used hereafter to designate genotypes of matings.

Frequency of Crossing Over

In Bridges' experiment (Table 10-4), recall that in parts 3 and 4, which involved crossing over in the female, a certain number of parental types and a certain number of recombinations were obtained. It might be asked if it is possible to calculate the frequency of new combinations. In fact, it is a simple matter to do so. In Cross 3, for example, the frequency of new combinations is found by dividing the new combinations (270+270) by the total number of progeny (5849) times 100. This equals 9.2 %. That is, there was 9.2 % recombination; and, more importantly, if the Morgan hypothesis is correct in stating that each recombination represents a crossover gamete, then it can be stated that there was 9.2 % crossing over. In other words, recombination percentage equals crossover percentage, at least in this example. The reader is forewarned, however, that recombination percentage does *not* always equal crossing over percentage. This point will be explained later. Therefore, in calculations of crossing over, it is always important to be sure that the equating of recombinations to crossovers is justified.

In the second crossing over experiment (Cross 4 in Table 10-4), the percentage of recombination is calculated as

$$\frac{265+234}{4740} \times 100 = 10.5\ \%$$

recombinations. It can again be demonstrated that, accepting the Morgan hypothesis, recombinations equal crossovers. The difference between 9.2 and 10.5 does not represent a difference in crossing over percentage but is due to differential viability of the different phenotypic classes and depends, to some extent, on the design of the experiment. It can be concluded that there is about 10 % crossing over between the genes for purple and vestigial.

In a different cross in Drosophila, two sex-linked genes were analyzed: a body-color mutant, yellow, **y**, and an eye-color mutant, white, **w**. Both mutants are recessive. Table 10-5 shows the crosses and the results obtained. The percentage of recombination in this experiment is calculated by

$$\frac{93+81}{14{,}939} \times 100 = 1.2\ \%$$

Notice that the recombination value, which equals crossover value, is much lower than the 10 % calculated in the purple vestigial experiment. The difference between 1.2 % and 10 % is real and statistically significant. Is it logical to expect such differences when different genes are tested? Morgan and his group concluded that it was, and Sturtevant, in particular, formulated the following hypothesis.

Table 10-5. A Cross Involving Two Sex-Linked Genes in Drosophila[a]

P: $\frac{\mathbf{y\,w}}{\mathbf{y\,w}}$ ♀ × $\frac{+\,+}{\rightharpoondown}$ ♂

F_1: $\frac{\mathbf{y\,w}}{+\,+}$ ♀ × $\frac{\mathbf{y\,w}}{\rightharpoondown}$ ♂

Progeny:	+ +	+ w	y +	y w	Total
	8093	81	93	6672	14,939

DATA FROM DEXTER. 1912. *BIOL. BULL.* **23**.

[a]Note that the male is hemizygous and has no genes on the Y chromosome. The headings for the progeny are phenotypes and not genotypes. Thus **+ w** symbolizes white and **+ +** equals wild-type, etc. Finally, **y** = yellow body, **w** = white eye.

Mapping Genes

If a chromosome is represented as a single line with three points, 1, 2, and 3 (Figure 10-4), simple physical considerations suggest that it is more probable that a break and exchange will occur between 2 and 3 than between 1 and 2, if the assumption is made that the probability of a break and exchange occurring is equal and random throughout the length of the chromosome.

Such assumptions lead to the interesting consequence that genes can be *mapped* according to their crossover frequency. Distance between genes is measured in *map units*, with the definition of a map unit provided by Sturtevant in 1913. He said: "Let 1 % crossing over equal one map unit." By 1 % crossing over, he

Figure 10-4. A representation of Sturtevant's proposal that the frequency of crossing over between any two genes will be a function of the distance between the genes.

meant that, of every 100 gametes, one will be a crossover gamete between a specific pair of genes. Note that Sturtevant's definition is based on crossing over percentage and not recombination percentage.

In the purple vestigial cross, there were 10 % crossovers, which means that, of every 100 gametes, 10 were crossover gametes. Therefore the genes purple and vestigial are 10 map units apart. The genes white and yellow are 1.2 map units apart.

Table 10-6. A Partial List of Some Sex-Linked Genes in Drosophila and Their Map Distances[a]

MUTANTS	MAP DISTANCE
y–l_1	0.8
y–w	1.5
y–ct	17.7
y–v	34.5
y–s	42.9
y–B	47.9

[a]The construction of such maps led to the compelling conclusion that the genes are arranged on the chromosome in linear order. The genes are: y = yellow body; w = white eye; ct = cut wing; v = vermilion eye; s = sable body; B = Bar eye.

A partial list of some sex-linked genes of Drosophila and their map distances is given in Table 10-6. Such results greatly influenced the conclusion drawn by Morgan and his group, which was that genes are arranged on the chromosome in linear order. Early investigators used the metaphor "beads on a string" to describe this linear arrangement of genes. This analogy, like many others in science, while picturesque, is not a completely accurate description, as will be shown later. However, it should be emphasized that the hypothesis that emerged from Morgan's research is that the genes are, in fact, arranged in linear order on the chromosome, each gene having its own place or *locus* (plural, *loci*).

In the examples so far analyzed only two genes were examined in a single experiment. Such experiments are often designated as a *two-point* cross. In another two-point cross in Drosophila, the sex-linked recessive gene scute (**sc**, a bristle mutant) was shown to be 8.9 map units from the recessive sex-linked gene echinus (**ec**, an eye mutant). In a second experiment involving echinus and a third sex-linked recessive gene, crossveinless (**cv**, a wing vein mutant), a map distance of 9.7 units was obtained. The results of these two experiments raise several interesting questions. What is the map relationship between scute and crossveinless? Can experiments be done with three genes at a time, a *three-point* cross? The answer to the second question is yes; it is possible to do a three-point cross. Crossover experiments involving more than three genes at one time can also be performed.

Three-Point Cross

For example, a cross involving three sex-linked genes—scute (**sc**), crossveinless (**cv**), and echinus (**ec**)—was performed. Table 10–7 shows the mating, beginning at the F_1 generation, and the results of this mating. To save time here and in subsequent examples, the experiment will be shown beginning with the F_1 female. The reader should be able to determine the parental genotypes of the cross that produced such an F_1 female. To map these genes, it is simplest to analyze the experiment two genes at a time, ignoring for the moment the third gene. For example, take the genes scute and crossveinless first, forgetting echinus for now. First, observe the parental arrangement of these two genes (Table 10-7). Examination of the F_1 female shows that **sc** and **cv** occurred in one parent and **+ +** in the other parent. Next, examine the progeny to determine the number of times where the parental arrangements are not observed—that is, where a recombination has occurred. In Table 10-7 the first group of progeny clearly consists of parental types with respect to the two genes scute and crossveinless because **sc** and **cv** are in the same progeny and **+ +** are together as they were in the original parents. The second group of progeny, however, represents a recombination of scute and crossveinless because **+** and **cv** are together, as are **sc** and **+**. Thus 150 recombinants are recorded. (See Figure 10-5.)

Table 10-7. A Three-Point Cross in Drosophila Involving Sex-Linked Genes[a]

F_1: $\dfrac{\mathbf{+\,+\,ec}}{\mathbf{sc\;cv\,+}}$ ♀ × $\dfrac{\mathbf{sc\;cv\;ec}}{\rightharpoondown}$ ♂

Progeny: (phenotype)		
+ + ec	810	1638
sc cv +	828	
+ cv +	88	150
sc + ec	62	
+ cv ec	89	192
sc + +	103	
	1980	

DATA FROM BRIDGES AND OLBRYCHT. 1926. *GENETICS* **11**.

[a]sc = scute; cv = crossveinless; ec = echinus.

The third group of progeny, with respect to the genes scute and crossveinless, also represents a recombination between these two genes because again the parental association of genes is not found in the progeny (**+ w** and **sc +**). Thus there are 192 more recombinants or a total of 342 recombinants between scute and crossveinless.

Next, turn to two other genes, scute and echinus, ignoring crossveinless for the time being. Note that the parental arrangements of these two genes is **+ ec** and **sc+**. Examine the progeny and notice that the first two are not recombinant between the genes scute and echinus. The second pair of progeny, however, is recombinant between these two genes in that now the association is **++** and **sc ec**. Therefore there are 150 recombinants between these two genes. The third pair of progeny represents the parental arrangement of these two genes. Therefore there are a total of only 150 recombinants between scute and echinus (Figure 10-5).

Finally, examine the two genes crossveinless and echinus and ignore scute for the moment. The F_1 shows that the parental arrangement is **+ ec** and **cv +**. In examining the progeny, notice that the first pair of progeny represents the parental

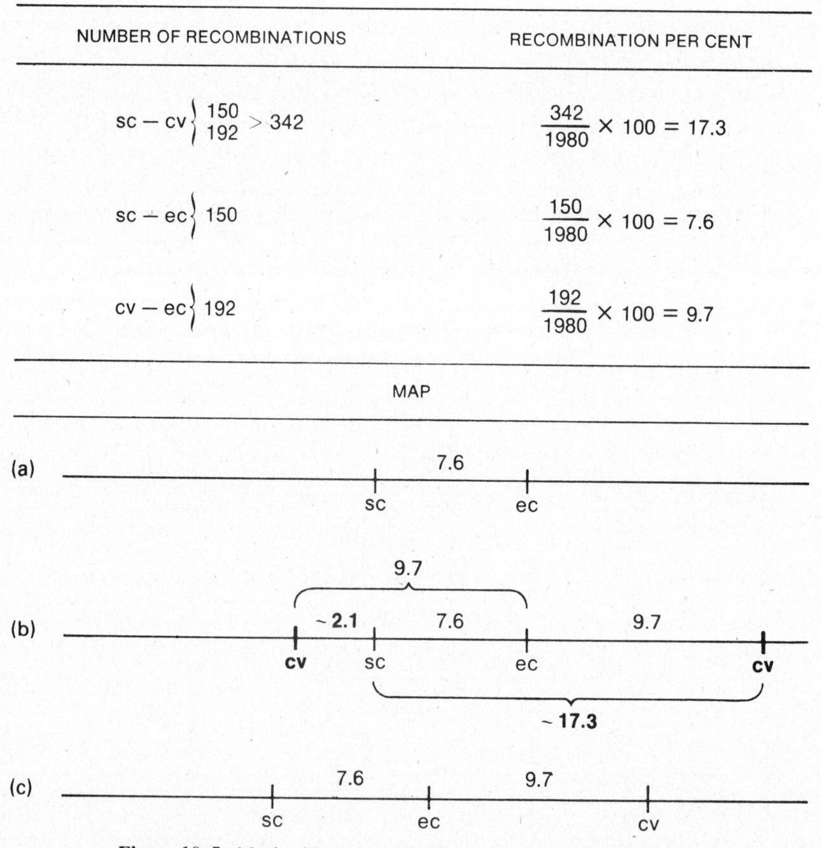

Figure 10-5. Method for mapping genes. Data from Table 10-6.

arrangement for crossveinless and echinus. It is worth pointing out, if the reader has not already noticed it, that clearly these first two progeny, which represent the majority of flies, are, in fact, the parental types. Their genotype is the same as the two parents originally involved in the cross. The second pair of progeny does not represent a recombination between the genes crossveinless and echinus. The third group of progeny, however, does represent a recombination between these two genes. Therefore there are 192 recombinants between these two genes (Figure 10-5).

These data may be summarized and the sums expressed as percentages by dividing by the total, 1980, ×100. This gives recombination percentages of 17.3 % for scute and crossveinless, 7.6 % for scute and echinus, and 9.7 % for crossveinless and echinus (Figure 10-5).

How, then, are the genes mapped? First, a chromosome is drawn and, as a rule, those two genes with the lowest recombination percentage are placed on the map. It will be shown later that this rule has a logical rationale behind it. In this experiment the smallest recombination percentage is 7.6, between scute and echinus, and so scute and echinus are placed on the map 7.6 units apart (Figure 10-5a). Determine the next smallest recombination percent which, in this instance, is 9.7 % between crossveinless and echinus. The questions now is: On which side does crossveinless go? Is it to the left or to the right of echinus (Figure 10-5b)? If crossveinless were to the left, the recombination percent between crossveinless and scute would be about 2 %, the difference between 9.7 and 7.6. The assumption is that the map is linear. If crossveinless is on the right, then the distance between scute and crossveinless should be roughly the sum of the map distances, or 17.3. The recombination percentage observed between scute and crossveinless is 17.3 %. Thus crossveinless must be to the right of echinus (Figure 10-5c).

Notice that the correct order of the genes is scute, echinus, crossveinless and not scute, crossveinless, echinus, as first written. Echinus is the gene in the middle. From these data, it is not certain which gene is left and which is right. Only the middle locus can be determined with certainty in a three-point cross. Why was not the gene order given correctly the first time? Because it was not known. The purpose of doing the experiment was to map the genes, and the genotypes were originally written in an arbitrary way without any insight as to what the correct order was.

In Figure 10-6 the cross is reshown now with the correct order of genes. The regions between the genes are numbered 1 and 2, respectively. Examination of the progeny indicates that the first two progeny, + ec + and sc+cv, are parental types and no crossing over need be assumed. The second set of progeny written in correct order, + + cv and sc ec +, can be simply explained by assuming a crossover in region 1. If we accept the Morgan hypothesis, no other assumptions are needed to explain these progeny. In a similar fashion, the third group of progeny, + ec cv and sc + +, can be explained by assuming a crossover in region 2. What were called recombination percentages are, in fact, crossover percentages. The map was correctly calculated using crossover percentages as required by Sturtevant's definition.

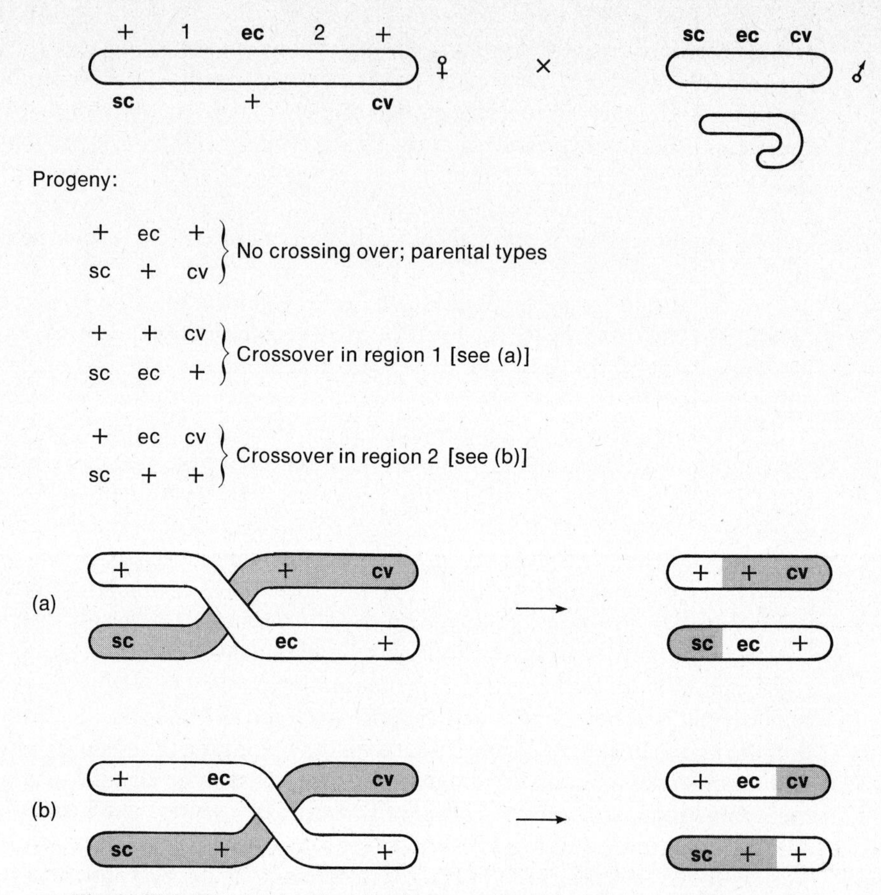

Figure 10-6. The three-point cross in Drosophila involving the sex-linked genes scute, echinus, and crossveinless. The progeny are designated as parentals or crossovers and the crossovers are represented diagrammatically. For simplicity, the chromosomes are drawn unduplicated.

Double Crossing Over

Another three-point cross in Drosophila was performed and involved three recessive sex-linked genes, crossveinless, cut (**ct**, a wing mutant), and vermilion (**v**, an eye-color mutant). The F_1 female from this cross was **+ + ct/cv v +**, and she was test crossed to the recessive **cv v ct** male. (See Table 10-8.) The object, again, is to map these three genes with the same procedure as before. Analyze only two genes at a time, crossveinless and vermilion. Examine the progeny to determine whether these two genes have recombined. The parental order of these two genes (Table 10-8) is **cv v** and **+ +**. The first pair of progeny does not represent a recombination between these two genes. The second pair does represent a recombination between crossveinless and vermilion, and so 153 recombinants are indicated. (See Figure 10-7.) The third pair of progeny also represents a recombination between crossveinless and vermilion, another 298 flies. The fourth pair of progeny does not represent a recombination between crossveinless and vermilion.

Table 10-8. A Three-Point Cross in Drosophila Involving Sex-Linked Genes[a]

F_1: $\frac{\mathbf{+\ +\ ct}}{\mathbf{cv\ v\ +}}$ ♀ × $\frac{\mathbf{cv\ v\ ct}}{\rightarrow}$ ♂

Progeny: (phenotypes)		
+ + ct	759	1525
cv v +	766	
+ v +	73	153
cv + ct	80	
+ v ct	140	298
cv + +	158	
+ + +	3	4
cv v ct	1	
	1980	

DATA FROM BRIDGES AND OLBRYCHT. 1926.

Crossveinless and cut are the second pair of genes to be analyzed. The parental order of these two genes is **+ ct** and **cv +**. The first pair of progeny does not represent a recombination between **cv** and **ct**. It should be noted that these progeny are the parental types, and as parental types they will not represent recombination between any pair of genes. The second group of progeny does represent a recombination between crossveinless and cut because now **cv** and **ct** are on the same chromosome, as are **+** and **+**. The third set of progeny does not represent a recombination between crossveinless and cut. The fourth set of progeny does represent a recombination between crossveinless and cut, so four are added to the 153 from the second group.

Vermilion and cut are the final pair of genes to be analyzed. The parental combination of vermilion and cut is **v +** and **+ ct**. The second pair of progeny does not represent a recombination between these two genes. The third pair of progeny does represent a recombination between vermilion and cut, as does the fourth set of progeny, thus resulting in a total of 302 (298 + 4). Figure 10-7 summarizes the data and expresses them in percentages.

In order to construct a map, the two genes with the smallest recombination percent are placed first on the chromosome—in this case, crossveinless and cut—at a distance of 7.9 (Figure 10–7a). The next smallest recombination is 15.3 between

NUMBER OF RECOMBINATIONS	RECOMBINATION PER CENT
cv — v { 153, 298 } > 451	$\frac{451}{1980} \times 100 = 22.8$
cv — ct { 153, 4 } > 157	$\frac{157}{1980} \times 100 = 7.9$
v — ct { 298, 4 } > 302	$\frac{302}{1980} \times 100 = 15.3$

MAP

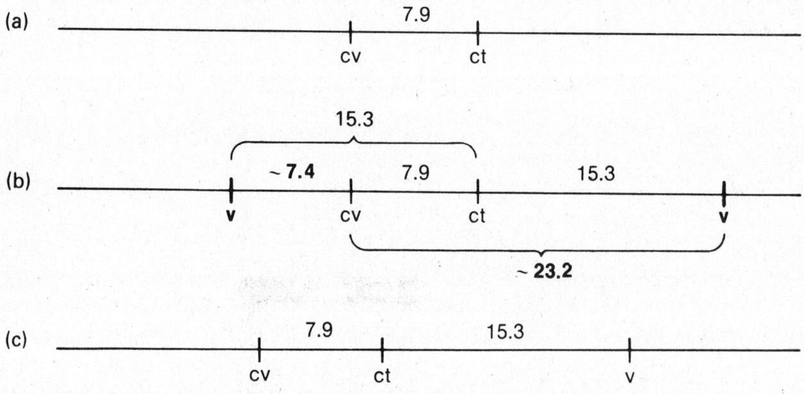

(a) cv —7.9— ct

(b) 15.3 (bracket) ; v —~7.4— cv —7.9— ct —15.3— v ; ~23.2 (bracket)

(c) cv —7.9— ct —15.3— v

Figure 10-7. Method for mapping genes. Data from Table 10-8.

vermilion and cut. Vermilion can be placed on the map either to the right of cut or to the left of cut. The latter would mean a map distance between crossveinless and vermilion of about 7.5 map units. Placing vermilion to the right of cut would suggest that the distance between them is about 23 units (Figure 10-7a). The recombination percentage between these two genes of 22.8 % easily decides the correct positioning of the genes (Figure 10-7a).

The map of these three genes shows that the original arbitrary positioning of the genes was incorrect. The correct order places cut in the middle instead of vermilion.

Figure 10-8 shows the cross with the correct order of the genes. The interval between crossveinless and cut is designated region 1 and that between cut and vermilion region 2. A reexamination of the results to determine the mechanism that resulted in the progeny shows that the first pair of progeny, + ct + and cv + v, represents the parental types. The second group of progeny, + + v and cv ct +, can

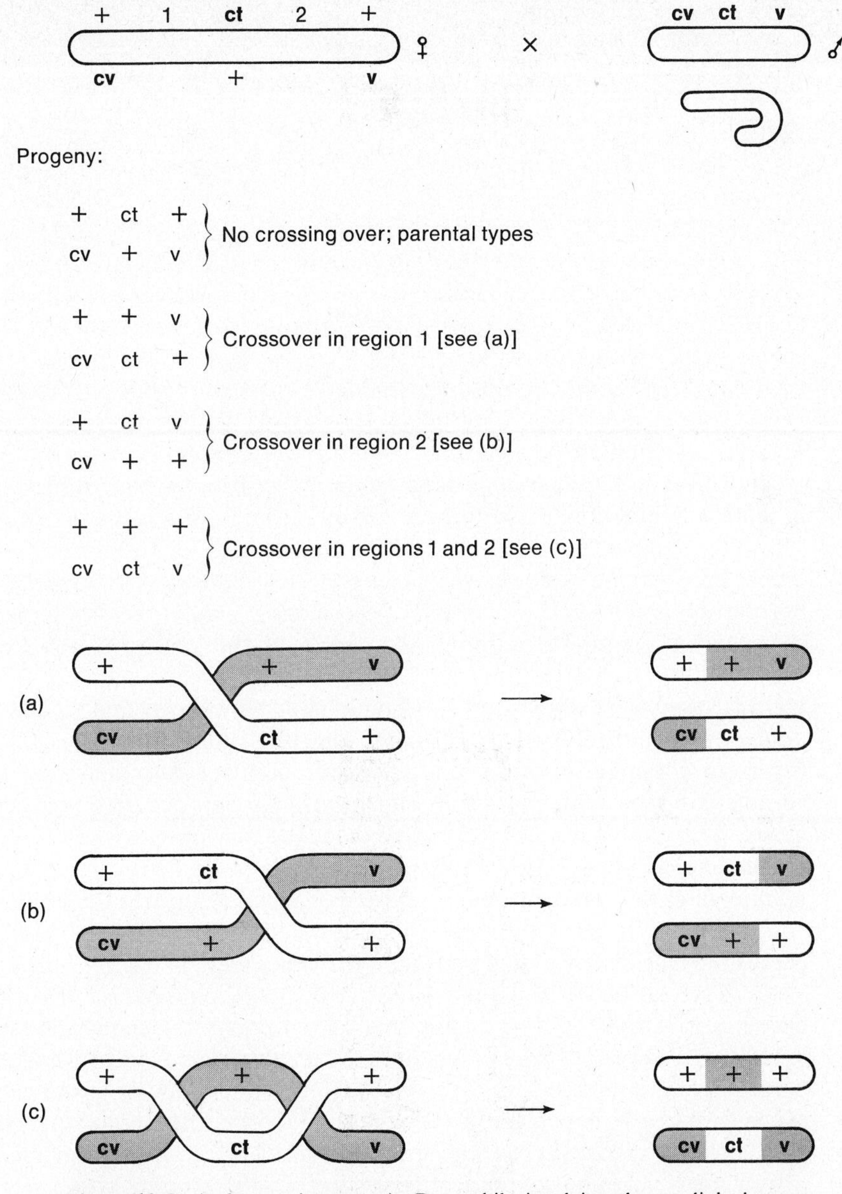

Figure 10-8. A three-point cross in Drosophila involving the sex-linked genes crossveinless, cut, and vermilion. The progeny are designated as parentals or crossovers and the crossovers are represented diagrammatically. For simplicity the chromosomes are drawn unduplicated.

be explained by a crossover in region 1. The third set of progeny, + ct v and cv + +, can be explained by a crossover in region 2. Finally, the fourth pair of progeny, + + + and cv ct v, can be explained by assuming a simultaneous crossover in region

1 and region 2. This last group is very interesting. It represents not one crossover but two and hence is called a *double crossover.* Double crossovers will reveal the important difference mentioned earlier between crossover and recombination percentage.

The analysis of this three-point cross gave a recombination percentage between crossveinless and vermilion of 22.8 %. What were the chromosomes that gave this 22.8 %? Clearly, they were + + v and cv ct +, 153, and + ct v, cv + +, 298. The total of 451 divided by 1980 and multiplied by 100 gave 22.8 %. Note that these four chromosomes, in fact, do represent a crossover gamete with the crossover occurring between crossveinless and vermilion. However, they are not the only chromosomes with crossovers between these two genes. The progeny + + + and cv ct v also represent chromosomes that have crossovers between crossveinless and vermilion and, moreover, represent two crossovers between these genes and not just one. So even though these chromosomes do not represent a recombination between crossveinless and vermilion because the double crossover has restored the original parental combination, they do represent crossovers—and *two* crossovers, not one.

Remember that Sturtevant said to let 1 % crossing over = 1 map unit. It was emphasized at the time that, in order to map the genes correctly, crossover percentages, not recombination percentages, must be employed. In the crossveinless-vermilion example, therefore, all the crossover gametes must be accounted for, including those represented by the double crossovers, in spite of the fact that they were not scored as recombinants between **cv** and **v**. The four (3+1) double crossovers represent 0.2 % of the total ($4 \div 1980 \times 100$). This 0.2 % does not represent a gamete in which one crossover occurred between crossveinless and vermilion; it represents gametes in which *two* crossovers occurred between these two genes. Therefore it is necessary to double the 0.2 % in order to indicate correctly the crossover percentage between the two outside genes. If the 0.4 % (which represents additional crossovers between **cv** and **v** that were not previously accounted for) is added to the 22.8 % recombination already noted, the final figure of 23.2 ($22.8 + 0.4$) equals the true distance obtained when the genes were mapped ($7.9 + 15.3$).

The danger of mapping genes only by recombination percentages now becomes apparent. When a double crossover occurs between two genes, no recombination is observed between them, but, in fact, two crossovers took place. Thus the percentage of double crossovers must be doubled and added to the percentage of single crossovers between the two genes in order to obtain the correct map distance. The reader can use this method as a check on the original mapping. The procedure of mapping genes described in the two three-point cross experiments will automatically account for the discrepancy caused by double crossovers. It should be emphasized that the correct map distance of 23.2 was obtained without knowing what double crossovers were. But a useful check on the calculations can be obtained by adding twice the frequency of doubles to the recombination percentage obtained between the outside genes. This sum should be exactly the same as the sum obtained by adding the separate map distances.

One rule for mapping genes was to start with those genes that show the least recombination. The logic for this procedure was revealed in the early Drosophila work. Double crossovers do not usually occur over short map distances. In Drosophila, the evidence is that if the genes are closer than 10 or 15 units apart, no double crossovers will occur between them. Therefore, by always starting the map with genes that are close together, the concern about doubles is minimized, particularly if the distance is smaller than 10 to 15 units. To find the true map distance between two genes that are far apart, design the experiment with intermediate genes between the two that are far apart so that the majority of double crossovers can be accounted for.

Determining Gene Order The occurrence and recognition of double crossovers make it possible to determine the gene order without actually mapping the genes. Examine the second cross as it was originally written (Table 10-8). It has been observed that a crossover is an infrequent event, and it would be logical to assume that a double crossover occurs more rarely. From the results of the three-point cross, it can be inferred that the class of progeny that occurs least frequently represents the double crossovers. In the crossveinless–vermilion–cut experiment it would be the + + + and cv v ct progeny that occurred only four times out of almost 2000 progeny that are the double crossovers. When a double crossover occurs, it is the two end genes—that is, the two genes that are the farthest apart—that end in the same combinations that they began in. In other words, no recombination occurs between these two genes when a double crossover takes place. Thus only the gene pair that is in the middle recombines or exchanges position. An analysis of the experiment to see which of the three gene pairs has recombined in the double crossover will show which gene pair is in the middle. Figure 10-9(a) illustrates how this works. It becomes clear that, of the three genes, the only one that can be recombined or exchanged to get the double crossover progeny is the cut gene. Therefore the cut gene pair is in the middle, giving the order identical to that obtained when the genes were mapped.

Return to the first three-point example that was presented (Table 10-7). Only six genetically different progeny are listed. Why not eight? The reason is simply that one class of progeny did not occur—the class + + + and sc cv ec. Stated another way, these progeny occurred with a frequency of zero. Then it is logical to conclude that this class represents the group that would have resulted from a double crossover. Making this assumption, we can ask: Which gene pair is the only one that could recombine to give the theoretical double crossover progeny? Examination of Figure 10-9(b) will show that only the echinus pair will satisfy this requirement, and hence echinus must be the middle gene pair, as was determined when the genes were mapped.

Why did the doubles not occur? Recall that double crossovers do not occur when the genes are relatively close together. In this example the total distance between the two genes is 17.3 units, and in an experiment involving only several

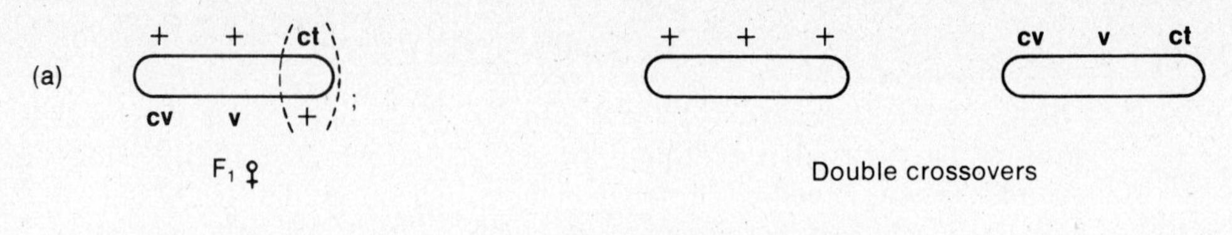

Only recombining the cut gene pair will transform

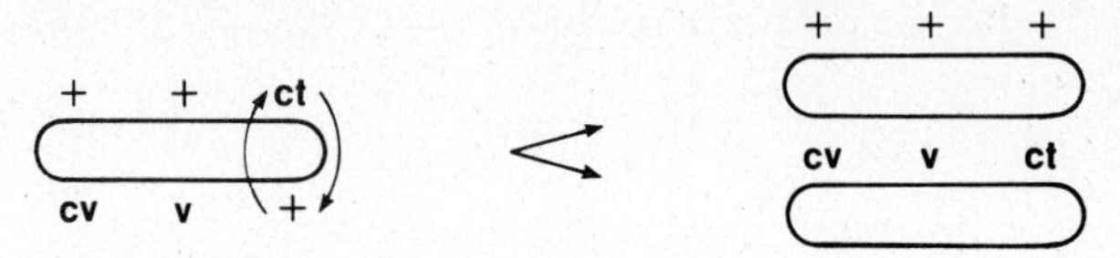

Therefore cut must be the gene pair in the middle.

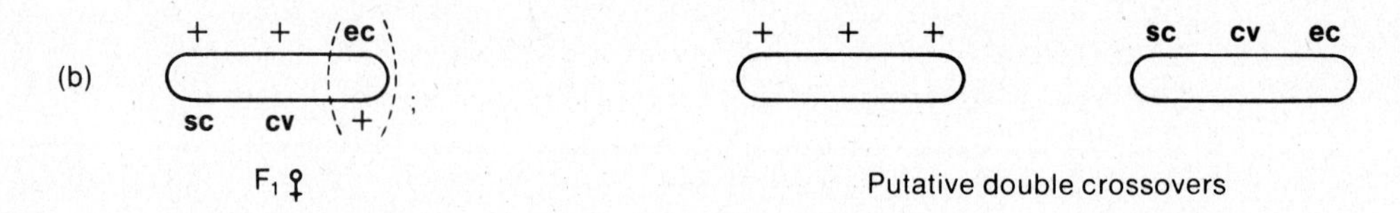

Only recombining the echinus gene pair will transform

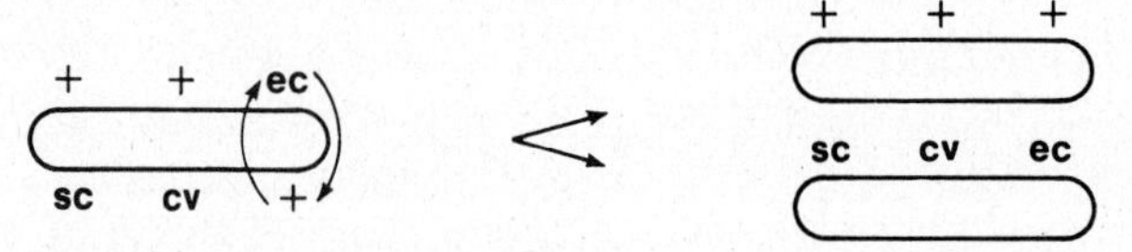

Therefore echinus must be the gene pair in the middle.

Figure 10-9. The determination of gene order in a three-point cross without mapping the genes. The least frequently occurring class of progeny must be the double crossover class, actual or potential. Since in a double crossover only the gene pair in the middle recombines, an examination of the original genotype and the double crossover genotype reveals which is the only pair of genes whose recombination gives the double crossover genotypes.

thousand flies it is possible that doubles have not occurred. Actually, when the experiment was done on a larger scale, a few doubles were produced.

Interference

It was shown that the map distance between crossveinless and cut is 7.9 and that between cut and vermilion 15.3. If the experiment were repeated, we would predict that crossovers between crossveinless and cut would be approximately 7.9 % and crossovers between cut and vermilion approximately 15.3 %, if these values are accepted as reasonably accurate and if errors of random sampling and other statistical vagaries are ignored for the present. It can be argued, therefore, that the probability of a crossover between crossveinless and cut is 7.9 % and that the probability of a crossover between cut and vermilion is 15.3 %. These probabilities are empirical because they are based on actual experimental data. Given all these assumptions, it may be asked: What is the probability of two crossovers occurring simultaneously between crossveinless and vermilion? In Chapter 3 a probability axiom was stated to the effect that if the probability of one event is a and the probability of a second independent event is b, then the probability that both events will occur at the same time is the product of the two, or $a \times b$. In our crossover example it can be argued that if the probability of a crossover between crossveinless and cut is 7.9 % and the probability of a crossover between cut and vermilion is 15.3 %, then the probability that two crossovers will occur at the same time between crossveinless and vermilion is the product of the separate probabilities, or 7.9 % × 15.3 %. Arithmetically, this would be 0.079 × 0.153, which equals 0.0121, or 1.21 %. Therefore *if the events are independent*, about 1.2 % double crossovers would be expected between crossveinless and vermilion. However, only 0.2 % was observed, which means that fewer doubles occurred than would have been expected due to chance. In other words, it appears as though one crossover between crossveinless and vermilion has decreased the chance of a second crossover occurring between these two genes. Such an observation is common. One crossover interferes with the probability that a second crossover will occur, a phenomenon that was given the name *interference* by H. J. Muller. Looked at in another way, the two events apparently are not independent, as was assumed in the calculation of probability.

In the first experiment with scute, crossveinless, and echinus, the expected percentage of doubles would be 7.6 % × 9.7 %, which is equal arithmetically to 0.74 %. Actually, there were 0 % doubles. Interference was absolute! It has been asserted that, in Drosophila no double crossing over is observed over short distances of about 15 units or less. The explanation, at least as proposed by Morgan, was thought to be due to mechanical reasons. Morgan postulated that when the chromosomes synapsed with each other, the coiling established certain tensions in the chromosomes. A break in one region presumably relieved the tension over a certain distance to the right or left, thus making it unlikely that a second break will occur. In recent years interpretations of interference at the molecular level have been proposed, although no completely satisfactory explanation has been suggested.

Coefficient of Coincidence

Muller, who was initially interested in this problem, proposed a method of quantifying the degree of interference, and he called the quantitative measure of interference the *coefficient of coincidence*, often referred to simply as *coincidence*. Muller proposed that the coefficient of coincidence is equal to the observed double crossovers in percent divided by the expected double crossovers in percent. What is observed when this formula is applied to the two three-point crosses discussed in this chapter? In the second example the coincidence would be equal to 0.2 % divided by 1.21 %, which equals 0.17. In the first example the coincidence is equal to 0 % divided by 0.74 %, which equals 0. Table 10-9 summarizes the theoretically possible coefficients of coincidence.

Table 10-9. The Theoretically Possible Kinds of Interference and the Corresponding Coefficients of Coincidence

COEFFICIENT OF COINCIDENCE	FREQUENCY OF DOUBLES	TYPE OF INTERFERENCE
0	If doubles = 0 %	absolute
<1	Doubles less frequent than expected	partial
1	Doubles frequency = expected frequency	none
>1	Doubles more frequent than expected	negative

No Crossing Over in Male Drosophila

Some readers may still be puzzled by the data of Bridges involving purple eye and vestigial wing in which only two classes of progeny occurred. In the cross of wild-type by purple vestigial, in which the F_1 male was mated with a purple vestigial female, only wild-type and purple vestigial progeny were obtained. There was no recombination in this experiment between the genes vestigial and purple. These results can be explained by the assumption that no crossing over occurs in the male of Drosophila. It is this postulate that Morgan and his group offered to explain the lack of recombination. It is, however, only an answer and not an explanation of why no crossing over occurs in Drosophila males.

It was suggested that perhaps no chiasmata occurred in male Drosophila, but cytological evidence indicates that meiotic figures in Drosophila males show what appear to be chiasmata similar to what is observed in females. Whether this fact indicates that breaks do or do not occur is not known. It does show that the seeming appearance of chiasmata cytologically does not necessarily mean that crossing over has occurred.

During the early part of this century the Japanese geneticist Tanaka was doing experiments with the silkworm *Bombyx mori.* Tanaka observed that there was no crossing over in the female of *Bombyx.* It is interesting to note that in the silkworm the female is the heterogametic sex, whereas in Drosophila the male is the heterogametic sex. And some believed that a reasonable generalization was that there is no crossing over in the heterogametic sex. However, this notion failed, for in most other organisms tested, crossing over has been found to occur in both sexes. The frequency of recombination is not always the same in both the male and the female, but crossing over is not, as a general rule, confined to one sex, as the early experiments would seem to have indicated.

Recent electron microscopy, particularly in Drosophila, has indicated meiotic differences between males and females. *Synaptinemal complexes* are present in females of Drosophila that are believed to play an important role in synapsis and recombination (Chapter 7). Such synaptinemal complexes are missing in males of Drosophila and in rather special strains of Drosophila in which crossing over is absent or reduced. The true significance of the synaptinemal complex will have to await further developments.

Factors Affecting Crossing Over

It turns out that crossing over, which seems reasonably simple, is actually a fairly complicated process and, more importantly, a very labile one. Many external factors are known to affect the frequency of crossing over, but the reason is not understood.

Morgan's laboratory performed many of the early experiments in which some of the variables affecting crossing over were analyzed. For example, Bridges did an experiment in which he tested the effect of the age of the mother in Drosophila on the frequency of crossing over and found that there was an effect. (See Figure 10-10.) The curve is not a simple one, nor is it easy to interpret the biological meaning of this variable.

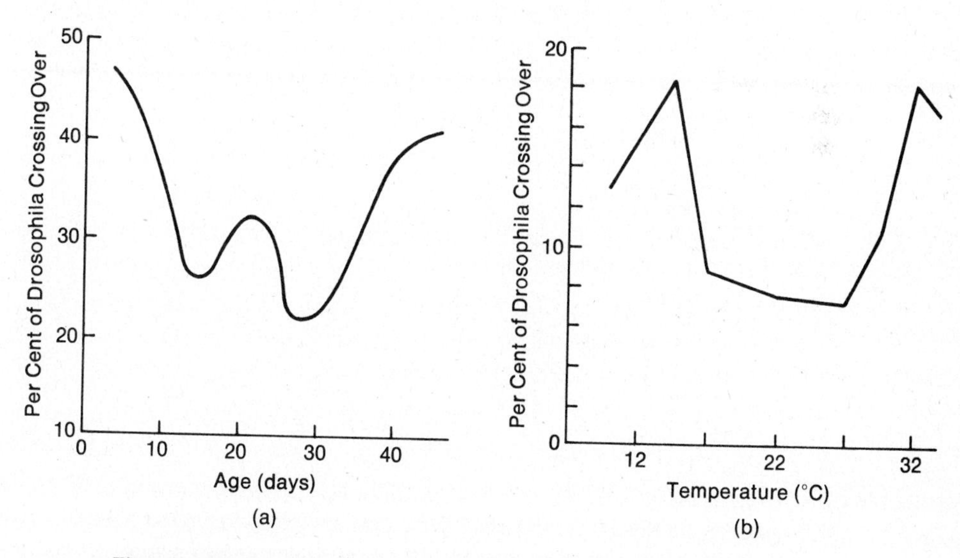

Figure 10-10. The effect of the age of the female (a) and the effect of temperature (b) on the frequency of crossing over in Drosophila. Age curve (a) redrawn from Bridges, 1929. *Carnegie Inst. Wash. Publ.* **399**. Temperature curve (b) redrawn from Plough, 1917. *J. Exp. Zoo.* **32**.

H. H. Plough, another student of Morgan, did a thorough analysis of the effect of temperature on crossing over in Drosophila and showed conclusively that this variable had a profound effect (Figure 10-10). The effect he observed is in no way linear nor even apparently simple, and easy explanations for this experiment are not forthcoming.

Other investigators tried many other things—chemicals, heat shock, radiation—on crossing over, and in most instances effects were observed. The fact that almost anything tried has an effect on the frequency of crossing over has made it difficult to understand what the mechanism may be if it is a simple one. Crossing over can be affected by external agents but in a manner that has led to no simple explanation as to how these various treatments work.

Finally, it can be said that if Morgan's hypothesis of the Chromosome Theory of Inheritance is correct, there should never be more linkage groups than there are pairs of chromosomes, for Morgan theorized that the genes are located on the chromosomes. There may, of course, be fewer linkage groups than chromosome pairs if not all the linkage groups have been experimentally detected. Table 10-10 shows a comparison between linkage groups and chromosome pairs to see whether this prediction holds up. It is seen that in no case does the number of linkage groups exceed the number of chromosome pairs, thus fulfilling an important requirement of the chromosome theory.

Table 10-10. The Correlation Between Linkage Groups and Chromosome Pairs[a]

ORGANISM	NUMBER OF LINKAGE GROUPS	NUMBER OF CHROMOSOME PAIRS
Drosophila melanogaster	4	4
Drosophila virilis	6	6
Drosophila pseudoobscura	5	5
Drosophila willistoni	3	3
Corn (*Zea mays*)	10	10
Sweet pea (*Lathyrus ordoratus*)	7	7

[a]The chromosome theory requires that the number of genetic linkage groups cannot exceed the number of chromosome pairs.

REFERENCES

BATESON, W., E. R. SAUNDERS, and R. C. PUNNETT. 1906. Experimental studies in the physiology of heredity. III. *Reports Evol. Comm. Roy. Soc.* **3**:1–53. (Excerpt available in *Classic Papers in Genetics*, edited by J. A. Peters. Prentice-Hall, Englewood Cliffs, N.J., 1959.) (This paper contains the early work with the sweet pea that showed exceptions to independent assortment.)

BRIDGES, C. B. 1929. Variations in crossing over in relation to age of female in *Drosophila melanogaster. Carnegie Inst. Wash. Publ.* **399**:63–89. (This paper contains the experiments that show the effect of age on crossing over in Drosophila.)

BRIDGES, C. B., and T. M. OLBRYCHT. 1926. The multiple stock "Xple" and its use. *Genetics* **11**:41–56. (The source of the data on crossing over in Drosophila involving the sex-linked genes scute, crossveinless, echinus, vermilion, and cut.)

DEXTER, J. S. 1912. On coupling of certain sex-linked characters in Drosophila. *Biol. Bull.* **23**:183–194. (The source of the data on crossing over in Drosophila involving the sex-linked genes white and yellow.)

JANSSENS, F. A. 1910. Spermatogénèse dans les Batraciens. 5. La théorie de la chiasmatypie. Nouvelles interprétation des cinèses de maturation. *La Cellule* **25**:387–411. (The important paper that demonstrated chiasmata during meiosis and suggested breakage and exchange.)

MORGAN, T. H. 1914. No crossing over in the male of Drosophila of genes in the second and third pairs of chromosomes. *Biol. Bull.* **26**:195–204. (The source of data on crossing over in Drosophila involving the genes purple and vestigial. Demonstrates clearly the absence of crossing over in male Drosophila.)

MULLER, H. J. 1916. The mechanism of crossing over. *Am. Nat.* **50**:284–305. (One of several important papers Muller wrote at this time on crossing over. This one gives his ideas on interference.)

PLOUGH, H. H. 1917. The effect of temperature on crossing over. *J. Exp. Zool.* **32**:187–212.) (This paper contains the experimental work that showed the effect of temperature on crossing over in Drosophila.)

PUNNETT, R. C. 1919. *Mendelism.* 5th ed., 219 pp. Macmillan and Co., London. (The source of the complete data involving flower color and pollen length in the sweet pea.)

STURTEVANT, A. H. 1913. The linear arrangement of six sex-linked factors in Drosophila as shown by their mode of association. *J. Exp. Zool.* **14**:43–59. (Available in *Classic Papers in Genetics*, edited by J. A. Peters. Prentice-Hall, Englewood Cliffs, N.J. 1959.) (The classic paper that shows that genes are linked, cross over, and can be mapped.)

QUESTIONS AND PROBLEMS

10-1. Define the following terms:

linkage
chiasma
two-point cross
three-point cross

crossing over
map unit
recombination
double crossover
interference
coefficient of coincidence

10-2. Discuss the significance of the early Drosophila work in Morgan's laboratory.

10-3. In the sweet pea experiment of Bateson and Punnett, if they made the following cross: purple long × red round and mated the F_1 × a red-round plant, what would be the distributions of phenotypes in 1000 progeny plants? (See Table 10-3.)

10-4. If, in the sweet pea, an F_1 plant from the cross of purple round × red long is test crossed to a red-round plant, what would be the distribution of phenotypes in 1000 progeny plants?

10-5. In Drosophila, white eye (w) is sex linked, recessive to red, curved wing (c), a second chromosome autosomal, recessive to straight, and speck (sp), another second chromosome autosomal recessive to nonspeck. In the female, c and sp undergo 25 % recombination. A female of the genotype $\frac{\text{w c +}}{\text{+ + sp}}$ is mated to $\frac{\text{w c sp}}{\text{w c sp}}$ male. What types of offspring and in what proportions will be produced? List genotypes and phenotypes.

10-6. What would be the results of the reciprocal cross: $\frac{\pm\ \text{c +}}{\text{+ sp}}$ male × $\frac{\text{w c sp}}{\text{w c sp}}$ female?

10-7. In Drosophila, a female of the constitution $\frac{\text{a b c}}{\text{+ + +}}$ is mated to an $\frac{\text{a b c}}{\text{a b c}}$ male. The following offspring are obtained:

a b c	+ + +	a + +	+ b c	a b +	+ + c
460	460	18	12	22	28

(a) What are the percentage frequencies of recombination between a and b; b and c; a and c?
(b) What is the coincidence value for double crossovers?

10-8. The genes a and b are linked autosomal recessives 10 map units from each other. The following crosses are made:

P: $\frac{\text{a +}}{\text{a +}}$ female × $\frac{\text{+ b}}{\text{+ b}}$ male

F_1: $\frac{\text{a +}}{\text{+ b}}$ female × $\frac{\text{a +}}{\text{+ b}}$ male

Show that the F_2 phenotypic ratios in this cross is performed in
(a) Drosophila
(b) Corn

10-9. In corn, a plant heterozygous for three gene pairs, A, a; B, b; C, c is test crossed to the triple recessive with the following results:

ABC	142	ABc	22
abc	148	abC	18
Abc	235	AbC	95
aBC	255	aBc	85

(a) What was the parental cross?
(b) What is the genotype of the heterozygous plant?
(c) Give the single and double crossover progeny.
(d) Determine the correct order of the genes without mapping them.
(e) Map the genes.
(f) Calculate the coefficient of coincidence.

10-10. In Drosophila, a fly that is a triple recessive for three sex-linked genes—ruby, vermilion, cut (rb, v, ct)—is crossed to a wild-type fly. The F_1 heterozygous daughters are crossed to triple recessive males. The offspring fall into the following eight classes:

+ + +	729	rb + ct	129
rb v ct	701	+ v +	151
rb + +	138	rb v +	13
+ v ct	122	+ + ct	17

(a) Map these genes.
(b) What is the coefficient of coincidence value?

10-11. In corn, the genes X, Y and Z are on chromosome 3 at loci 15, 21, and 24. What type of gametes and in what percentages will a plant produce which obtained X, Y, and z from one parent and x, y, and Z from the other?

10-12. Assuming m and n to be sex-linked recessive genes in Drosophila 40 units apart and r and s to be second chromosome recessives 20 units apart, determine the phenotypes and their frequencies from the cross:

$$\frac{++}{mn}\frac{++}{rs}\text{ female} \times mn\frac{rs}{rs}\text{ male.}$$

10-13. Sturtevant (1921) reported the following data from crosses made with *Drosophila simulans.* The genes were yellow body (y), carmine eyes (c), and forked bristles (f), all three recessive to wild-type and sex linked. Yellow carmine females were mated to forked males and the F_1 females were mated to yellow carmine forked males to produce the following offspring:

y c +	725	y c f	134
+ + f	719	+ + +	109
y + f	419	y + +	34
+ c +	383	+ c f	32

(a) Map these genes.
(b) What is the coincidence value?

10-14. The map distance between two genes, a and b, is 20 units. In the progeny from the cross $\frac{a\,b}{+\,+} \times \frac{a\,b}{a\,b}$ only 18 % recombinations were found. State the frequency of double crossovers in percent?

10-15. Assume that a number of experiments have been performed in Drosophila involving the following genes: curved wing (cur), dachs body (d), bent wing (bt), sepia eye (se), blackbody (b), claret eye (ca), scarlet eye (st), Star eye (S), purple eye (pr), and rotated abdomen (rt).
From these experiments, the following crossover values were obtained:

cur-st	14 %	se-st	18 %
d–b	18 %	b-pr	6 %
S-b	47 %	cur-rt	7 %
se-rt	11 %	d-pr	24 %
se-cur	4 %	rt-st	7 %
S-d	30 %		

All other combinations showed 50 % crossing over. Plot these genes on their chromosome or chromosomes.

10-16. In Drosophila, the recessive genes scarlet eye (st), spineless bristles (ss), and ebony body (e) are linked and their map distances from the end of the chromosome are 44, 58, and 70, respectively. If many flies are produced from the following cross

$$\frac{st\ ss\ e}{+\ +\ +}\text{ female} \times \frac{st\ ss\ e}{st\ ss\ e}\text{ male}$$

and if the coincidence value is 0.7, what phenotypes will be expected and in what percentages?

10-17. In Drosophila, a cross was made involving four sex-linked recessive genes: cut wing (ct), forked bristle (f), raspberry eye (ras), and yellow body (y). From a cross of an F_1 female heterozygous for all four genes to a male completely recessive, the following progeny were obtained:

+ + + +	2005	ct + ras y	555
ct f ras y	1825	+ f + +	629
ct + + +	69	ct + + y	317
+ f ras y	99	+ f ras +	386
ct f + +	35	ct f + y	81
+ + ras y	46	+ + ras +	94
ct f ras +	441	ct + ras +	169
+ + + y	603	+ f + y	185

(a) What was the parental cross?
(b) Give the single, double, and triple crossovers.
(c) Map the genes.
(d) Calculate coincidence values.

11/Linkage and recombination

PART II

Four-Strand Crossing Over

In one of Bridges' early experiments with nondisjunction in Drosophila, he crossed a female heterozygous for two sex-linked eye-color genes, eosin (**w**^{e}) and vermilion (**v**), by wild-type males (Figure 11-1). He obtained a few unusual progeny that proved to be homozygous for vermilion and heterozygous for eosin. This result is not easy to explain, and at first a logical possibility might be that it is some genetic mistake or mutation. Then the sperm might contain, instead of the wild-type gene for vermilion, a mutation to the recessive allele, **v**. However, mutation cannot account for the observed exception. Also, as will be shown in Chapter 14, mutation rates are very low, and the exceptions observed by Bridges, while infrequent, were not that rare.

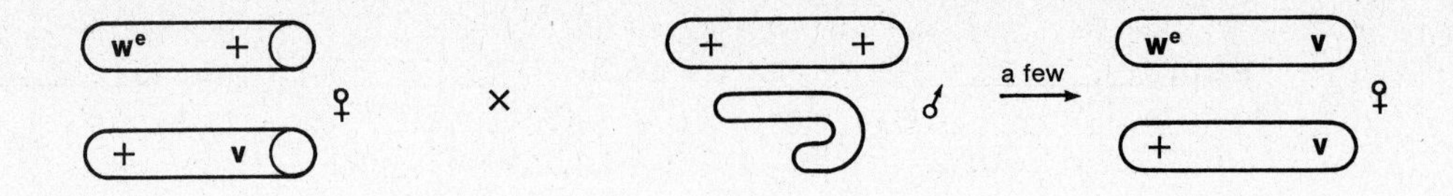

(a) Two-strand crossing over:

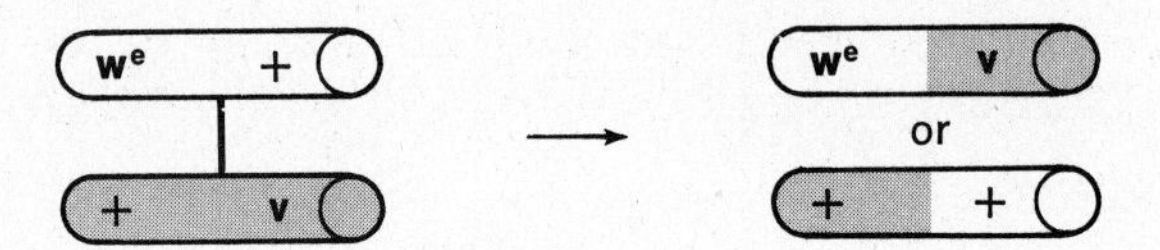

Neither gamete can account for the rare flies.

(b) Four-strand crossing over:

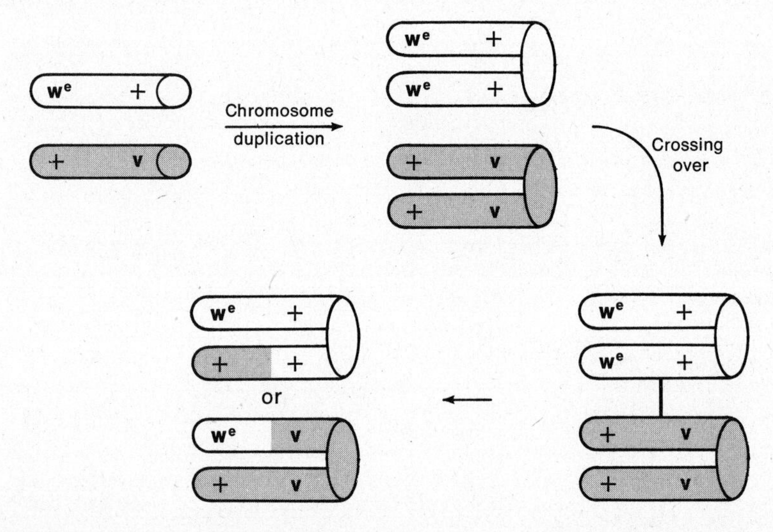

Subsequent nondisjunction and fertilization by a Y-bearing sperm can account for the rare flies.

Figure 11-1. Evidence for four-strand crossing over in Drosophila. The rare females at the top right cannot be explained by mutation or by crossing over at the two-strand stage. Only if crossing over occurs after chromosome duplication with an accompanying nondisjunction can Bridges account for the rare females. $\mathbf{w^e}$ = eosin, **v** = vermilion.

Another possible explanation is an hypothesis that includes crossing over. However, any attempt to provide an explanation based on crossing over with only two unduplicated chromosomes, as was illustrated in Chapter 10, will fail. (See Figure 11-1a.) Bridges, however, thought that perhaps crossing over occurs during the four-strand, or tetrad, stage, after chromosome duplication. If true and if there was a subsequent nondisjunction, then the few unusual females could be accounted for. (See Figure 11-1b.) On the basis of this evidence, Bridges in 1916 postulated that crossing over must occur when the chromosomes are in the four-strand stage. Although no one seriously doubted this particular explanation, it should be apparent that Drosophila is not the most suitable organism to show the stage at which crossing over occurs. The reason is that a single fly that might be scored as a crossover represents only one chromosome from the meiotic tetrad. The other three chromosomes are distributed to the polar bodies and are not available for analysis. More favorable material is necessary to test the four-strand crossing-over hypothesis.

Neurospora

Morgan left Columbia University in 1927 to become head of the new biology division at the California Institute of Technology. He took with him two of his students, Sturtevant and Bridges, and they soon attracted a new active group of geneticists on the West Coast. One of the projects undertaken in this new laboratory was a genetic analysis of the fungus, *Neurospora crassa*, a bread mold.

Life Cycle of Neurospora

Neurospora reproduces both asexually and sexually. It is the sexual reproduction that is of most interest to the geneticist. The vegetative stage of reproduction involves filamentous growths called *hyphae* (singular, *hypha*), which grow by simple mitotic division. The hyphae form a network of strandlike growths known as a *mycelium* (plural, *mycelia*). If, in a mycelium, two hyphal strands of proper mating type come close together by chance, their nuclei can fuse, thereby forming a diploid nucleus or zygote. This zygote immediately undergoes meiosis in a structure called an *ascus* (plural, *asci*). Following two meiotic divisions, there is a mitotic division, resulting finally in eight nuclei. These divisions proceed along a longitudinal axis such that, at the conclusion of the division, the eight nuclei are arranged in a longitudinal row. Heavy walls form around these nuclei, so that eventually they are contained in eight ascal spores called *ascospores*. The divisions take place within the ascus, and so there is no random shifting of the nuclei and the final eight nuclei are ordered by the divisions. Because of this type of division all four chromatids from one tetrad can be analyzed. This is known as a *tetrad analysis*. (See Figure 11-3.)

Segregation in Neurospora

For the hyphal strands to have a nuclear fusion, they must be of the proper mating types. It was learned that a gene pair controls mating types. The two mating types are under genetic control of two alleles designated **A** and **a**. Figure 11-2 illustrates the segregation of a pair of chromosomes following nuclear fusion. On each chromosome is an allele for mating type. Meiosis and subsequent segregation result in asci with two possible arrangements of the genes.

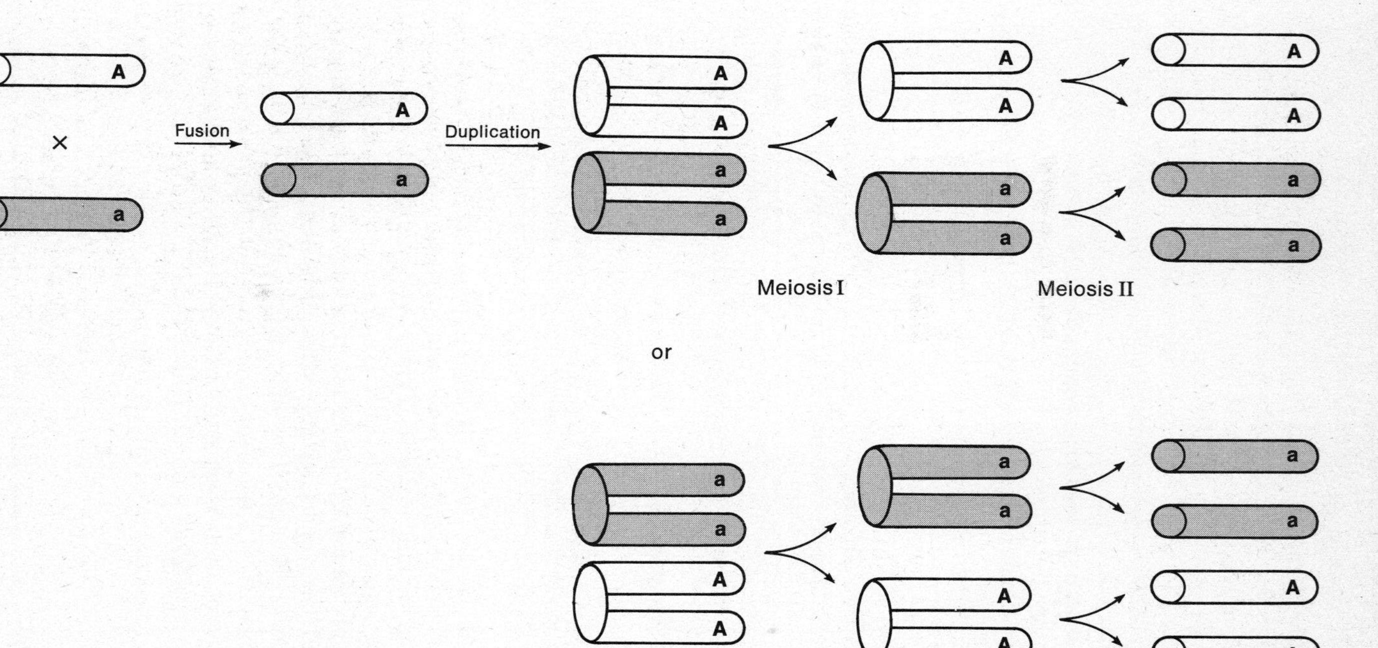

Figure 11-2 Segregation of the mating pair alleles, **A**, **a**, in Neurospora. Because of the ordered meiosis in the ascus only two arrangements of the genes are possible. The **A** genes are always together and the **a** genes are always together. Because the mitotic division does not alter the genetic situation, it has been omitted for simplicity.

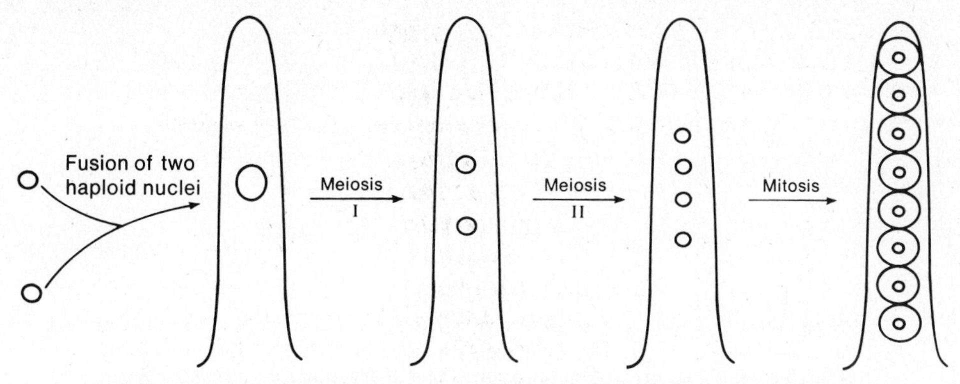

Figure 11-3. Sexual reproduction in the bread mold, *Neurospora crassa.* Two haploid nuclei, if the proper types, can fuse to form a diploid nucleus. This diploid nucleus, in a structure called an ascus, undergoes two meiotic divisions forming four haploid nuclei. Each of these divides mitotically forming eight haploid nuclei which become spores known as ascospores. During the divisions the nuclei maintain their position in the ascus and are *not* positioned randomly.

Crossing Over in Neurospora

However, occasionally the eight spores would be detected in an ascus in a different arrangement. (See Figure 11-4.) In this case, one possible explanation is

A		A		a		a
a		a		A		A
	or		or		or	
A		a		a		A
a		A		A		a

Figure 11.4. Alternate arrangements of the alleles for mating type in Neurospora. For simplicity the mitotic division has been omitted.

that crossing over accounts for these alternate arrangements. But it will soon be apparent that the alternate ascus arrangements cannot be explained if crossing over occurs at the two-strand stage. Try it! However, if crossing over has occurred at the four-strand stage, then it is possible to see how these alternate arrangements arise. (See Figure 11-5.)

Neurospora geneticists sometimes use the term *segregation* to describe the possible events that might occur in a tetrad. When the two products of the meiotic division contain *unlike genetic components,* it is said that segregation has occurred. If this process occurs as the result of the first meiotic division, it is called *first division segregation.* (See Figure 11-6.)

If, however, a crossover occurs between the gene and the centromere at the four-strand stage, segregation will not occur at the first division (Figure 11-6). The two daughter cells have identical genotypes. But at the second division, for either one of these nuclei, a segregation has occurred. This is an example of *second division segregation.* Whenever a crossover occurs, there is second division segregation, resulting in one of the four alternate arrangements shown in Figure 11-4. Thus by

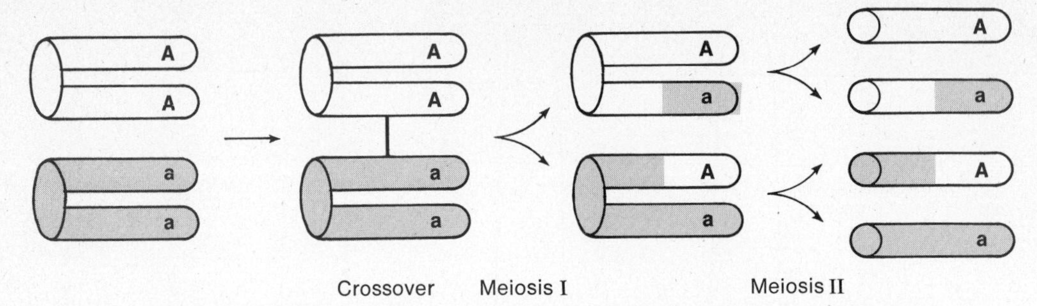

Figure 11-5. Crossing over at the four-strand stage in Neurospora to account for alternate gene arrangements occasionally observed in the ascus. The crossover is represented as a vertical line between two of the four tetrad chromosomes. When the chromosomes move toward the poles the centromere always goes first with the arms trailing after. In analyzing exchanges, therefore, always begin at the centromere. For simplicity the mitotic division is omitted.

serially dissecting out the spores of an ascus, it can be determined whether first or second division segregation has occurred by the order of the spores and also whether crossing over has occurred. Figure 11-7 shows asci with both first and second division segregation.

Mapping Genes in Neurospora

Assume that, in an experiment, 20 % of the asci show second division segregation for a particular pair of alleles. What does this mean in terms of crossing over? Can a map distance be calculated from such a Neurospora experiment? Since 20 % of the asci showed second division segregation, then 20 % showed crossing over and 80 % did not. If there had been 100 asci, which represent 400 chromatids (four in each tetrad), and 80 asci did not have crossing over, then 320 chromatids were noncrossover chromatids.

The 20 asci that showed crossing over represent 80 original chromatids. However, in any one tetrad in which there was a crossover, *only two of the four* chromatids are involved in the exchange. Therefore only half the 80 strands, or 40, were crossover chromatids and half (40) were noncrossover chromatids. Summing up, the noncrossover chromatids of the original 100 tetrads are represented by 320 + 40 and the crossovers by 40, so that, out of the total of 400 chromatids present as tetrads in the 100 asci, 40 of them, or 10 %, were crossover chromatids. Therefore there was 10 % crossing over between the gene and the centromere; and, applying Sturtevant's definition, there are ten map units between the gene and the centromere. It is interesting to note that, in Neurospora, a single gene pair can be mapped, since the centromere behaves as a gene for purposes of mapping.

No More Than 50 % Recombination

This example of mapping genes in Neurospora provides additional insight into the process of crossing over. No more than 50 % recombination should ever be observed even if there is a chiasma in *every* tetrad involved. In other words, if each tetrad has a

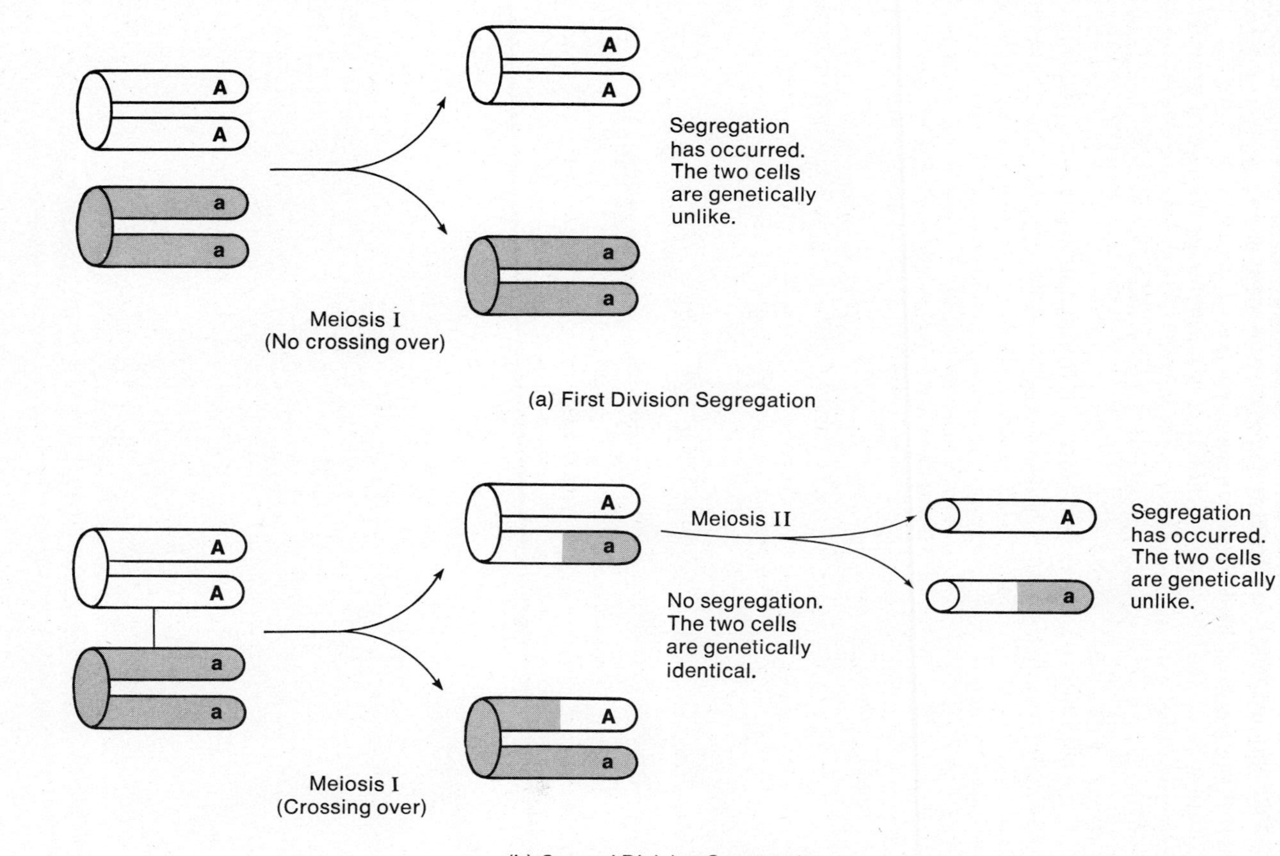

Figure 11-6. Segregation in Neurospora. If crossing over does not occur, segregation occurs at the first division. If a crossover does occur, then segregation occurs at the second division. Therefore, the four alternate gene arrangements shown in Figure 11-4 are examples of second division segregation and are an indication that crossing over has taken place.

crossover between the centromere and the gene, only 50 % recombination occurs because each crossover involves only two of the four chromosomes of each tetrad.

This statement follows from the evidence in Neurospora involving a two-point cross (a gene and the centromere) and when no more than one crossover per tetrad is observed. It is fair to ask, however, whether the presence of double or multiple crossovers will affect the statement that 50 % recombination is the upper limit obtainable. An examination of double crossing over in Neurospora will answer this question. (See Figure 11-8.)

If the recombinant and nonrecombinant chromatids are summarized (Figure 11-8, last column), eight recombinant and eight nonrecombinant chromatids are

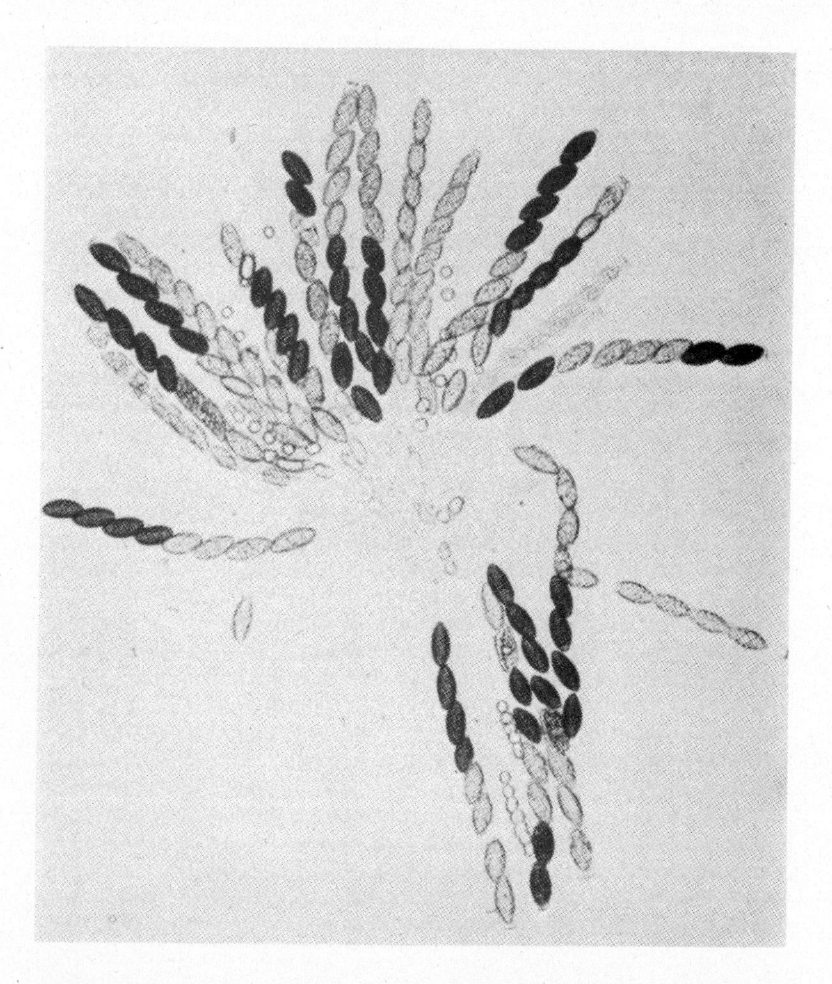

Figure 11-7. A photomicrograph of Neurospora asci showing nine first division segregation asci and five second division segregation asci. (After Stadler. 1956. *Genetics* **41**A.)

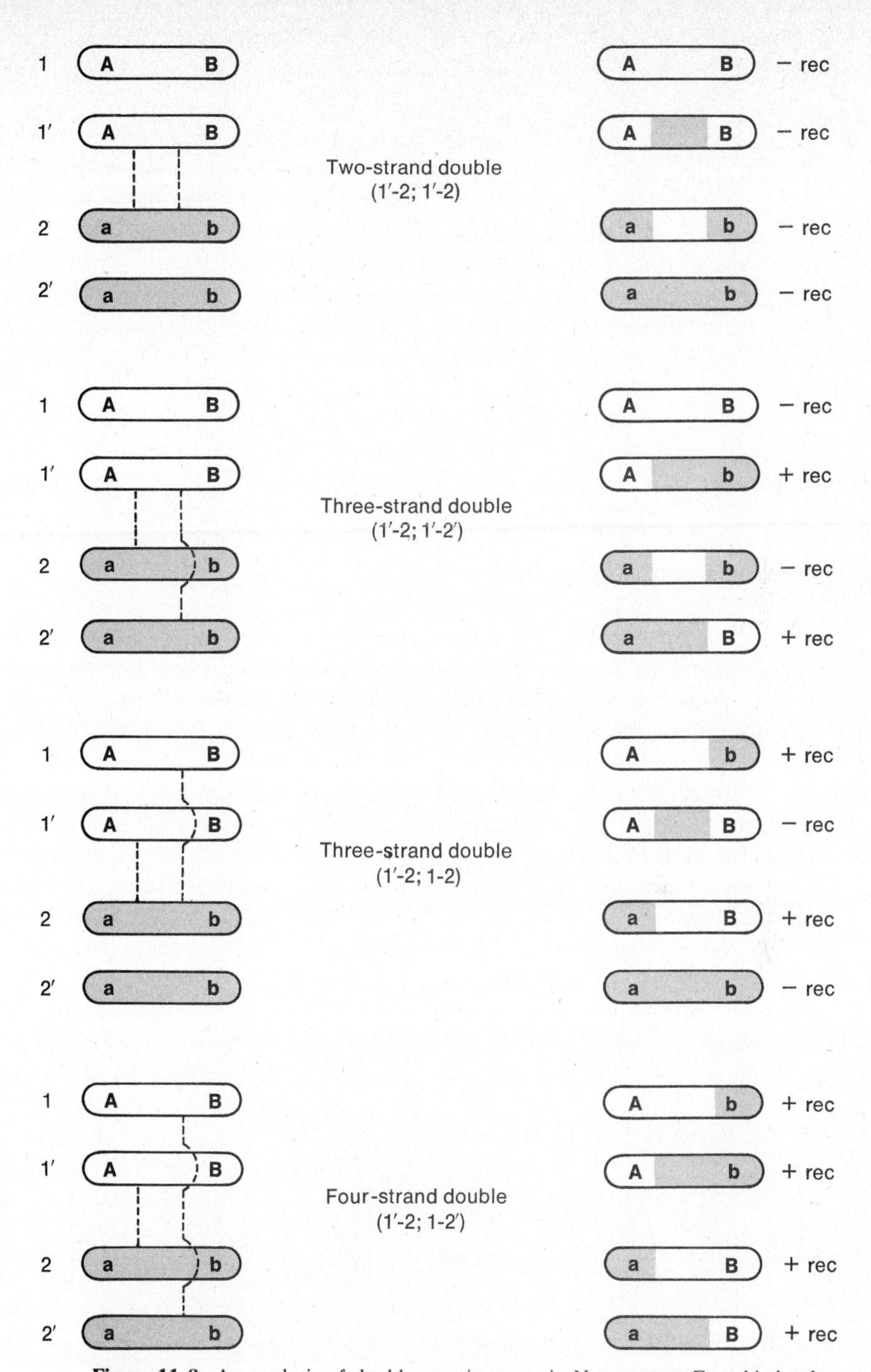

Figure 11-8. An analysis of double crossing over in Neurospora. Four kinds of doubles are possible: a two strand, a four strand and two kinds of three strand doubles. The strands are numbered and the doubles indicate which strands are involved in the exchanges. The first exchange is always between strand 1′ and strand 2. The centromere is assumed at the left. The + signifies recombination occurred between **A** and **B**. The − signifies no recombination occurred.

observed with respect to the two genes involved in this experiment. The conclusion is that even though there is double crossing over, more than 50% genetic recombination is not observed. This assumes that the three kinds of double crossovers (*two strand*, involving only two of the four chromosomes in the tetrad, *three strand*, and *four strand*) occur in the expected frequency of 1 : 2 : 1. This assumption has been experimentally tested in Drosophila and other organisms, and it appears to be true that the three kinds of doubles occur in a frequency of 1 : 2 : 1. So even the presence of double crossovers will not result in more than 50% recombination.

It can be shown that triple crossovers or higher multiples will not affect the general argument, given the assumption that the various possible crossovers occur randomly.

Cytological Proof of Crossing Over

In his explanation of the chromosome theory Morgan postulated that genetic recombination is due to a breakage and exchange of parts of homologous chromosomes. Now it is time to consider whether there are any alternative hypotheses to explain genetic recombination and, if so, whether there is any way to decide among the various possibilities.

In fact, there are other possibilities beside the Morgan hypothesis of exchange of parts of homologous chromosomes. Richard Goldschmidt in 1917 proposed a recombination theory that involved not an exchange of chromosome parts but an exchange of alleles. Goldschmidt assumed that the gene is a separate entity from the chromosome and that at metaphase the genes can detach from the chromosome. Later, during meiosis, the genes would be reabsorbed on the chromosome at the same place or possibly in a different place. Figure 11-9 indicates how Goldschmidt imagined that genetic recombination could occur. Sturtevant recognized that the Goldschmidt hypothesis had certain intrinsic theoretical difficulties,

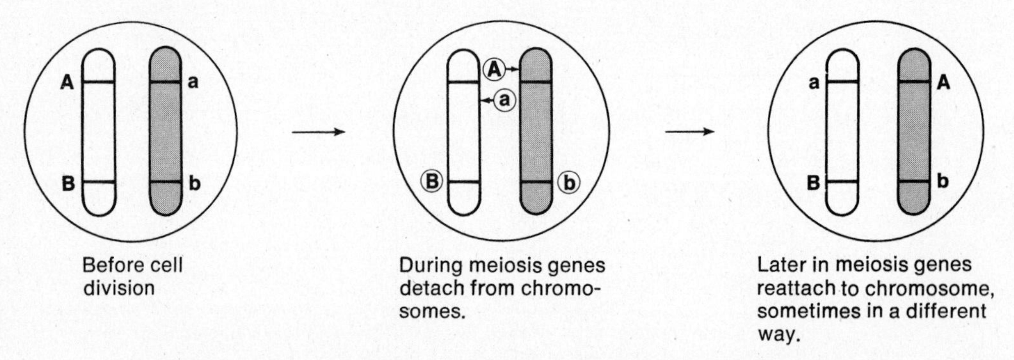

Figure 11-9. Goldschmidt's explanation for genetic recombination. His hypothesis does not involve an exchange of chromosome parts.

and in a paper published in direct response to Goldschmidt he refuted, to the satisfaction of almost everyone, including Goldschmidt, the interesting hypothesis that Goldschmidt had advanced. However, it should be remembered that there can be, at least hypothetically, recombination without an exchange of chromosome parts.

In 1930 the German scientist Winkler also proposed a recombinational mechanism that did not involve exchange of chromosome parts. He called his hypothesis *gene conversion*, a term used later by geneticists to explain a somewhat different phenomenon. What Winkler had in mind was that during gene replication, particularly if two homologous chromosomes were in intimate contact, the wrong allele might be copied. If this situation occurs for only one gene pair on the chromosome, it would appear as though a crossover had occurred. (See Figure 11-10.) If the mechanism proposed by Winkler is true, then genetic recombination could result without exchange of chromosome parts.

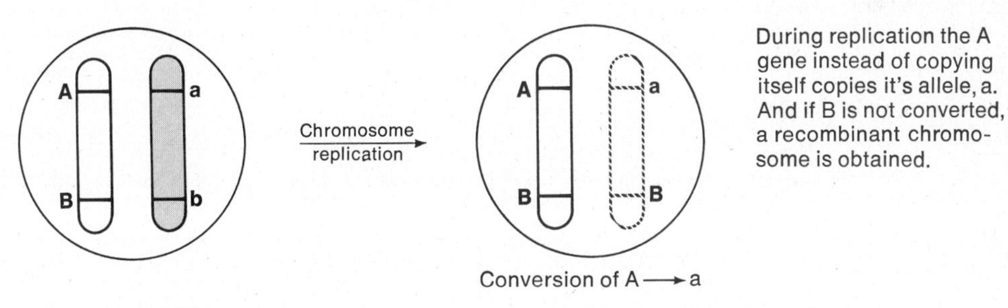

Figure 11-10. Winkler's theory of gene conversion. The hypothesis can account for genetic recombination without an exchange of chromosome parts.

The question now is which of the preceding hypotheses to select. Unlike the schematic presentations in this book, the chromosomes are not differentially colored or marked, and the two members of an homologous pair look exactly alike. However, it is known, as pointed out in Chapter 8, that occasionally a heteromorphic chromosome pair occurs in which one member of the pair differs from the other by one or more characteristics. Would such a heteromorphic pair of chromosomes be sufficient to test our hypotheses? Unfortunately not! For even if the chromosomes exchange parts, as Morgan postulated, exactly the same chromosomes would be observed following the exchange as were observed before. (See Figure 11-11a.) Several people recognized that if there were a doubly heteromorphic pair of chromosomes, in which one homolog differs from the other by two morphological characteristics, the problem could be attacked (Figure 11-11b).

As often happens in science, this problem was solved independently in two different laboratories. In 1931 the American botanists Barbara McClintock and Harriet Creighton performed the experiment in corn, while, independently, Curt Stern, in Germany, did it in Drosophila. The Stern experiment will be discussed, but the two laboratories share equal credit for their brilliant analysis.

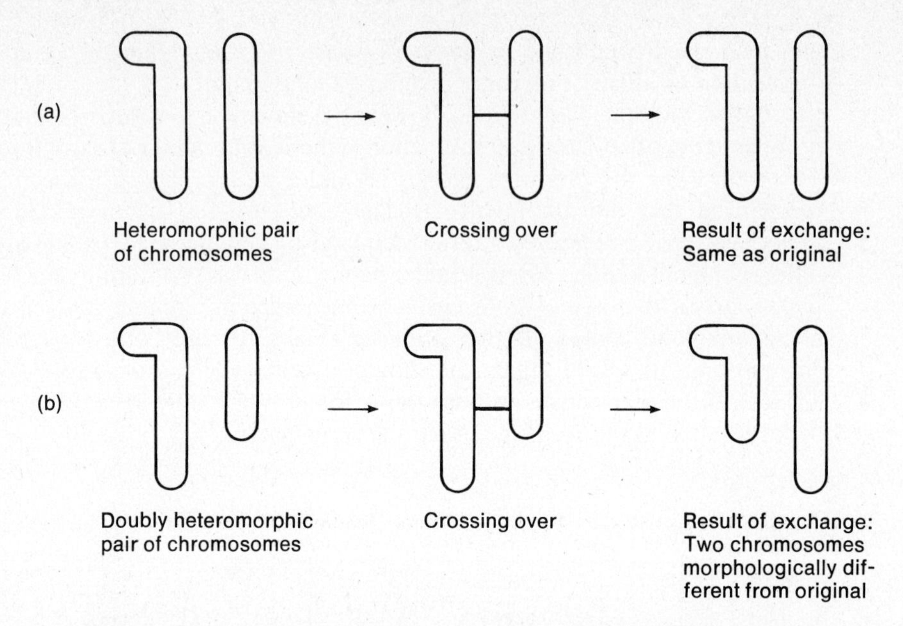

Figure 11-11. Experimental ingredients necessary to test the hypothesis that crossing over involves an exchange of homologous chromosomes. There must be a pair of homologous chromosomes that differ from each other in two ways.

By means of a *translocation*, which refers to the fact that a piece of one chromosome has broken off and become attached to another chromosome (see Chapter 13), Stern obtained a stock of Drosophila in which a portion of the Y chromosome had broken off and become attached to the X chromosome. Stern also had a second stock of Drosophila in which another translocation involving the X resulted in an X chromosome that appeared to be broken in two. Stern then experimentally "placed" genes on these translocated chromosomes. On the "broken" X chromosome there were the recessive eye-color gene carnation (**car**) and the dominant eye-shape gene Bar (**B**). On the unbroken X chromosome with the translocated piece of the Y were the wild-type alleles of each of these genes. He crossed this female parent by a male that had a normal Y and a normal X chromosome marked with the recessive gene for carnation and the wild-type gene for Bar. (See Figure 11-12.) From this mating, Stern obtained females that were carnation Bar and females that were wild type. These females could easily be explained as nonrecombinant or noncrossover females. Stern also obtained, however, females that were carnation, not Bar, and females that were Bar but not carnation. These females were clearly recombinations, and the question is whether they were crossovers in the Morgan context.

Stern reasoned that if Morgan's hypothesis was correct, the two classes of flies classified as crossovers should show a chromosomal rearrangement cytologically. In

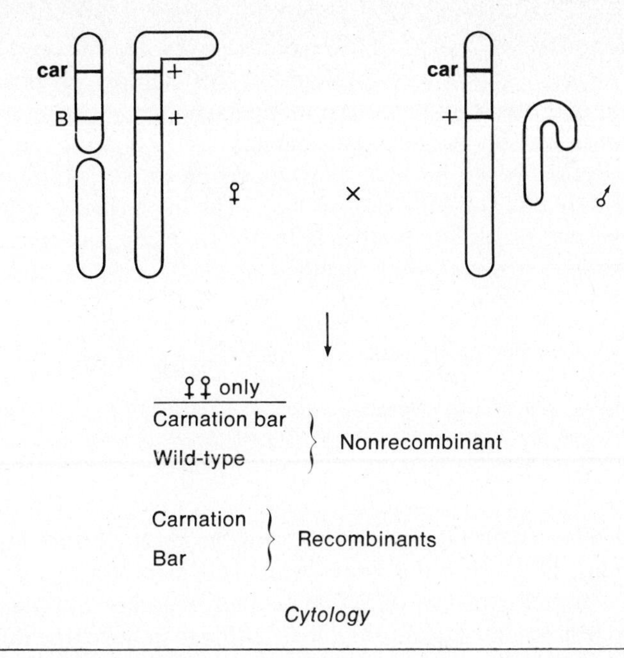

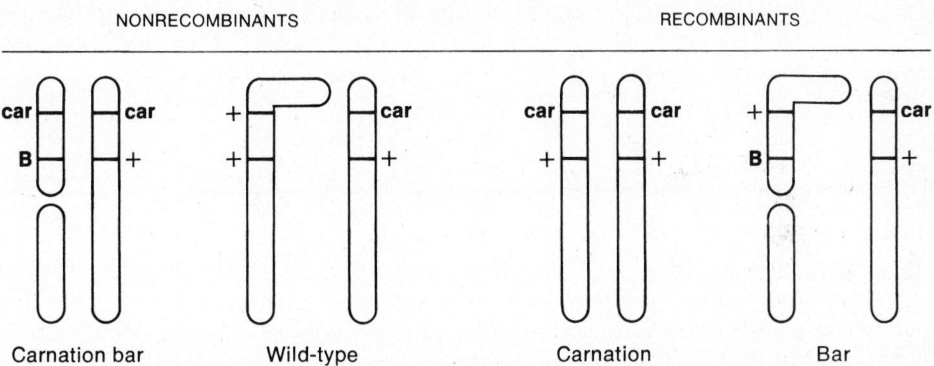

Figure 11-12. Stern's proof of Morgan's hypothesis that genetic recombination is accompanied by an exchange of chromosome parts. The parental female has an X chromosome with part of the Y translocated to it and an X chromosome translocated to the fourth chromosome and giving the appearance of an X chromosome broken in the middle. **B** = dominant gene for Bar eye, **car** = recessive gene for carnation eye.

examining the chromosomes, Stern observed exactly what he had predicted. A genetic recombination was accompanied by a chromosomal exchange of parts. (See Figure 11-12.) Thus Creighton and McClintock, on the one hand, and Stern on the other had provided the cytological proof of crossing over by showing that genetic recombination was accompanied by an exchange of chromosome parts. This discovery excluded both the Goldschmidt and Winkler hypotheses.

The Belling Hypothesis

However, the story does not end here, for an American cytologist, John Belling, proposed another theory of crossing over that took into account an exchange of chromosome parts but not the Morgan idea of breakage and exchange.

Belling based his hypothesis partly on an interpretation of the chromosome that he derived from his study of lilies. The lily has very large chromosomes and is cytologically favorable material. In the lily chromosomes, Belling observed a considerable structure that included a linear array of "beads," which he called

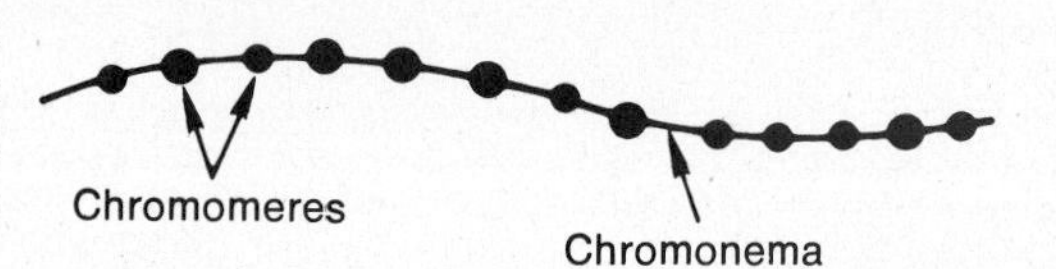

Figure 11-13. Belling's view of the chromosome. He observed a linear array of chromomeres which were characteristic for a given pair of homologues but varied from chromosome to chromosome.

chromomeres. (See Figure 11-13.) The chromomeres are believed to be localized thickenings of the chromosome strand. The linear arrangement and their constancy within a homologous pair of chromosomes, together with their variability from one pair of chromosomes to another, led Belling to believe that the chromomeres might, in fact, represent the genes. In any case, Belling's idea of recombination differed from Morgan's. (See Figure 11-14.) He supposed that the chromomeres first

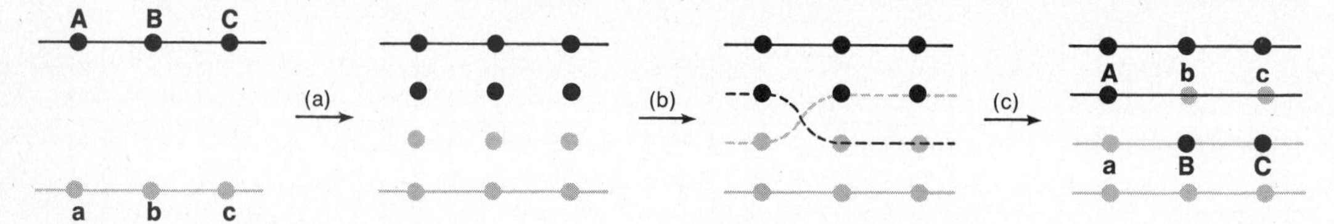

Figure 11-14. Belling's hypothesis of crossing over based on his observations of lily chromosomes. He proposed that first the chromomeres replicate themselves (a). Second the chromonemata are resynthesized and occasionally they may link up a new order of chromomeres (b) leading to recombinant chromosomes (c).

replicated themselves. Later the chromonemata were synthesized and connected to the new chromomeres. If, by chance, one of the strands did not go through the chromomeres in the original order but "crossed over" and picked up another chromomere, the end result would be a genetic exchange. Neither the Stern nor the Creighton and McClintock experiments could distinguish between this hypothesis and that of Morgan's breakage and exchange. As a matter of fact, in the 1950s many geneticists, particularly microbial geneticists, were intrigued with the Belling hypothesis for recombination, and they gave it a name, *copy choice.* Many felt an intuitive distrust that such an important and exact mechanism as crossing over could have evolved as an essentially destructive mechanism involving breaks in the chromosome. In Chapter 23 it will be shown that further sophisticated tests designed to distinguish between breakage and exchange, on the one hand, and copy choice on the other have all been resolved in favor of breakage and exchange; and there is little evidence now that copy choice plays an important role in genetic recombination.

REFERENCES

BELLING, J. 1931. Chromomeres of Liliaceous plants. *Univ. Calif. Publ. Bot.* **16**:153–170. (A description of an alternate hypothesis to Morgan's breakage and exchange. Belling's hypothesis enjoyed a rebirth among microbial geneticists in the 1950s.)

CREIGHTON, H. B. and B. MCCLINTOCK. 1931. A correlation of cytological and genetical crossing over in *Zea Mays*. *Proc. Natl. Acad. Sci. U.S.* **17**;492–497. (Available in *Classic Papers in Genetics*, edited by J. A. Peters. Prentice-Hall, Englewood Cliffs, N.J., 1959.) (One of the two papers that showed the cytological proof for crossing over.)

STERN, C. 1931. Zytologisch-genetische Untersuchungen als Beweise für die Morgansche Theorie des Faktorenaustauschs. *Biol. Zentralbl.* **51**:547–587. (This paper should be looked at as the companion to the Creighton-McClintock work. The diagrams clearly indicate the results even if the reader does not understand German.)

QUESTIONS AND PROBLEMS

11-1. Define the following terms:

ascus
segregation
tetrad analysis
first division segregation
second division segregation
doubly heteromorphic chromosomes
three-strand double
copy choice

11-2. Suppose that two genes, a and b, are 20 units apart on a chromosome.

(a) What is the percentage of crossing over between the two?

(b) What is the percentage of chiasmata formed during meiosis in this region of the chromosome?

11-3. What is meant by the expression "crossing over at the four-strand stage?"

11-4. What is the evidence that crossing over occurs at the four-strand stage?

11-5. In Neurospora, a gene, x, shows second division segregation in 18 % of the asci.

(a) Can this gene be mapped?

(b) If so, what is the second marker used?
(c) What is the map distance between x and the marker?

11-6. In Neurospora, a gene, y, is known to be 11 map units from the centromere. How many asci will show second division segregation for this gene?

11-7. In Neurospora, the linkage arrangement for three genes and the centromere is: centromere, M, N, O. From a hybrid (M N O)/(m n o), asci with the following types of spores were found. (The mitotic division has been omitted.)

(a) MNO, MNo, mnO, mno
(b) MNO, MnO, mNo, mno
(c) MNO, MNO, mno, mno
(d) Mno, MNo, mNO, mnO
(e) Mno, MNO, mNo, mnO

For each of these asci, indicate whether crossing over has occurred and, if so, what kind.

11-8. In Neurospora, two genes, A and B, are linked at a distance of a few map units. After a cross of AB × ab, which of the following types of asci would you expect?

(a) Asci with 4 AB and 4 ab spores
(b) Asci with 6 AB and 2 ab spores
(c) Asci with 4 Ab and 4 aB spores
(d) Asci with 2 AB, 2 Ab, 2 aB and 2 ab spores?

Explain your answers.

11-9. In Neurospora, a heterozygote of the constitution RS/rs, with the centromere lying between the two loci, produces the following types of asci from double crossing over. (The mitotic division has been omitted.)

(a) RS rs Rs rS
(b) RS rs RS rs
(c) rS Rs RS rs
(d) Rs rS Rs rS

For each ascus, indicate whether a two-, three-, or four-strand double crossover is responsible.

11-10. In Neurospora, the genes X and Y are linked on the same chromosome with the centromere between them. The following cross is made: XY by xy. Indicate the genetic constitutions of the spores in an ascus resulting from the following situations. (Assume only four spores in an ascus by ignoring the final mitotic division.)

(a) No crossing over
(b) A crossover between X and the centromere
(c) A crossover between Y and the centromere
(d) A two-strand double, with one crossover between X and the centromere and the second between Y and the centromere

(e) A three-strand double, with one crossover between X and the centromere and the second between Y and the centromere

(f) A four-strand double with one crossover between X and the centromere and the second between Y and the centromere

11-11. What is the significance of Stern's experiment in Drosophila with doubly heteromorphic chromosomes?

11-12. Does Stern's experiment rule out Belling's hypothesis? Explain.

12/Linkage and recombination

PART III

Geneticists speculated whether recombinational mechanisms were present in all living organisms. They had obviously been looked for in only a few. Plants, both higher and lower, and animals, particularly Drosophila, had shown recombination; but it seems safe to generalize that recombination that involves a Morgan-type exchange, is widespread. The question now is: Are combination mechanisms found in lower organisms, such as bacteria and viruses, and, if so, can anything be learned about the mechanism for such recombination?

Recombination in Bacteria

The organism used to determine recombination in microorganisms is the common colon bacterium *Escherichia coli*, and the experiment was done by a young geneticist,

Joshua Lederberg, in association with his professor at Yale, Edward Tatum. The first account of the experiment occurred in 1946, and the fuller report was the publication of Lederberg's Ph.D. thesis in 1947.

Mating in E. coli

Lederberg had determined that it was possible for the wild-type or normal *E. coli* to be cultured on a *minimal medium*. This minimal medium contains only the basic growth requirements for *E. coli*, which are glucose, KH_2PO_4, K_2HPO_4, sodium citrate, magnesium sulfate, and ammonium sulfate. These simple inorganic requirements are all that are required for *E. coli* to synthesize all the necessary compounds needed for growth. Of course, it is possible to grow *E. coli* on a *complete medium*, which contains not only the minimal growth requirements but also all other known constituents, such as amino acids, vitamins, and nitrogenous bases. A fuller description of the technique of minimal medium growth appears in Chapter 15.

If a mutation occurs in *E. coli* that involves, for example, the synthesis of an amino acid or a vitamin, such a strain will *not* grow on the minimal medium unless the particular substance that it lacks is added to the medium. It will grow on a complete medium, since the latter, by definition, contains the substance that it is lacking.

Lederberg began with two strains of *E. coli* that were unable to grow on the minimal medium because of one or more mutations in genes controlling nutritional requirements. The first strain can be symbolized $\mathbf{B^- M^- T^+ L^+ B_1^+}$, where **B** stands for the vitamin biotin, **M**, **T**, and **L** represent the amino acids, methionine, threonine, and leucine, and $\mathbf{B_1}$ represents vitamin $\mathbf{B_1}$ or thiamine. The $^-$ refers to the fact that it is unable to synthesize the nutrient in question and thus will not grow on a minimal medium. The $^+$ means that it is able to synthesize the nutrient in question. The second strain is symbolized $\mathbf{B^+ M^+ T^- L^- B_1^-}$. Neither strain will grow on a minimal medium.

When both strains were plated together on a minimal medium, growth was observed in a very small number of cells. Lederberg theorized that they would have the genotype $\mathbf{B^+ M^+ T^+ L^+ B_1^+}$. This wild-type or completely normal strain is known as a *prototroph*, and any biochemically mutant strain derived from it is known as an *auxotroph*. Thus the mixture of two auxotrophs, $\mathbf{B^- M^- T^+ L^+ B_1^+ \times B^+ M^+ T^- L^- B_1^-}$, yielded prototrophs at the rate of about 1 for every 10^6 or 10^7 cells plated. These results could be explained on the assumption that two cells, one from parent strain A and the other from parent strain B, when mated formed a zygote from which parental and recombinant cells emerged.

Lederberg was puzzled as to why the frequency of recombination was so low. Simplistically, there might be two explanations. The first, based on Lederberg's assumption that there is some kind of sexual fusion of cells, is that the fusion is limited and occurs only rarely among two cells. A second explanation would be that fusion occurs rather commonly, but, for one reason or another, recombination after fusion between conjugating cells is rare. At this time it was not possible for Lederberg to determine which of these two hypotheses might be correct or whether other explanations might be responsible for the low frequency of recombination.

During the 1950s genetic experiments with bacteria became widespread, and the understanding of recombination was further refined. The British investigator Hayes treated potentially recombing strains of *E. coli* with streptomycin, an antibiotic. Somewhat to his surprise, he observed that when he treated one strain (by strain is meant an auxotroph as used in Lederberg's experiment) with streptomycin, recombination did not follow. On the other hand, if the second strain was treated with streptomycin, recombination proceeded as though streptomycin had not been introduced at all. Hayes concluded that one strain plays only a transitory role in the recombination mechanism, perhaps analogous to the role played by a male in fertilization in a higher organism, and this transitory role is not impaired by streptomycin. To get recombinants, however, cell division must occur,and Hayes postulated that it is this division process that is interfered with by streptomycin. Thus Hayes viewed genetic recombination in bacteria as a one-way transfer of genetic material from a "male type" to a "female type."

High-Frequency Recombination

Hayes and, independently, the Italian geneticist Cavalli-Sforza subsequently discovered a strain of *E. coli* in which recombination occurred a hundred to a thousand times more frequently than in the Lederberg experiment, giving recombinations of the order of 2×10^{-4}. Such a strain is referred to as an Hfr strain, where Hfr stands for high-frequency recombination.

The Mode of Genetic Transfer

The mode of genetic transfer in *E. coli* was worked out in an elegant experiment by two French geneticists, Francois Jacob and Eli Wollman. They reasoned, from the experiments of Hayes and others, that genetic transfer may, in fact, occur from what would be a "malelike" donor to a "femalelike" recipient and that this genetic transfer may involve a linear arrangement of genes. To test their hypothesis, they wondered whether conjugation could be interrupted by some physical means. If so, they might be able to vary the time of the interruptions to see whether recombination was also varied. They discovered that they could interfere with conjugation by placing the bacteria in a food blender. When the motor was turned on, the conjugating pairs of bacteria were physically disrupted and separated but without any cellular damage. (See Figure 12-1.) Following conjugation, the two separated cells are called *exconjugants.*

The results of this "interruption" experiment are interesting. Jacob and Wollman noticed that recombination of certain genes depended on the length of time the bacteria were allowed to mate before separation. If mating proceeded for 10 minutes, recombination for gene **A** was detected. After 15 minutes, genes **A** and **B** recombined. After 20 minutes, genes **A**, **B**, and **C** recombined and so on. (See Figure 12-2.) These results were best interpreted by the assumption that the genes of *E. coli* were arranged on a "chromosome" in linear order. During mating or conjugation this chromosome is passed to the recipient cell, and the order or passage is constant for any given strain of *E. coli* as though the chromosome has an origin that is always passed first.

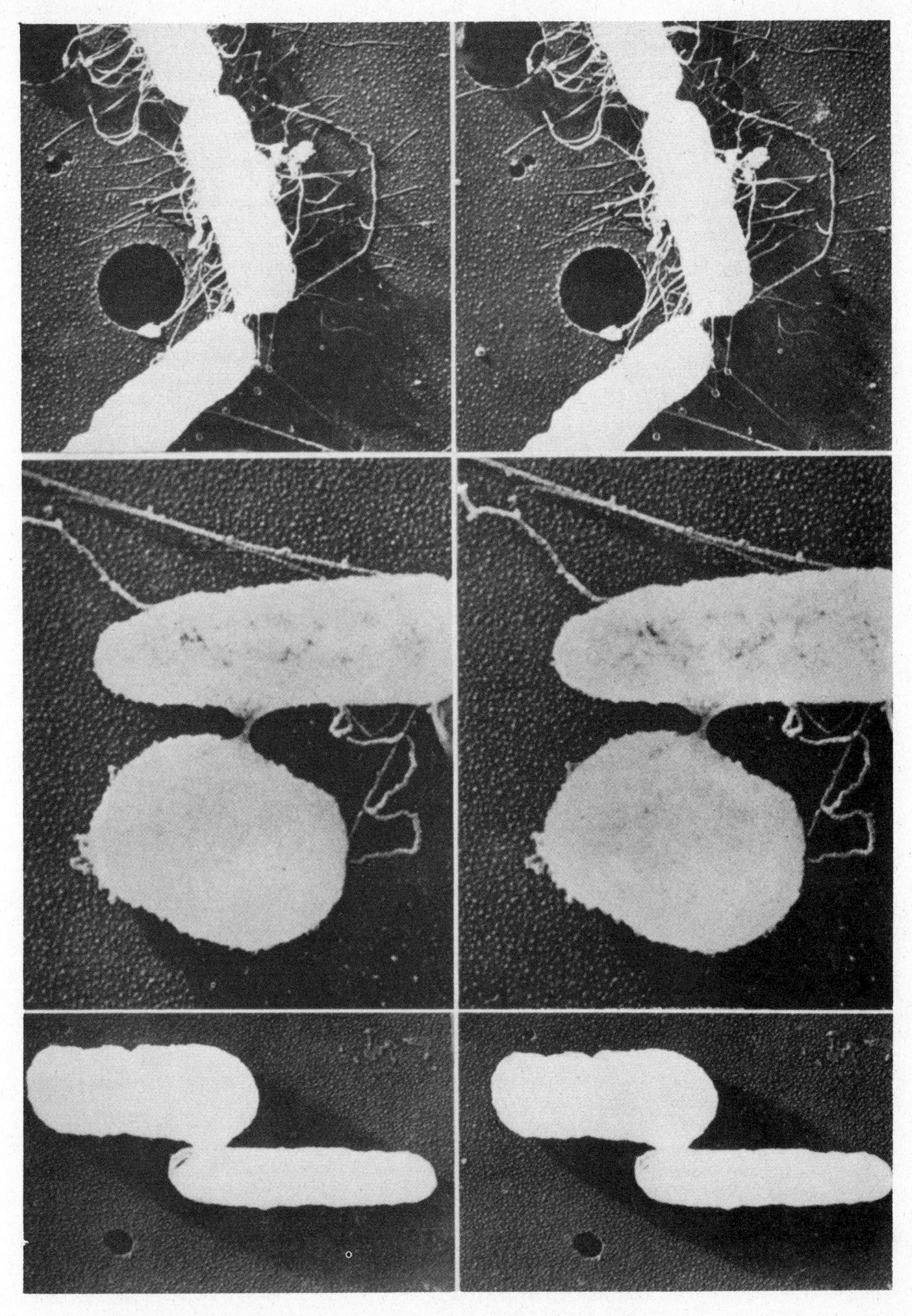

Figure 12-1. Stereoscopic electron micrographs of conjugating *E. coli.* (From Wollman, Jacob, and Hayes. 1962. *C. S. H. S. Quant. Biol.* **21**: 152.)

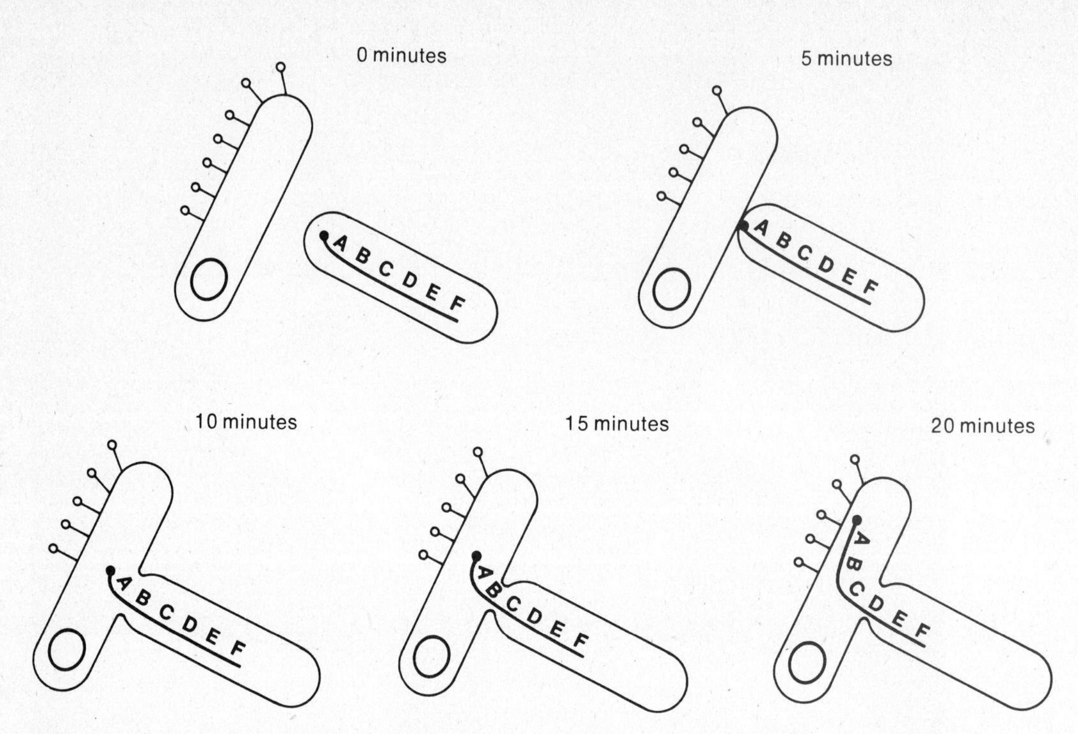

Figure 12-2. Jacob and Wollman's explanation of recombination in *E. coli*. The "female" or recipient cell is on the left and is marked with the adsorption of bacterial viruses. The "male" or donor cell is on the right. The time intervals are the time the cells are allowed to mate before being separated in the food blender. The interpretation is that the donor chromosome is passed to the recipient at a constant rate and in the same orientation. Therefore recombination is time-dependent. (After Jacob and Wollman. 1961. *Sexuality and the Genetics of Bacteria*. Academic Press.)

The Circular Chromosome

When others repeated the Jacob and Wollman experiment, often using different strains of *E. coli* as donors, another interesting fact was revealed. Although, for any donor strain, the order in which the genes entered the recipient was constant, it was not the same from donor to donor. In one strain, gene A might enter first, whereas in a second strain gene D might be first, and in a third strain gene F, and so on. However, careful analysis of the results showed that although the order in which the genes entered varied from strain to strain, the linkage relationship among the genes was constant. This fact led Wollman and Jacob to the conclusion, bizarre though it may seem, that the chromosome in *E. coli* is *circular*! When transferred, it is linear, and the point of break may be different from one strain to another. (See Figures 12-3, 12-4, and 12-5.)

STRAIN OF HFR *E. COLI*	ORDER IN WHICH GENES ENTER RECIPIENT CELL BASED ON JACOB-WOLLMAN EXPERIMENT (FIG. 12-2)
1	← A B C D E F G H I J K L M N O P
2	← D E F G H I J K L M N O P A B C
3	← F E D C B A P O N M L K J I H G
4	← I H G F E D C B A P O N M L K J
5	← M N O P A B C D E F G H I J K L

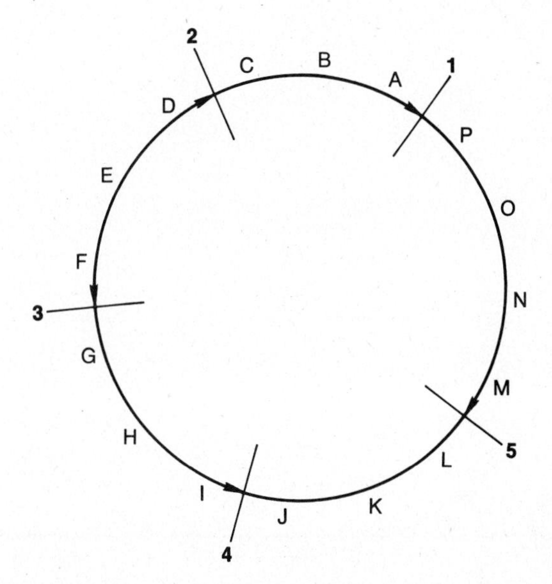

Figure 12-3. Evidence that led Jacob and Wollman to postulate a circular chromosome for *E. coli.* Above are represented five different Hfr donor strains with the linkage arrangement of the genes. The ← represents the origin. Below is the interpretation of such evidence. The chromosome is viewed as circular with a given gene order. In each strain the break occurs in a different place opening up the circle. The small arrow on the circle indicates the origin.

Fertility Strains

Subsequent experiments by these investigators and others revealed additional complexity that was unanticipated in what would seem to be an organism as simple as the bacterium, *E. coli.* Sometimes recombination experiments using strains that had been successful previously resulted in no recombinations. Only matings involving certain strains were successful. This situation led to the postulate that there are two mating types in *E. coli.* One, the donor, possesses an autonomous fertility factor, F, and is known as the F^+ strain. The second lacks the F factor and is the F^- strain. Matings between two F^+ or two F^- strains are sterile and no recombination can be obtained.

After mating an $F^+ \times F^-$, it was observed that all the exconjugants were F^+. Somehow, during the conjugation, the F^- recipients became F^+. It was con-

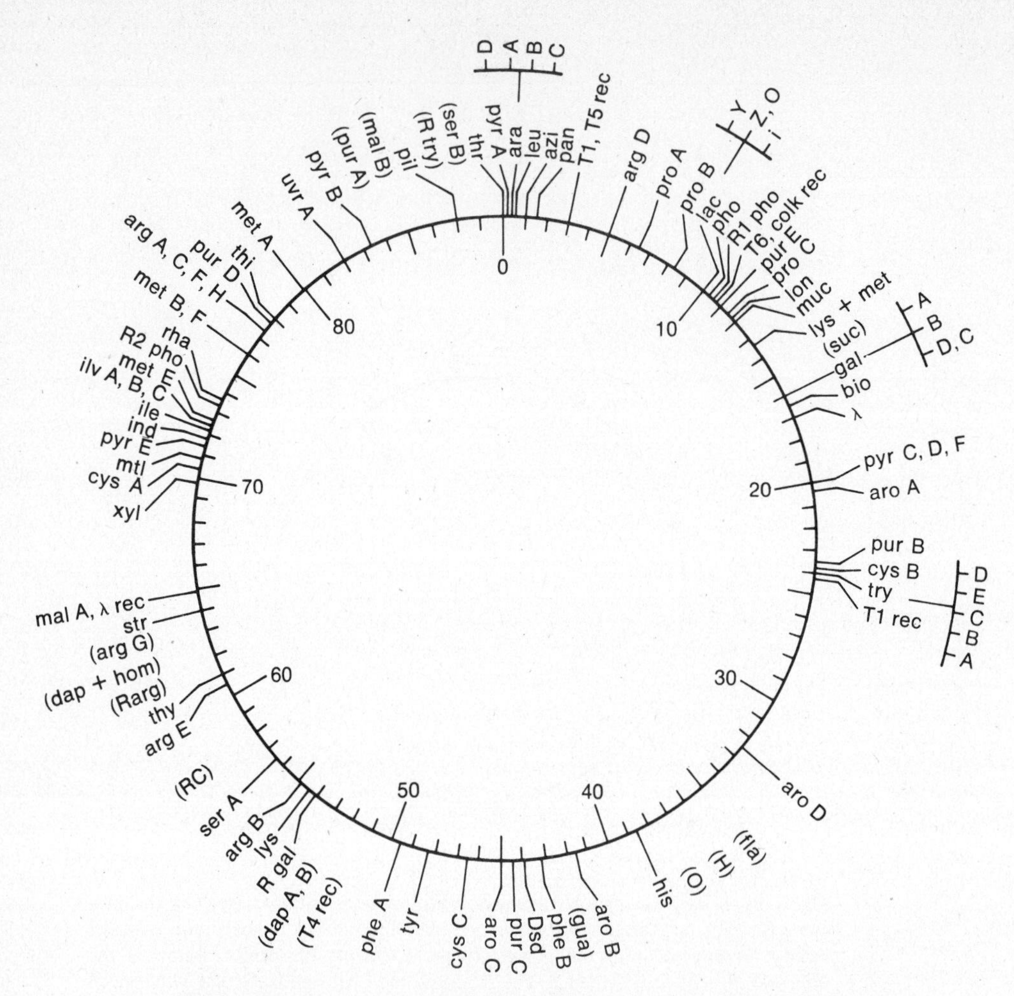

Figure 12-4. Genetic map of *E. coli* drawn to scale. The map is graduated in one-minute intervals (89 minutes total) and numbered at 10-minute intervals to facilitate computation of interlocus distances. (After Taylor and Thoman. 1964. *Genetics* **50**: 667.)

cluded that F^+ *E. coli* are characterized by the presence of an F^+ agent, which is a self-replicating particle and which during conjugation between F^- and F^+ is passed through the cytoplasmic bridge from F^+ to F^- such that the F^- strain is transformed into an F^+ bacteria.

Further experiments revealed that the mating of an $F^+ \times F^-$ usually results in no genetic recombination. However, from such a mating recombination is sometimes observed, and it was concluded that this recombination is a consequence of an F^+ becoming an Hfr. In other words, it is the mating of an F^- by another donor, Hfr, that results in genetic recombination. As a general rule, from the mating of an

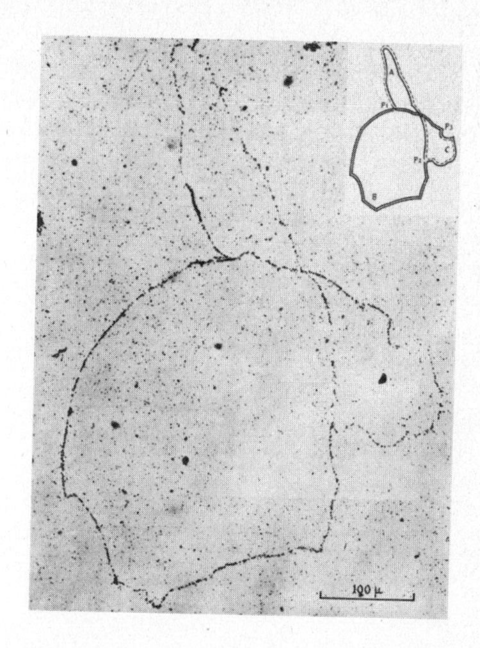

Figure 12-5. Autoradiograph of the chromosome of *E. coli* K 12 Hfr. (After J. Cairns. 1963. *C.S. H. S. Quant. Biol.* **28:** 44.)

F^- × Hfr, the exconjugants are F^- and Hfr. Whatever it is that makes an *E. coli* Hfr is usually not transmitted during conjugation.

Jacob and Wollman suggested that the F^+ strain became an Hfr whenever the former "lost" its self-replicating F factor from the cytoplasm. Moreover, they proposed that the F factor, in this instance, somehow becomes integrated into the bacterial chromosome, converting the F^+ to Hfr. It is thought that this integration opens up the chromosome from a ring to a rod so that transfer can occur. Apparently, in different strains of *E. coli*, the homology between F and the chromosomes varies. That is, the region on the chromosome where the F factor pairs and becomes integrated is different from strain to strain. Because of this variation, different strains donate their genes in different orders, as shown in Figure 12-3. Finally, the F factor that has become integrated passes into the recipient last, so that, in general, the recipient never receives it and thus remains F^-. The reason is that conjugation is usually interrupted spontaneously before chromosome transfer is completed. A suggested model for the conversion of F^+ to Hfr is shown in Figure 12-6.

Episome

Jacob termed a factor such as F that can exist in two alternate conditions, one free in the cytoplasm and the other integrated in the chromosome, an *episome*. By this definition, the F factor is an episome.

It can be concluded that recombination does take place in microorganisms and that the mechanism may be similar to that in Drosophila. Once the donor material is transferred, the process of recombination that occurs in the recipient can be visualized as a breakage and exchange phenomenon. Breakage and exchange is also the mechanism proposed for the integration of the F factor into the bacterial chromosome. Finally, it should be mentioned that in the Lederberg and Tatum experiment, which yielded recombinants in very low frequencies, it is now understood that the matings involved $F^- \times F^+$. The few recombinants obtained were due to the conversion of F^+ to Hfr.

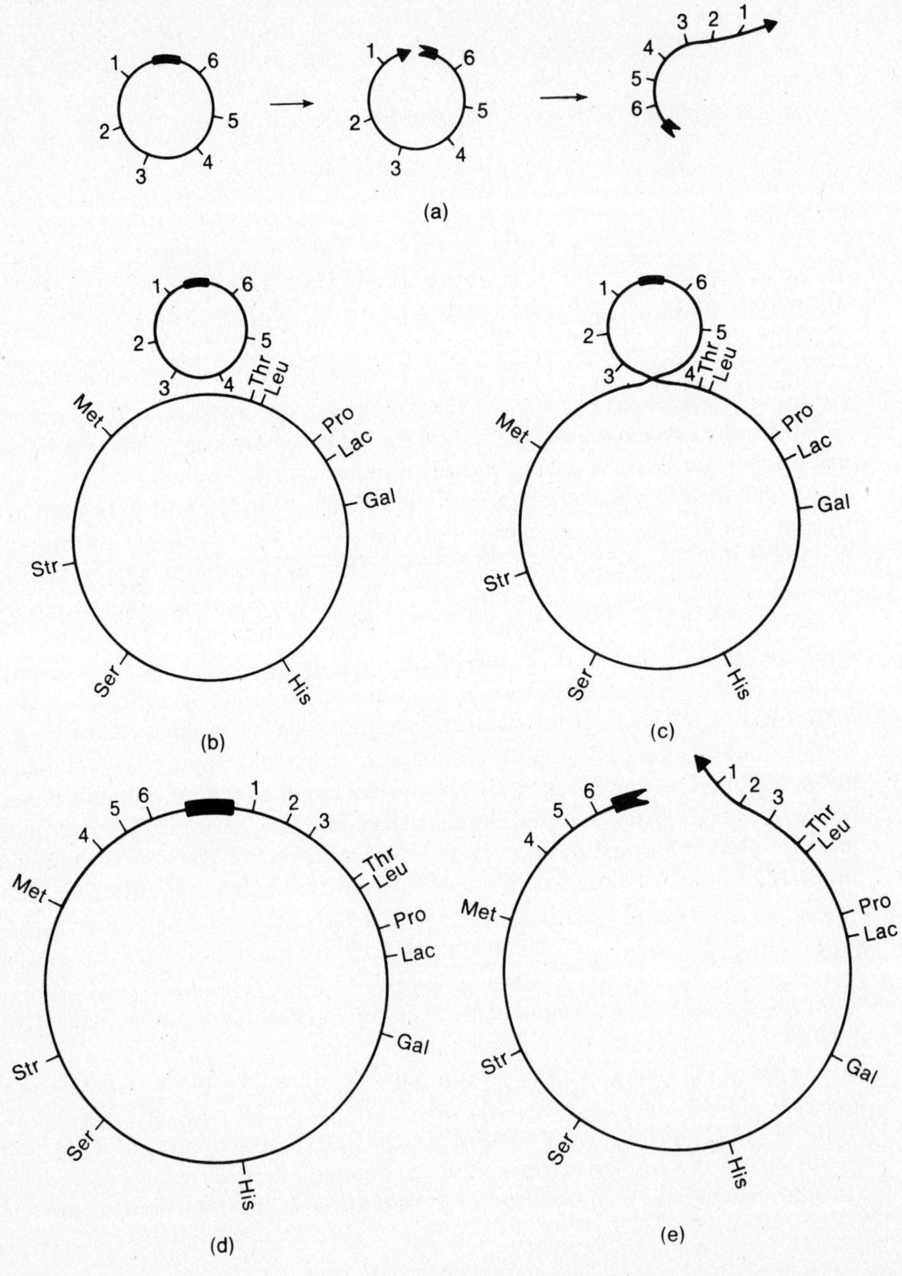

(a)

(b)

(c)

(d)

(e)

Recombination in Phage

The next question is whether recombinational mechanisms occur in viruses. Many kinds of viruses are known, including the so-called animal viruses, such as influenza, mumps, and polio virus, and a large group of viruses whose normal host is a bacterium. These viruses are known as *bacteriophages, phage* for short, and it is the phages that geneticists investigated to the exclusion of most other viruses.

The study of phage genetics is both interesting and complicated, but a detailed description is beyond the scope of this book. We shall, however, outline the life cycle of phage because it is needed in order to understand phage genetics.

Phage Life Cycle—Virulent Phage

The life cycle can be conveniently divided into two parts. The first involves *virulent* phage and results in the *lysis*, or destruction, of the bacterium. Because they are capable of being lysed, the bacteria are referred to as *lytic* bacteria. The phage are adsorbed onto the bacterium tail first and inject their genetic material inside the bacterium. (The evidence for this assertion will be given in Chapter 18.) (See Figure 12-7.) For a period of about 20 minutes, phage proteins are produced and autonomous multiplication of phage genetic material begins. The bacterial chromosome disintegrates and disappears. Materials from the bacteria will be incorporated into the phage genetic material during its synthesis. Finally, proteins are organized around the newly synthesized phage and phage particles are formed. An enzyme will be produced by the phage that will hydrolyze the bacterial wall, and the bacterium will be lysed, thus liberating 100 to 200 phage progeny into the medium. These progeny will then repeat the cycle with neighboring bacteria. In time a petri plate that has bacteria on it will show clear areas, known as *plaques*, which are the result of the lysis of bacteria by phage. (See Figure 12-8.)

Temperate Phage

In some instances, after adsorption of phage and injection of the genetic material, reproduction of phage is not started, and the bacterium survives the infection. These survivors multiply normally, but now each bacterium possesses and perpetuates the potentiality of producing phage particles in the absence of infection. Such bacteria are called *lysogenic* bacteria, and these nonvirulent viruses are called *temperate* phage. The phage

---

Figure 12-6. The breakage and transfer of the Hfr chromosome as a consequence of its integration with F. (a) F is shown as a circle, which has a special site of breakage between markers 1 and 6. Breakage is followed by transfer (during conjugation), such that F markers penetrate the recipient in the order 1–2–3–4–5–6. (b) F has paired with a chromosomal site between Met and Thr. (c) A crossover in the region of pairing integrates the two circles. (d) Same as (c), but redrawn as a single circle. (e) Breakage at the special F site leads to transfer causing F markers 1–2–3 to enter the recipient first, followed by chromosomal markers Thr, Leu, Pro, Lac, and so on; F markers 4–5–6 enter last. Recombinants will be males if they receive all six F markers; thus the terminal end of the chromosome must be transferred to produce an Hfr recombinant. (After Stanier, Doudoroff, and Adelberg. 1970. *Microbial World*, 3rd. Ed. Prentice-Hall.)

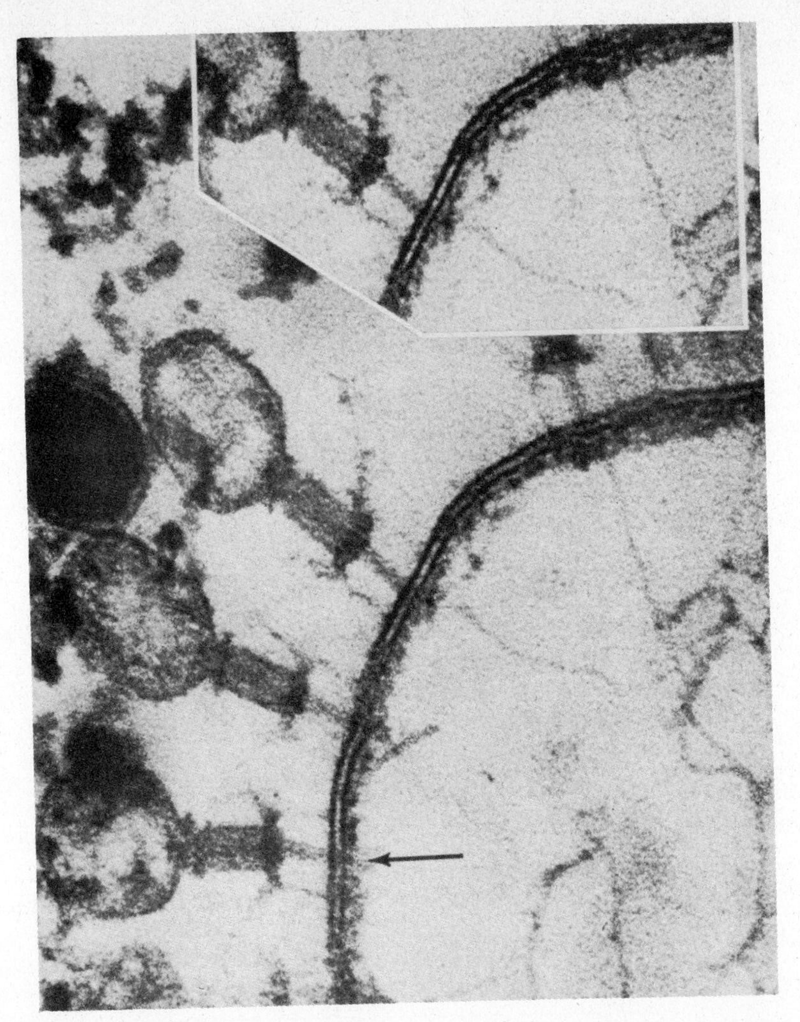

Figure 12-7. T_4 phages adsorbed on an *E. coli* B cell wall, as seen in a thin section. The phages are bound to the bacterial surface by short tail fibers extending directly from their baseplates to the cell wall. The needle of one of the phages can be seen to penetrate just through the cell wall (arrow). The inset shows a second micrograph of the same field take $0.3\,\mu$ underfocus. (After Simon and Anderson. 1967. *Virology* **32**: 294.)

genetic material, which has been injected inside the bacterium, instead of producing many daughter progeny at the expense of the bacterium, becomes incorporated into the bacterial chromosome. When this process occurs, it is known as a *prophage*. This prophage is the genetic material of the phage, and it carries the information for the production of the phage particles. Generally speaking, there is only one prophage per bacterial chromosome, and the site of incorporation of some prophage is specific but that of others is not.

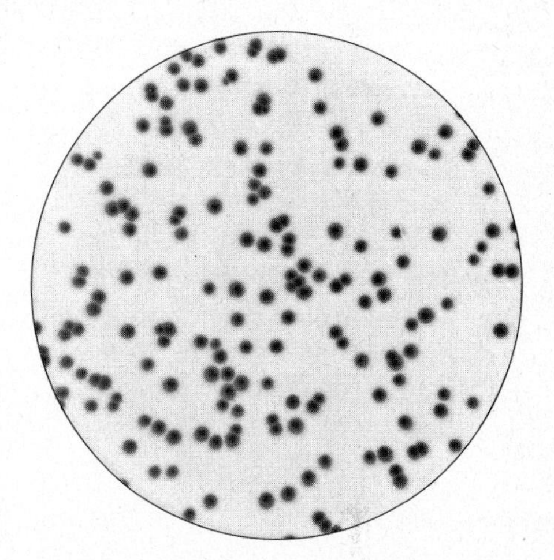

Figure 12-8. A petri plate showing growth of a lawn of *E. coli* bacteria on which T_2 phages have formed plaques. (After G. S. Stent. 1963. *Molecular Biology of Bacterial Viruses.* W. H. Freeman and Company.) (After G. Stent. 1971. *Molecular Genetics.* W. H. Freeman and Company.)

As long as the prophage remains incorporated in the chromosome, the bacterium is incapable of being lysed by subsequent infections with the same phage. In other words, it possesses an *immunity* to this phage. It can, however, be lysed if infected by a virulent phage of a different kind.

Spontaneously, and with a frequency that varies but is on the order of 10^{-5}, such lysogenic bacteria become lytic bacteria. The prophage detaches from the bacterial chromosome and goes through the lytic cycle described above, lysing the bacterium and releasing daughter progeny into the medium. (See Figure 12-9.) It is known that this detachment of prophage, called *induction*, can be faciliatated by ultraviolet radiation. It should be apparent now that such bacteria are called lysogenic because they are capable of being lysed if induction takes place. The behavior of this phage/prophage is reminiscent of the F factor described earlier in *E. coli.* In fact, the definition of an episome—namely, a particle that can exist in two independent states, one free in the cytoplasm, the other incorporated into the chromosome—would include the temperate phage.

Recombination in T_2 Phage

With this understanding of the life cycle of bacteriophage, the question originally asked about recombinational mechanisms in viruses can be considered. The answer came initially from the laboratory of A. D. Hershey at Cold Spring Harbor. Hershey was working with a strain of phage designated T_2. He discovered that there were differences among T_2 phage with respect to the number of bacteria lysed per unit time. The mutant phage lysed many more bacteria per unit time than did the wild-type strain, clearing out a larger plaque in the process. Hershey designated it an **r** mutant, which stood for rapid lysing. The **r** mutant forms a larger plaque with a clear, sharply defined outline.

A second phage mutant was discovered because of its property of infecting some strains of bacteria but not others. If *E. coli* is infected with T_2 bacteriophage, it is possible to isolate some *coli* that are not infected by the phage. These resistant

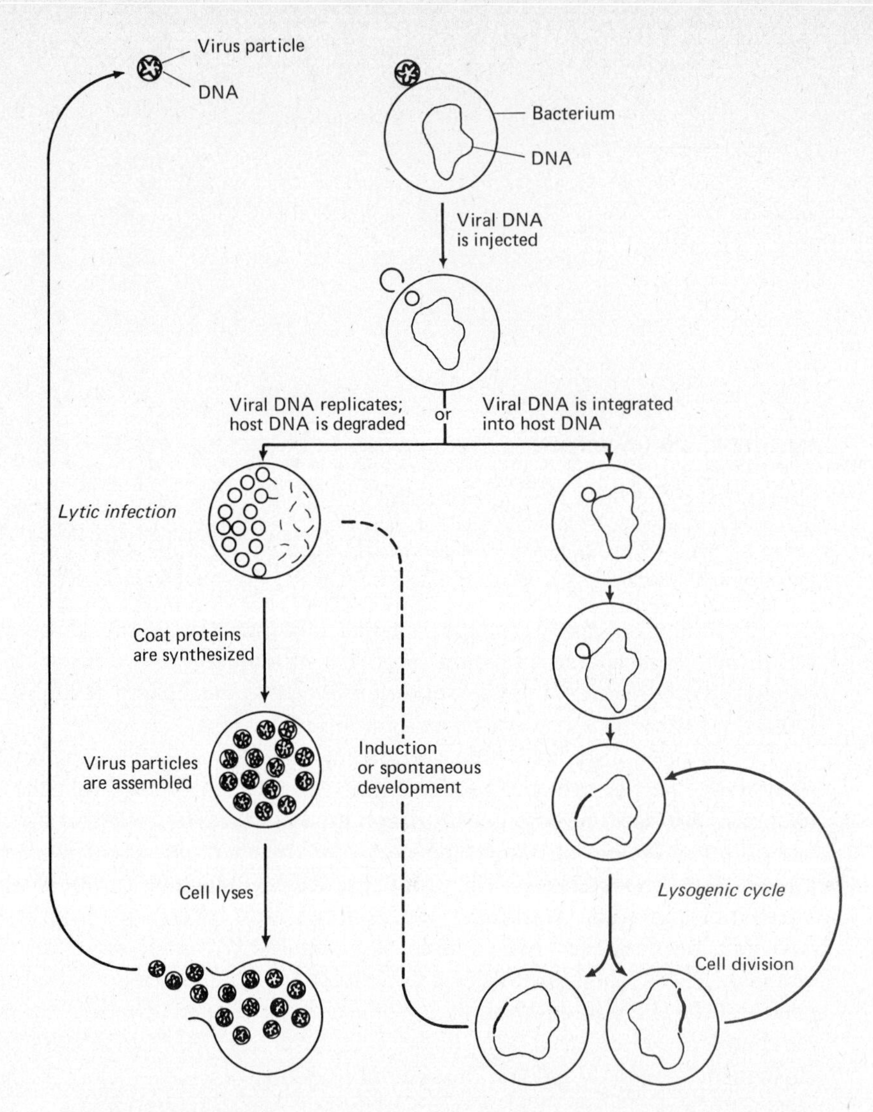

Figure 12-9. Consequences of infection by a temperate bacteriophage. The alternatives upon infection are integration of virus DNA into the host DNA (lysogenization) or replication and release of mature virus (lysis). The lysogenic cell can also be induced to produce mature virus and lyse. (From Brock, T. D. *Biology of Microorganisms.*

bacteria are designated as strain B/2 in contrast to strain B, which are sensitive to T_2 phage. Continued experiments with B/2 *coli* and phage T_2 uncovered a mutant phage that now can infect B/2 and form plaques. It can also be shown that the newly mutant phage can form plaques on the originally sensitive B strain. This mutant phage, designated **h** for host range, forms plaques on either B or B/2 bacteria. However, the wild-type phage, $\mathbf{h}^+$, forms plaques on B but *not* on B/2.

This distinction permitted Hershey to distinguish $\mathbf{h}^+$ from **h**. If he infected a mixture of bacterial strains B and B/2 with phage $\mathbf{h}^+$ or **h**, it could be shown that **h** results in clear plaques because it lyses both strains but that $\mathbf{h}^+$ results in turbid plaques because B/2 is not lysed by it. With the additional ability to tell **r** from $\mathbf{r}^+$ phage, Hershey now was able to distinguish among four possible phenotypes. (See Figure 12-10.)

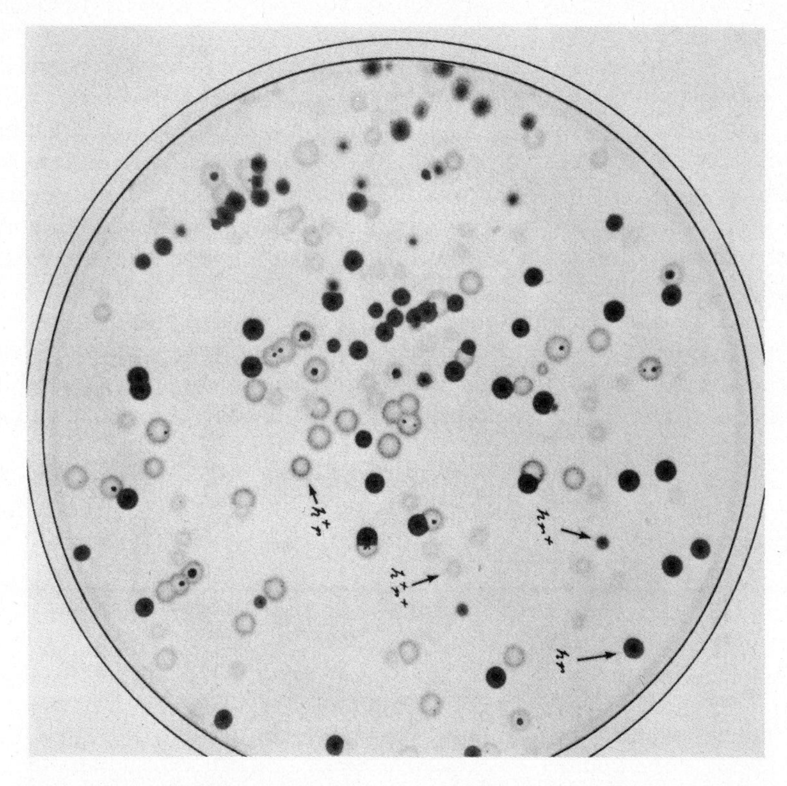

Figure 12-10. The four types of plaques formed by $T_2h^+r^+$, T_2h^+r, T_2hr^+, and T_2hr on $Tto^2 + Tto^r$ mixed indicators. (After G. S. Stent. 1963. *Molecular Biology of Bacterial Viruses.* W. H. Freeman and Company.) (After G. S. Stent. 1971. *Molecular Genetics.* W. H. Freeman and Company.)

The experimental procedure is relatively simple. Bacteria are infected with two genotypes of phage, $\mathbf{h}^+\mathbf{r}$ and $\mathbf{hr}^+$. The number of phage particles used is sufficiently high so that it is statistically almost certain that each bacterium is infected by at least two phages. After lysis occurs, the lysate that contains the progeny phage is analysed by reinfecting more bacteria. This time a mixture of B and B/2 is used to distinguish between $\mathbf{h}^+$ and **h**.

The results of the experiment showed that, in addition to the parental types, $\mathbf{h}^+\mathbf{r}$ and $\mathbf{hr}^+$, recombinants $\mathbf{h}^+\mathbf{r}^+$ and **hr** were recovered in approximately equal frequency. The frequency of recombinants was less than that of the parentals. The reciprocal cross, in which the two parental strains were $\mathbf{h^+r^+}$ and **hr**, was made. Again, all four genotypes were recovered—parental and recombinant. This result is best explained by assuming that recombination took place between the two original phage particles after infection. It is assumed that, in a single bacterium, it is possible for two or more phage to enter, and if the two phage are different genetic types, then during their replication the possibility of genetic recombination exists and apparently occurred.

It seems safe to conclude now that most organisms from the highest, such as man and Drosophila to the lowest, including bacteria and viruses, possess recombination mechanisms. Actually, it would be an exception if an organism were discovered in which no recombinational mechanisms were present. In Chapter 29 some of the arguments on why recombinational mechanisms are so widespread in living systems will be presented, but the reader might think of this question now and try to imagine what the selective advantages might be for recombinational mechanisms.

As the microbial geneticists examined genetic mechanisms in bacteria and their viruses, several cases of what appeared to be rather bizarre events were discovered. It is now understood that these events are almost certainly recombinational, and so they will be considered in this chapter, which deals with recombination in microorganisms.

Transformation

In 1928 the bacteriologist Griffith was studying various strains of the bacterium *Streptococcus pneumononiae*, also known as "pneumococcus." Griffith knew that certain strains of pneumococcus were *virulent*; that is, they were capable of causing infection. He could easily detect such virulent strains because their injection into a mouse resulted in its death in about 24 hours. Griffith also knew that the virulent strains possessed a polysaccharide capsule, and it was apparently this capsule in which the property of virulence resided. If such *encapsulated* strains were cultured on an agar plate, the outline of the growing colony was rather regular or smooth in appearance, and so such strains became known as smooth, or *S-type*, pneumococcus. Finally, Griffith also knew that the polysaccharide capsule was responsible for antigenic specificity, and, by using standard serological techniques, it was possible to tell, antigenically, one particular strain of encapsulated pneumococcus from another. These antigenically distinctive strains were designated as SI, SII, SIII, and so on. Later it became known that most of the properties that reside in the polysaccharide capsule, such as the antigenic specificity, are genetically determined.

S-types spontaneously mutate to strains known as R strains. R strains are *avirulent*—do not possess the polysaccharide capsule—and their colonies, when cultured on agar, give a rough, irregular appearance. Because of the lack of capsule, these R strains of pneumococcus cannot be typed antigenically, but if the cultures are properly sorted and identified in the laboratory, it is possible to tell the serological S strains from which they mutated.

With this information, Griffith performed the following experiment. He took an R strain of pneumococcus, which had originated by mutation from an SII virulent strain, and he injected it into a mouse, together with heat-killed SIII pneumococcus. The mouse was dead in a day and from it he recovered virulent SIII pneumococcus. This is a surprising result, for it would be thought that the R strain could not cause death, since it is avirulent, and the heat-killed SIII should not cause infection because it was heat-killed. Perhaps there was some kind of contamination by living SIII, and one way to test for this factor would be to increase the time needed to heat-kill. Heating was done by placing the bacteria in a test tube of water and boiling for a period of time. Extending the time of boiling had no effect on the experiment. When Griffith performed the control experiments, which included infecting the mouse with R pneumococcus alone and (a separate experiment) infecting the mouse with a heat-killed SIII alone, no infection resulted. It was necessary to have the combination of both R pneumococcus plus the heat-killed SIII in order to recover SIII. Griffith imagined that the R strain was transformed in some way by something in the heat-killed SIII and thus resulted in virulent SIII pneumococcus. This phenomenon became known as *transformation.* A few years later it was shown that transformation could be effected with just a cell-free extract of SIII, and scientists began to ask whether all the extract was needed for transformation or whether one or more parts were sufficient. These scientists began to look for this *transforming principle*, if it existed.

Transformation by DNA

It was some years later, in 1944, that the so-called transforming principle was, in fact, found. This work, which is now recognized as a milestone in genetics, was done by three investigators at the Rockefeller Institute, Avery, MacCleod, and McCarty. Avery and his colleagues attacked this problem in the way that would occur to biochemists. They asked whether they could extract various chemical moieties from the pneumococcus and determine whether any one or a combination of several moieties together could cause the transformation that Griffith had obtained with the whole organism. By utilizing standard biochemical techniques, Avery and his coworkers succeeded in finding the transforming principle. They discovered that it was possible to transform by adding not the entire SIII nor a cell-free extract from it but only one particular chemical fraction. This chemical fraction was *deoxyribonucleic acid, DNA*. Now the Griffith transformation experiment can be done by injecting into a mouse an R strain from SII plus DNA, and only DNA, extracted from SIII. From the mouse will be recovered virulent SIII. Several questions must be asked of the Avery experiment. First, is it certain that DNA alone is responsible for the transformation? The reason for this question is that in the original extraction procedures it was not possible to isolate pure DNA. The transforming fraction always contained a small amount of protein. Is it possible that this small amount of protein was responsible for the transformation? It seems unlikely, for Avery was able to do the following experiments. If he treated the DNA, plus its small protein contaminant, with DNase, an enzyme specific for DNA and one that results in its breakdown, no transformation occurs. If this same experiment is repeated by substituting RNase, an enzyme specific for RNA, a different nucleic acid, transformation still occurs. Finally, if the enzyme used is protein specific, transformation is still observed. Such evidence makes it unlikely that the protein contaminant is responsible for transformation; rather, it clearly implicates DNA.

The induced change in transformation is permanent, and, as mentioned earlier, many of the capsule properties in pneumococcus are inherited. Avery's experiment, therefore, becomes important not only because it shows what the transforming principle itself is but also, and what is more significant, because it largely implicates DNA as the genetic material. This question will be discussed in greater detail in Chapter 18. As the historical development of genetics proceeds from the descriptive or classical to the molecular, questions about the nature of the genetic material and how it functions will inevitably arise. One of the paramount questions is: What, chemically, is the gene? The transformation experiments, for the first time, suggest that the gene might be the nucleic acid, DNA.

Subsequent experiments showed that transformation is not confined to those particular capsule properties in pneumococcus nor to pneumococcus itself. It has been demonstrated in other strains of bacteria involving a variety of different phenotypes. It is not absolutely clear exactly what occurs during transformation, but the most reasonable explanation involves the concept of recombination. The introduced genetic material undoubtedly pairs with the homologous genes of the host, and recombination, probably involving breakage and exchange, ensues.

Transduction

In 1952 Zinder and Lederberg reported another kind of recombination in the bacterium, *Salmonella typhimurium.* Their experiment was conducted in a U-shaped tube. (See Figure 12-11.) In the U-shaped tube was a liquid minimal medium. In the bottom of the tube was a glass filter with holes sufficiently small so that the Salmonella could not pass from one side of the tube to the other. Into the left arm of the tube Zinder and Lederberg placed strain 22A, which was auxotrophic for tryptophan. It was unable to grow in a minimal medium because it could not synthesize this amino acid. In the right arm of the tube was a different auxotrophic strain, 2A, that was unable to synthesize the amino acid histidine.

After introducing these auxotrophic strains into the tube, Zinder and Lederberg recovered prototrophs. However, the prototrophs were only recoverable in the left arm of the tube. No prototrophs were recovered in the right arm! One possible explanation would be that the prototrophs were mutations and that the tryptophan locus is more mutable than histidine. But this interpretation was incorrect. The most telling evidence against a mutation explanation resulted when Zinder and Lederberg repeated the experiment but placed the same auxotroph in

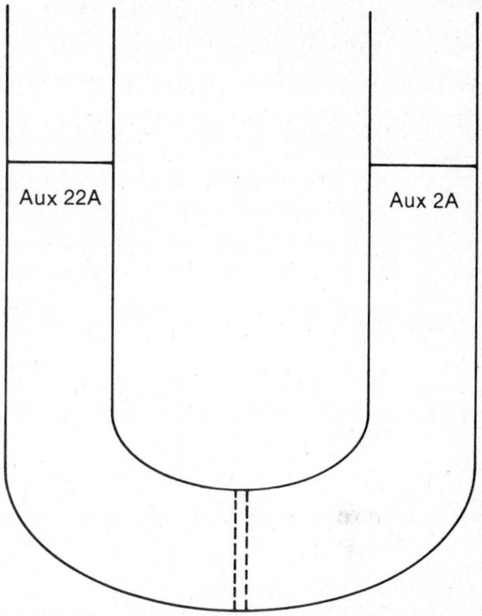

Figure 12-11. U-shaped tube used for recombination experiments in Salmonella by Zinder and Lederberg. In both arms of the tube is minimal medium. A filter prevents bacteria from crossing from one arm to the other. Aux 22A is a strain auxotrophic for tryptophan. Aux 2A is auxotrophic for histidine. Prototrophs are recovered only in the left (Aux 22A) arm.

each arm of the tube. Prototrophs were not recovered with the frequency observed in the experiment in which one arm had 22A and the second 2A. If mutation were the explanation, it should make no difference what strain is in the other arm. It therefore seemed that both bacterial strains were necessary, suggesting possibly that some kind of recombinational mechanism is involved.

Zinder and Lederberg next discovered that 22A was a lysogenic strain of bacteria, harboring a prophage. Occasionally the prophage detached and the bacterium was lysed, releasing daughter phage into the medium. These daughter phage would occasionally find their way into the other arm of the tube. Since 2A is a nonlysogenic bacterium, adsorption, infection, and lysis occur. Zinder and Lederberg postulated that only rarely during phage replication in 2A do the phage genome pick up a small piece of bacterial genome and includes this bacterial genetic material within the phage protein coat. If such a phage makes its way back to the other side of the tube and injects its genetic material into the lysogenic 22A, the small fragment of bacterial genome will also be injected. Should this bacterial gene fragment contain the wild-type gene for tryptophan, which is present in the 2A strain, then it would follow that the bacterium could become a prototroph. This hypothesis of the phage vector carrying bacterial genes was verified in subsequent experiments, and the phenomenon was named *transduction.* Further experiments indicated that phage could carry any particular bacterial gene but that, as a general rule, only one bacterial gene at a time could be carried. In other words, the amount of bacterial genome that could fit into the head of the phage was limited. Although only one gene *at a time* is usually transduced, later experiments have shown that transduction is not limited to just one or two genes in the bacterial genome.

Restricted Transduction

Later, Lederberg and his coworkers discovered that some transducing phage—for example, λ phage in *E. coli*—are very specific in that they can transduce only one gene. This results from the fact that the site of attachment of the λ prophage is very specific in the bacterial genome. Lambda phage transduces only the galactose locus of *E. coli.* Such transduction, limited to a specific gene, is referred to as *restricted* transduction, in contrast to *generalized* transduction described first.

The evidence is quite persuasive that transduction involves recombinational mechanisms. A piece of bacterial genome, when introduced into the lysogenic bacterium, apparently is incorporated into the bacterium by a process analogous to, if not identical with, crossing over. For this reason, transduction is included in a chapter on recombinational mechanisms. Whether such recombination occurs in nature is not known.

REFERENCES

AVERY, O. T., C. M. MACCLEOD, and M. MCCARTY. 1944. Studies on the chemical nature of the substance inducing transformation of pneumococcal types. I. Induction of transformation by a deoxyribonucleic fraction isolated from pneumococcus type III. *J. Exp. Med.* **79**:137–158. (Available in *Classic Papers in Genetics*, edited by J. A. Peters. Prentice-Hall, Englewood Cliffs, N.J., 1959.) (The classic paper that focused attention on DNA as the genetic material.)

HAYES, W. 1968. *The Genetics of Bacteria and Their Viruses.* 2nd ed., 925 pp. John Wiley & Sons, New York. (A comprehensive and very readable book covering bacterial and viral genetics.)

HERSHEY, A. D., and M. CHASE. 1951. Genetic recombination and heterozygosis in bacteriophage. *C.S.H.S. Quant. Biol.* **16**:471–479. (Available in *Papers on Bacterial Viruses*, 2nd ed., edited by G. Stent, Little, Brown and Company, Boston and Toronto, 1965.) (This paper describes the experiments that showed recombination in bacterial viruses.)

JACOB, F., and E. WOLLMAN. 1961. *Sexuality and the Genetics of Bacteria.* 374 pp. Academic Press, New York. (An excellent book describing recombination and genetics in bacteria by two scientists who did much of the work.)

LEDERBERG, J. 1947. Gene recombination and linked segregations in *E. coli. Genetics* **32**:505–525.

LEDERBERG, J., and E. L. TATUM. 1946. Gene recombination in *E. coli. Nature* **158**:558 (Available in *Classic Papers in Genetics*, edited by J. A. Peters. Prentice-Hall, Englewood Cliffs, N.J., 1959.) (These two papers demonstrated that recombination occurs in *E. coli* and provided great stimulus to microbial genetics.)

ZINDER, N., and J. LEDERBERG. 1952. Genetic exchange in Salmonella. *J. Bacter.* **64**:679–699. (Available in *Classic Papers in Genetics*, edited by J. A. Peters. Prentice-Hall, Englewood Cliffs, N.J., 1959.) (A description of transduction in bacteria.)

QUESTIONS AND PROBLEMS

12-1. Define the following terms:

complete medium	episome
minimal medium	virulent phage
prototroph	temperate phage
auxotroph	lysis
exconjugant	lytic bacteris
immunity	lysogenic bacteria
induction	transduction
transforming principle	restricted transduction
plaque	prophage

12-2. Why was it important that Lederberg and Tatum use doubly auxotrophic strains of *E. coli* in their experiment?

12-3. What is the evidence that recombination in bacteria involves a union of two bacteria and a transfer of genetic material?

12-4. What is the evidence that the "chromosome" in *E. coli* is circular?

12-5. In the following crosses in *E. coli*, indicate what the phenotype of the exconjugants will be and whether recombination can occur.

(a) $F^+ \times F^+$.
(b) $F^+ \times F^-$.
(c) $F^- \times F^-$.
(d) Hfr $\times F^+$.
(e) Hfr $\times F^-$.

Briefly explain each answer.

12-6. Compare and contrast the life cycle of lytic phage and temperate phage.

12-7. It appears that recombinational mechanisms exist in many different organisms. Why, would you suppose, have so many different organisms evolved recombination mechanisms?

12-8. What is the significance of transformation in the history of genetics?

12-9. In the Zinder and Lederberg experiment, what other explanation for the appearance of prototrophs might be suggested?

12-10. What experiments and what evidence might be brought to bear to rule out other explanations?

12-11. What is the distinction between generalized and restricted transduction?

12-12. What is the logic for including a discussion of transformation and transduction in a chapter on recombination?

13/Chromosomal aberrations

In Chapter 10 the way in which Morgan's group developed their hypothesis of crossing over, principally from genetic inference, was described. Although corroborating cytological evidence existed, almost all the details of the theory came from genetic experiments. It would be interesting to ask: Is it possible to test the idea of linear arrangement of genes by other techniques? This chapter will discuss the attempt to localize genes by cytological methods.

In order to attempt gene mapping by cytological methods, use will be made of *chromosomal aberrations.* A chromosomal aberration is a change in chromosome structure or number. It has been known for some time that such aberrations can occur spontaneously, although in a very low frequency. It is also possible to increase the frequency of these aberrations with radiation of several kinds.

There are many kinds of chromosomal aberrations, and it is possible to organize them according to their relationship to the normal chromosome complement. For example, there are changes in the *number of chromosomes.* First are changes involving *entire sets* of chromosomes. As pointed out in Chapter 7, the normal chromosome complement consists of pairs of chromosomes, called the diploid condition. It is possible to have an organism in which each chromosome is represented only once. This situation is called *haploidy*. Finally, each chromosome in the set may be represented by more than two homologs, and the general name for this unusual event is *polyploidy*. Specifically, *triploidy* refers to three chromosomes each, four chromosomes each would be designated as *tetraploidy*, and so on. Chapter 30 will treat the evolutionary significance of polyploidy.

Not only are there chromosome changes involving the entire set, but there are also changes involving the *number of chromosomes in a set.* For example, if one chromosome of the entire genome is lost, then it is spoken of as *monosomic* for that particular chromosome. Conversely, if there is one extra chromosome in the set, it is called *trisomic* for that particular chromosome. Recall that trisomics were discussed in Chapter 8 in Blakeslee's experiment with Datura, where he showed a qualitative difference in the number of chromosomes. As with the changes involving the entire set, this chapter will not discuss monosomics or polysomics in detail.

Deficiency

The aberrations of concern now are those that involve a change in *number or arrangement* of gene loci within a chromosome. Discussed first will be changes involving the *number* of gene loci. An obvious aberration would be one in which there has been a loss of one or more genes from the chromosome. This aberration is called a *deficiency*.

How can a deficiency be detected from genetic experiments? Imagine that a cross is made of **AA** by **aa**, where **A** stands for a dominant phenotypic characteristic. All the F_1 from such a cross should be phenotypically A. If on rare occasions however, an individual with "a" phenotype occurs, one explanation would be a deficiency or loss of the **A** gene from one parent. Although other interpretations are possible and would need to be investigated experimentally, the occurrence of a rare recessive phenotype where one is not expected suggests a deficiency. (See Figure 13-1.)

P: AA × aa

F_1: All Aa and phenotypically "A"

If "a" phenotype occurs, one possible explanation is the loss of the A gene from one parent.

Figure 13-1. A genetic experiment designed to detect deficiencies for the A locus.

Synapsis Involving Deficiencies

A second question that should be asked is: How will the homologous chromosomes pair if one is deficient for a few genes? Chapter 7 stated that during meiosis the homologous chromosomes pair, and it is known that this pairing is orderly and apparently occurs in a precise manner, gene with homologous gene. What will happen to the pairing if one homolog is deficient for one or more gene loci present in its normal partner? Cytological observations, particularly with large plant chromosomes, show that in this situation the normal homolog with the entire genetic set will form a little loop or buckle at the site where its homolog is deficient. (See Figure 13-2.)

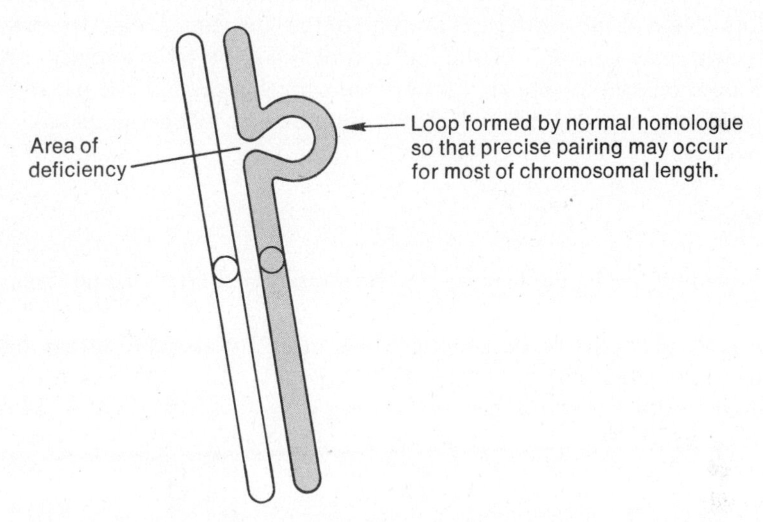

Figure 13-2. Cytological picture observed in synapsing chromosomes when one is deficient for a few genes. For purposes of the diagram, the size of the deficiency and the corresponding loop has been exaggerated.

Viability of Deficiencies

With genetic evidence suggesting a deficiency as indicated in Figure 13-1, a cytological examination of the chromosomes can verify the hunch. The observation of a loop or buckle provides persuasive evidence that a deficiency has occurred. The hypothesis is reinforced by taking into account what is known about the viability of organisms that have deficiencies. In Drosophila—and it is probably true, in general, for most diploid animals—only individuals heterozygous for *short* deficiencies are viable. In addition, they sometimes, but not always, also show some phenotypic abnormality. Homozygous deficiencies in Drosophila are usually lethal. This situation suggests, of course, that most genes are indispensable, at least in single dose, for the development of a viable organism, a fact that follows logically from Boveri's earlier experiments in sea urchins (Chapter 8).

Plants with deficiencies for entire chromosomes or even pairs of homologous chromosomes may survive. Monosomics in plants are not rare. This difference between plants and animals probably lies in the different evolutionary history of these two groups and in the fact that development in plants is less complicated than in animals. It can be also asserted that because many present-day plants are probably polyploid, the loss of one chromosome would not be as detrimental to development of the organism as it would in a nonpolyploid animal.

It was mentioned that in animals such as Drosophila a deficiency is often accompanied by a phenotypic effect. A well-known example in Drosophila involves a sex-linked deficiency known as *Notch*. In this phenotype, there are short notchings on the wing of the fly; and cytological examination has revealed that there is often a loss of one or more loci near one end of the X chromosome. It is possible to obtain a rough approximation of the location of this gene by cytological means, by noting where the loop or buckle occurs. Later in this chapter the use of deficiencies as a means of localizing genes in a more refined way will be discussed.

Duplication

A counterpart to a deficiency is a chromosomal aberration known as *duplication*, the addition of one or more genes to the normal genome. Here the organism now carries a particular part of its genetic constitution in triplicate rather than in the usual duplicate situation.

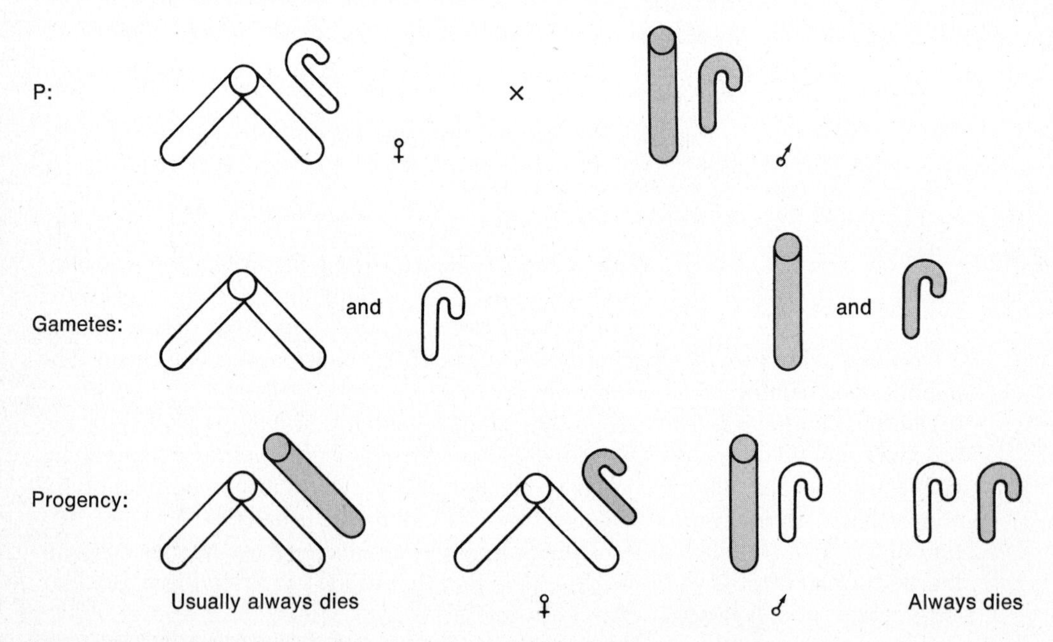

Figure 13-3. The genetics of attached-X females in Drosophila. Normally the only surviving progeny are females that receive both X-chromosomes from the mother and, therefore are *matroclinous* and males that receive the X-chromosome from the father and are *patroclinous*.

The detection of a duplication genetically is not as simple as in the case of a deficiency. Without making the example overly complex, an experiment can be analyzed that might suggest a duplication in Drosophila.

In the 1920s Mrs. T. H. Morgan discovered a female Drosophila in which the two X chromosomes were attached to a single centromere. It turned out that these females normally have a Y chromosome also. Such flies are known, naturally enough, as *attached-X* females. At meiosis these females produce two kinds of eggs. One has the two attached-X chromosomes. The other has the Y chromosome. If the latter is fertilized by a Y sperm, the zygote dies due to lack of an X. If the attached-X egg is fertilized by the X sperm, the zygote dies most of the time. Usually the only surviving progeny are females that obtain their attached-X from the mother and the Y from the father and males that obtain their X from the father and the Y from the mother. (See Figure 13-3.)

Before we discuss the testing for duplication, we should point out that other explanations are possible and need to be investigated experimentally. (See Figure 13-4.)

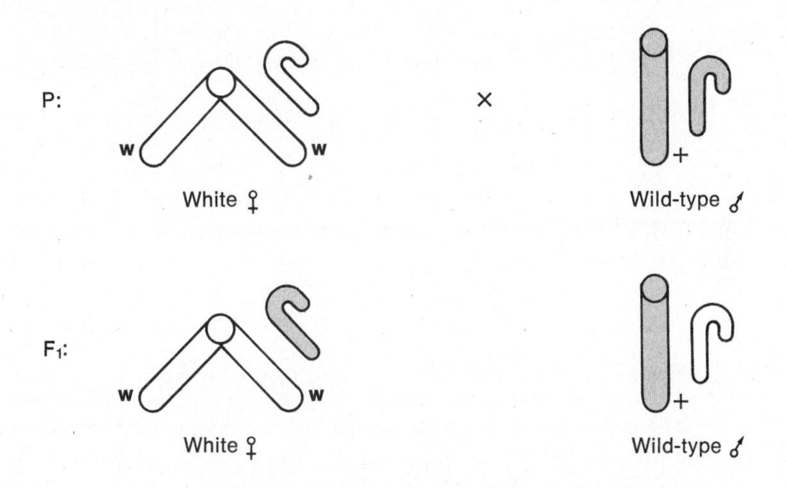

If however the + gene is "detached" from the parental male chromosome and passes to the daughter:

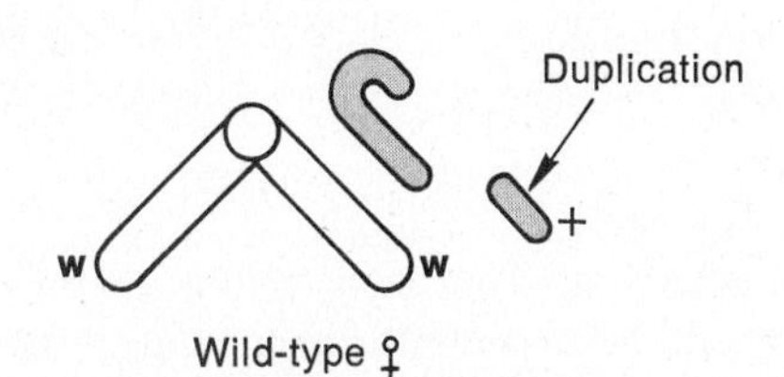

Figure 13-4. The use of attached-X females in Drosophila to detect duplications. The duplication is suggested by the occurrence of a wild-type female when only white-eyed females were expected. For simplicity the duplication is pictured as if it were free in the nucleus. Actually it most likely is inserted into another chromosome in the genome. **w** = white eye.

To test for duplication, an attached-X female that is homozygous for a recessive sex-linked gene, like white eye, is mated by a wild-type male. All the female progeny from this cross should be white eyed, since the X chromosomes remain attached and are passed only to the female progeny. Occasionally and rarely, however, a red-eyed or wild-type female will occur. One possible explanation is the presence of a duplication. It is supposed that the normal gene for white eye is now present in these exceptional females. As we mentioned, other explanations besides duplication are possible and would need to be investigated experimentally. A duplication is suggested by the occurrence of a dominant phenotype when only recessives are expected.

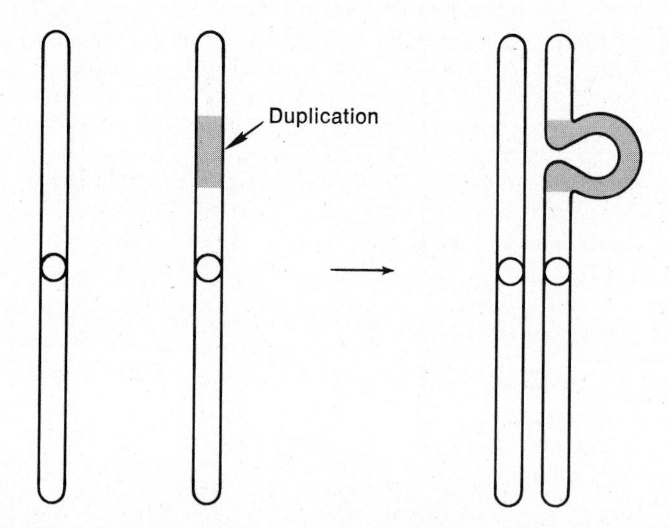

Figure 13-5. The cytological picture observed in synapsing chromosomes when one has "duplicate" genetic material not found in the homologue. For purposes of the diagram, the size of the duplication and the corresponding loop is exaggerated.

Synapsis Involving Duplications

A duplication is detected cytologically by examining the chromosomes. The cytological picture predicted is the same as that observed for a deficiency. That is, one chromosome has a small extra piece of genetic material present that is not found on the homolog. This extra material is the duplication. When these two homologs pair, it would be expected that the chromosome with the extra piece of genetic material will form a loop or buckle at the point of insertion, since this extra material has no counterpart on the normal homolog. (See Figure 13-5.)

Translocation

The final group of chromosomal aberrations involves changes in the *arrangement* of gene loci. There are two kinds of arrangements and the first is known as *translocation.* A translocation is an exchange of parts between nonhomologous chromosomes. Although a translocation need not be reciprocal—that is, involving two different chromosomes—most are, and the discussion in this chapter refers only to reciprocal translocations. (See Figure 13-6.)

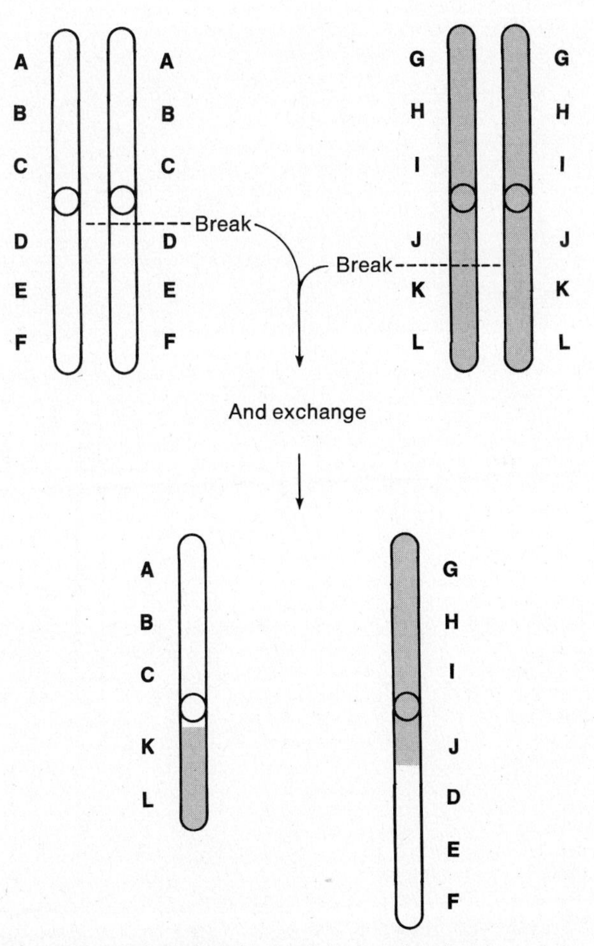

Figure 13-6. A reciprocal translocation involving break and exchange between two nonhomologous chromosomes. Note that the genes **K** and **L** which were originally independent of **A**, **B**, and **C** are now linked to them and **K** and **L** now assort independently of **G**, **H**, **I**, and **J** to which they were originally linked. Similarly **D**, **E**, and **F** now have new linkage arrangements.

How are translocations suspected genetically? Actually, this question is relatively easy to answer. Since genes have moved from one homologous chromosome to a nonhomologous one, and reciprocally, new linkage arrangements could be expected to occur. That is, if the genes **A**, **B**, **C**, **D**, **E**, and **F** are linked on one chromosome and the genes **G**, **H**, **I**, **J**, **K**, and **L** are linked on another nonhomologous chromosome, a reciprocal translocation occurring between these two chromosomes would establish new linkage groups. Genes that once assorted independently are now linked, and genes that were originally linked now assort independently (Figure 13-6). The occurrence of new linkage arrangements not found in the original stock would suggest a reciprocal translocation.

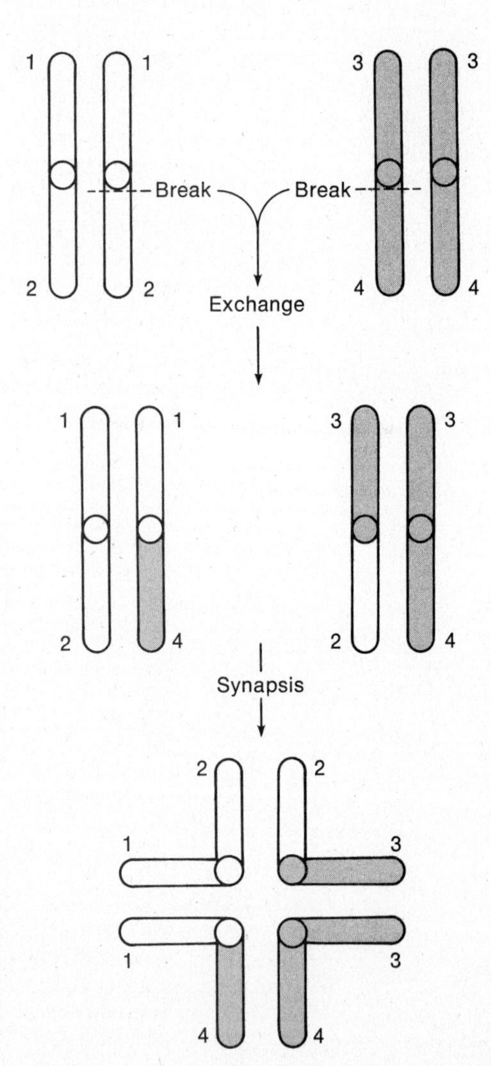

Figure 13-7. Synapsis of two pair of chromosomes involved in a reciprocal translocation. The appearance of cross-like configurations is suggestive cytological evidence for reciprocal translocations. For simplicity the breaks are drawn to occur near the centromere in each case.

Synapsis Involving Translocations

An even more interesting question is how the two pairs of chromosomes that have exchanged parts synapse during meiosis. Evidently when chromosomes pair, they do so in a rather exact fashion. Can this exact pairing be accomplished with reciprocal translocations? The answer is yes. During pairing, reciprocal translocations will form a crosslike configuration. (See Figure 13-7.) This quite characteristic cytological picture makes detection of translocations relatively easy, particularly in plants which often have large chromosomes.

Semisterility

This pairing arrangement raises additional interesting questions about what happens in meiosis. For example, how will the chromosomes segregate to the poles? There are four centromeres; and if the two pairs of chromosomes were not involved in a reciprocal translocation, then two centromeres would go to one pole and two would go to the other. If, in the translocation, it is assumed that the centromeres segregate in a random fashion—two to one pole and two to the other—then there are, a priori, three possible segregations. (See Figure 13-8.) Note that only in the first combination do the gametes have a complete chromosome for each of the two pairs, even though, in one case, the parts are rearranged in the translocation. However, notice that in all the other arrangements there are duplication-deficiency gametes. That is, one chromosome is present in excess, whereas the other chromosome is deficient for one of its arms. In plants, particularly in the male plant, such duplication-deficiency arrangements will cause the pollen to abort and be nonfunctional. According to the figure, if the chromosomes segregate at random, about 67% sterility would be expected. Actually, because the centromeres do not segregate at random, about 50% sterility, due to aborted pollen that do not contain an intact set of chromosomes, is observed. In other words, in plants, another clue that suggests reciprocal translocations is the presence of semisterility, particularly in the male line.

Does semisterility caused by reciprocal translocations also occur in animals? The answer is no, because duplication-deficiency arrangements in general are not inviable in the gametes of animals. However, such duplication/deficiencies will not survive in the zygote. Consequently, some interesting segregations that suggest the presence of a reciprocal translocation will occur.

This situation can be demonstrated with an experiment involving the second and third chromosome in Drosophila. A female, homozygous for the second chromosome recessive gene for brown eye (**bw**) and for the third chromosome recessive gene for scarlet eye (**st**), is mated to a male with a normal second chromosome that has **bw** and a normal third chromosome that has the **st** gene, plus a translocated second and third chromosome with the wild-type alleles for these two genes. If a male heterozygous for two independently assorting pairs of genes is mated to a female homozygous for both recessives, four kinds of progeny in equal frequency would be expected. Because of the reciprocal translocation, however, some of the zygotes produced will have duplication-deficiency chromosomes and will not survive. The only sperm that will produce viable zygotes are the ones with nontranslocated chromosomes containing brown and scarlet or the ones with

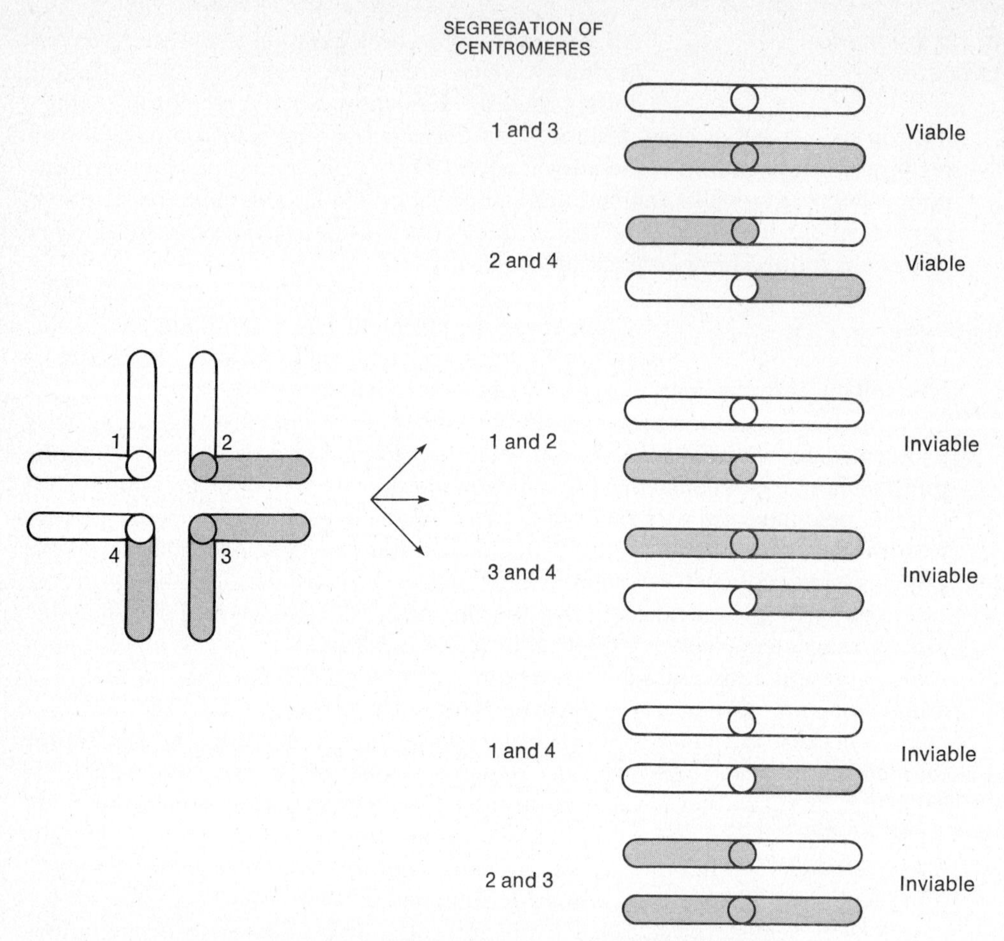

Figure 13-8. The consequence of centromere segregation after synapsis of two pair of chromosomes involved in a reciprocal translocation. If the centromeres segretate at random three possibilities exist, two of which result in duplication-deficiency gametes and only one third of which give viable gametes. Because segregation of centromeres is not random only about 50 % of the gametes are duplication-deficiency gametes.

translocated chromosomes containing the wild-type alleles for both genes. (See Figure 13-9.) From a cross that should have produced four kinds of progeny in equal frequency, only two kinds of progeny, brown-scarlet and wild type, are observed. This unexpected genetic result is another indication that a translocation has occurred.

It is possible, even in an organism like Drosophila in which the mitotic chromosomes are very small, to make a rough approximation of gene location by observing translocations. If a translocation has occurred between one of the large

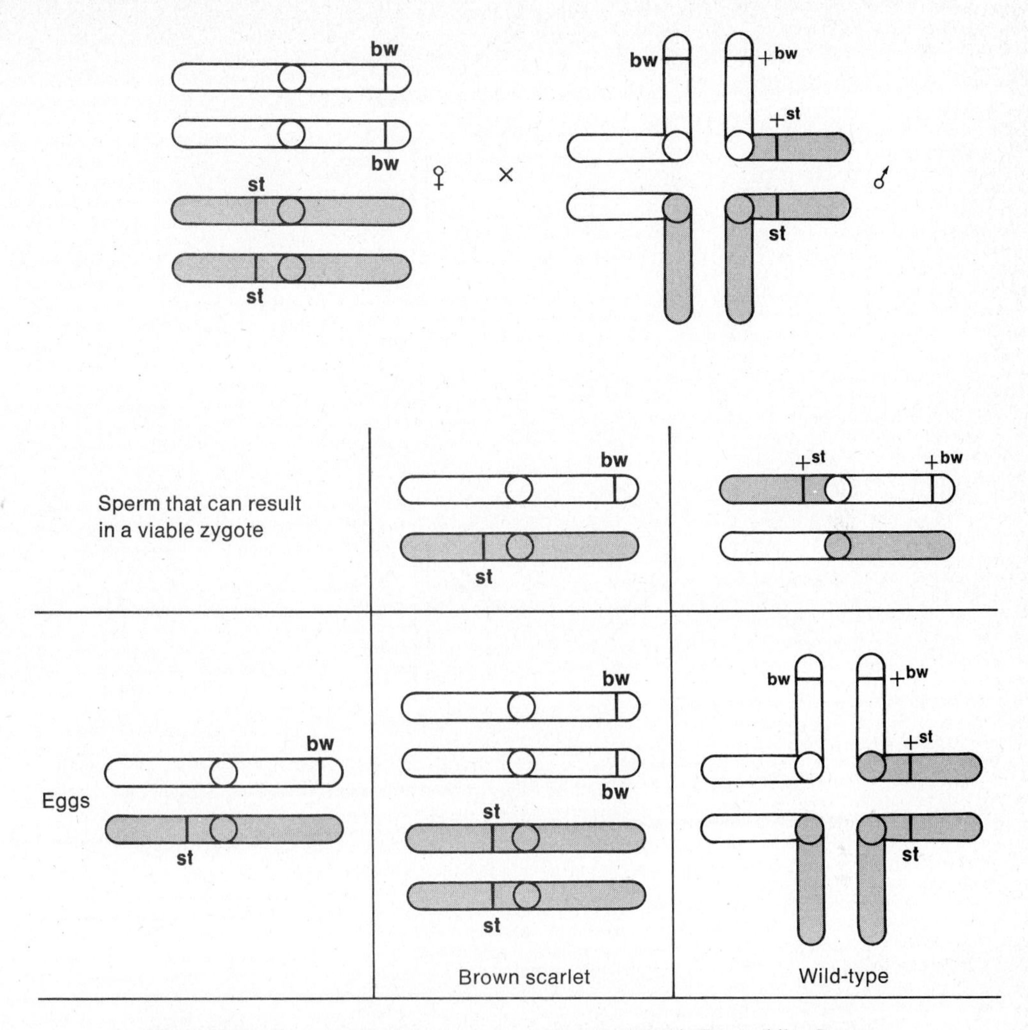

Figure 13-9. Genetic detection of a reciprocal translocation in Drosophila. Because duplication-deficiency zygotes are inviable, only two zygotes survive giving two phenotypes instead of the four predicted by independent assortment. **bw** = brown eye; **st** = scarlet eye; + = wild type.

V-shaped chromosomes and the small-dot fourth chromosome, it should be possible to correlate the cytologically observed breakpoints with the changed genetic linkage arrangements. Although the correlation is only approximate, because of the difficulty of working with Drosophila chromosomes, reasonable cytological maps have been constructed by this method. (See Figure 13-10.)

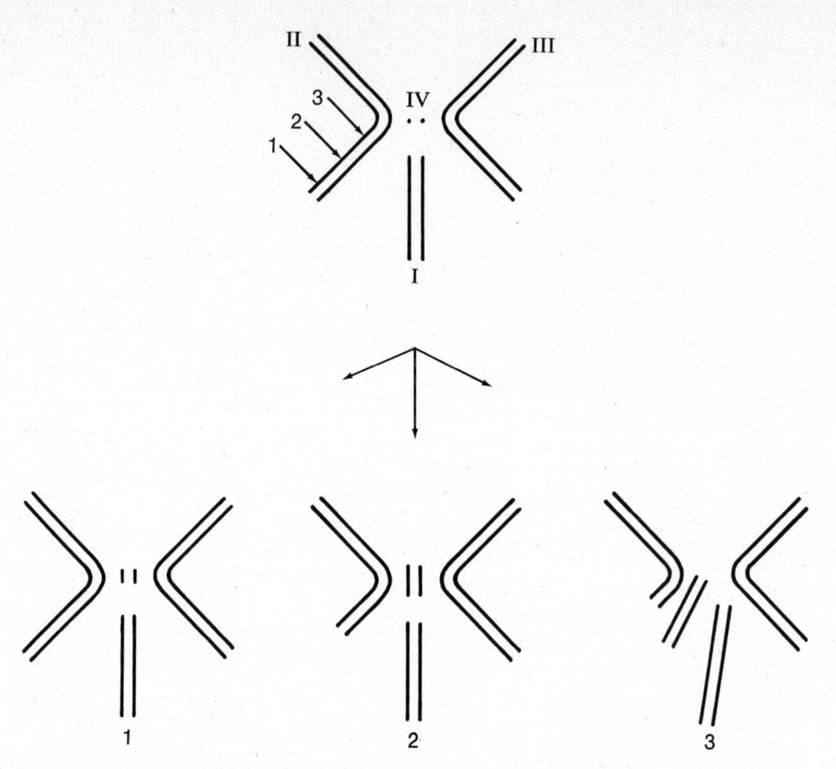

Figure 13-10. Cytological use of translocations between the second and fourth chromosome in Drosophila. The translocations are drawn as though they were homozygous. Even with the small somatic chromosomes of Drosophila, it is possible to obtain a reasonable approximation of where the break on the second chromosome occurred.

Inversion

The final chromosome aberration also involves a new arrangement of linked genes. Imagine a chromosome that is simultaneously broken in two places. Imagine also that the broken piece comes out, turns around in some way, and is then reinserted. If this process occurs, the chromosome possesses an *inversion*. (See Figure 13-11.)

An examination of Figure 13-11 should make it apparent how inversions are detected genetically. Notice that, in the original chromosome, the genes **B** and **C** were next to each other and the genes **B** and **F** were relatively far apart. This should be reflected in crossing-over experiments in which **B** and **C** should show little recombination, whereas **B** and **F** should give many recombinants. In the inverted chromosome, however, it would be revealed that **B** and **F**, which are now cytologically close to each other, give little genetic recombination. Similarly, **B** and **C**, once closely linked, would now appear to be loosely linked. In summary, a rearrangement

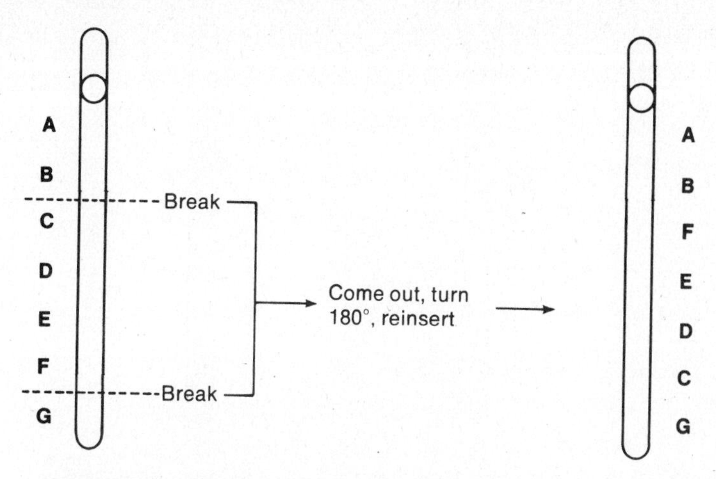

Figure 13-11. Diagram of a chromosomal inversion. A large segment of the chromosome has come out and turned 180° before reinserting. Note that genes once closely linked, **B** and **C**, are now relatively far apart while genes once far apart, **B** and **F**, are now closely linked.

of known linkage relationships wherein genes once closely linked appear to be far apart and genes far apart appear to be closely linked is strongly suggestive of an inversion.

Synapsis Involving Inversions

How would a pair of homologous chromosomes synapse if one has a normal sequence of genes and the other an inverted sequence? The solution is not easy to visualize. In order for the chromosomes to synapse in a more or less exact way when one homolog has an inverted region, they must be arranged in a loop. (See Figure 13-12.) The detection of such inversion loops is relatively easy, especially with large plant chromosomes, and would tend to confirm the genetic inference that an inversion has occurred.

Crossing over in Heterozygous Inversions

What occurs if crossing over takes place within an inversion loop? (See Figure 13-13.) Notice that chromatids 1 and 3 are uninvolved in the exchange resulting in nonrecombinant chromatids. However, chromatid 2 is involved in the exchange and, consequently, forms a chromosome that has two centromeres. Such a chromosome is called a *dicentric*, and at subsequent cell divisions it will break and eventually become lost. A remaining fragment is unaccounted for in the exchange. This fragment has no centromere at all. A chromosome without a centromere is called an *acentric* fragment and, as a rule, is immediately lost at cell division. Such an analysis leads to the conclusion that the only viable products recovered from crossing over within an inversion heterozygote are the nonexchange

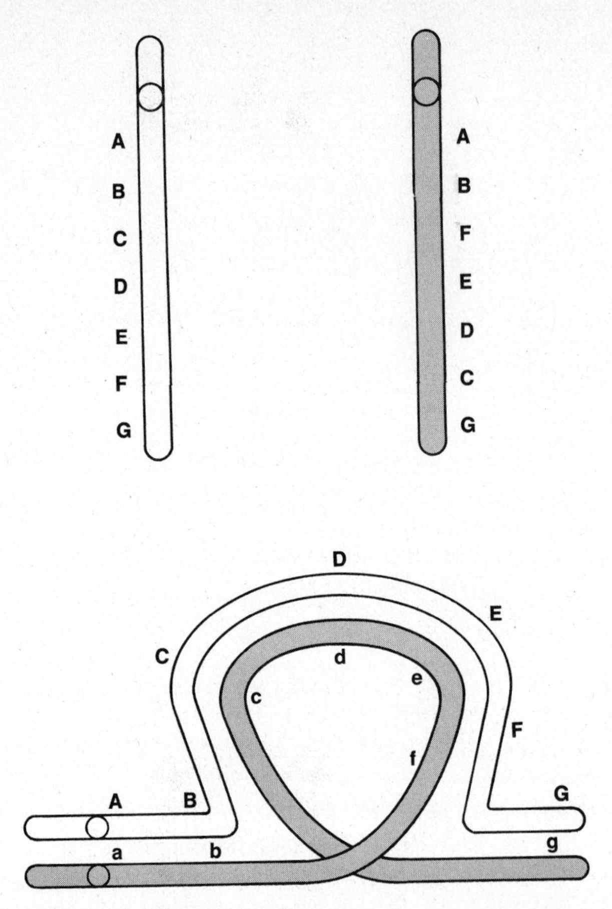

Figure 13-12. Synapsis of homologous chromosomes when one homologue has an inverted segment. For simplicity the synapsed chromosomes are drawn unduplicated and greatly attenuated.

chromosomes. In other words, if there is a heterozygous inversion, there will be, in theory, no genetic recombination. In fact, there will be some recombination due to multiple exchanges, particularly if the inversion is large. A two-strand double, for example, would result in four viable chromosomes and would show some genetic recombination. The net result, however, is that genetic recombination within heterozygous inversions is almost reduced to zero; and with a complex series of breaks in which there are inversions within inversions, crossing over will be effectively reduced to zero. This point is important, and its evolutionary significance will be discussed in Chapter 29. Although the example in Figure 13-13 shows an inversion heterozygote in which the centromere is not included in the inversion, the same conclusions can be drawn about inversion heterozygotes in which the centromere is included in the inversion. Only the details are somewhat different.

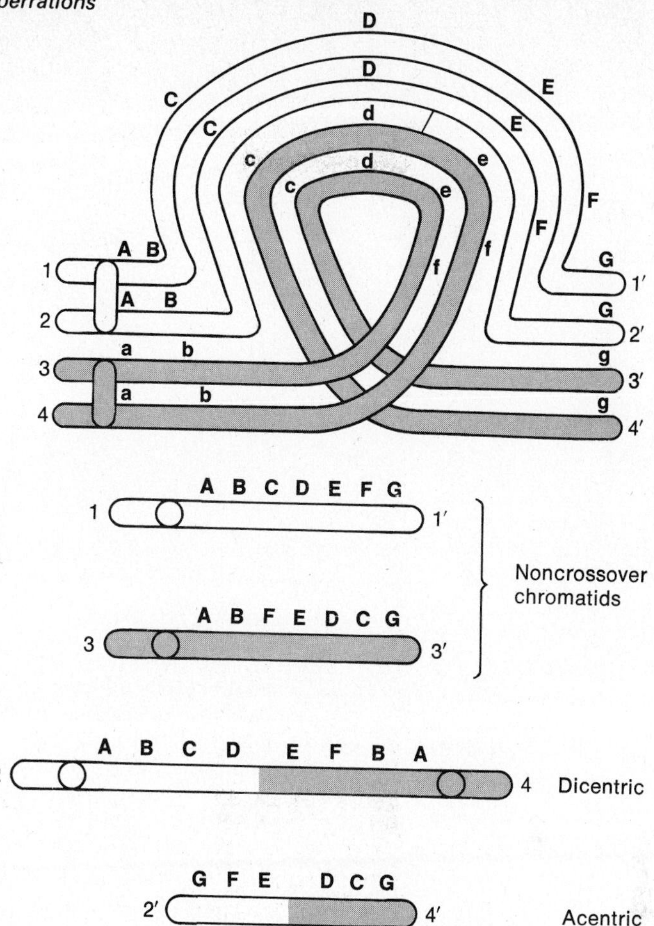

Figure 13-13. Crossing over within a heterozygous inversion. Only chromatids uninvolved in the exchange survive and are nonrecombinants. The chromatids involved in the exchange result in either a dicentric duplication-deficiency chromosome or an acentric duplication-deficiency fragment. Neither survives so that the net result is a great reduction in recombination in heterozygous inversions. For simplicity both ends of the chromosomes are numbered. As a rule of thumb, after the crossover proceed along the new strand in the same direction as obtained on the old strand.

Gene Mapping with Chromosomal Aberrations

In the beginning of the chapter we stated that one aim in studying cytological aberrations was to map genes by a method other than simple genetic experiments, in order to see whether the same conclusion could be independently reached. This method of analysis was done in Drosophila, mainly by using translocation analyses, as shown in Figure 13–10. An example of the results obtained is seen in Figure 13-14.

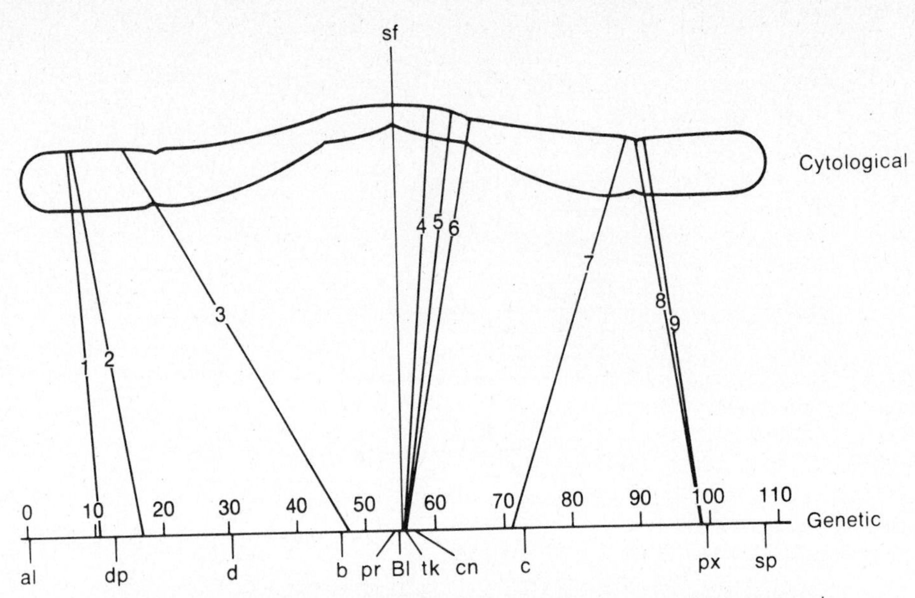

Figure 13-14. The genetic and the cytological maps of the second chromosome in *Drosophila melanogaster.* Letters in the lower part of the drawing indicate loci of the various genes of the second chromosome. sf = the locus of the spindle-fiber attachment. (After Th. Dobzhansky. 1930. Cytological Map of the Second Chromosome of *Drosophila melanogaster. Biolog. Zentral.* **50**.)

The first conclusion is that the linear order of genes, as mapped by cytological techniques, is exactly the same as the linear order of genes mapped genetically. Note that in Figure 13-14 none of the lines cross each other. Independent means have shown that the genes are arranged in linear order on the chromosome, an inference drawn earlier from Morgan's genetic analysis.

One of the assumptions implicit during the discussion of Morgan's work was that the probability of a break and exchange is equal and random throughout the length of the chromosome. This assumption is not substantiated by the cytological investigation. For example, note that genes that appear to be tightly linked genetically (**Bl**, **tk**, **cn**) do not appear to be so tightly linked on the cytological map. Conversely, genes that appear to be relatively far apart (**px** and **c**) on the genetic map appear to be closer together on the cytological map. The result seen for the second chromosome in Drosophila is also true for the third and X chromosomes, and the conclusion can be generalized. In the area near the centromere, genetic recombination is low, considering the actual distance apart of the two genes, whereas, out in the arms of the chromosomes, genetic recombination is relatively high, taking into account the true spatial relationship of the genes.

Cytologists had demonstrated that the chromosome does not always stain as though it were homogeneous throughout. It was found that very often whole chromosomes or parts of chromosomes stained differently at a given time from other chromosomes or other parts of the chromosome. This differently staining material was called *heterochromatin*, in contrast to *euchromatin*, the "true" chromatin material of the chromosome. It can be generalized that the heterochromatic regions of the chromosome contain fewer genes than the euchromatic regions and that genes in heterochromatin recombine less frequently than those in euchromatin. Since the centromere in Drosophila chromosomes is in heterochromatin, genes near the centromere are also in heterochromatin. Because the probability of exchange in heterochromatin is less than in euchromatin, such genes appear, from genetic experiments, to be tightly linked.

So far a crude approach to the cytological localization of genes, based primarily on translocations, has been used. It is possible to obtain a much more refined cytological picture of the chromosomes.

Salivary Gland Chromosomes

It has been known since 1881, through the investigations of Balbiani, that chromosomes of gigantic size were present in the cells of salivary glands of larvae of some species of Diptera, the two-winged flies. These chromosomes are not only extremely large, but they also show a wealth of structural detail. (See Figure 13-15.) The significance of this discovery, however, was not appreciated until 1933, when the nature of these chromosomes was clarified independently by Heitz and Bauer in Germany, and particularly Painter in the United States, who indicated their use in studying genetical problems. These giant chromosomes are called, appropriately, the *salivary gland chromosomes.*

In the Dipterans, the homologous members of each pair of chromosomes tend to lie side by side in the cell nucleus. That is, there is *somatic pairing* in this order of insects. The salivary gland chromosomes are believed to represent an interphase condition in which the chromosomes are very long and attenuated. Moreover, they lie close together and have replicated as many as ten times, giving rise to a structure of 1000 or more laterally arranged strands. As a consequence, these chromosomes contain a pattern of darkly staining disks or bands along their length. The pattern of bands is characteristics for each particular chromosome. Such detail allows for the recognition of one chromosome from another in Drosophila. More importantly, specific regions of one chromosome can be unambiguously distinguished from other regions of the same chromosome. Bridges, in the mid thirties, devoted himself to making salivary chromosome maps of Drosophila in which he carefully studied the banding patterns and drew the standard maps. It is possible, therefore, to study the map and learn the normal banding configuration. With this information, cytological aberrations can be studied in great detail. In theory, it should be possible to detect a deficiency as small as a single band, although technical problems sometimes prohibit it.

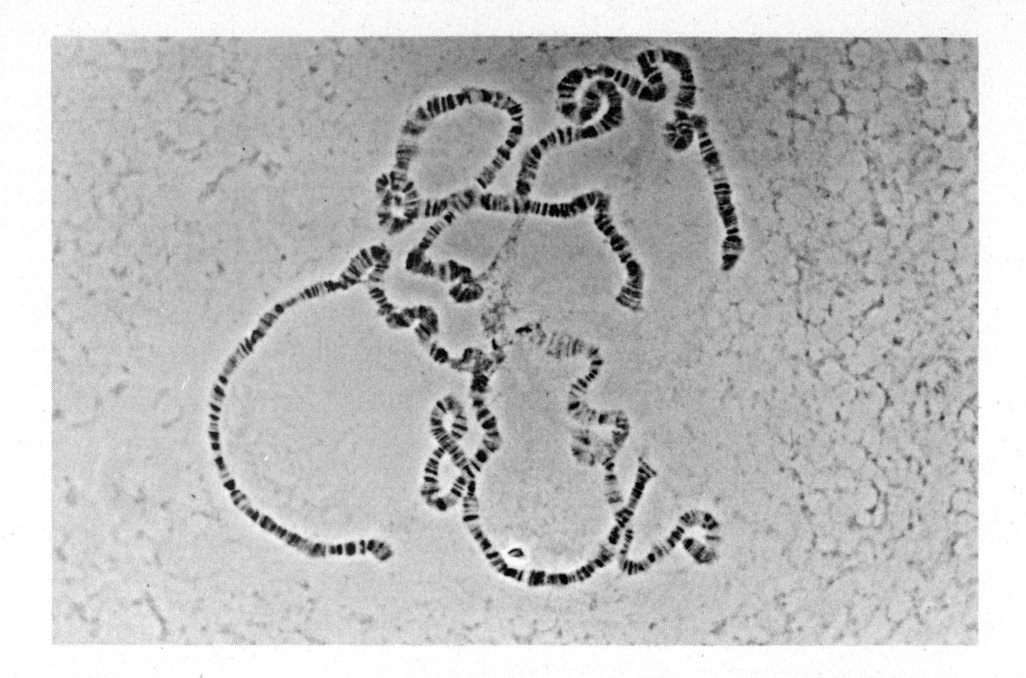

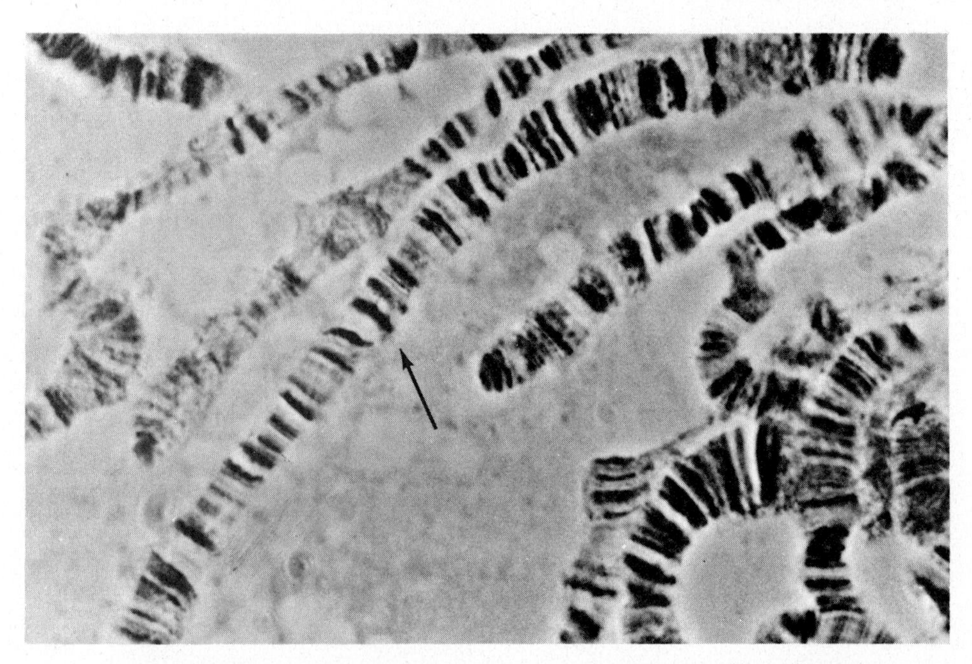

Figure 13-15. Giant chromosomes compared with "normal." The main figure is a camera lucida drawing of the elements found in a salivary gland nucleus of a female larva, showing the chromosomes just as they lay on the slide. (After T. S. Painter. 1934. Salivary Chromosomes and the Attack on the Gene. *J. Hered.* **25**.)

Gene Localization with Salivary Gland Chromosomes

It has been possible, however, to make a refined localization of the genes by studying the banding pattern in the salivary band chromosomes. In the earlier discussion of deficiencies mention was made of one particular deficiency in Drosophila, Notch, which is characterized phenotypically by notches in the wing. It turned out that many of the early Notch mutants were, in fact, deficiencies; and investigators were interested in whether it would be possible to localize the Notch deficiency to any one particular region of the chromosome.

Figure 13-16 shows some results of such an analysis. Notice that a deficiency may be of varying length and still result in the Notch phenotype. A careful examination of the results indicated that every one of the Notch deficiencies has one missing band in common—namely, the band 3C7. The conclusion is inescapable

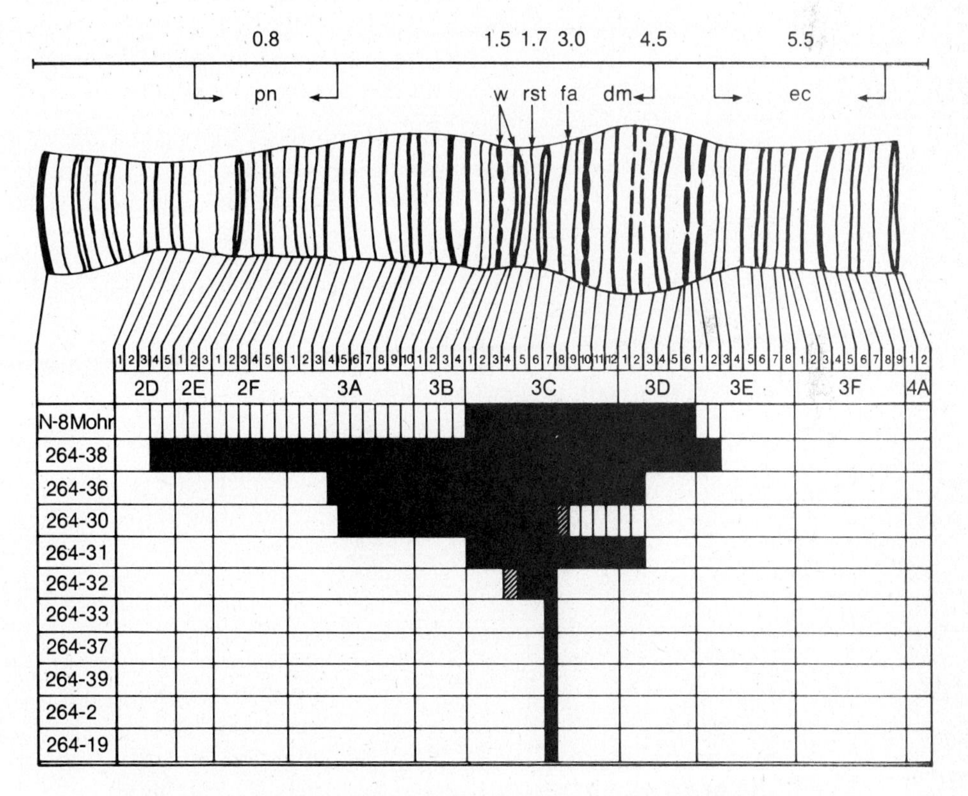

Figure 13-16. Genetic and salivary chromosome maps of the Notch region of the X chromosome and a diagram showing results of the study of some deficiencies of that region. Black areas represent deficient segments and shaded areas indicate sections for which it is uncertain whether or not they are deficient. The symbolism between the salivary chromosome and the representation of the deficiencies is Bridges' coordinates for the map. (After H. Slizynska. 1938. Salivary Chromosome Analysis of the White Facet Region of *Drosophila melanogaster. Genet.* **23**.)

that the Notch phenotype is determined by band 3C7 or its immediate interband neighborhood. In similar ways, other genes in Drosophila have been localized on the cytological chromosome.

It is a tempting hypothesis that a 1 : 1 correspondence exists between the genes and the bands on the salivary gland chromosomes, but this hypothesis is unproven.

REFERENCES

DOBZHANSKY, TH. 1930. Cytological map of the second chromosome of *Drosophila melanogaster*. *Biol. Zentralbl.* **50**:671–685. (An example of the use of chromosomal aberrations, in this case, translocations, to map genes cytologically.)

PAINTER, T. S. 1933. A new method for the study of chromosome rearrangements and the plotting of chromosome maps. *Science* **78**:585–586. (Available in *Classic Papers in Genetics*, edited by J. A. Peters. Prentice-Hall, Englewood Cliffs, N.J., 1959.) (The important paper that suggested the significance of the salivary gland chromosomes for genetics.)

SLIZYNSKA, H. 1938. Salivary chromosome analysis of the white-facet region of *Drosophila melanogaster*. *Genetics* **23**:291–299. (An example of the use of the salivary gland chromosomes to localize a particular deficiency, Notch.)

SWANSON, C. P. 1957. *Cytology and Cytogenetics*. 596 pp. Prentice-Hall, Englewood Cliffs, N.J.

SWANSON, C. P., T. MERZ, and W. J. YOUNG. 1967. *Cytogenetics*. 194 pp. Prentice-Hall, Englewood Cliffs, N.J.

(Two excellent sources for additional information. The former is comprehensive, the latter more up to date but not as detailed.)

QUESTIONS AND PROBLEMS

13-1. Define the following terms:

chromosomal aberration
haploidy
polyploidy
monosomic
trisomic
deficiency
euchromatin
heterochromatin
duplication
attached-X chromosomes
translocation
inversion
dicentric chromosome
acentric chromosome
somatic pairing

13-2. In Drosophila, a female homozygous for the sex-linked recessive genes yellow, cut, raspberry, and forked was mated to a male wild-type for all four genes. All the F_1 females were wild-type with one exception, which was cut but wild-type for the other three genes.

(a) What is a possible explanation involving a chromosomal aberration for this one exception?

(b) How would you verify your hypothesis?

13-3. In Drosophila, a female was found that had nicks in her wings. It was suspected that this phenotype was *Notch*, a known deficiency on the X chromosome.

(a) How might the hypothesis of Notch deficiency be tested genetically?

(b) How might the hypothesis of Notch deficiency be tested cytologically?

13-4. From a previous experiment in Drosophila, it was known that a, b, c, d, e, f, and g were all sex-linked recessive genes. In a later experiment, a female with a, b, c, d, e, f, and g on one X chromosome and all the wild-type alleles on the other yielded male progeny of only two phenotypes: either abcdefg or wild-type.

(a) Explain this result.

(b) How might the explanation be verified?

13-5. Does crossing over occur in a pair of homologous chromosomes if one of the homologs is inverted with respect to the other?

13-6. If crossing over does occur, why is there a marked reduction in recombination?

13-7. Although there is a marked reduction in inversion heterozygotes, rarely is recombination reduced to zero. Why?

13-8. Can you think of any practical application for the fact that, in inversion heterozygotes, recombination is markedly reduced?

13-9. In maize, there are three strains with the following gene orders for a given chromosome:

a b c i h g k j d e f l
a b c i h g f e d j k l
a b c d e f g h i j k l

If it is assumed that the last chromosome represents the original order, what can be said about how the other two chromosomes originated?

13-10. In an organism, chromosome 2 has the following gene order: A B centromere C D E F G H I. Chromosome 3 in the same organism has the following gene order: J K L M N centromere O P Q R S. A break occurs in chromosome 2 between D and E, and simultaneously a break occurs in chromosome 3 between L and M with the result that a reciprocal translocation is produced between chromosomes 2 and 3. Draw the chromosome configuration that would result during meiotic synapsis in this plant with the reciprocal translocation.

13-11. In Drosophila, the recessive genes a, b, and c are located on the X, second, and third chromosomes, respectively. From the mating (+ / +) (+ / +) (+ / +) female × (a / a) (b / b) (c / c) male, a single wild-type male was mated to a (a / a) (b / b) (c / c) female. The progeny from this cross were as follows:

FEMALES	MALES
$\frac{1}{2}$ + + +	$\frac{1}{2}$ a b +
$\frac{1}{2}$ + + c	$\frac{1}{2}$ a b c

What is the explanation for the genetic behavior of the single wild-type F_1 male?

14/Mutation

In the discussion of sex-linkage in Chapter 9 it was pointed out that Morgan found one fly, a male, that had white eyes. This white-eyed male arose through mutation. For some reason or other, the wild-type allele, instead of making a faithful copy of itself, changed to a different form, which during the course of development became manifest as a white eye. Such an event is called a *mutation* and is defined simply as any change from the wild-type that is inherited.

Mutations may be of several kinds. The white-eyed mutation is referred to as a *point mutation* or a *gene mutation.* It is a change that apparently involves only one locus on the chromosome. However, chromosomal aberrations are also mutations in the sense that they are inheritable changes, and, as was learned in the last chapter, there are many possible kinds of chromosomal mutations. In this chapter the discussion is concerned essentially with point or gene mutations.

Spontaneous Mutations

Detection

How can experiments be designed to detect the presence of mutations? This question was asked in the 1920s by an American geneticist, L. J. Stadler, who worked with corn and barley. He reasoned that it should be possible to detect recessive mutations (see Figure 14-1);

P: AA♀ × aa♂

F_1: All Aa show the dominant "A" phenotype

Occasionally an "aa" phenotype occurs. Such an occurrence suggests a mutation in the parental ♀ of A→a. The F_1 recessive must be bred to verify that the change is inherited and to rule out other possible causes.

Figure 14-1. Method for determining recessive mutations in plants.

and when he designed a large-scale experiment in corn utilizing this experiment design, he did, in fact, observe mutations. (See Table 14-1.)

Table 14-1. Spontaneous Mutations for Eight Loci in Corn

GENE	NUMBER OF GAMETES TESTED	NUMBER OF MUTATIONS	FREQUENCY OF MUTATION PER MILLION GAMETES
R	554,786	273	492
I	265,391	28	106
Pr	647,102	7	11
Su	1,678,736	4	2
C	426,923	1	2
Y	1,745,280	4	2
Sh	2,469,285	3	1
Wx	1,503,744	0	0

FROM L. J. STADLER. 1942.

Two important conclusions may be drawn from Stadler's results. First, mutations are recurrent. For the most part, they occur more than once. Secondly, the rate of mutation is low. Ignoring the genes R and I, which later research showed were of a special nature, it is evident from Stadler's experiment that spontaneous mutation rates are on the order of 10^{-6}. Another conclusion of lesser importance can be made. It appears that different genes mutate at different rates, although this may be a consequence of sampling error. It should be apparent to the reader that it was essential for Stadler to test all the phenotypically recessive plants obtained in order to be sure that what is called a mutation was, in fact, an inherited gene change.

Reverse Mutations

Having established that mutations can occur spontaneously, several general questions about mutation should be raised. First, is there such a thing as reverse mutation? In the experimental design in Figure 14-1 the mutation rate of **R** → **r** was being measured. Does the reverse event from **r** → **R** also occur? This particular question was investigated in several organisms but extensively in Drosophila. Thoroughly studied was the white-eye mutant to determine whether mutation from **w** → + could be observed as had been the mutation from + → **w**. The results indicate that not only could + mutate to **w** but also that **w** could mutate to +. The frequency of forward mutation seems higher than for reverse mutation. Further complicating the interpretation was the fact that mutation might proceed in steps through an intermediate allele. For example, white might mutate to eosin, another allele at the white locus. In a subsequent experiment this eosin allele mutated to wild-type. For many years there was considerable confusion about the reality of reverse mutations. However, present evidence, largely obtained with microorganisms, makes it quite clear that genes may mutate in either direction. The rates of such mutations often vary. The mutation rate from wild-type to mutant generally is higher than from mutant to wild-type. In Chapter 23 mutation will be considered at the molecular level, and a clearer understanding of why rates of mutation vary will be given.

Somatic Mutations

A second question asked about mutations was: Where in the life cycle do they occur? For obvious reasons, a geneticist usually is studying mutations involved in the meiotic cycle. These mutations are technically much easier to study and, from the standpoint of transmitted characteristics, are the only mutations worth talking about. However, it is also of interest to ask whether mutations can occur in somatic cells that do not give rise to gametes but constitute the great majority of cells in most organisms.

In Drosophila, it was discovered that flies sometimes occur that are *mosaic* with respect to eye color. A mosaic consists of two or more types, and an eye color mosaic simply means that two or more eye colors are present. Such a mosaic pattern could vary, but normally it involves a white patch in an otherwise red eye. At its most extreme, there could be one red eye and one white eye. The question is: What is responsible for these mosaics? The most reasonable answer would appear to be mutation from the wild-type allele to white. This hypothesis is supported by the observation that such eye mosaics generally occur in males. This result would be expected of a mutational hypothesis, since the male is hemizygous and mutation of only one gene is necessary. Testing for somatic mutation is, by definition, difficult, since the cells involved are not passed on to the next generation. However, it is possible, in some instances, to verify the mutation hypothesis in Drosophila by breeding the fly to determine whether the mutation included cells that gave rise to the gonads. Such experiments have been done; and in a few cases of bilateral mosaics where one eye was white and the other red, the fly bred as though it were heterozygous for red and white. Heterozygosity cannot be explained easily in a hemizygous male without postulating that one testis was derived from the cell in which the mutation occurred and gave rise to sperm that carried the white allele, whereas the other testis derived from the cell in which no mutation occurred and gave rise to sperm that contained the wild-type allele. Such a male would appear to be heterozygous in spite of the fact that he can only be hemizygous in the genetic sense.

Occasionally there are humans who have one blue eye and one brown eye. The most likely explanation for this mosaicism is a somatic mutation. It is probable that the zygote was heterozygous for the brown and blue alleles. Since brown is dominant to blue, such a person would have brown eyes. But if, early in development, a mutation occurred from brown to blue in a cell that gives rise to the cells of one eye, such a person would have one brown and one blue eye.

Mutation in Microorganisms

The Stadler experiment in corn shows us that there is a significant technical problem in the study of mutation. Spontaneous mutation rates are low; and with experiments in such organisms as Drosophila, corn, or mice, the task of revealing the nature of mutation is almost insurmountable. For this reason, people interested in mutational studies turned their attention to the microorganisms, particularly bacteria such as *E. coli.* It is possible to culture a large number of bacteria on a single Petri plate, and the following kind of experiment can be visualized. It is known that many bacteria are sensitive to various antibiotics, such as streptomycin. Plate a large number of bacteria on a petri plate on medium that contains streptomycin. It is expected that most of the bacteria would be killed; but when the experiment is performed, a few bacteria survive and give rise to colonies resistant to streptomycin that can be maintained as resistants for many generations.

It might appear at first that the resistant colonies arose by mutation and that bacteria represent the organism in which mutation studies can be easily performed. However, there are other possible explanations for the occurrence of resistant colonies on the streptomycin plate. Until the various hypotheses can be sorted out, bacteria cannot be exploited as a genetic organism.

What are these different explanations? First, it can be argued that *adaptation* has occurred. It is possible that some organisms are able to adapt their physiology to the presence of the antibiotic streptomycin, which raises the corollary question of why the adaptation persists so long. For it can be shown that the newly arisen resistant cells maintain their resistance for many generations, even in the absence of streptomycin. If adaptation is rejected as an explanation and mutation is considered as an alternative, there are still two possibilities. Is it a *directed* mutation that is caused by the influence of an external agent? Or is it a *spontaneous, preexisting* mutation? Did the mutation actually occur before the application of the external agent, suggesting that bacteria, like higher organisms, mutate at a certain low but spontaneous rate? Once the mutation has occurred, of course, it is quickly selected for in a streptomycin environment. The question is which hypothesis is correct: adaptation or mutation? And if mutation, is it directed or spontaneous?

Fluctuation Test

The answer was provided in an elegant experiment by Luria and Delbrück in 1943. The experiment is called the *fluctuation test.* It is historically interesting to point out that the fluctuation test represents one of the beginnings of molecular biology, for it clearly established the utility of microorganisms for genetic experiments. In their experiment Luria and Delbrück subjected bacteria to T_1 bacteriophage, but the general principle is the same as outlined in the hypothetical example with streptomycin resistance. Luria and Delbrück inoculated *E. coli* into a number of test tubes in which they had placed liquid culture medium. The tubes were incubated for a given period of time. After incubation, equal samples were plated on Petri plates that contained a large number of T_1 phage. In a second part of the experiment *E. coli* was placed in one test tube, and this single tube was incubated. At the end of incubation, equal samples were taken from this one tube and plated on plates with T_1 phage. Luria and Delbrück reasoned as follows. If the phenomenon is adaptation or directed mutation, then the number of mutant colonies on the first set of plates should be roughly the same from plate to plate. Adaptation to the external agent, T_1 phage, or induced mutation by it should cause about the same number of resistant cells in each tube within errors of random sampling. The same argument would apply for the second set of plates, which are merely random subcultures from a single tube. On the other hand, Luria and Delbrück reasoned that if the phenomenon is preexisting spontaneous mutation, it would be expected that a mutation could occur at any generation in the first set of tubes that were incubated over a period of time. In some tubes a mutation to resistance might occur very early, giving rise to a large number of phage-resistant cells. In other tubes the mutation might have occurred about midway during the incubation period. Finally, in some other tubes the mutation, if it occurred, may have occurred very late in the incubation period, so that there are very few phage-resistant cells. In other words, if it is spontaneous, preexisting mutation, the prediction would be that the mutation can occur at any time during the incubation period. As a result, the number of resistant cells from tube to tube, and therefore on the plates cultured from each tube, should vary greatly. In some tubes where mutation occurred early there would be a large number of resistant cells, whereas in other tubes where the mutation, by chance, occurred late during the incubation period, there would be very few. On the second set of plates that came from the single tube, it would be argued that the number of resistant cells from plate to plate should be relatively the same, for all represent random samples from a single tube. Whatever number of resistant cells was present in the single tube should be distributed randomly among the plates.

Luria and Delbrück measured the number of resistant cells on each plate in the two parts of the experiment and applied a statistical test known as a *variance analysis* to the results. Without bothering to explain the statistical analysis in detail, it can be asserted that if there is little difference in the number of resistant cells from plate to plate, the variance will be low, whereas if there are large differences from plate to plate, the variance will be large. Variance analysis is simply a statistical measure of variability. Luria and Delbrück predicted that if they were dealing with adaptation or directed mutation, the variance would be relatively low in both sets of plates, whereas if they were dealing with preexisting mutation, the variance would be high in the first set of plates and low in the second set of plates that represent random subcultures from a single tube. The results of their experiment are shown in Table 14-2. In the experiment with individual cultures, the variance is very high. In contrast, the plates that came from a single tube gave a low variance. These results led Luria and Delbrück to conclude that resistance to T_1 phage was due not to adaptation or directed mutation but to preexisting mutation, random and spontaneous. Thus they established the validity of such organisms as *E. coli* for genetic analysis.

Table 14-2. The Results of Luria and Delbrück's Fluctuation Test[a]

Subsamples from Single Tube		*Individual Culture*	
SAMPLE NUMBER	NUMBER OF RESISTANT COLONIES	SAMPLE NUMBER	NUMBER OF RESISTANT COLONIES
1	14	1	10
2	15	2	18
3	13	3	125
4	21	4	10
5	15	5	14
6	14	6	27
7	26	7	
8	16	8	17
9	20	9	17
10	13		
Average number of resistant colonies	16.7		26.8
Variance	15		1217

FROM LURIA AND DELBRŨCK. 1943.

[a]Although the mean number of resistant cells is similar in the two tests, the variance is low in the single-tube experiment and very high in the individual cultures. This led to the conclusion that resistance to T_1 phage in *E. coli* comes about through spontaneous mutation.

Spreading Test

Six years later the Canadian geneticist, Newcombe, in a similar but more straightforward and easily understood experiment, reaffirmed the concept of preexisting mutation in *E. coli*. His technique was as follows. On a large number of Petri plates, he spread *E. coli* and allowed them to incubate for a given period of time. After the incubation period, he redistributed the cells on every other plate by spreading the bacteria with a mechanical spreader. It should be noted that the number of cells per plate will not be changed by this mechanical spreading; only the distribution of cells will be altered. After spreading every other plate, Newcombe introduced the external agent, T_1 phage, to each plate, spread and unspread.

His argument is: If the phenomenon is adaptation or directed mutation, then the number of resistance colonies should be the same on each plate, for the mechanical redistribution should not, in any imaginable way, affect adaptation or induced mutation. On the other hand, if the mutation is spontaneous and preexisting, then it can be argued that on some plates mutation occurs early in the incubation and gives rise to a colony with a large number of resistant cells. When the phage is introduced to this unspread plate, although there are a large number of resistant cells because of the early mutational event, only one resistant colony will grow on the plate due to the one mutational event. On the other hand, if the cells on the plate are mechanically spread and redistributed, then a large number of resistant cells from one colony has been spread about the plate. The addition of phage to this spread plate would lead to the expectation that there are many resistant colonies on the plate, as a result of the redistribution of the resistant cells in a colony by mechanical spreading. Therefore Newcombe predicted that if adaptation or directed mutation was the cause of resistance to phage, he would expect to find approximately the same number of resistant colonies on the spread and unspread plates. On the other hand, if spontaneous, preexisting mutation was the cause of resistance to phage, then he should find many more resistant colonies on the spread plates than on the unspread plates. An examination of Table 14-3 shows that the latter hypothesis is correct. There were many more resistant colonies found on the spread plates, reaffirming what Luria and Delbrück had already shown. There can be no doubt that mutations in bacteria are random and spontaneous and apparently of the same nature as in higher organisms.

Table 14-3. Resistance of *E. coli* to T_1 Phage on Spread and Unspread Plates[a]

		Number of Resistant Colonies	
INCUBATION PERIOD (IN HOURS)	FACTOR INCREASE IN GROWTH	UNSPREAD	WIDESPREAD
5	5,100	28	353
6	54,900	240	12,638

AFTER NEWCOMBE, 1949. *NATURE* **164**.

[a]If resistance is due to adaptation or directed mutation the number of resistant colonies should be the same on the spread and unspread plates. The fact that this is not so leads to the conclusion that resistance is due to spontaneous preexisting mutations.

Replica Plating

Some people remained unconvinced by the Luria-Delbrück and Newcombe experiments. The skeptics argued that regardless of how the experiment was designed, addition of the external agent was necessary to screen for resistant cells, and hence some kind of adaptation or direct mutation could not unequivocally be eliminated as a hypothesis. Although such skepticism seems unwarranted in light of the two experiments just described, it is possible, due to the cleverness of Lederberg, to make even this criticism untenable.

Lederberg recognized that velveteen has a very stiff nap and that it can be used to inoculate bacteria from one plate to another, because the velveteen will substitute for a "thousand" inoculating needles. With this technique, it is possible to make a number of exact replicas of a plate of bacteria. (See Figure 14-2.) Lederberg spread some *E. coli* onto a Petri plate that had no phage added to it. He then

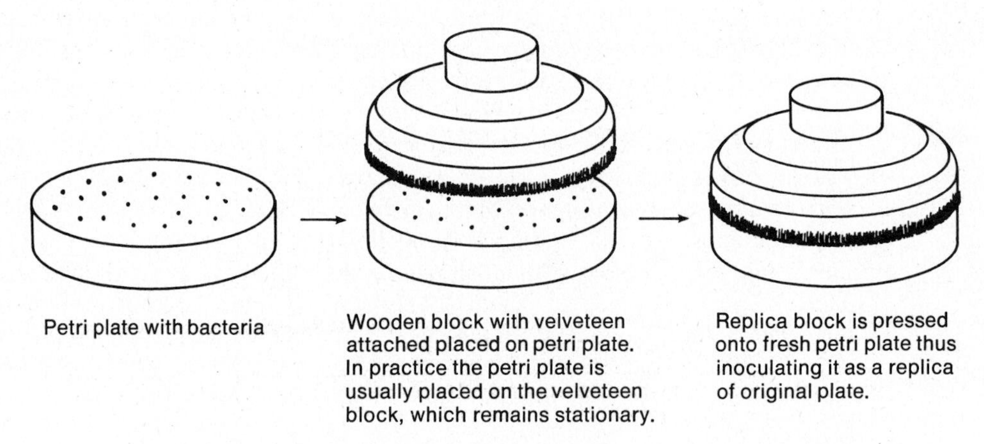

Figure 14-2. The technique of replica plating. The velveteen acts as many inoculating needles in transferring bacteria from one plate to another. Each plate should be given a constant reference mark so that exact replicas can be obtained.

replicated this plate several times on new plates with phage added. A few resistant colonies were noted on the phage plates, always in the same places, indicating that the mutations were already present on the master plate to which no phage had been added. Returning to the original plate, he picked a few cells from the area where resistant colonies showed up on the phage plates. He recultured these cells and spread them, in lower concentration, on a nonphage plate from which he again made several replicas onto plates with phage added. By repeating this process several times, Lederberg ultimately obtained a culture of *E. coli* that was completely resistant to phage, although none of the cells had been exposed to the external agent! (See Fig. 14-3.) Such an experiment effectively rules out adaptation or directed mutation as the source of phage-resistant *E. coli*.

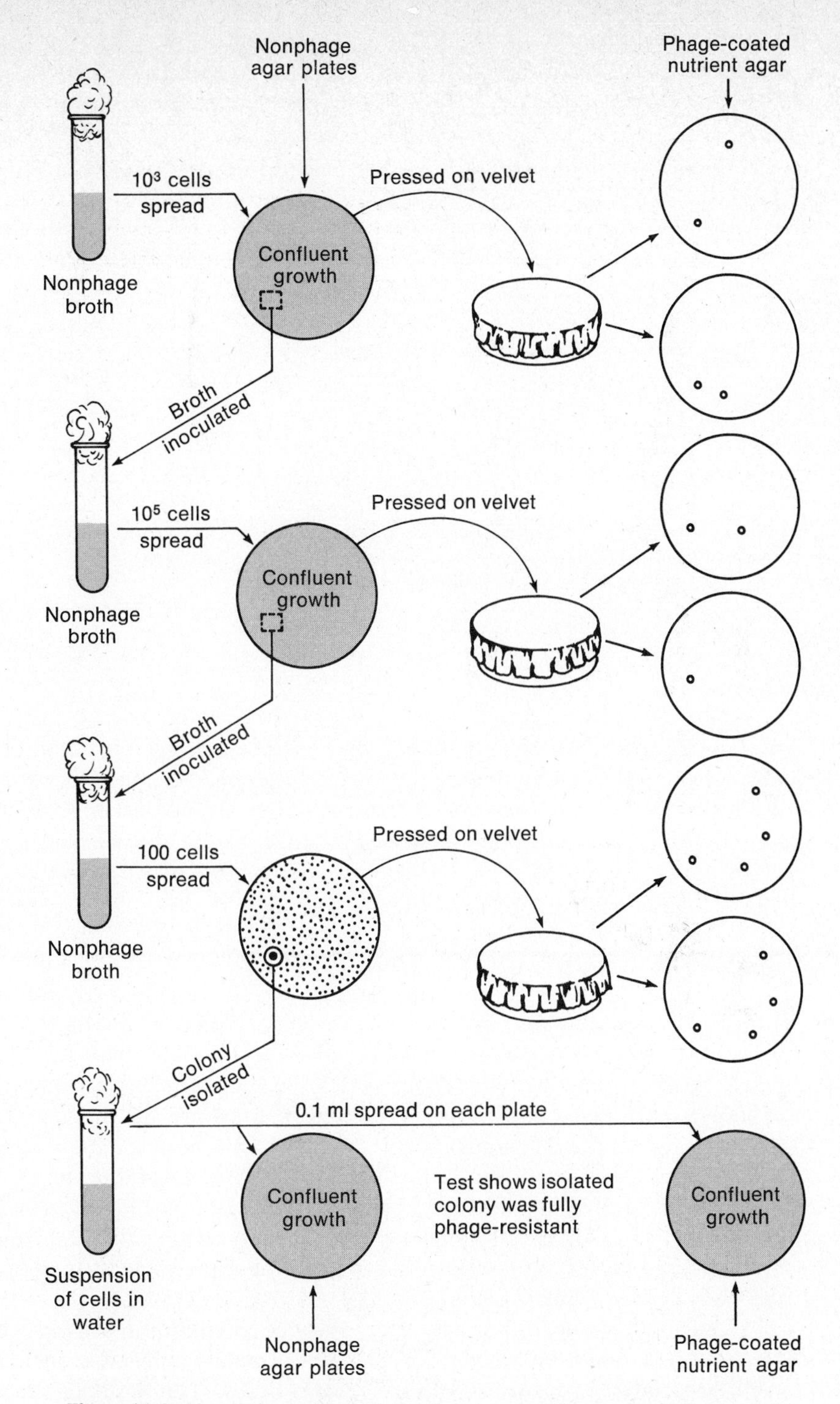

Figure 14-3. Use of the replica plate method to prove the nondirected nature of bacterial mutation. The bacteria are successively transferred between nonphage broth and nonphage agar (left side of diagram). At each stage, replica plates are made on the phage agar (right side of diagram). The final suspension of bacteria is found to be completely phage resistant, as is indicated by confluent growth on the two plates at the bottom of the diagram. (After Stanier, Doudoroff, and Adelberg. 1970. *The Microbial World*, 3rd ed. Prentice-Hall, Englewood Cliffs, N.J.)

Once it became established that bacteria mutate similarly to higher organisms, it was possible to measure spontaneous mutation rates. (See Table 14-4.) Bacterial studies reveal the same general picture as observed by Stadler with corn. Spontaneous mutations are recurrent and occur with low frequency.

Table 14-4. Some Spontaneous Mutation Rates in Bacteria

BACTERIUM	SCREENING AGENT	RATE OF MUTATION
E. coli	resistance to T_1 phage	1×10^{-3}
E. coli	resistance to T_3 phage	1×10^{-8}
E. coli	resistance to T_6 phage	1×10^{-7}
E. coli	resistance to radiation	1×10^{-5}
Staphylococcus aureus	resistance to penicillin	1×10^{-7}
Staphylococcus aureus	resistance to sulfathiazole	1×10^{-9}

Induced Mutations

As soon as geneticists understood that the genetic material was capable of mutation, they looked for ways to influence the rate of mutation. The most successful and one of the first in this search was one of Morgan's original students, H. J. Muller. Muller left the Morgan group in 1919 and went to Texas, where he began a series of experiments that ended in 1927 with publication of his classical study of the influence of ionizing radiation on the genetic material.

The ClB Technique

Muller, a Drosophila geneticist, realized that there were two major difficulties in studying mutation rate. First, mutation rate is low and therefore he needed an efficient method for detecting mutations. Secondly, Muller realized that discriminating among several similar phenotypes is subjective. If eye colors or bristle lengths in Drosophila are the object of study, some phenotypic changes will be clear to anyone, but there might be other, more subtle changes that would be scored one way by one observer and a second way by another. Since Muller wanted the best possible chance of observing mutations, he did not want to eliminate the marginal phenotypic changes; but he realized that their classification was slow and difficult. Consequently, Muller decided to pick lethal mutations for his test material; and to simplify the experiment further, he restricted the analysis to *sex-linked recessive lethals.*

As with many other important genetic breakthroughs, the elegance of Muller's experiment lay in its design. It took him a number of years to obtain the stock he needed, a strain of Drosophila called *ClB*. The **B** stands for Bar eye, a semidominant sex-linked mutation that affects the size of the eye. The **l** stands for a recessive lethal on the X chromosome, and the **C** symbolizes an inversion of the X chromosome. In Chapter 13 it was pointed out that, in heterozygous inversions, there is little, if any, recombination. Figure 14-4(a) and (b) shows the use of the ClB technique in detecting sex-linked recessive lethals in Drosophila. Several points should be noted about this experiment. First, it is essential that in mating the F_1 ♀, only a single female is used in each vial. Mating only one female per vial means that each F_2 vial represents a test of one X chromosome from the parental male. Note also the importance of the semidominant gene Bar. From the parental cross, two kinds of F_1 females are produced; but it is the female with the ClB chromosome, containing the sex-linked recessive lethal and the inversion, that must be used as the female parent. The marker Bar makes possible the distinction between this female and her wild-type sister. Finally, note that, in the F_2, half the males die because of the presence of this ClB chromosome—specifically, the sex-linked recessive lethal that it carries. Thus a 2 : 1 ratio of females to males is observed under normal breeding conditions.

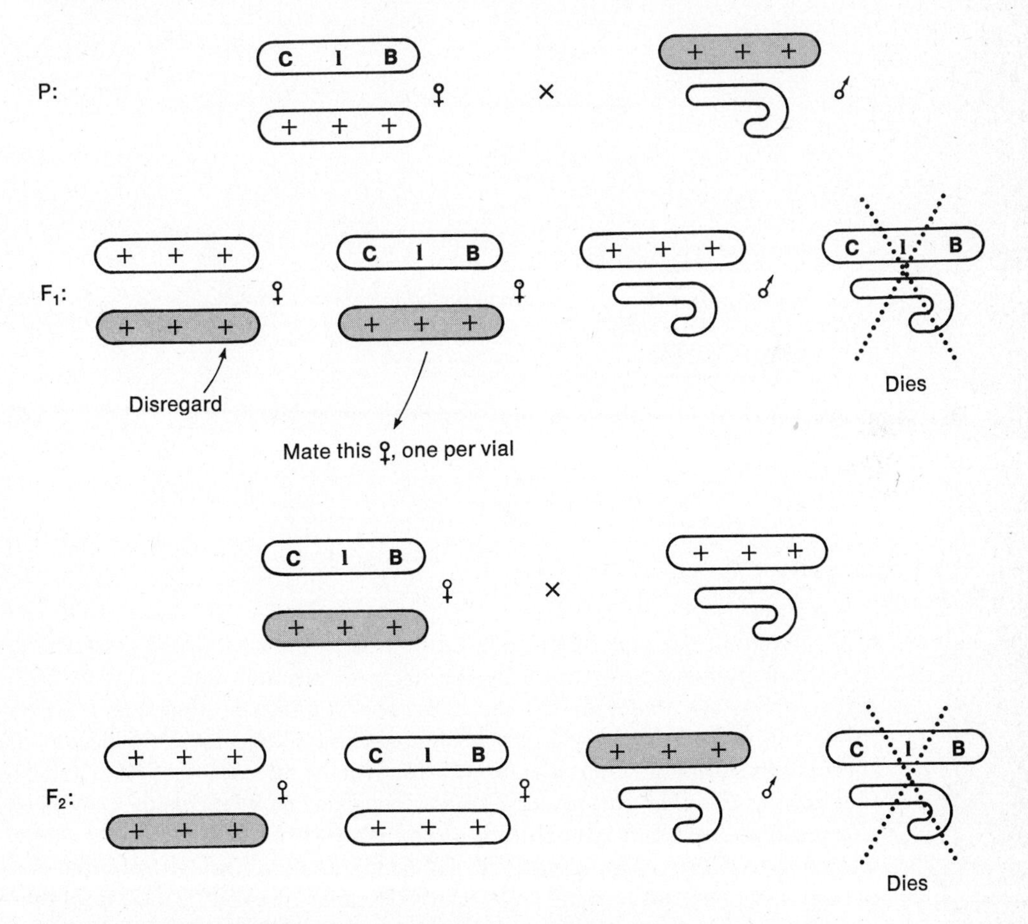

Figure 14-4(a). Muller's ClB technique for detecting sex-linked recessive lethals in Drosophila. If no lethal is produced, the F_2 sex ratio is 2♀ : 1♂.

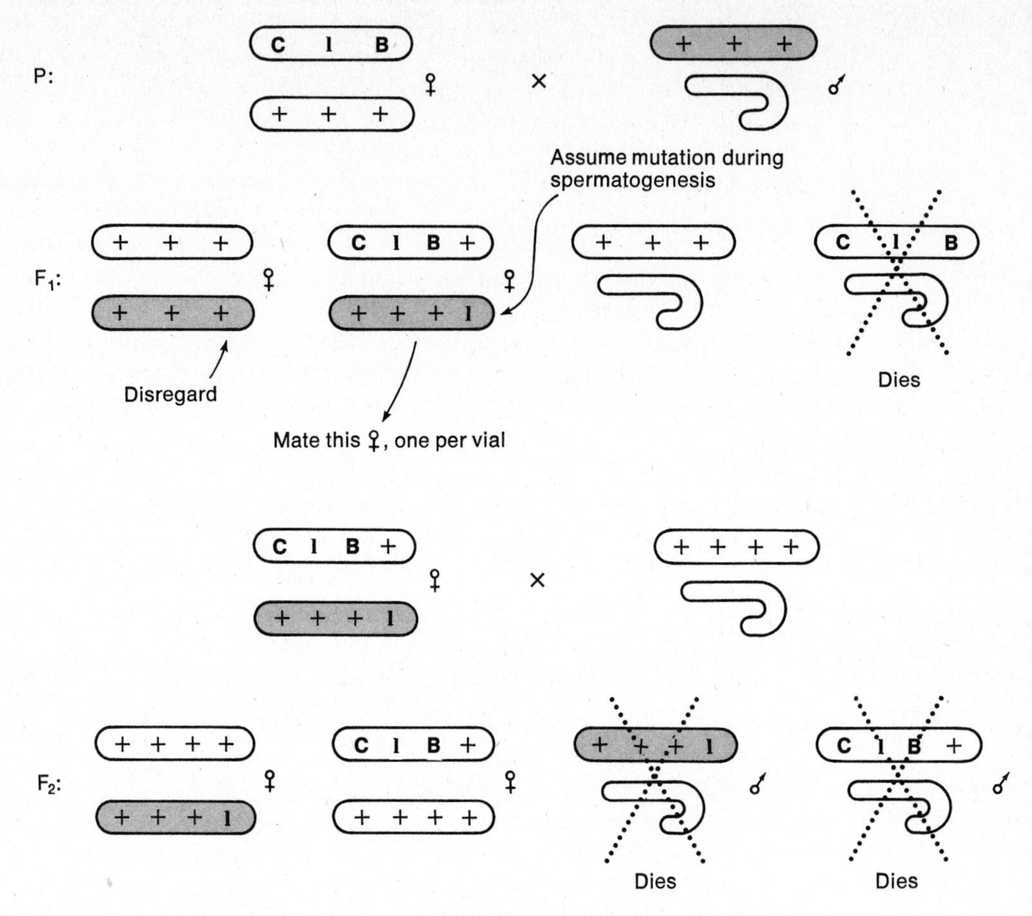

Figure 14-4(b). Muller's ClB technique for detecting sex-linked recessive lethals in Drosophila. If a lethal occurs during sperm production of the parental male, it will be detected by the presence of an F_2 culture without male progeny.

Examine the experiment now with one further assumption. Assume that, in the production of spermatazoa by the original parental male, a mutation occurs occasionally such that a few of the many spermatazoa produced contain a newly formed sex-linked recessive lethal. If the assumption that a lethal has occurred in the sperm of the original parental male is correct, all the males in the F_2 die: half because they receive the ClB chromosome and half because they receive the newly mutated chromosome from their grandfather. (See Figure 14-4b.) Therefore Muller had an objective test for scoring mutations. He simply looked for vials in which no males occurred. It also turns out to be a relatively rapid method; Muller could screen a large number of vials in a short time, thus accomplishing the objectives he set for himself with this experimental design.

The significance of the crossover suppressor should now be evident. It is essential that, in the F_1 female, the chromosomes retain their original integrity and that the original order not be lost through crossing over. The test relies on the fact that one lethal chromosome is the original ClB chromosome and the other lethal chromosome comes from the original father. If free recombination existed in the F_1 female, these chromosomes could easily recombine and the test would fail because some wild-type males could be reproduced by recombination.

Muller's ingenious experimental design allowed him to test the frequency of spontaneous mutations in Drosophila. Muller, who maintained a lifelong interest in the process of mutation, also designed other stocks that enabled him to test the frequency of other mutations besides sex-linked recessives. Table 14-5 contains the results not only from his ClB experiment but also from other experiments on spontaneous mutation rates in Drosophila. The data, which are not mutation rates per locus but for an entire sperm, suggest that on the order of 58 new mutations per 1000 sperm are produced spontaneously in Drosophila. To express the data in a slightly different way, approximately 4 to 8 % of all sperm have a new mutation in them, which probably represents an underestimate, since all possible types of mutations are not examined. This is an interesting figure to reflect on, particularly if it were true for human populations in roughly the same order of magnitude.

Table 14-5. Spontaneous Mutations in Drosophila[a]

MUTATION	FREQUENCY PER THOUSAND SPERM
Sex-linked recessive lethals	2
Autosomal recessive lethals	10
Autosomal viability genes	36
Visible genes	< 10
	58 = 5.8 %

[a] At face value, the data indicate that from 4 to 8% of all sperm produced have a new mutation.

Effect of X rays

As noted, one question that geneticists were interested in, Muller in particular, was whether or not external agents, such as ionizing radiation, would have any influence on the mutation rate. The design of the ClB experiment made it a relatively simple task for Muller to test this hypothesis. He used the ClB method with the experiment divided into (1) a "control," in which the parental male was untreated and (2) a "treated," in which the original parental male was exposed to ionizing radiation. He then compared the number of mutations obtained in the two experiments by analyzing the data to see whether any difference that occurred was statistically significant.

Muller observed that ionizing radiation did increase the frequency of mutation. He obtained sex-linked recessive lethals at a rate of 20 mutations per 1000 sperm tested if the original male was exposed to 1000 roentgens (r) of X ray. This compares to a rate of two sex-linked recessive lethal mutations per 1000 sperm in the control (Table 14-5).

Linearity of Mutation Curve

Muller and his collaborators undertook an extensive experimental program that lasted many years, one in which the additional effects of radiation on the genetic material were tested. One of the first items of interest to them was the mutation curve they obtained when the dosage of radiation was varied. How many mutations would be observed at 1000 r, 2000 r, 3000 r, and so on? The results of many

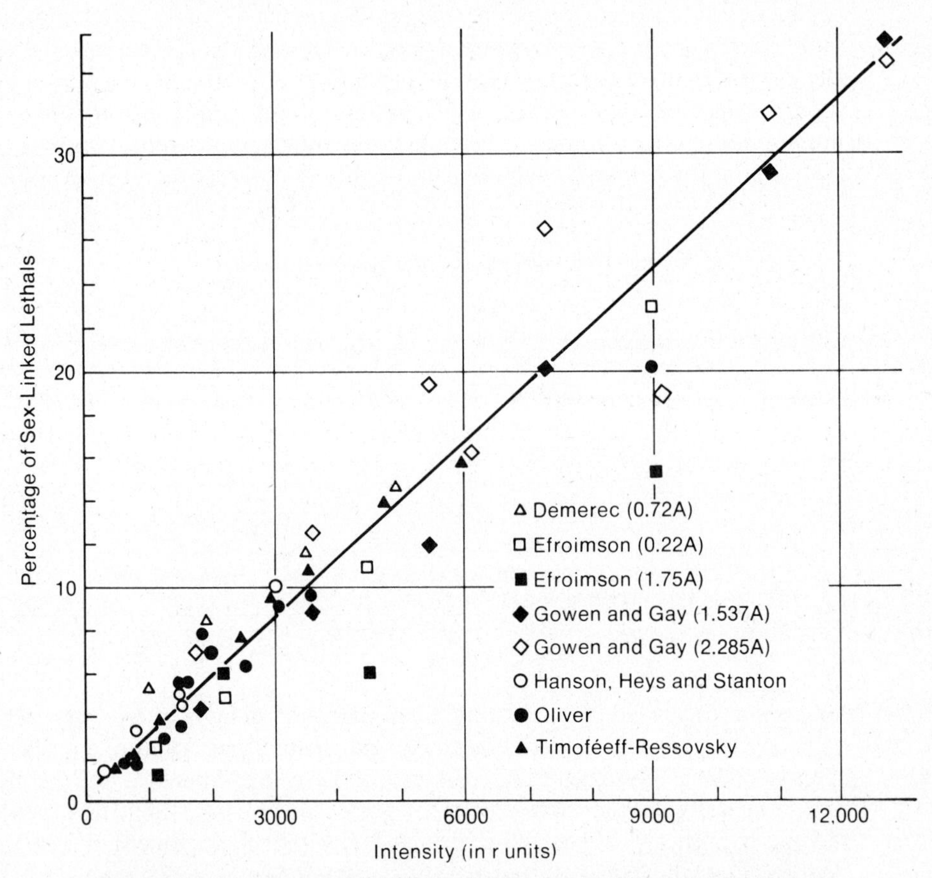

Figure 14-5. Percentage of lethals in the X chromosome of *Drosophila melanogaster* in relation to the intensity of radiation. (After J. Schultz. 1936. Radiation and the Study of Mutation in Animals. In Duggar, ed. *Biological Effects of Radiation.* McGraw-Hill. p. 1029–1261.)

investigators revealed the same simple conclusion. (See Figure 14-5.) The curve is linear, which means that the frequency of sex-linked recessive lethal mutations is proportional to the given dosage of radiation. If the curve were extrapolated down to the vertical axis, it would be seen that the curve does not go through the origin. This difference between the horizontal axis and the point of intercept of the extrapolated curve represents the spontaneous mutation rate.

The ending of World War II with the explosion of two atomic bombs in Japan marked the beginning of the atomic age and an increasing awareness of radiation. Geneticists aware of the linear curve became interested in the effects of low-dosage irradiation. Does the curve remain linear as the dose falls to 100 r, 50 r, 25 r, and less? It should be apparent to the reader that when the experiment is performed at lower doses, the size of the test needed to distinguish a statistically significant difference becomes large. However, the importance of the question made it necessary for the experiment to be performed; and Figure 14-6 presents the results. It is clear that linearity is still the rule to doses as low as 25 r. Other investigators have extended this generalization to as low as 5 r. During the years immediately after World War II, both the United States and the Soviet Union conducted tests of atomic and hydrogen devices in the atmosphere that resulted in radioactive fallout over many populated areas. Many geneticists were concerned about the mutagenic effects of the fallout even though the dosages were considerably lower than 1 r. The linearity of the dose-rate curve convinced them that the fallout, regardless of how small, represented an additional mutation rate and hence a hazard. A few scientists argued the contrary, suggesting that at such low dosages there might be a threshold effect for genetic damage, as had been shown to be the case for physiological effects of radiation, and therefore concern over mutagenic effects was unwarranted. Since the curve has proved linear every time a lower-dose experiment was tried, it would seem that the burden of proof falls on those who believe in a threshold effect. It would be more prudent to assume that the curve is linear for any dosage of irradiation, although this point has not been experimentally proved.

Chronic Versus Acute Radiation

Another interesting question asked by Muller and others was whether there is any difference in mutation rate if the method of providing the irradiation is altered. Muller wondered whether the 2% sex-linked recessive lethals obtained with a dose of 1000 r would be the same if the radiation were given in 10 equal doses of 100 r over 10 days rather than all at once. Stated another way, is the mutation rate following *acute radiation*, the dose given all at once, the same as the mutation rate following *chronic irradiation*, the same dose broken up over a period of time? When Muller tested this in Drosophila, he found that the mutation rate was the same regardless of how the dose was given. A thousand roentgens gave approximately 2% sex-linked recessive lethals per sperm whether given all at once or chronically. This finding is significant because it assumes that all radiation, regardless of how low the dose or how it is given, results in more mutations. Moreover, most geneticists assert

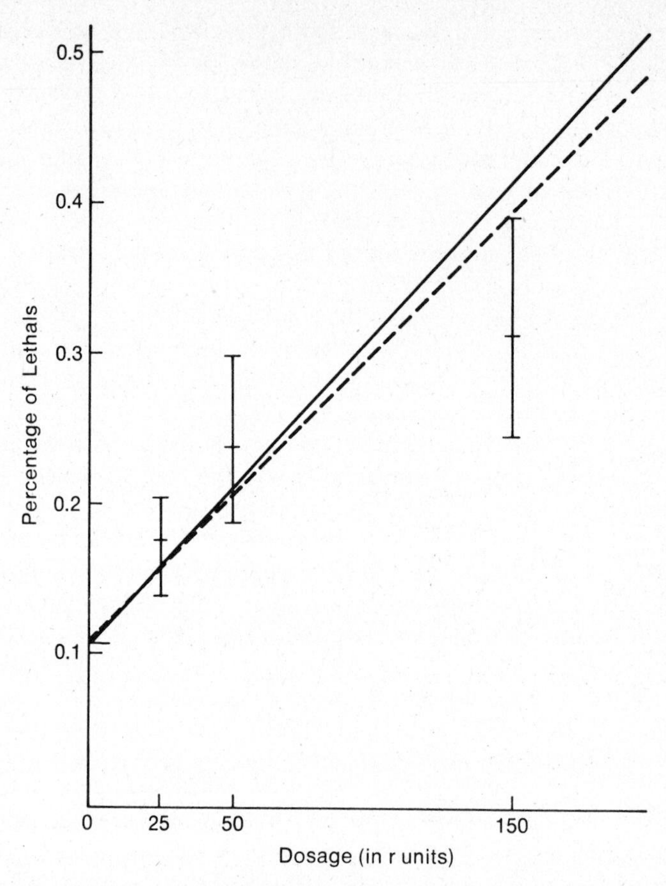

Figure 14-6. The linear relation between r-dose and lethal mutation rate. The solid line is based on all data in Table 14-3; the broken line is based on the control through 1000 r data, omitting the 2000, 3000, and 4000 r data. The data points for control, 25, 50, and 150 r are shown as small central horizontal marks. The ends of the vertical lines represent values of which the observed would be the upper and lower 5 % fiducial limits. (After W. P. Spencer and C. Stern. 1948. *Genetics* **33**.)

that the overwhelming majority of mutations is deleterious. (Some of the arguments for this assertion are in Chapter 29.) The danger of an increased number of mutations was expressed by many geneticists during the period of atomic testing when radioactive fallout of very low magnitudes was experienced almost every day. It was projected that a newborn infant who was subjected to low chronic radiations might, 20 to 30 years later at the reproductive age, have accumulated a rather significant dose. And if it is true in humans for low dosages as it is in Drosophila at higher dosages—that the net effect of radiation is the same regardless of how the dose is given—mankind then faces a question of considerable magnitude.

It is difficult to be too dogmatic about the effects of radiation, for different organisms have different responses and experiments have been performed in the mouse that seem to indicate a difference between chronic and acute radiation, with acute radiation giving the higher mutation rate. It is not possible here to discuss all the variables involved in the effects of ionizing radiation; the reader is referred to the literature for a more extensive treatment.

Chemical Mutagenesis

It should also be pointed out that other kinds of radiation, in particular, ultraviolet, have been thoroughly tested for genetic effects. All affect mutation rate by increasing it. During World War II Charlotte Auerbach made the important discovery that chemicals might also act as mutagens. She tested the effect of mustard gas, but soon other investigators showed that numerous substances, including formaldehyde and hydrogen peroxide, were effective. Depending on the substance and how it was administered, both point mutations and chromosomal aberrations might result.

The discovery of chemical mutagenesis has important implications. As our society becomes more technological and increasingly reliant on drugs and chemicals as medicinal aids and food preservatives, to give only two examples, it must be ever-vigilant against genetic side effects, for the number of chemicals known to act as mutagens is large. Secondly, the discovery of a way to alter the genetic material chemically excited geneticists, for it offered hope in solving the important questions of what the genetic material is made of and how it functions (Chapter 23).

REFERENCES

AUERBACH, C. 1967. The chemical production of mutations. *Science* **158**: 1141–1147. (Available in *Papers on Genetics*, edited by L. Levine. C. V. Mosby Co., St. Louis, 1971.) (A review of chemical mutagenesis, with a molecular interpretation by the person who first developed this field of study.)

LEDERBERG, J., and E. M. LEDERBERG. 1952. Replica plating and indirect selection of bacterial mutants. *J. Bact.* **63**:394–406. (The introduction of an invaluable technique for microbiologists and also the convincer that bacteria have spontaneous mutation.)

LURIA, S. E., and M. DELBRÜCK. 1943. Mutations of bacteria from virus sensitivity to virus resistance. *Genetics* **28**:491–511. (Available in *Papers on Bacterial Genetics*, 2nd ed., edited by E. A. Adelberg. Little, Brown and Company, Boston and Toronto, 1966.) (One of the first articles published on molecular biology, this paper made possible most of the microbial genetics that followed.)

MULLER, H. J. 1927. Artificial transmutation of the gene. *Science* **66**:84–87. (Available in *Classic Papers in Genetics*, edited by J. A. Peters.

Prentice-Hall, Englewood Cliffs, N.J., 1959.) (This paper, for which Muller was awarded the Nobel Prize in 1946, demonstrated that ionizing radiation affects gene mutation.)

MULLER, H. J. 1950. Our load of mutations. *Am. J. Hum. Genet.* **2**:111–176. (A very important paper summarizing radiation genetics and particularly pointing out the hazards of radiation for humans.)

NATIONAL ACADEMY OF SCIENCES. 1960. *The Biological Effects of Atomic Radiation.* 90 pp. National Academy of Sciences, Washington, D.C. (A good, short, readable review of the possible hazards of radiation for man.)

NEWCOMBE, H. B. 1949. Origin of bacterial variants. *Nature (London)* **164**:150. (Another approach that demonstrates that spontaneous mutations occur in bacteria.)

SCHULTZ, J. 1936. Radiation and the study of mutation in animals. In *Biological Effects of Radiation*, edited by B. M. Duggan. McGraw-Hill, New York, pp. 1209–1261. (Although this paper was written many years ago, it is still an excellent review of the early and exciting work on mutation in Drosophila.)

SPENCER, W. P., and C. STERN. 1948. Experiments to test the validity of the linear r-dose/mutation frequency relation in Drosophila at low dosage. *Genetics* **33**:43–74. (This paper tested the hypothesis that the dose-rate curve is linear even at low dosages and found the hypothesis true.)

STADLER, L. J. 1942. Some observations on gene variability and spontaneous mutation. *The Spragg Memorial Lectures (Third Series)*, Michigan State College, pp. 3–15. (A review of spontaneous mutation in corn with a summary of the results Stadler obtained in nearly 20 years' work.)

QUESTIONS AND PROBLEMS

14-1. Define the following terms:

point mutation
reverse mutation
mosaic
somatic mutation
fluctuation test
replica plating
induced mutation
acute radiation
chronic radiation

14-2. Assume that you have available two strains of homozygous plants, AA and aa, affecting flower color. Design an experiment for efficient detection of mutations from

(a) A to a.
(b) a to A.

14-3. Assuming the detection of such mutations (Problem 14-2), how would you calculate the rate of mutation from

(a) A to a?
(b) a to A?

14-4. When early embryonic stages of male Drosophila from homozygous red-eyed females are X rayed, the adults occasionally show large areas of their eyes not red but white in pigmentation. When later embryonic stages are irradiated, mosaic eyes of the adults show only small white spots. What is the cause of

(a) the appearance of white areas?
(b) the difference in size of white areas?

14-5. A certain Drosophila male is the offspring of a homozygous wild-type (gray body color) female and a wild-type male. Assume that a mutation to the recessive sex-linked allele yellow-body color had occurred in the fertilized egg nucleus with which the male started its development. What would be the phenotype of the male?

14-6. Had the mutation occurred in one of the first two cleavage nuclei, what would you expect to be the phenotype of the male?

14-7. If the male were mated to a homozygous yellow female, what would you expect regarding the phenotypes of the offspring in

(a) Problem 14-5?
(b) Problem 14-6?

14-8. In Problem 13-2, an F_1 female exception had cut wings but was wild-type for the other three genes.

(a) What is another possible explanation (different from the one suggested in Problem 13-2) for this exceptional female?
(b) How would you verify this hypothesis?

14-9. What is the significance for geneticists of Luria and Delbrück's fluctuation test?

14-10. Assuming that bacteria mutate in the same way as other organisms (that is, that newly arising mutant colonies are not due to adaptation or directed mutation), design an experiment in bacteria using Lederberg's replica-plating technique to detect new mutations.

14-11. In Muller's ClB technique for the detection of sex-linked recessive lethals in Drosophila, what is the significance of the following?
(a) C.
(b) l.
(c) B.

14-12. If a recessive lethal mutation is produced on the X chromosome of a sperm that fertilizes an egg that has a ClB chromosome, will the resulting individual live? Explain your answer.

14-13. If a recessive lethal mutation is produced in a sperm at exactly the same locus as the sex-linked recessive lethal contained on the ClB chromosome, would not the experiment be ruined or the results compromised? Why or why not?

14-14. How would you design an experiment in Drosophila for the detection of dominant lethals?

14-15. In Drosophila, an X-ray dose of 1000 r given over a period of 5 minutes induces about 2 % sex-linked lethal mutations in the sperm. What percent of mutations are induced if

(a) the same total radiation is given within one minute?
(b) 10 equal fractions of 200 r, each one hour apart, are given?
(c) a total dose of 1 r is given, distributed over a whole hour?

14-16. Although there is general agreement that mutation rate varies linearly with dose for dosages of, say, 50 r or higher, there is still some disagreement whether linearity applies for very low dosages, say less than 1 r. Why, do you think, unequivocal data for or against linearity at low dosages are not available?

14-17. If one assumes that the atomic bombs dropped on Japan had a mutagenic effect on the exposed survivors, what effect on the sex ratio might be observed in the progeny of

(a) exposed women?
(b) exposed men?

15/Gene action

Influence of Environment

It was not long after the discovery of Mendelism that geneticists, having satisfied themselves about the mechanics of genetics, became interested in the action of genes. As shall be seen, many of the early investigators had shrewd insights into the way genes acted, but it was many years before definitive evidence was available. Early students of gene action focused on the general field of chemical reactions. There were many reasons for this focus, but one, in particular, derived from early work in the gene action field in which the end product of a particular gene or genes was altered by various environmental agents. A look at some of these environmentally controlled experiments may indicate if any conclusions can be drawn from them.

Temperature Effect in Flowers

It should not be surprising that one of the most easily controlled environmental agents is temperature. Also unsurprising is the information that many of the early experiments involved flowers, because so much was known about the genetics of flower color. In the primrose, flower color is controlled by a pair of alleles, **A** and **a**. The former results in red flowers, whereas the homozygous recessive gives white flowers. This is true in plants raised at a temperature of 20 to 25°C. However, if the plants are raised at a higher temperature of 30°C, with accompanying high humidity, then, regardless of the genotype, **A** or **a**, the plants have white flowers. It would seem that the action of genes can be modified by temperature.

Temperature Effect in the Himalayan Rabbit

An intriguing experiment that attempted to gain some insight into gene action was done on the pigmentation pattern of the Himalayan rabbit by a German pediatrician named Schultz and others. It was pointed out in Chapter 6, in the discussion of multiple alleles, that the Himalayan rabbit is essentially an albino rabbit that has pigmented ears, nose, tail, and paws. Aside from the question of gene action, there is an interesting problem in differentiation, since theory demands that every cell of the Himalayan rabbit be genetically identical. The question is: Why do some cells give rise to white pigmentation and others to dark? Nor is that the only problem. Why are the darkly pigmented cells in a particular place?

Schultz reasoned that since all the dark markings were at the tips of the animal, it is possible that the temperature of the extremities is cooler because more rapid evaporation is occurring at these surfaces. He made a rather ingenious although crude test of this hypothesis. He shaved off a surface of the rabbit's body, placed the animal in a very cold cellar, and allowed the fur to grow back in that environment. He noted that the hair grew back pigmented. In a reciprocal experiment he shaved the fur from the paws and then bandaged them to keep them warm. Now the fur grew back unpigmented in a normally black region.

Schultz concluded from these experiments that his hypothesis was essentially correct. There is a temperature differential, and the genotype $\mathbf{c^hc^h}$ responds differently to the different temperatures. In a warmer temperature the genotype results in no pigmentation, whereas in the cooler extremities the fur is pigmented. Other workers performed sophisticated experiments involving thermocouples to obtain precise measurements of temperature and confirmed Schultz's general hypothesis.

Temperature Effect in Drosophila

In Drosophila, the sex-linked semidominant mutant Bar, mentioned first in the description of Muller's ClB technique (Chapter 14), is also affected by temperature. The normal eye of Drosophila has somewhat over 700 individual facets that make up the compound eye of this insect. If flies that have the genotype Bar are raised at a high temperature, the eye will have approximately 60 facets. If the same genotype is raised at low temperatures, the eyes will have 200 facets. The wild-type gene is unaffected by variations in temperature.

Does it make any difference when the fly is subjected to the high or low temperature during the developmental period? That is, in order to obtain a small number of facets in the eye, must the fly be raised at a high temperature thoughout its developmental period or will raising it at this temperature for only a short period of the developmental cycle have the same effect? It was shown experimentally that the latter interpretation is correct. If Drosophila are raised at the high temperature for only a very small part of the developmental period, the same phenotype will be observed that was obtained if the fly was raised at the high temperature throughout the larval period. Experiments with Drosophila involving other mutants and similar temperature shocks have verified this general finding, indicating that, in many cases, there is a special period during larval development, known as the *temperature-sensitive period* or *temperature-effective period*, during which the temperature has its effect.

Because Drosophila is so easily manipulated genetically, other interesting hypotheses can be tested. If flies of the Bar genotype are raised at a constant temperature for a long time and then suddenly raised at a quite different temperature, what phenotype would be observed? If, for example, a strain of flies is raised at low temperature for 60 generations and second strain of flies for the same 60 generations is raised at high temperature, what would the eye size be after these 60 generations, if both strains are raised at the same temperature? When this experiment was done, it turned out that both strains, when raised at the same temperature, after many generations at different temperatures, had the same eye size even though, for the previous 60 generations, one strain averaged 200 facets and the other only 60. The thesis that acquired characteristics can be inherited, which was proposed by the French evolutionist J. B. Lamarck around 1800, receives no support from the temperature experiments with Bar eye in Drosophila.

Humidity Effect in Drosophila

Many other examples of temperature effect on genes during development could be cited, but instead the final example of environmental effects on gene action will be another experiment in Drosophila involving a different environmental factor. The experiment was performed by Morgan and involved a sex-linked dominant trait in Drosophila known as Abnormal abdomen. In this mutant strain, there were no well-formed chitinized bands in the abdominal region of the fly. (See Figure 15-1.) Instead the bands are irregularly formed; and although the expression varied considerably from extreme disarrangement to only minor differences in the chitinous bands, the abdomen usually takes on a quite abnormal twisted appearance.

When Morgan made counts of crosses involving Abnormal abdomen genes, he obtained some unusual ratios. If he started his count as soon as the flies began to hatch, he obtained a ratio of wild-type to Abnormal abdomen that he would have expected on the basis of the genetic hypothesis. But as he continued to count on successive days, he observed in the later counts more wild-type flies than expected. Morgan eventually concluded that the expression of the gene Abnormal

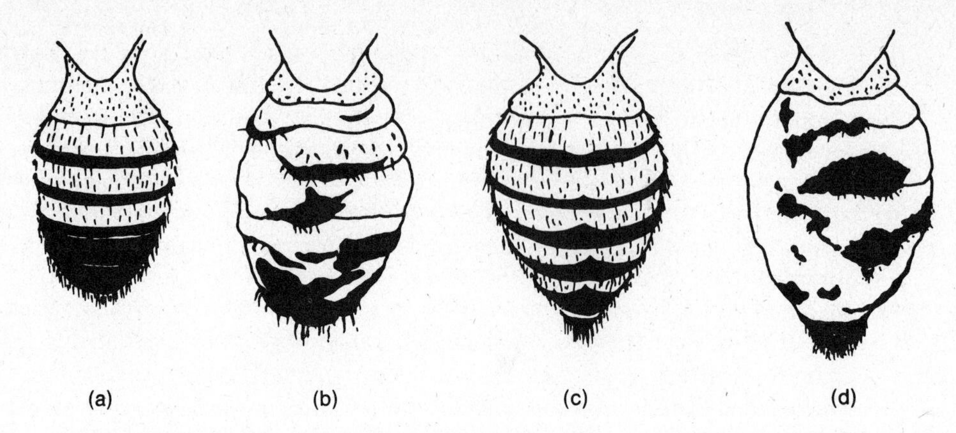

Figure 15-1. Normal and abnormal abdomen of *Drosophila melanogaster.* (After T. H. Morgan. 1919. *The Physical Basis of Heredity.* J. P. Lippincott Co.)

abdomen depended on the amount of moisture in the culture; and it was well known to Drosophila geneticists that the older the culture bottle, the drier the culture medium.

Morgan concluded that the wild-type allele gave a normal phenotype whether the conditions of the culture were wet or dry. The mutant allele, however, gave an abnormal abdomen when the culture conditions were relatively wet but gave normal or nearly normal flies as the culture dried. This result explained the unexpected ratios that Morgan observed in the later counts. As the culture medium dried, more wild-types than expected were seen.

Some years later another gene that gave approximately the same abnormal abdomen phenotype was discovered. This gene, however, was not an allele of the dominant sex-linked Abnormal abdomen gene just described but proved to be an autosomal recessive gene. Interestingly, it was found that humidity affected the expression of this gene also but in the opposite way. Flies that were homozygous for the autosomal recessive mutant, when grown under dry conditions, gave abnormal abdomen. Mutant flies raised under relatively moist conditions gave a phenotype that approached wild-type, and the abnormal abdomen phenotype became undetectable.

Presence or Absence Hypothesis

During the early years of the twentieth century the geneticist William Bateson, in addition to championing Mendelism (Chapter 4), was also thinking about how genes acted. Recall that many of the early genetic experiments involved the pigmentation of flowers. In addition, the dominant allele often gave a pigmented flower, whereas the recessive genotype resulted in no pigmentation or white flowers. From this and similar observations, Bateson postulated that the gene **A**, which makes for red flowers, represents a "presence," but its allele, **a**, which results in white flowers, is an "absence." This formulation became known as the *presence or absence theory* and was often invoked early in the century as one explanation for gene action.

The hypothesis received support in some of the early reports from Drosophila geneticists in which it was reported that strains of flies were recovered that were homozygous for very small deficiencies involving the tip of the X chromosome. It also turned out that such flies had yellow body color. It was known from mapping experiments that the yellow locus was near the tip of the X. Such an observation supported the notion that a mutant might represent an absence, since the phenotype of the true deficiency mimicked that of the homozygous recessive.

Multiple Alleles

However, it soon became apparent that the presence or absence hypothesis was inadequate to explain the observed facts. The discovery of multiple alleles in rabbits, which indicated that at least three alleles were involved, **C**, $\mathbf{c}^h$, and **c**, made untenable Bateson's hypothesis, which visualized only two allelic alternatives. It is not easy to imagine more than one kind of absence for a given locus resulting in similar but still different phenotypes.

Gene Dosage

A direct attack on this theory was performed in a clever experiment by Curt Stern, who used the bobbed gene in Drosophila. In Chapter 9 it was pointed out that bobbed, a mutant that results in reduced bristle length, is present on both the X and the Y chromosome. Because of its presence on both sex chromosomes, it is reasonably simple for a Drosophila geneticist to design crosses that will give flies with different numbers of the bobbed allele. For example, Stern was able to obtain males that were XO and that had a bobbed allele on the X chromosome. He was also able to obtain males that were XY with a bobbed allele on both the X and the Y. And, finally, he could get a stock of males that were XYY in which there was a bobbed allele on each of these chromosomes. Thus Stern was able to obtain flies with doses of 1, 2, and 3 bobbed alleles, respectively. If the Bateson hypothesis is correct, the phenotype of these males should be nearly identical because one absence should equal two absences should equal three absences. If, however, the bobbed allele is not an absence in the functional sense but is a functioning gene (although it need not be stated at this moment how), then it is predicted that flies with different dosages of these "functioning" alleles might well give different phenotypes. Figure 15-2 shows the results that Stern obtained. It is evident that bobbed does not represent an absence of function. A more plausible hypothesis is that bobbed is a functioning allele with respect to bristle formation but that it functions less efficiently than the wild-type allele. For example, the following model might be hypothesized. Suppose that the alleles, wild-type and mutant, are responsible for the production of some units of bristle substance, and further suppose that, in order to obtain a wild-type bristle, at least 20 units of bristle substance are needed. Finally, assume that the wild-type allele produces 20 units of bristle substance and that the bobbed allele produces only

CHROMOSOME COMPLEMENTS	NUMBER OF BOBBED ALLELES	PHENOTYPE OF BRISTLES
XO	1	Very short /
XY	2	Short
XYY	3	Almost wild-type

Figure 15-2. The results of Stern's experiment with the dosage of bobbed alleles in Drosophila. It was clear that the more bobbed genes present, the more nearly normal were the bristles. This result is inconsistent with the presence and absence hypothesis and suggests that bobbed is a functioning gene.

6 units of bristle substance. Given this *hypothetical* model, Stern's results can be easily explained. A heterozygous fly that has one wild-type and one bobbed allele should produce 26 units of substance and have wild-type bristles. A fly with only one bobbed allele would have only 6 units of substance and have a very short bristle. A fly with two bobbed alleles would make 12 units of substance and produce an intermediate-sized bristle; and, finally, a fly with three bobbed alleles would produce 18 units of substance and give a bristle nearly wild-type. It must be emphasized that there is no evidence that these genes produce units of bristle substance. What was presented is simply a useful model to account for the results Stern obtained. In any event, it is clear that bobbed cannot represent an absence of function.

Deficiencies

If a mutant truly represents an absence of function, then there should be no difference in phenotype between a heterozygote of wild-type and mutant (**+/m**) and a heterozygote of wild-type and deficiency (**+/Def.**). On the other hand, if a mutant represents a less-efficient allele than the wild-type, it is possible that **+/Def.** would show a mutant phenotype. In fact, if the recessive allele functions nearly as efficiently as the wild-type, the substitution for it by a deficiency may cause the **+/Def.** to show a greater deviation from wild-type than does the homozygous mutant. A hypothetical model similar in kind to the model proposed for bobbed can be constructed to illustrate this situation. Suppose that, for normality, 20 units of a hypothetical substance are needed. Further suppose that the wild-type allele produces 12 units of substance and that the mutant allele, only slightly less efficient, produces 9 units of substance. Given these assumptions, it follows that the homozygous wild-type produces 24 units of substance and has the wild-type phenotype. The heterozygote **+/m** has 21 units of substance and a wild-type phenotype. The homozygous mutant **m/m** gives 18 units of substance and has, therefore, a mutant phenotype. Finally, the heterozygote **+/Def.** gives only 12 units of substance and produces a phenotype more extreme than the homozygous mutant itself. Such examples of deficiencies that show exagerrated phenotypic effects are known in Drosophila, leading to the conclusion that many mutant genes are not absences of function but represent less-efficient functioning genes than their wild-type alleles.

Muller's Terminology

In an important paper presented before the Sixth International Congress of Genetics at Ithaca, New York, in 1932, H. J. Muller, taking into account the previously described experiments as well as his own, introduced some new terms into the genetic language to describe the action of mutant genes. The term most commonly used is *hypomorph*, which Muller proposed to describe genes that do the same thing as the wild-type allele but do it less efficiently. Bobbed would be an example of a hypomorph, and the evidence suggests that most mutant genes are hypomorphs. There are, however, other possibilities. Muller proposed the term *amorph* for a gene that has no function at all. This would be analogous to what Bateson was proposing in his original presence and absence hypothesis. A *hypermorph* would be a gene that does the same as the wild-type but does it more efficiently. In an hypothetical model, a hypermorph would produce more substance than the wild-type allele. Theoretically, there could be an allele, called by Muller an *antimorph*, that has an effect opposite to that of the wild-type allele. Evidence for such genes has not been firmly established. Finally, Muller proposed the term *neomorph* to describe those genes that may do something quite different from what is done by the wild-type allele.

Gene-Controlled Processes

Gene-Controlled Processes in Ephestia

In the flour moth, Ephestia, a mutant gene has been found that affects pigmentation in various parts of both the caterpillar and the adult. The homozygous recessive, **aa**, has the generalized effect of reducing pigmentation compared to the wild-type allele, **A**. Figure 15-3 gives specific details of the phenotypes in various parts of the caterpillar and adult.

The geneticist E. Caspari performed a number of experiments with this organism, which involved transplanting parts from one animal to another and then observing the results of the transfer on both the transplant and the host animal. Caspari transplanted testes from an **AA** animal to a mutant host. He noticed that the grafts retained their dark color and in addition, caused certain of the characters of the **aa** host to develop the **A** phenotype. For example, **aa** adults with

AFFECTED PART	WILD-TYPE, A—	MUTANT, aa
Adult eyes	Black	Red
Adult testis	Brown	White
Adult brain	Brown	Pale red
Caterpillar skin	Red	Pale
Caterpillar ocelli	Pigmented	Colorless

Figure 15-3. The action of the gene pair A/a in the flour moth, Ephestia. The recessive aa has the general effect of reducing pigmentation.

AA testis implants had black instead of red eyes. The testis and brain of the host resembled more those of the wild-type. If the transplantation was made into early **aa** caterpillars, the larval skin and larval ocelli in later molts assumed the **AA** phenotype. Similar results were obtained by implanting ovaries or brains. In the reciprocal type of transplantation where **aa** testis is implanted into the **AA** host, the host retains the wild-type appearance, but the transplants assume the **A** phenotype. To explain these results, Caspari and his associates postulated that a substance contained in the implanted tissues from the wild-type donor could produce its characteristic effects in host tissues that did not contain the **A** gene. Since the substance was assumed to be diffusible and could be extracted, it was called a *hormone.*

In an interesting variation of the experiment, Caspari transplanted an **A** testis into **aa** females, which were then mated by **aa** males. Since the progeny from such a cross are all **aa**, the caterpillars would be expected to show the **aa** phenotype. They did not! Eggs from **aa** mothers implanted with **A** testis produced caterpillars with the **A** phenotype. The interpretation is that the postulated **A** hormone diffused into the eggs and persisted in the egg cytoplasm long enough to result in larvae that were phenotypically **A** although genotypically **aa**. With successive molts, the substance originally present in the eggs was dissipated, and the caterpillar eventually assumes the phenotype expected of its genotype. The **A** hormone has been biochemically identified as kynurenine.

Gene-Controlled Processes in Drosophila

The work in Ephestia was a beginning, limited in scope, in the area of gene-controlled processes in animals. It did, however, prove a stimulus for the next step, taken a few years later in the mid 1930s by two geneticists, George Beadle and Boris Ephrussi, working both in Morgan's laboratory at the California Institute of Technology in Pasadena and in Ephrussi's laboratory in Paris.

Beadle and Ephrussi began an analysis of eye-color pigments in Drosophila. It is known that the wild-type eye in Drosophila contains two pigments, a brown and a red, which together give the dull red wild-type eye color. Figure 15-4 gives some details on these pigments in a selected group of mutants in Drosophila.

Beadle and Ephrussi knew that, in the Dipteran insects, many of the structures of the adult, such as the compound eyes, legs, wings, and antennae, develop from small disks that are formed in the late embryo or early larval stages. These are known as *imaginal disks.* In Drosophila, the imaginal disks of the compound eye can be transplanted from one larva to another; and if the transplant is successful, after metamorphosis the adult host will have a third eye in the abdominal cavity. This third or transplanted eye can be examined and its phenotype determined, usually directly through the abdominal wall of the host.

They performed many transplantation experiments that exploited the large number of eye-color mutants in Drosophila. They found that, in almost all cases, disks from mutant genotypes, such as white, peach, pink, raspberry, and carmine,

GENOTYPE	PIGMENT		PHENOTYPE
	BROWN	RED	
Wild-type (+ +)	+	+	Dull red
Vermilion (**v v**)	−	+	Bright red
Cinnabar (**cn cn**)	−	+	Bright red
Brown (**bw bw**)	+	−	Brown
Cinnabar brown (**cn cn**; **bw bw**)	−	−	White
White (**w w**)	−	−	White

Figure 15-4. The presence and/or absence of the two eye color pigments in several genotypes of Drosophila. Presence of pigment is symbolized by + and its absence by −.

when transplanted into wild-type larvae gave transplanted eyes that were the phenotype expected for the genotype of the transplanted disk. The color of the transplanted eye was not affected by the genotype of the host. In the reciprocal experiment the same result was observed. The transplanting of wild-type disks into various mutant hosts resulted in eyes whose phenotypes were predictable according to their genotype. Development is *autonomous*, in that the eye develops according to its own genotype.

However, Beadle and Ephrussi found two exceptions to autonomy. If an eye disk from a vermilion donor (**v** disk) is transplanted into a wild-type host, the disk does *not* behave autonomously and give the bright red phenotype expected of vermilion. Instead the abdominal eye is wild-type! They tentatively concluded from this experiment that the host produced a $\mathbf{v}^+$ substance that diffused into the transplanted disk and was responsible for the normal pigmentation. The analogy to the Ephestia work is evident. The wild-type makes $\mathbf{v}^+$ substance, whereas the mutant vermilion produces no $\mathbf{v}^+$ substance.

In a second experiment Beadle and Ephrussi transplanted a cinnabar (**cn** disk) into a wild-type host and, again, contrary to the expected autonomous development, the transplanted eye was wild-type. It might be tentatively concluded that the wild-type is producing a substance that is diffusing into the cinnabar disk, to produce the wild-type transplanted eye. Such an assumption immediately suggests the following question: Are the **cn** flies and the **v** flies lacking the same substance?

In a third experiment a **cn** disk was transplanted into a **v** host. The transplanted eye did *not* have a wild-type phenotype but retained the cinnabar phenotype. From this experiment it could be concluded that the host vermilion is not supplying what is lacking in the **cn** transplant.

A fourth experiment provided information, which together with information from the other experiments, made an understanding of gene action in this system possible. Beadle and Ephrussi transplanted a vermilion disk into a cinnabar host,

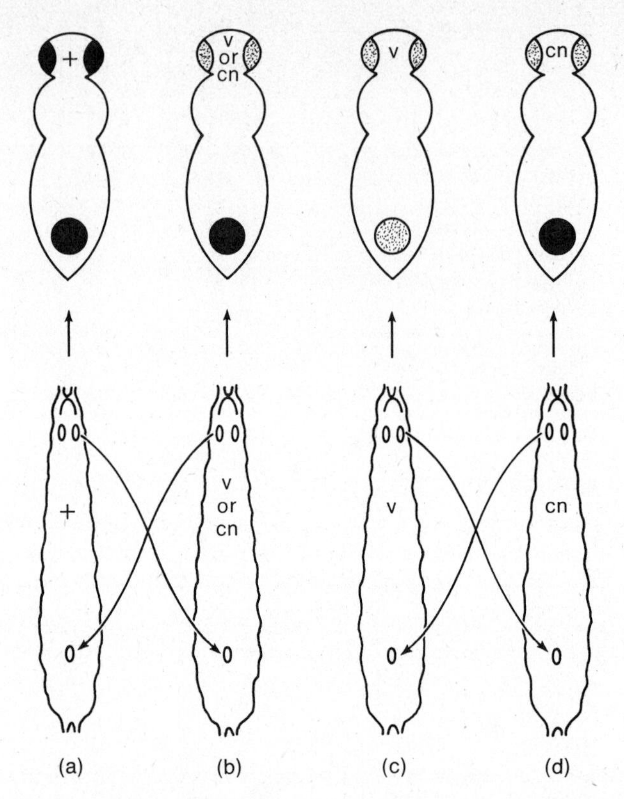

Figure 15-5. Results of transplantations involving wild-type, vermilion, and cinnabar flies. Below, host and donor larvae. Above, adult flies containing differentiated implants. **+** = wild-type; **v** = vermilion; **cn** = cinnabar. (After B. Ephrussi. 1942. *Quart. Rev. Biol.* **17**.)

and the transplanted eye was not bright red but wild-type. The conclusion seems inescapable than the **cn** host is supplying what is lacking in the vermilion disk. (See Figure 15-5.)

SEQUENTIALLY RELATED REACTIONS. It appears, then, that two substances, one known as $\mathbf{v}^+$ and the second as $\mathbf{cn}^+$, are lacking in the vermilion genotype. Only one of them, the $\mathbf{cn}^+$ substance, is lacking in cinnabar. Both the $\mathbf{v}^+$ and $\mathbf{cn}^+$ substances are present in the wild-type. Beadle and Ephrussi also deduced from these experiments and others that the second of the substances, the one that cinnabar lacks, is produced only if the first substance is present. In other words, one substance is a precursor in a chain of reactions for the other. Beadle and Ephrussi have, therefore, demonstrated two related links in a sequence of reactions leading to the development of wild-type eye color. This sequence is apparently interrupted at an early point by mutation to vermilion and at a later point by mutation to cinnabar.

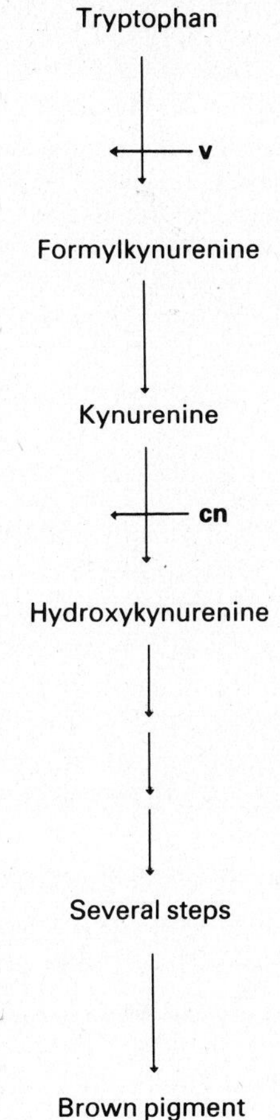

Figure 15-6. Part of the biochemical sequence leading to the production of brown pigment in Drosophila. The mutant *v* blocks the reaction tryptophan ⟶ formylkynurenine. The mutant *cn* blocks the reaction kynurenine ⟶ hydroxykynurenine.

When biochemists analysed this system, it was learned that these two mutants are involved in a biochemical sequence beginning with tryptophan and going through several steps, leading eventually to the formation of brown pigment. (See Figure 15-6.) The **v**$^+$ substance was identified as formylkynurenine, since the vermilion mutant blocked the reaction from tryptophan to formylkynurenine. The **cn** mutant blocks the reaction from kynurenine to hydroxykynurenine, so that the **cn**$^+$ substance must be hydroxykynurenine. Figure 15-7 summarizes the relationship of these genes and the substances involved.

Gene-Controlled Processes in Neurospora

The experiments with eye color in Drosophila led Beadle to further important investigations. He realized that Drosophila was not an ideal organism to undertake coordinated biochemical and genetic experiments. Drosophila, on the one hand, is a diploid organism, which makes the detection of biochemical mutants more difficult

Figure 15-7. A summary of the relationship between vermilion and cinnabar genes in Drosophila. Although the vermilion mutant lacks both **v**$^+$ and **cn**$^+$ it is capable of producing **cn**$^+$ substance if provided with **v**$^+$ substance. Thus, a **v** disc in a **cn** host gives a wild-type eye.

GENOTYPE	SUBSTANCE	
	v$^+$	**cn**$^+$
Wild-type (+/+)	+	+
Cinnabar (**cn/cn**)	+	−
Vermilion(**v/v**)	−	−

v$^+$ substance is formylkynurenine.
cn$^+$ substance is hydroxykynurenine.

than in a haploid organism, in which mutations are expressed in the first generation. Secondly, Drosophila is a fairly complicated and highly evolved organism containing many complex reactions and complicated substances. A biochemical analysis was not likely to be simple, particularly in 1940, when many of the sophisticated biochemical techniques that are routine today were not available. Beadle, therefore, looked for another organism in which these difficulties were not present. He selected the bread mold, Neurospora, to continue his work with gene controlled processes. In Chapter 11 it was pointed out that the genetics of Neurospora had been investigated some years earlier in Morgan's laboratory at the California Institute of Technology, and so the choice of this organism seemed natural. In collaboration with Edward Tatum, Beadle began a series of investigations at Stanford University that provided a much clearer answer to gene action.

BIOCHEMICAL MUTANTS. Beadle and Tatum adopted the following assumptions as the rationale for their experiment. They asked first whether it was possible to grow wild-type Neurospora on a medium containing chemically simple constituents. They found that Neurospora could be cultured on what they called a *minimal medium.* For Neurospora, the only requirements were some inorganic nitrates, various other inorganic salts, a simple sugar, and only one chemically complex molecule, biotin, a vitamin. Even though Neurospora is grown on such simple substances, the wild-type organism consists of many vitamins, 20 amino acids, a variety of nucleic acids, and other organic molecules. Thus Beadle and Tatum assumed that Neurospora is able to synthesize complex organic molecules from simple building blocks. They further reasoned that if these syntheses were genetically controlled, it might be possible to obtain mutant strains that were unable to perform some of these syntheses and thus could not grow in a minimal medium. Such mutants that were unable to grow in a minimal medium might be sustained by supplying the nutrient that the strain was unable to produce, for they assumed that growth could occur by providing the necessary substance exogenously. Therefore an important feature of their experimental design was a *complete medium* to culture the supposedly nutritionally deficient mutants. A complete medium is a minimal medium to which is added all necessary vitamins, amino acids, nucleic acids, and other constituents.

With these assumptions in mind, Beadle and Tatum designed their experiment in the following way. They knew from the work of Muller that the spontaneous mutation rate is low and that if they were to obtain mutations, it would be desirable to utilize radiation. They irradiated spores and crossed them to wild-type of the opposite mating type. From the ascus, they cultured the ascospores on a complete medium. From those colonies that grew on the complete medium, they made a subculture of each onto a minimal medium, carefully marking the strains so that they knew the origin of each one. Many of the subcultures grew on the minimal medium, but a few did not—the important ones. Returning to the original cultures on complete medium, they subcultured again into a series of tubes that had minimal medium plus all available constituents of a certain class of substances. For example, they would place spores in a tube that contained minimal medium plus all the known vitamins. A second tube contained minimal medium plus all known amino acids. A third subculture had minimal medium plus all nucleic acid derivatives and so on. If the culture grew in the tube with minimal plus all vitamins but not in the other subcultures, the assumption was that the strain had lost its ability to synthesize a certain vitamin. If so, additional subcultures were made in which the strain was cultured in tubes of minimal medium plus individual vitamins. For example, minimal medium plus thiamine, minimal medium plus riboflavin, minimal medium plus niacin, and so forth. If the growth was observed in one tube but not in the others, the assumption was that a nutritionally deficient strain that was unable to synthesize a specific vitamin had occurred. (See Figure 15-8.)

It is an interesting historical sidelight that Beadle and Tatum, although they certainly had faith in their assumptions and rationale, did not know whether the experiment would work and arbitrarily decided to start 1000 cultures. If no mutants were obtained in 1000, they would abandon this experiment as a good idea but a lost cause. It turned out that in culture #299 they obtained their first mutant and in culture #1085 their second. Encouraged by these positive results, they continued the experiment on a much larger scale and were able to verify their assumptions by accumulating, relatively quickly, a large number of biochemically deficient strains.

Because the genetics of Neurospora was well understood, Beadle and Tatum were able to verify that the strains of biochemically deficient Neurospora were simple genetic mutants with 1 : 1 ratios of mutant to wild-type spores in a given ascus. After accumulating a large number of mutants, they were able to map them on the chromosome, obtaining linkage groups similar to those obtained 25 years earlier in Drosophila. There seemed no doubt in their minds that the biochemical mutants that they had isolated in Neurospora were, genetically speaking, the same as the morphological mutants that the Drosophila workers had been using for two decades.

SEQUENTIALLY RELATED REACTIONS. The discovery of biochemical mutants by Beadle and Tatum led to additional experiments that gave considerable insight into gene action. A number of mutants were obtained that were unable to grow on a

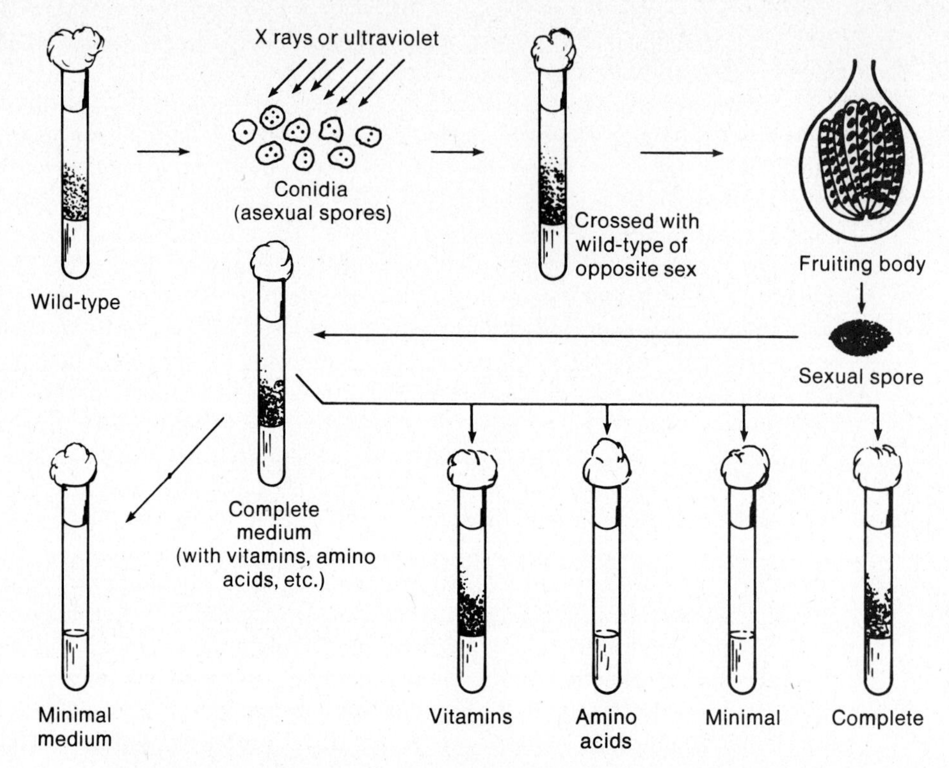

Figure 15-8. The experimental design of Beadle and Tatum in Neurospora for the detection of biochemical mutants. (After G. W. Beadle. 1947. In *Science in Progress.* Yale University Press, New Haven.)

Ornithine		Citrulline		Arginine
NH_2		NH_2		NH_2
$(CH_2)_3$		$C=O$		$C=NH$
$CHNH_2$	$\xrightarrow[+NH_3]{+CO_2}$	NH	$\xrightarrow{+NH_3}$	NH
$COOH$		$(CH_2)_3$		$(CH_2)_3$
		$CHNH_2$		$CHNH_2$
		$COOH$		$COOH$

Figure 15-9. Biochemical synthesis of arginine from citrulline which in turn is synthesized from ornithine. These reactions form part of the urea cycle in man.

minimal medium unless the amino acid arginine was added. Such mutants are called *arginineless.* The biochemical geneticist adds the suffix *less* to the nutritional requirement to designate a biochemical deficiency. The biochemical pathway involving arginine had been worked out by biochemists, and it was known that arginine comes from citrulline, which, in turn, is synthesized from ornithine. (See Figure 15-9.) Of the seven arginineless mutants that had been accumulated independently of each other, it was shown that four of the strains would grow on a minimal medium if supplied with ornithine, or citrulline, or arginine; two would grow if either citrulline or arginine were added; and one strain would grow only if supplied with arginine. Since there is good evidence of the sequential synthesis of arginine from ornithine through citrulline, it is probable that these different mutants represent blocks at different steps. (See Figure 15-10.) Note that the strain will grow only by adding the requirement that comes *after* the genetic block.

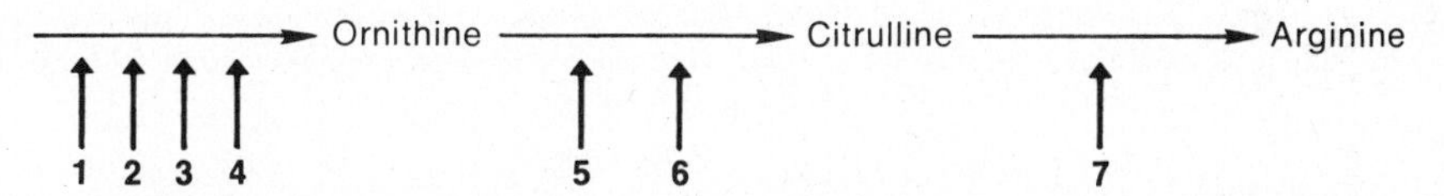

Figure 15-10. Seven arginineless mutants in Neurospora and their supposed site of action. Mutant 7 which blocks the conversion of citrulline to arginine will grow only in a minimal medium to which arginine is added. In a similar way, mutants 5 and 6 grow on a minimal medium if either citrulline or arginine is added, whereas mutants 1, 2, 3 and 4 grow with the addition of any one of the three substances.

In a second example involving biochemical mutants in Neurospora, a strain was obtained that would not grow on a minimal medium without the addition of the amino acid adenine. In addition, it was observed that this adenineless strain also produced a purple pigment. The assumption is that, in this mutant, the precursor, whatever it may be, is not converted to adenine but instead accumulates in the cell and, in this particular instance, results in a purple pigment. Sometime later a second mutant was discovered that also was unable to grow on a minimal medium without adenine but in which no purple pigment is produced. The hypothesis is that this second mutant blocks the formation of the precursor. (See Figure 15-11).

The Neurospora investigations reinforced the conclusion obtained with eye pigments in Drosophila that genes control sequentially related reactions. If this hypothesis is true, biochemical mutants are not expected to grow on minimal medium without the addition of a substance that comes after the genetic block. In addition, the hypothesis predicts the accumulation of the immediate precursor. These assumptions have been repeatedly tested and proven to be true. Such information proved invaluable not only to the geneticist but also to the biochemist because in many instances it became possible to work out the specific details of certain biochemical syntheses by obtaining mutants that accumulated precursors that could be identified biochemically. Some scientists say that the Beadle-Tatum

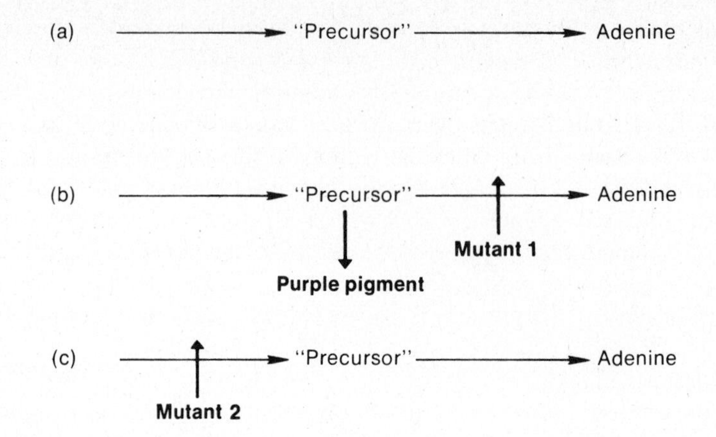

Figure 15-11. An analysis of two adenineless mutants in Neurospora. Adenineless 1 is associated with a purple pigment assumed to be the pile-up of the precursor to adenine. Adenineless 2 is not associated with purple pigment leading to the assumption that the genetic block is prior to the precursor.

experiment started biochemical genetics and, together with Luria and Delbrück's experiment on mutations in bacteriophage, forms the true beginning of molecular biology.

One Gene–One Enzyme

But the significance of Beadle and Tatum's work goes beyond the isolation of biochemical mutants and the verification of genetically controlled sequential reactions. They were interested in the more fundamental question of why it is that the mutations do not permit the syntheses to occur.

Beadle and Tatum had a hunch, for they knew of the work of Sir Archibald Garrod, who in the early 1900s had written a book called *Inborn Errors of Metabolism.* In it Garrod talked about four diseases of man—alcaptonuria, albinism, pentosuria, and cystinuria. At the turn of the century some of the biochemical pathways involved in these various diseases were known, and Garrod proposed that the reason for the biochemical deficiencies involved in these diseases was the absence of an enzyme necessary to catalyze the reaction. Beadle and Tatum hypothesized that enzymes were involved in the Neurospora work, but now the biochemical techniques were at hand to test this hypothesis. It was possible to perform biochemical assays on the genetically mutant strains of Neurospora and show that in many cases specific enzymes were absent, at least in the functional sense. From studies of this kind came the hypothesis of *one gene–one enzyme*, which, simply stated, says that the function of a given enzyme is under the control of a specific gene. If the gene mutates, the enzyme will not function. This one gene–one enzyme hypothesis proved an important stimulus for the work in biochemical genetics following World War II. Although it is now known (see Chapter 19) that this hypothesis is oversimplified, it is essentially correct and proved to be of great significance to workers in the field. For their work, Beadle and Tatum were awarded in 1958, along with Joshua Lederberg, the Nobel Prize in medicine and physiology.

REFERENCES

BEADLE, G. W., and E. L. TATUM, 1941. Genetic control of biochemical reactions in Neurospora. *Proc. Natl. Acad. Sci. U.S.* **27**:499–506. (Available in *Classic Papers in Genetics,* edited by J. A. Peters. Prentice-Hall, Englewood Cliffs, N.J., 1959.) (This important paper was the first work published on biochemical genetics.)

MORGAN, T. H. 1915. The role of environment in the realization of a sex-linked Mendelian character in Drosophila. *Am. Nat.* **49**:385–429. (One of the earliest papers that demonstrated an influence of environment on gene action.)

MULLER, H. J. 1932. Further studies on the nature and causes of gene mutation. *Proc. 6th Int. Congr. Genet.* (*Ithaca*) **1**:213–255. (In this paper Muller reviewed his ideas on the nature of the gene, gene action, and gene mutation.)

ROBINSON, R. 1958. Genetics of the rabbit. *Bibliographica Genetica* **17**:229–558. (A very good review of the physiological genetic studies on the rabbit, particularly the temperature experiments with the Himalayan rabbit.)

STERN, C. 1929. Über die additive Wirkung multipler Allele. *Biol. Zentralbl.* **49**:261–290. (This paper gives the evidence that the bobbed gene in Drosophila functions but not as efficiently as the wild-type allele. This work helped to discredit the presence-or-absence hypothesis.)

QUESTIONS AND PROBLEMS

15-1. Define the following terms:

temperature-effective period	autonomous
presence or absence hypothesis	one gene–one enzyme
hypomorph	imaginal disk

15-2. Does the knowledge that the same genotype may result in two or more different phenotypes, if subjected to different environments, give you any insight into how genes act?

15-3. Does the knowledge that there are temperature-effective periods give you any insight into how genes act?

15-4. Discuss briefly the different lines of evidence that indicate that many mutant genes do not represent an absence of function.

15-5. For each of the following experiments in Drosophila, indicate whether or not the implanted disk forms brown pigment.

(a) **v** implant → +host (b) v implant → cn host
(c) **cn** implant → +host (d) cn implant → v host

15-6. Discuss briefly the experimental evidence indicating that genes may control sequential chemical reactions in
(a) Drosophila.
(b) Neurospora.

15-7. All that is known about a newly discovered mold is that it bears a morphological resemblance to Neurospora and apparently has a similar life cycle. Outline the experimental steps necessary to study the biochemical genetics of this new mold.

15-8. Neurospora spores are irradiated. Outline briefly the experimental procedures to find mutations that block the synthesis of any amino acid.

15-9. You have a strain of Neurospora that is biochemically deficient for histidine. You wish to obtain a strain of Neurospora that is simultaneously deficient for two biochemical syntheses, one of which is histidine. Outline the experimental procedures to obtain such a strain.

15-10. In Neurospora, you wish to obtain a strain that is deficient for the synthesis of histidine and *no other* deficiencies. Outline the most efficient procedure to obtain such a strain.

15-11. In Neurospora, it is known that a chain of biochemical reactions links a number of substances as follows:

$$A \rightarrow B \rightarrow C \rightarrow D \rightarrow E$$

A normal strain of Neurospora grows well without the presence in the culture medium of any of the substances A, B, C, D, or E. Three mutant strains, 1, 2, and 3, do not grow on minimal medium. Strain 1 grows if C, D, or E is present. Strain 2 grows only when E is present. Strain 3 grows when any one of A, B, C, D, or E is present. Where do mutants 1, 2, and 3 fail to act in the above-listed normal sequence of events?

15-12. If you were asked in the mid-1940s (shortly after the publication of Beadle and Tatum's paper), what do genes do, what would your reply be?

16/The nature of the gene

The discussion of the early Drosophila work in Chapter 10 emphasized that Morgan and his group presented compelling evidence that the genes are arranged on the chromosome in linear fashion. It was pointed out that in his public lectures Morgan often used the metaphor that genes were like "beads on a string." It is time to examine this metaphor more carefully to see whether it is a valid description of reality.

Position Effect

To take the metaphor literally, that genes are like beads on a string, is to say that if a bead is taken off the string and placed elsewhere, it is still the same bead, unaltered in any way. If a red bead from the left end of the string is moved to the right end of the string or anywhere else, it will still be a red bead. Is this true of the genetic material?

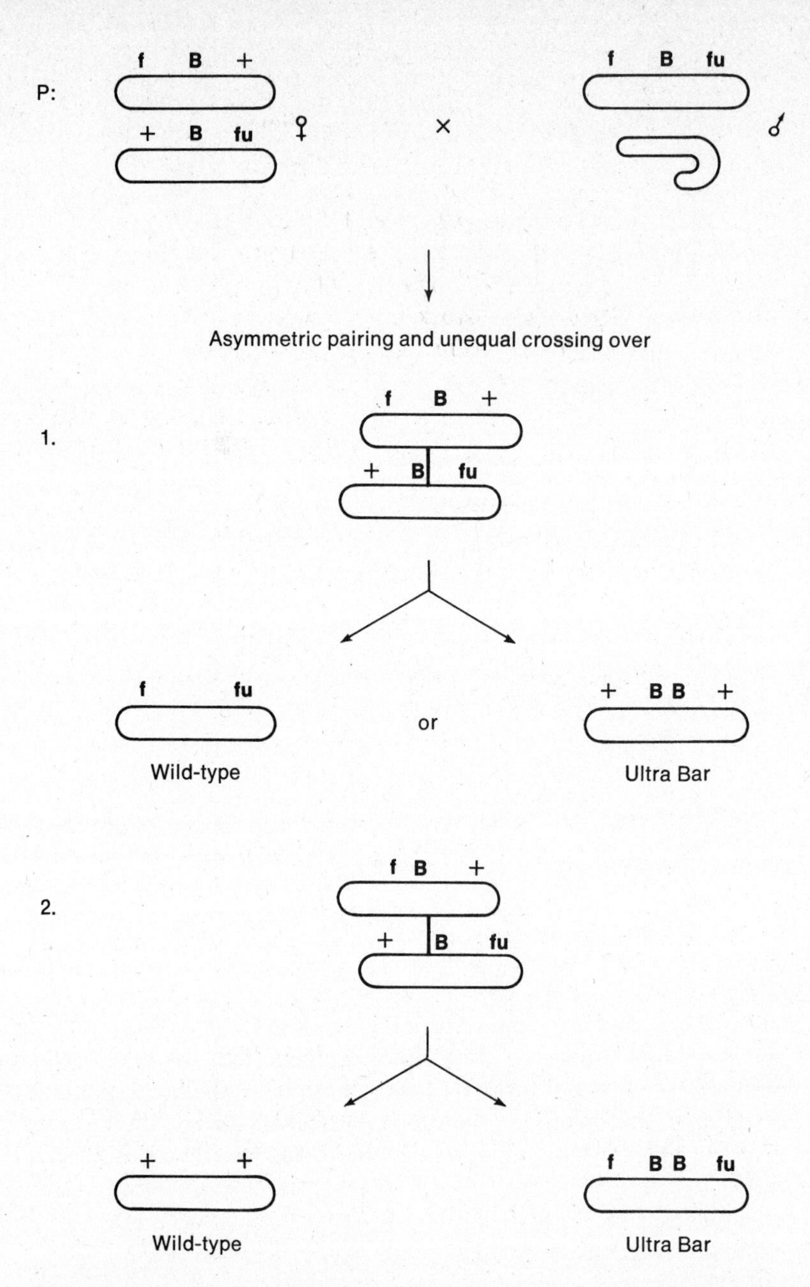

Figure 16-1. Unequal crossing over as a consequence of asymmetric pairing at the Bar locus in Drosophila. Sturtevant proposes that the asymmetric pairing may occur in either direction, (1) or (2). Note that the outside markers are different for each case so that verification of these assumptions is straightforward. **B** = Bar eye; **f** = forked bristles; **fu** = fused veins.

Bar Eye in Drosophila

This question was first answered in 1925 in an elegant paper by Sturtevant, who made an analysis of the Bar eye mutant in Drosophila. It had been known that, in cultures homozygous for Bar, there occurred "mutations" to wild-type in a frequency of about 1 in 1600. Also, but in lower frequency, a more extreme eye type with very narrow eyes was observed. This phenotype was given the name *ultra-Bar.*

Sturtevant reasoned that the frequency of these unusual events was much higher than one would expect from spontaneous mutation, and so he sought another explanation. He designed stocks in which the females were homozygous for Bar and in which the Bar locus was marked on either side by other genes closely linked to Bar. In this instance, he used the marker genes, forked (**f**) and fused (**fu**), a recessive bristle mutant, and a recessive wing vein mutant, respectively. (See Figure 16-1.) By using this experimental design, Sturtevant could detect whether crossing over was occurring in this region.

The experimental results quickly confirmed his intuition that all the "mutants" at the Bar locus were accompanied by recombination of the outside markers, forked and fused. Sturtevant suggested that perhaps, during synapsis, the pairing was not exact and that the Bar genes did not pair exactly with each other. If true and if, simultaneously, a crossover occurred between the two Bar genes, a chromosome might be recovered that had no Bar gene and that was wild-type. Or the complement might be recovered, one having two Bar genes and that was ultra-Bar. This illegitimate or asymmetrical pairing could occur in either direction, as indicated in Figure 16-1(a) and (b). Recombination of the outside markers should serve as a verification of his hypothesis. The data indicated the essential correctness of Sturtevant's hypothesis. All wild-types and ultra-Bars were associated with recombination of the outside markers. Also, the two ways that asymmetric pairing was supposed to occur did occur, and in equal frequency.

The explanation for the occurrence of wild-types and ultra-Bar was achievement enough, but Sturtevant made another observation that was of greater significance. He noticed that females that had one Bar gene on each homolog gave the phenotype Bar, whereas females that had two Bar genes on one homolog and none on the other were ultra-Bar, a phenotype different from Bar. Since the compound eye of Drosophila is composed of many individual facets that can be counted, it was possible for Sturtevant to quantify his observations. He showed that Bar females had 68.1 ± 1.1 facets, whereas the ultra-Bar females had only 45.4 ± 0.2 facets. The difference between these two sizes is statistically significant.

Sturtevant concluded that two Bar genes on the same chromosome produce a stronger effect on the developing eye than the same two genes would produce when on different homologs. Thus Sturtevant was the first to demonstrate the phenomenon called *position effect.* By it he meant that the position of the genes has an effect, at least in this case, on their genetic activity. The metaphor of beads on a string is not true in the literal sense. It would appear that the spatial relationships of genes to one another makes a difference.

This theory of asymmetric pairing and unequal crossing over, which was largely a theoretical postulate by Sturtevant, was fully confirmed some ten years later by a cytological examination of the salivary gland chromosomes when their significance was discovered. Bridges and Muller, independently of each other, examined the salivary chromosomes of wild-type, Bar, and ultra-Bar and provided the cytological verification for Sturtevant's hypothesis. The cytological studies added one refinement to Sturtevant's analysis. It turned out that Bar is a duplication of the wild-type gene. There are two sections of chromosome adjacent to each other that make the Bar mutant. Since these two sections are duplicates of each other, it is easy to understand how asymmetric pairing might occur. Unequal crossing over gives a chromosome with either one section (wild-type) or three (ultra-Bar). The position effect is now thought of in terms of four sections that are either arranged two and two or three and one. (See Figure 16-2.)

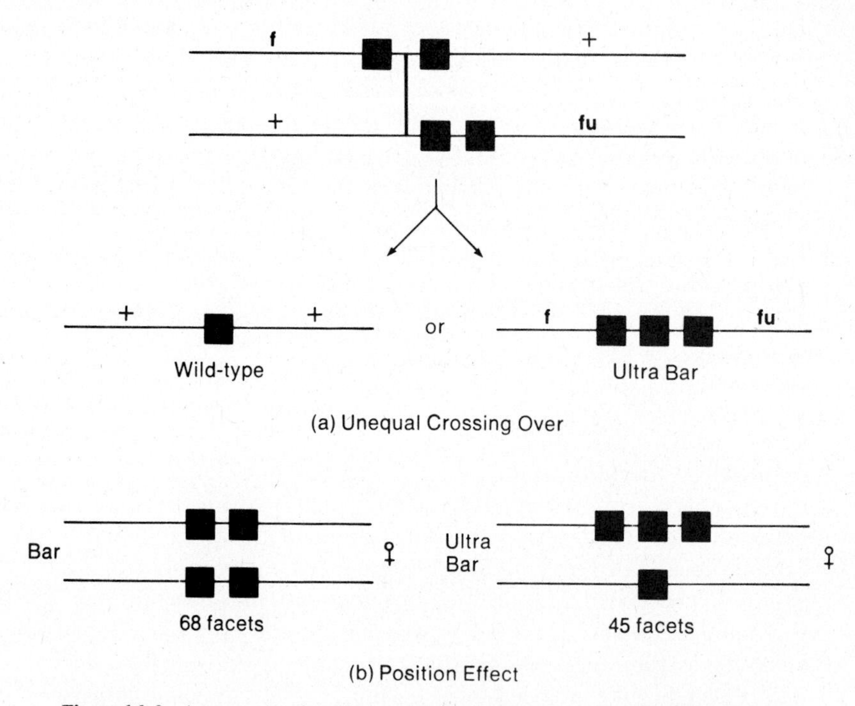

Figure 16-2. A reexamination of unequal crossing over and position effect at the Bar locus in Drosophila based on the cytological discovery that Bar is a duplication.

Variegated Position Effect Whereas Sturtevant's analysis of Bar eye in Drosophila was the first demonstration of position effect, many more dramatic examples have been obtained, primarily in Drosophila but in other organisms as well. One of the most interesting in Drosophila involves the mutant white eye. Recombination analysis has shown that white is located near the tip of the X chromosome at the end opposite from the centromere. In Chapter 13 it was indicated that the region around the centromere in Drosophila is heterochromatin, and the region away from the centromere is largely euchromatin. The white-eye mutant in Drosophila is in a region of euchromatin nowhere near a heterochromatic region.

It is possible to obtain X chromosomes in Drosophila that contain a large inversion. In some cases, the breakpoints are such that the locus for white eye is now near the region of heterochromatin. (See Figure 16-3.) It should be emphasized

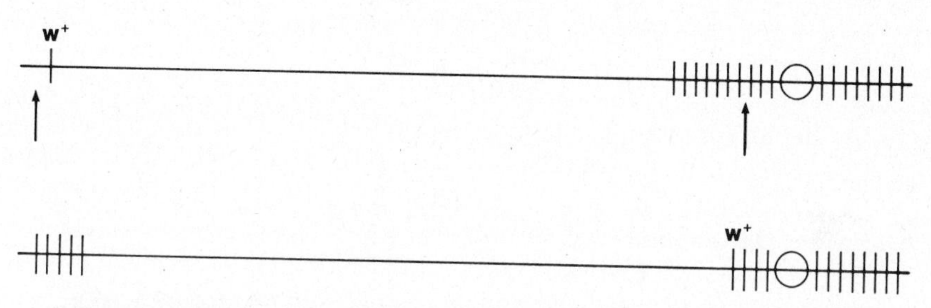

Figure 16-3. Example of variegated position effect in Drosophila involving the white locus. If breaks, indicated by arrows, occur so that an inversion is produced which moves the wild-type allele for white near a region of heterochromatin, a mottling of the eye will occur in males which contain the inversion or in females which have the inversion in one X and the recessive allele **w** in the other.

that the strain of Drosophila in which the inversion occurs has the wild-type allele for white and not the mutant. The assumption is that the inversion is not accompanied by a simultaneous mutation from wild-type to white but is simply a physical rearrangement of the X chromosome. The net consequence of this rearrangement, however, is a mottling of the eye in which some of the facets are wild-type and some are white, thus giving a *variegated* appearance.

The apparent reason for this variegation is the placement of a gene in a new environment and, in this specific instance, the placement of the wild-type allele for white into or adjacent to a heterochromatic region.

Additional experiments have shown that the assumption that there was not a simultaneous mutation is correct. It has also been shown that variegated position effects in Drosophila can occur by moving genes located in the heterochromatin into areas of euchromatin. Such dramatic examples clearly confirm Sturtevant's notion of position effect and indicate beyond a reasonable doubt that the function of a gene may well be altered if the position of the gene is changed.

Variegated position effects are relatively easy to demonstrate, but the underlying causes are not clearly understood. The degree of variegation can often be altered by the amount of heterochromatin present in the fly. For example, in the case of the white-mottled described first, if the rearrangement occurs in a male without a Y chromosome, more mutant facets result such that the amount of variegation is increased. On the other hand, if extra Y chromosomes are added to a male containing the rearrangement, the eye becomes more normal as the variegation is suppressed. The Y chromosome itself is almost entirely heterochromatin, and clearly the variegated position effects are sensitive to additions or subtractions of heterochromatin. In the case of variegated position effects where a gene normally in heterochromatin is moved near euchromatin, the addition of extra Y chromosomes will enhance the variegation, the inverse of the effect on the mottling of white. It has not been possible to derive any consistent hypothesis to explain these experiments.

Pseudoallelism

In Chapter 6 a multiple allelic series was defined. It was stated that two mutants are true alleles if they produced a mutant phenotype when crossed and if they never occurred together in the same gamete. The second part of that definition now needs further examination.

There is a sex-linked recessive eye mutant in Drosophila, lozenge (**lz**), which results in a smaller, darker, and more eliptically shaped eye. Many different alleles of this mutant have occurred independently, and all share this eye phenotype. They are, however, not identical, and it is possible to distinguish among several lozenge alleles on the basis of the severity of the effect on the eye and various other phenotypic effects in the fly. In 1940, in some preliminary experiments, Oliver recovered wild-type flies from crosses involving females that were heterozygous for two different lozenge alleles. The frequency of occurrence for these exceptional flies was low, one in several thousand, but much higher than would be expected from spontaneous mutation. The lozenge locus was thoroughly analyzed some years later by a student of Oliver, M. M. Green.

Green designed his experiment as Oliver had but on a larger scale. He mated females that were heterozygous for two different lozenge alleles to lozenge males. As expected, the progeny were lozenge except for an occasional wild-type fly.

Green, recalling the analysis by Sturtevant of the Bar locus, considered the possibility of recombination as the mechanism generating these unusual wild-types. He designed the experiment in such a way that the lozenge locus was marked on either side with other mutants, so that if the wild-type fly was due to crossing over the outside markers would recombine. He observed that, in the few wild-type flies recovered, the outside markers had recombined. (See Figure 16-4a.) Unlike the Bar example, however, the markers always recombined in a given way. That is, the two mutant lozenge alleles behaved as though they were in a definite spatial relationship to each other.

Green concluded that the two lozenge alleles were not true alleles of each other. They did not occupy the exact same position on the two homologous chromosomes but were actually separable by a small physical distance. As a result of this separation, a crossover can occur between them. (See Figure 16-4b.) Because

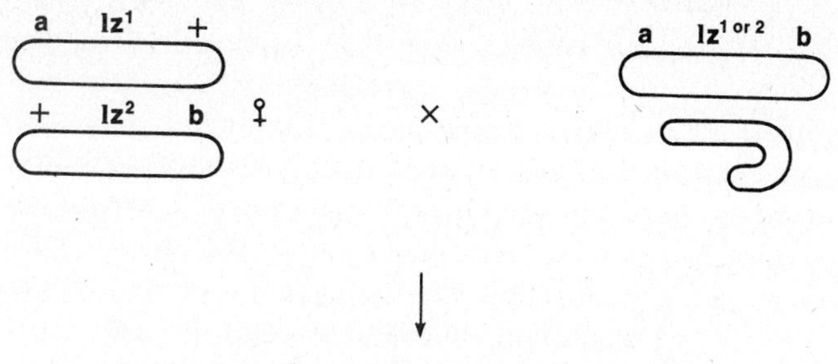

(a) Green's Observation

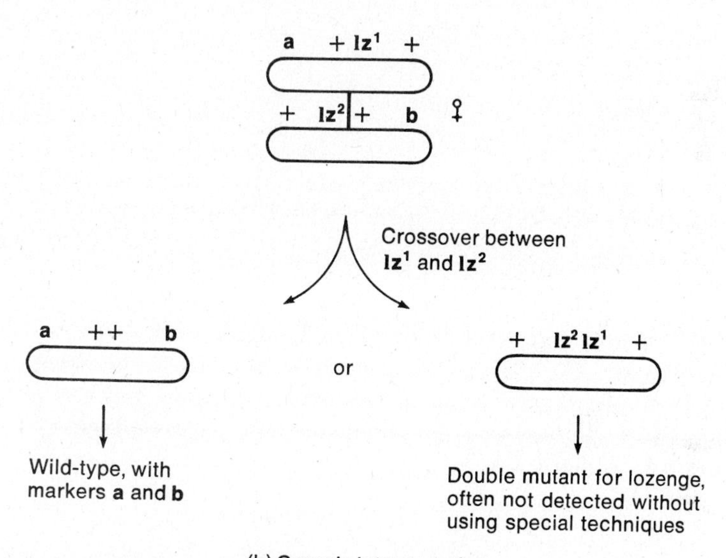

(b) Green's Interpretation

Figure 16-4. Green's analysis of pseudoalleles at the lozenge locus in Drosophila. The occurrence of wild-types associated with a fixed recombination of outside markers led Green to suppose that **lz**' and **lz**² were not true alleles but were spatially distinct with **lz**² to the left of **lz**'. In most experiments the double mutant is not detected because it resembles phenotypically the single mutants. However, the occurrence of double mutants has, in some experiments, been verified. **lz**' and **lz**² = two lozenge alleles. **a** and **b** = two different recessive markers.

the lozenge alleles did not satisfy both criteria for allelism (both alleles could occur in the same gamete), a redefinition of allelism was necessary. The lozenge mutants are now referred to as *pseudoalleles*, "false" alleles. Many examples of pseudoalleles are known in Drosophila and a host of other organisms.

Position Pseudoallelism

Pseudoallelism represents another kind of position effect. It should be emphasized that the female heterozygous for the two lozenge alleles was phenotypically lozenge. Additional experiments with lozenge and other pseudoalleles have permitted the recovery of the other arrangement of two pseudoalleles, in which the two mutant alleles are on one homolog and the two wild-type alleles are on the other. The first arrangement is known as a *trans* configuration and the second as a *cis* configuration. (See Figure 16-5.) This is, by definition, a position effect because the two arrangements involving the same genes give two phenotypes.

Figure 16-5. Position pseudoallelism in Drosophila, a position effect. The genes involved are the same but their spatial relationship to each other is changed. In the left example, the fly is wild-type. In the right example, the fly is mutant. Adopting the language of the chemist, the case of the two mutants on the same homologue is sometimes called the *cis* configuration. When the two mutants are on separate homologues, it is defined as the *trans* configuration. Geneticists may now speak of a *cis/trans* position effect.

An interesting question is: Why a mutant phenotype in the first case and not in the second? The question is especially intriguing when it is realized that both lozenge mutants are recessive to wild-type. Several hypotheses have been offered in explanation, but the most influential one was proposed by E. B. Lewis. (See Figure 16-6.) In addition to providing an explanation for position pseudoallelism, Lewis

Figure 16-6. The hypothesis of E. B. Lewis to explain position pseudoallelism. The assumptions are: The recessive pseudoalleles **a** and **b** control sequentially related reactions; the site of the reactions is on or near the chromosome; the products of the reaction are nondiffusible. If the wild-type alleles are present at both loci on the *same* homologue, the reaction goes optimally and sufficient B is produced to result in a wild-type phenotype (upper left). If, on the other hand (upper right), mutant **a** is present, insufficient product A is produced and even in the presence of + for **b**, insufficient B is produced. On the other homologue, sufficient A is produced but, because of **b**, again insufficient B is produced. The net result is that neither homologue produces sufficient B, and so a mutant phenotype is produced.

also discusses the evolution of genetic systems and the possible role played by pseudoalleles in the evolution of genes.

An analysis of pseudoallelic loci in Drosophila provides further insight into the nature of the gene. Pseudoallelism raises the possibility that a gene is not simply a single unit of function, mutation, and recombination, as described by T. H. Morgan in the mid-1920s, but suggests that the gene may have a *fine structure*. In fact, there may be a semantic problem as to exactly what is called a gene when it is shown that there are several lozenge mutant sites where once only one gene was visualized.

Genetic Fine Structure

Genetic Fine Structure in Bacteria

M. Demerec, working with the bacterium Salmonella, obtained more evidence about genetic fine structure. Salmonella is an organism in which transduction experiments are easily carried out (Chapter 12). It was indicated that, as a general rule, only one gene at a time can be transduced. The conclusion drawn was that the transducing fragment carried by the phage vector is very small. Demerec decided to turn the argument around. He reasoned that it should be possible to do a fine-structure analysis by transduction because the only recombination obtained would be for genes linked very closely to each other. In this way, he could perform an experiment similar in design to the lozenge experiment in Drosophila but utilizing the special techniques of transduction.

Over a period of time Demerec and his coworker Hartman had accumulated a number of independently arising tryptophanless mutants in Salmonella. These mutants could not grow on a minimal medium without the addition of tryptophan. They arbitrarily symbolized the mutants as tryptophan A, tryptophan B, tryptophan C, and tryptophan D. They did transduction experiments involving several combinations of these mutants as both donor and recipient bacteria. The tryptophan locus was marked on either side with other genes so that they could tell whether recombination had occurred. Through these experiments, Demerec & Hartman were able to separate the tryptophan locus in Salmonella into four different sites. The recombination of the outside markers indicated exactly what the linkage arrangement of these four tryptophan sites was. (See Figure 16-7.)

The reaction sequence for the biochemical synthesis of tryptophan is well known to biochemists (Figure 16-7). Demerec and coworkers determined which particular enzymatic blocks were involved with the four tryptophan mutants, A, B, C, and D. It turned out that tryptophan A blocked the synthesis prior to anthranilic acid. Tryptophan B involved an enzyme that converted anthranilic acid into indole glycerol phosphate. Tryptophan C was involved in the biochemical sequence of the third step, and tryptophan D controlled an enzyme that permitted the conversion of indole to tryptophan (Figure 16-7). In other words, the sequence of reaction in the synthesis of tryptophan was in the same order as the linkage map of the four mutants that controlled this sequential reaction. This result created considerable excitement

Linkage of tryptophan genes in Salmonella:

try A try B try C try D

Biochemical synthesis of tryptophan:

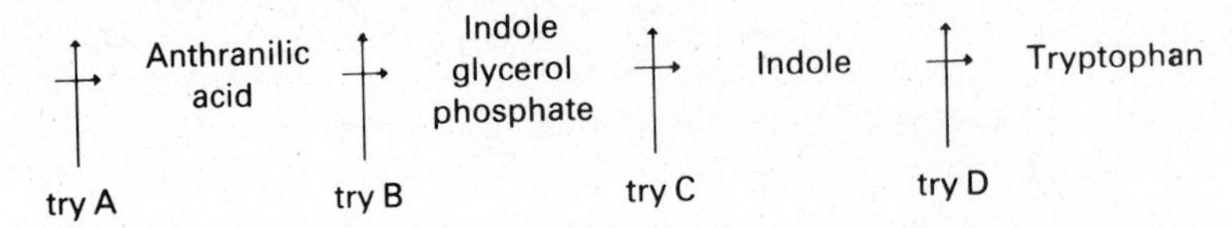

Figure 16-7. Demerec's analysis of the tryptophan locus in Salmonella. By transduction tests he separated the locus into four sites with the linkage arrangement as indicated above. The genes controlling each step in the biochemical synthesis of tryptophan are linked in exactly the same order as the sequence of steps they control.

when first announced, and geneticists and microbiologists began to construct hypotheses on why an organism would evolve a system where a linkage arrangement paralleled the biochemical sequence it controlled. It turns out that, as microbiologists looked for similar pathways in other organisms in which a close correlation between genetic linkage and biochemical sequence might exist, the tryptophan story in Salmonella is an exception. Although other examples were observed in which at least some relationship between the genetic linkage and biochemical sequence is shown, in most cases a 1 : 1 correlation was not found in other organisms. Moreover, the same biochemical pathways may show quite different relationships in different organisms.

Genetic Fine Structure in Phage

What is now recognized as the ultimate in genetic fine-structure analysis was carried out by Seymour Benzer. Benzer used as his experimental organism T_4 phage of *E. coli*. Over a period of time a large number of rapid lysing (**r**) mutants had been accumulated in T_4 phage. As indicated in Chapter 12, the **r** mutants make large, sharp plaques on *E. coli*, strain B. The wild-type allele, $\mathbf{r}^+$, makes small fuzzy plaques. Genetic analysis reveals that several independent **r** loci are involved; but the one that Benzer analyzed is known as the $\mathbf{r}_{II}$ locus, and the discussion will be confined entirely to $\mathbf{r}_{II}$ mutants of T_4 phage. It was discovered that another strain of bacterium known as *E. coli*, strain K reacted somewhat differently to the $\mathbf{r}_{II}$ mutants. The $\mathbf{r}_{II}$ mutants, when plated on *E. coli* K, make no plaques at all. They infect the cell, multiply, and kill the bacterium; but no lysis occurs and no phage are released. The wild-type, however, makes the typical wild-type plaque on K as well as on B. (See Figure 16-8.) Benzer recognized that he had a discriminating system for differentiating genetically separable mutants of $\mathbf{r}_{II}$ with a high resolution. As his experiment will show, he was able to screen with high efficiency events that might occur only rarely.

	Strain of *E. coli*	
	B	K
r_{II}^{+}	Small, fuzzy plaques	Small, fuzzy plaques
r_{II}	Large, sharp plaques	No plaques

Figure 16-8. Plaque formation of $\mathbf{r}_{II}$ mutants and the wild-type allele on two different strains of *E. coli.*

COMPLEMENTATION TEST. For his first experiment, Benzer made a mixed infection of $\mathbf{r}_{II}^{+}$ by $\mathbf{r}_{II}$ on strain K. He noticed that the K cells were lysed and that in half an hour a large clearing of the plate was observed. This is not a surprising result. Even though $\mathbf{r}_{II}$ will not lyse K, the wild-type will; and when bacteria are infected with a multiplicity of phage such that there are about five times as many phage particles as bacteria, then each bacterium is likely to be infected with two phages. The $\mathbf{r}_{II}^{+}$ is capable of lysing the cell, and the $\mathbf{r}_{II}$ that had infected and multiplied will be released.

Benzer now tried a similar experiment, but instead of using an $\mathbf{r}_{II}$ mutant and a wild-type for his double infection, he used two different $\mathbf{r}_{II}$ mutants, arbitrarily designated $\mathbf{r}_{II}^{x}$ and $\mathbf{r}_{II}^{y}$, and plated them on *E. coli* K. In many cases, as would be predicted, no lysis occurred and no plaques were formed, since $\mathbf{r}_{II}$ forms no plaques on *E. coli* K. However, Benzer did observe lysis on K in some cases involving two $\mathbf{r}_{II}$ mutants. As the experiments continued and many different combinations of two different $\mathbf{r}_{II}$ mutants were plated together on K, it became clear that the $\mathbf{r}_{II}$ mutants fell into two separate groups, which Benzer symbolized as A and B. After sorting out the mutants in these two groups, it was possible to state that the cross of an $\mathbf{r}_{II}$ mutant from Group A by another $\mathbf{r}_{II}$ mutant from Group A gave no plaques on K. Similarly a cross of a mutant from Group B by another mutant from Group B gives no plaques on K. However, the cross of an $\mathbf{r}_{II}$ mutant from Group A by an $\mathbf{r}_{II}$ mutant from Group B resulted in lysis on *E. coli* K!

Benzer concluded from this result that the two groups of mutants, A and B, were functionally different from each other. All the mutants that resided in Group A share a common functional deficiency. Similarly, all the mutants that resided in Group B share a common functional deficiency but one that is different from Group A. When a mutant from Group A is plated with another mutant from Group A, no lysis occurs because they have the same functional deficiency. But when a mutant from Group A is plated with a mutant from Group B, then, argues Benzer, they can make up for each other's functional deficiency. The two mutants complement each other! A can do what B cannot, and B can do what A cannot; and the combination gives a functionally wild-type result—lysis and the formation of plaques on *E. coli* K. This phenomenon, in which two different mutants each provide the function lacking in the other, is called *complementation.*

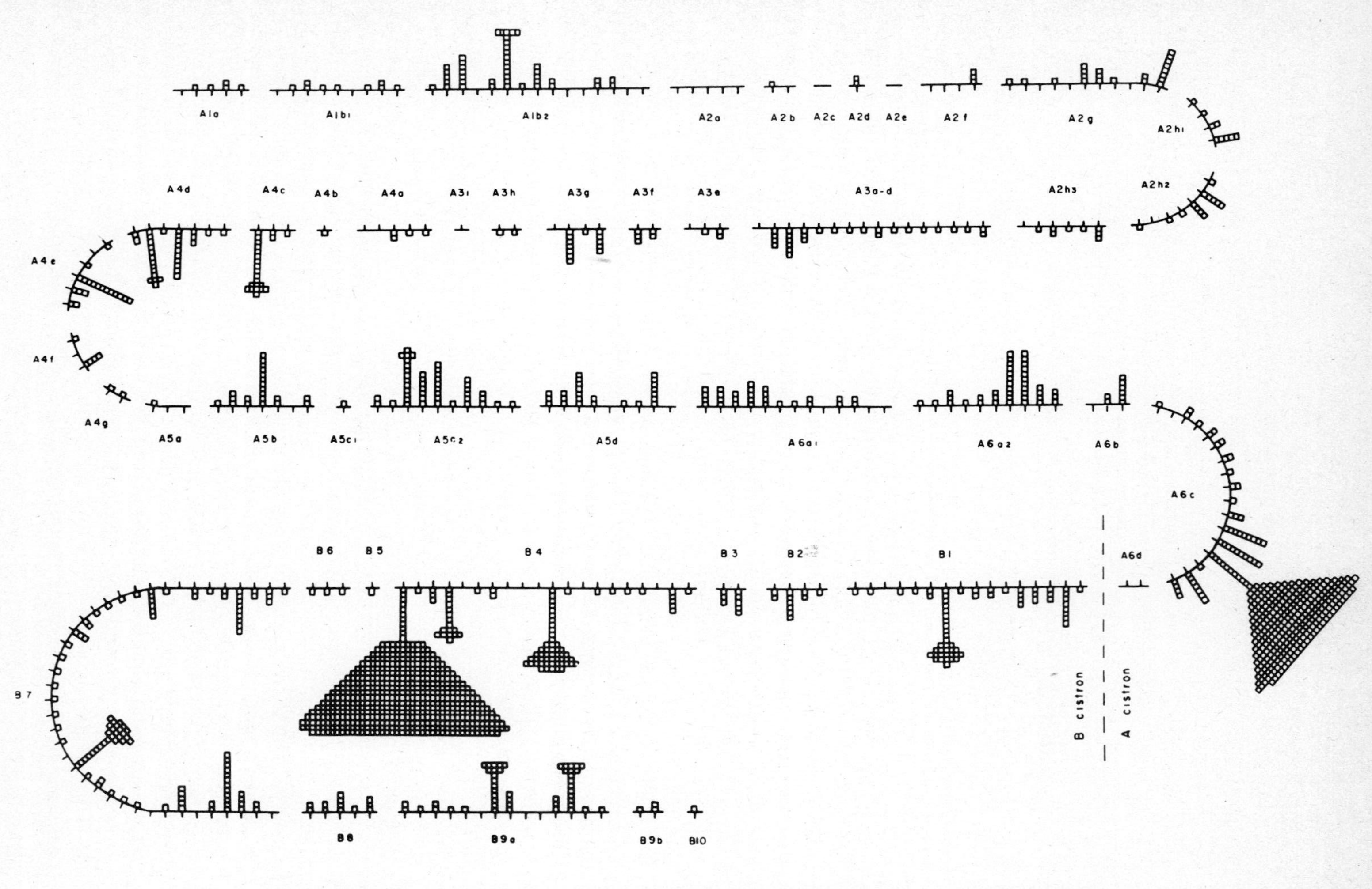

Figure 16-9. A fine structure map of the r_{II} region of T_4 phage. Within the two cistrons A and B each square represents an independent mutational event. Some sites represent highly mutable areas and are known as *hot spots.* (After S. Benzer. 1961. *PNAS* **47**.)

RECOMBINATION TEST. Benzer next asked whether the collection of mutants that fell in a single complementation group—for example, the A mutants—were all the same or were genetically separable. To answer this question, Benzer did a recombination experiment. From the A group of mutants, two independent mutations, $\mathbf{r}_{II}^1$ and $\mathbf{r}_{II}^2$, were plated in mixed infection on *E. coli* B. Since $\mathbf{r}_{II}$ mutants lyse B, plaques occur, and from these plaques a lysate is obtained that will consist of all the phage released by the lysis of the cell. The lysate is then plated on K. Since it has been shown that $\mathbf{r}_{II}$ mutants from the same complementation group never lyse K, no plaques are expected to form. Benzer reasoned, however, that if $\mathbf{r}_{II}^1$ is not genetically identical to $\mathbf{r}_{II}^2$, then recombination might occur, generating among others, phage that have both wild-type alleles. If this situation should occur, plaques will form on K. Benzer's hypothesis proved to be correct. Most $\mathbf{r}_{II}$ mutants from Group A, when subjected to the experimental procedure described, recombined, as was evidenced by the formation of plaques on K. Similar results were observed when the mutants from Group B were analyzed. Control experiments ruled out other possible explanations, such as mutation. After many experiments, he found that most of the $\mathbf{r}_{II}$ mutants are genetically separable from each other. Genetic identity proves to be the exception. By using this procedure and several technical tricks, Benzer was able to test a large number of $\mathbf{r}_{II}$ mutants for recombination and to construct a genetic map. The result of Benzer's brilliant analysis was to show that not only can the $\mathbf{r}_{II}$ mutants be separated into two complementation groups, A and B, but within each group there is a large number, measuring in the hundreds, of separable mutant sites that can be recombined by genetic analysis as well. (See Figure 16-9.)

CISTRON, MUTON, RECON. In addition to demonstrating genetic fine structure, Benzer introduced several new terms, which, to a large extent, resolve some of the semantic problems facing geneticists when they talk about genetic systems. As a result of both the complementation and recombination tests, Benzer was able to obtain recombination phage that had two mutants from the same complementation group in one phage. If Benzer mixed this double mutant phage by multiple infection with a wild-type and plated on K, plaques were obtained. As pointed out earlier, the reason is that the wild-type permits lysis to occur on K and from the lysate both the wild-type and the double mutant phage are recovered. However, if the experiment of mixing two different mutants from Group A and placing them on K is performed, no lysis and no plaques are observed. (See Figure 16-10.) Benzer realized that this position effect was similar to the cis-trans position effect observed earlier with the lozenge pseudoalleles in Drosophila. (See Figure 16-5.) Because the cis-trans test resulted in different phenotypes, and because Benzer assumed that the reason was due to the functional differences of the complementation groups, he called the unit of function the *cistron* (from cis-trans). The $\mathbf{r}_{II}$ locus consists of two cistrons, which implies that there are two functional units.

Secondly, Benzer pointed out that within each cistron there is a large number of sites in which mutations can occur, each resulting in a $\mathbf{r}_{II}$ mutant. Benzer called such units of mutation—that is, the smallest element that, when altered, gives rise to

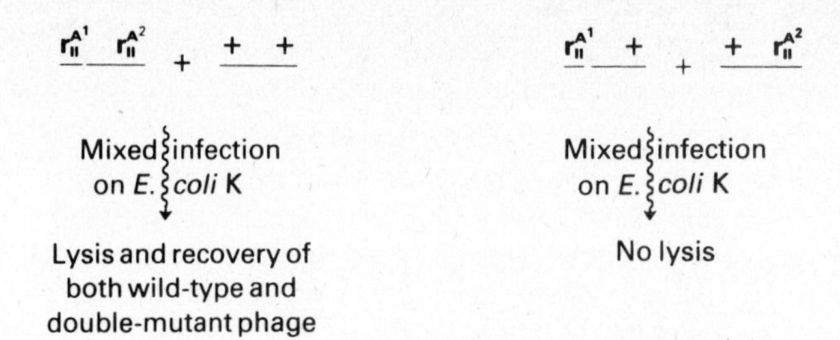

Figure 16-10. The cis-trans position effect at r_{II} locus of T_4 phage. Two different mutants from the same complementation group fail to give lysis on K because they are defective for the same function. Two mutants from different complementation groups give plaques on K because they are defective for different functions and complement each other. On this basis, plus the cis-trans position effect above, Benzer named the two groups *cistrons*, or units of function.

a mutant—a *muton.* And, finally, Benzer pointed out that there must be a smallest distance between which any two mutons can recombine. He named this smallest element of recombination, a *recon.* The term cistron as a genetic unit of function and muton as a unit of mutation have become useful, and these terms, particularly cistron, will be seen frequently in the literature. Do the terms cistron, muton, and recon help in answering the question: What is a gene? The answer is still no! The gene is, as Johannsen defined it, a unit that behaves in a Mendelian fashion. But more and more contemporary geneticists reserve for the gene its property of function, and so if any term is synonymous with gene, it is the term cistron.

The masterly analysis by Benzer of the $\mathbf{r}_{II}$ region of T_4 phage presents a picture of the gene that was unimagined a few years earlier. No longer can unitary systems be pictured. What used to be regarded as a simple gene is now visualized as a complex array of mutant sites residing in one or more functional components. More importantly, the genetic fine structure revealed by Benzer, Demerec, and others force our attention on the chemical nature of heredity. It appears as though genetic analysis has been pushed almost as far as it can go and that the final resolution of the nature of the gene must come from a study of the molecules that constitute the genetic system.

REFERENCES

BENZER, S. 1957. The elementary units of heredity. In *The Chemical Basis of Heredity*, edited by W. D. McElroy and B. Glass. Johns Hopkins Press, Baltimore, pp. 70–93. (An early review by Benzer of his work with the r_{II} mutants of T_4 phage. This paper contains a good description of Benzer's experimental techniques and the results he obtained.)

BRIDGES, C. B. 1936. The Bar "gene," a duplication. *Science* **83**:210–211. (Available in *Classic Papers in Genetics*, edited by J. A. Peters. Prentice-Hall, Englewood Cliffs, N.J., 1959.) (The cytological verification of Sturtevant's hypothesis about the Bar gene in Drosophila.)

DEMEREC, M., and P. E. HARTMAN. 1959. Complex loci in microorganisms. *Ann. Rev. Microbiol.* **13**:377–406. (A good review of the work on genetic fine structure in Salmonella.)

GREEN, M. M., and K. C. GREEN. 1949. Crossing over between alleles at the lozenge locus in *Drosophila melanogaster. Proc. Natl. Acad. Sci. U.S.* **30**:586–591. (The techniques and results that demonstrated pseudoallelism in Drosophila.)

LEWIS, E. B. 1951. Pseudoallelism and gene evolution. *C.S.H.S. Quant. Biol.* **16**:159–174. (An excellent review of the early pseudoallelic work with a theoretical discussion of how genes may have evolved.)

STURTEVANT, A. H. 1925. The effects of unequal crossing over at the Bar locus in Drosophila. *Genetics* **10**:117–148. (Available in *Classic Papers in Genetics*, edited by J. A. Peters. Prentice-Hall, Englewood Cliffs, N.J., 1959.) (A classic paper that showed position effect in Drosophila suggesting that genes may not be like beads on a string.)

QUESTIONS AND PROBLEMS

16-1. Define the following terms:

position effect
asymmetric pairing
variegated position effect
pseudoalleles
cis-trans position effect
cistron
muton
recon

16-2. What does the metaphor "genes are like beads on a string" mean?

16-3. What experimental evidence suggests that the metaphor is false?

16-4. In view of what was learned in Chapter 15 about gene action, does the knowledge of position effect surprise you? Why or why not?

16-5. White and apricot are two sex-linked mutants in Drosophila. Females that are heterozygous for these two genes are "apricotlike" in phenotype. Male progeny from such females are either white or apricot or wild-type, the latter occurring in a frequency of about one in 10,000.

(a) What do these results tell us about the relationship between white and apricot?

(b) What is the significance of this experiment for our understanding of the gene?

16-6. In Drosophila, a translocation is often accompanied by a visible phenotypic effect of a gene very near the site of one of the breaks. It is *not* thought that the gene in question has actually mutated. How would you explain this phenomenon and what insight does it give for gene action?

16-7. Demerec and Hartman's analysis of tryptophan synthesis in Salmonella indicates a gene order identical to the order of biochemical synthesis. What selective advantage might such an arrangement have?

16-8. Briefly discuss the significance of Benzer's analysis of the $\mathbf{r}_{II}$ locus of T_4 phage, particularly with respect to the nature of the gene.

16-9. What is the difference between a cistron and a gene?

17/ Proteins and enzymes

The work that led to the one gene–one enzyme hypothesis posed the interesting question of how genes can control enzymes. There are two ready answers: genes can direct enzyme synthesis or genes can produce a regulator of enzymatic activity. Before we can begin to choose between these alternatives, it is necessary to know what an enzyme is and how it acts. Without this information we cannot construct models to explain gene action. Consequently, we must digress briefly from a strictly genetic approach to consider facts about enzymes that the biochemists were uncovering simultaneously with the discoveries in genetics that have been discussed so far.

Perhaps the most important fact is that all enzymes are proteins. This point is of great significance, since proteins have a number of other functions and are the most common large molecules in cells, composing up to two-thirds of their dry weight. Many of the obvious cell structures are protein (spindle fibers, centrioles, cilia, and flagella), as are some of the not so obvious (cytoplasmic matrix, microtubules, and microfibrils). Others have a high protein content (chromosomes, mitochondria, plastids, membranes in general). Since the enzymes, the most important functional units of the cell, are also protein, it is evident that if genes can control protein synthesis or protein function, they can rule the cell.* And, of course, that is just what genes do. The hypothesis that genes controlled proteins had been around for some time, but there was no really compelling evidence to support it until the work of Beadle and Tatum. After that, it became more than a vague possibility, and many geneticists who initially were ill prepared for the idea had to make the same digression that we now make to consider the structure of proteins and their relation to cell function.

Primary Structure

The basic unit from which a protein is built is the amino acid. Figure 17-1 presents the structure of the three very different amino acids: glycine, glutamic acid, and phenylalanine. These amino acids, and all the ones involved in the formation of proteins, have one region in common—a terminal carboxyl group (COO^-) and an amino group (NH_3^+) attached to the carbon adjacent to that carboxyl group. The carboxyl group is an acid ($COOH \rightleftharpoons COO^- + H^+$), hence the name amino acid. The amino group is a base ($NH_2 + H^+ \rightleftharpoons NH_3^+$), and normally the amino acid exists as a doubly charged molecule with the proton (H^+) released by the carboxyl group being taken up by the amino group. One of the most important features of proteins is the charged character of the amino and carboxyl groups.

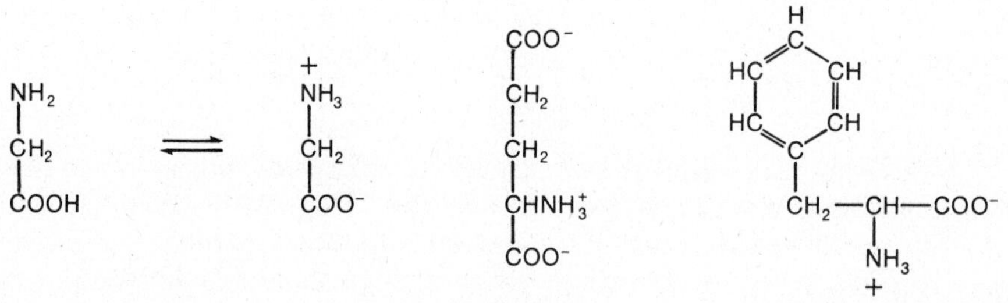

$$NH_2{-}CH_2{-}COOH \rightleftharpoons {}^{+}NH_3{-}CH_2{-}COO^-$$

$$^{-}OOC{-}CH_2{-}CH_2{-}CHNH_3^+{-}COO^-$$

$$C_6H_5{-}CH_2{-}CH(NH_3^+){-}COO^-$$

Glycine **Glutamic acid** **Phenylalanine**

Figure 17-1. Three amino acids. Glycine is the simplest amino acid. It consists of a carboxyl ($COOH$, COO^-) group and an amino group (NH_2, NH_3^+) attached to the carbon that is adjacent to the carboxyl group. All of the important amino acids have this form. The amino acids can exist in either the charged or uncharged forms, as indicated for glycine. Under physiological conditions, the charged form is the most common.

*The importance of proteins was recognized very early. In fact, their name was derived from the Greek word *proteios*, meaning primary, by Müller in 1838.

Glycine is the simplest amino acid. The others are larger than glycine and, in some cases, may have relatively large "side chains." Glutamic acid has a three-membered side chain, containing a second carboxyl group. The side chain of phenylalanine contains a six-membered ring. The character of these side chains is of great importance in determining the higher levels of structure of proteins. For the moment, while we are discussing the primary structure, the side chains can be ignored. In principle, it is possible to make a protein that is composed solely of glycine, and, in fact, silk is a protein with an extremely high glycine content.

The amino acids are joined together within the protein by peptide bonds. Figure 17-2 shows the formation of two such linkages between three amino

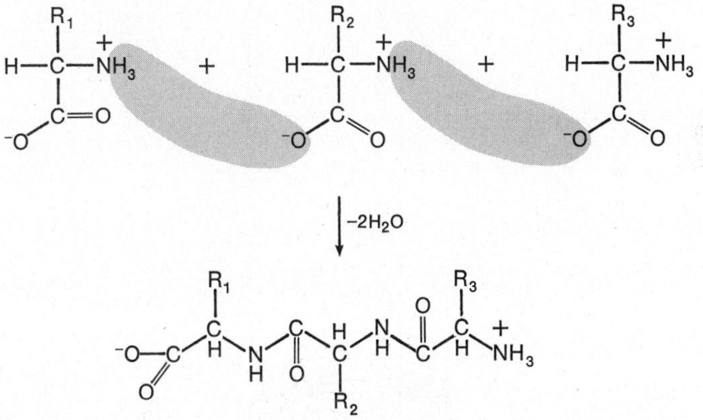

Figure 17-2. Peptide synthesis. The peptide backbone of a protein is formed with the loss of water in the reaction of the carboxyl (COO^-) and amino (NH_3^+) groups. R_1, R_2, R_3 represent the side chains of any of the 20-odd different amino acids. The illustration shows a tripeptide; proteins are polypeptides. Regardless of chain length, the basic structure is the same, with a single, free amino group at one end (N terminal) and a single, free carboxyl group at the other end (C terminal).

acid molecules. The basic structure of the protein, then, is a series of amino acids joined together by peptide linkages, having at one end a free carboxyl group (the C-terminal end) and at the other end a free amino group (the N-terminal end). The side chains, which are represented as R_1, R_2, R_3, are not involved in the peptide linkage and could be anything.

In one sense, a protein is a polymer of a large number of units joined together. On the other hand, in the strict sense, it is not a polymer, since the units may be very different indeed. Cellulose is a polymer composed of many glucose residues joined together. As such, it is a true polymer. In a protein, however, the only constantly repeated structure is the peptide linkage itself. In each case, the amino acids are joined by the same linkage, and thus the "backbone" of the protein is formed by a series of such peptide linkages. It is for this reason that proteins are named in terms of the peptide linkage: dipeptide, tripeptide, tetrapeptide ...

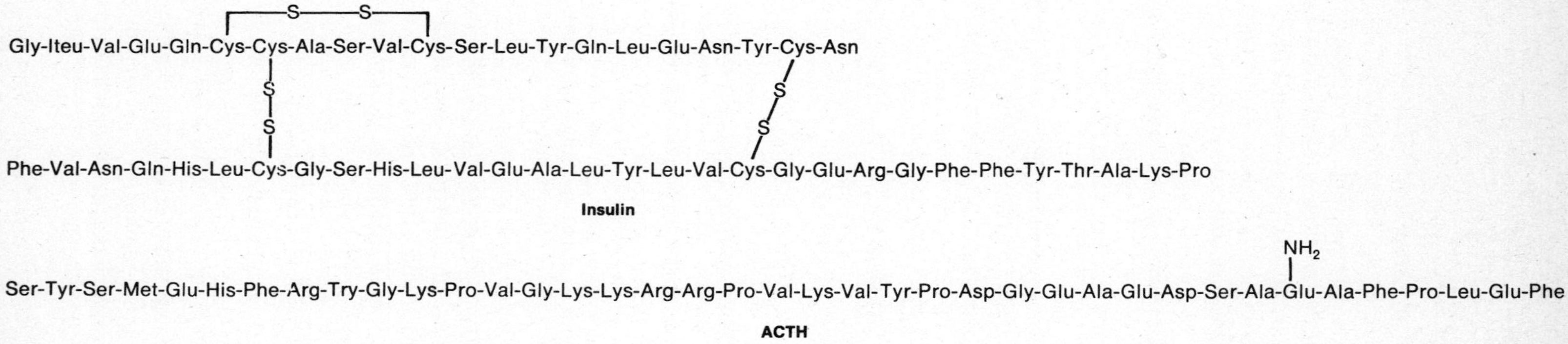

```
     ┌─────S─────S─────┐
Gly-Iteu-Val-Glu-Gln-Cys-Cys-Ala-Ser-Val-Cys-Ser-Leu-Tyr-Gln-Leu-Glu-Asn-Tyr-Cys-Asn
                         |                                                  /
                         S                                                 S
                         |                                                /
                         S                                               S
                         |                                              /
Phe-Val-Asn-Gln-His-Leu-Cys-Gly-Ser-His-Leu-Val-Glu-Ala-Leu-Tyr-Leu-Val-Cys-Gly-Glu-Arg-Gly-Phe-Phe-Tyr-Thr-Ala-Lys-Pro
```

Insulin

```
                                                                                                                 NH2
                                                                                                                  |
Ser-Tyr-Ser-Met-Glu-His-Phe-Arg-Try-Gly-Lys-Pro-Val-Gly-Lys-Lys-Arg-Arg-Pro-Val-Lys-Val-Tyr-Pro-Asp-Gly-Glu-Ala-Glu-Asp-Ser-Ala-Glu-Ala-Phe-Pro-Leu-Glu-Phe
```

ACTH

Figure 17-3. Amino acid sequence of two protein hormones. Both proteins are relatively small. ACTH is a single polypeptide; insulin is composed of two different polypeptides joined by covalent disulfide (S—S) linkages. In reality, both would have a three-dimensional configuration. The flat projection given here distorts the bond lengths of the disulfide bridges. The amino acids are shown according to the prevailing convention that uses the first three or four letters of their names (refer to Figure 17-5).

polypeptide. The names are slightly illogical, since the dipeptide contains only one peptide bond and so on, but they describe accurately the fact that the peptide linkage is the basic structural unit and the number of amino acids joined together by that linkage.

Most proteins are relatively large molecules and, as such, are best described as polypeptides. However, the terms polypeptide and protein cannot always be used interchangeably. They can whenever a protein is composed of one polypeptide chain only. However, in cases where two or more polypeptide chains join together to make a protein, the words have distinct meanings. Thus the A chain of insulin is a polypeptide, and the B chain is similarly a polypeptide. However, the insulin molecule is a protein composed of those two polypeptide chains. Figure 17-3 presents the amino acid sequence of two different proteins, ACTH and insulin. ACTH is a very small protein by usual standards, and even insulin is below average. Thus it should be evident that proteins are relatively large molecules.

The size of proteins makes analysis of their structure difficult, and it is only since the 1950s that the technical problems presented in such an analysis have been solved. However, it is now possible to break up polypeptide chains and analyze them for their constituent amino acids so that we can determine the exact sequence of amino acids in a protein. That is how we know the structure of the proteins presented in Figure 17-3. The details of this kind of analysis are far beyond the scope of this text. For our purposes, it is sufficient to know that, given enough time and enough money, any protein that can be purified can be analyzed for its amino acid sequence. What is critical to us is that no protein that has been so analyzed has been found to contain anything other than the twenty-odd amino acids characteristic of all living cells. While other molecules may sometimes be attached to the periphery of proteins, the polypeptide itself is the fundamental structure. Furthermore, different proteins do not seem to share any common sequences of amino acids. That is, each type of protein apparently has its own unique sequence of amino acids; but within any one protein type, the amino acid sequence is invariant. That is, every time insulin molecules are isolated from pig pancreas, the amino acid sequence determined is the same. Consequently, it is clear that the control of protein synthesis must involve a mechanism that can specify the exact placement of each amino acid in each polypeptide chain!

Higher Levels of Structure

It is far too simpleminded to think of a protein as just a string of amino acids. Most proteins have a complex three-dimensional structure that is the result of interactions between the various amino acid side chains. To understand this situation at even the most general level, it is necessary to consider what those side chains are. (A somewhat more detailed discussion is given in the appendix to this chapter.) Figure 17-4 presents the structure of the 20 amino acids most commonly found in proteins. They have been grouped together according to the similarities of their side

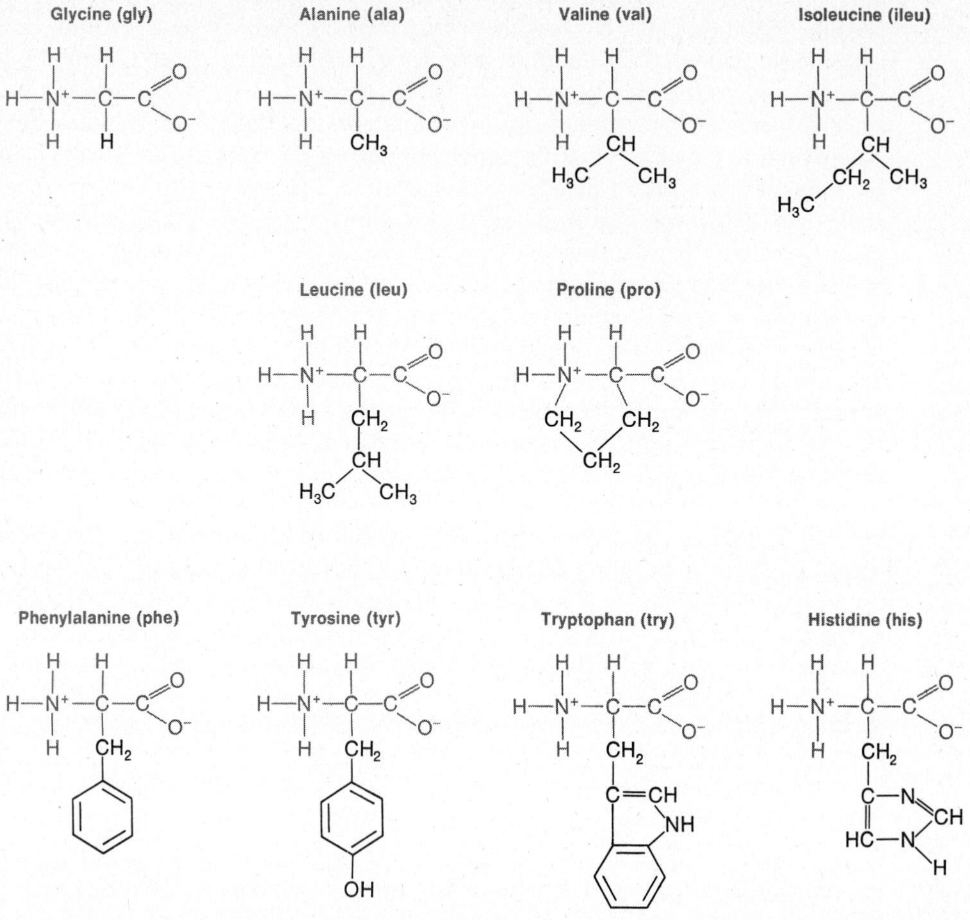

Glycine (gly) Alanine (ala) Valine (val) Isoleucine (ileu)

Leucine (leu) Proline (pro)

Phenylalanine (phe) Tyrosine (tyr) Tryptophan (try) Histidine (his)

chains. (Concentrate only on these; ignore the groups "buried" in the peptide linkages.) For example, glutamic acid and aspartic acid carry a negative charge. Since similar charges repel, they cannot come close to one another. On the other hand, the positive charge on lysine will attract the negatively charged glutamic acid. Should they come into contact as the polypeptide folds up, they will interact in such a way that they will remain together until some force is exerted to separate them.

Uncharged side chains can also interact. The side chains of leucine and valine are not soluble in water, but they do have a slight attraction for one another (van der Waals associations). The proteins in the cell are normally quite hydrated. Consequently, should two (or more) such groups happen to come into close contact, they will stick together. If they are grouped together tightly, no water will be present and they will be held by their mutual attraction. And so it goes. With the addition of the covalent disulfide bonds (S—S) that can form between the side chains of cysteine

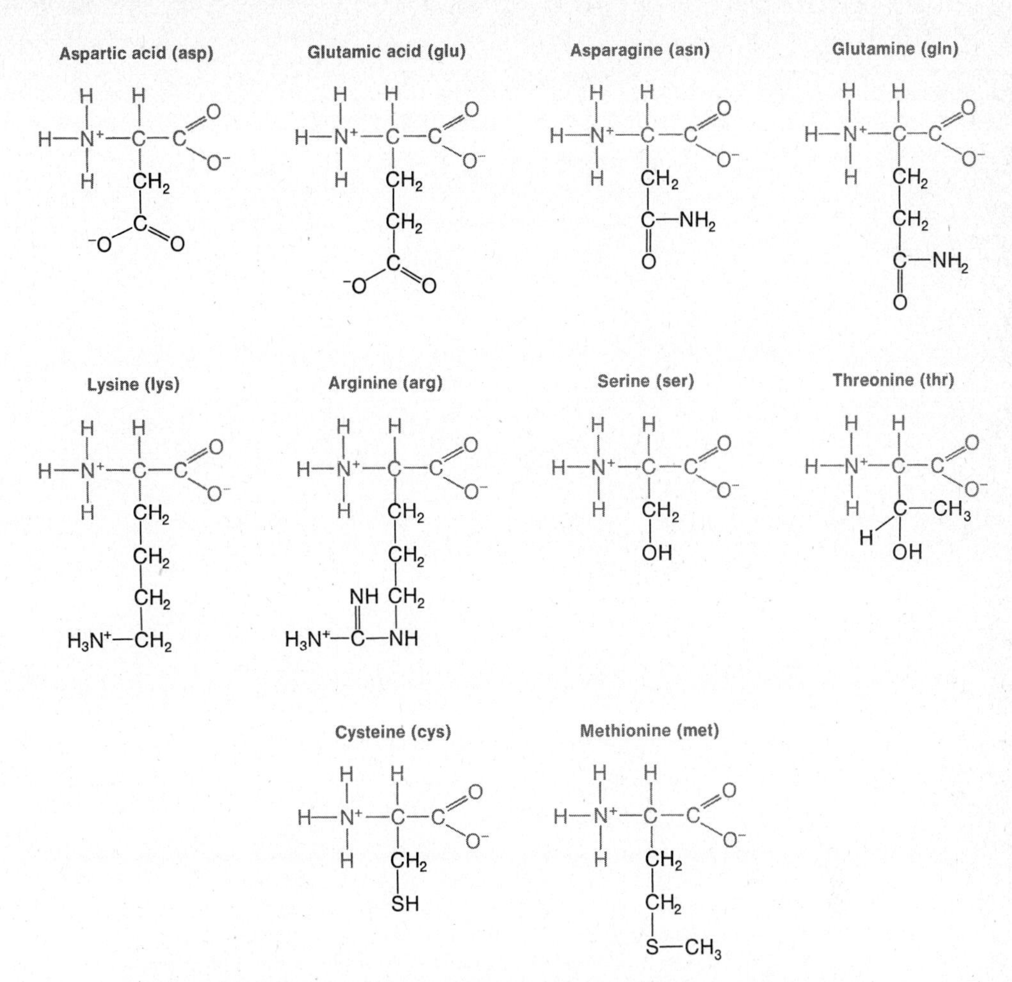

Figure 17-4. Twenty amino acids. The twenty amino acids shown here are the basic units from which all proteins are made. The simplest one is glycine. All the rest are derived from it to the extent that each can be seen to have the same carboxyl group (COO^-) and amino group (NH_3^+); only the side chains (indicated by the dashed lines) are different. It is the side chains, extending out from the peptide backbone of the protein, that give the bonding possibilities to develop higher levels of structure. The amino acids are grouped according to reactive similarity of the side chain. To make this more evident, the side chains are shown larger than the parts normally "buried" in the peptide linkage. The commonly used abbreviation for each amino acid is given with the name.

molecules and the ubiquitous hydrogen bonds, the associations that can be generated *within* a folded polypeptide are more than sufficient to provide a quite stable, highly convoluted, three-dimensional configuration (Figure 17-5).

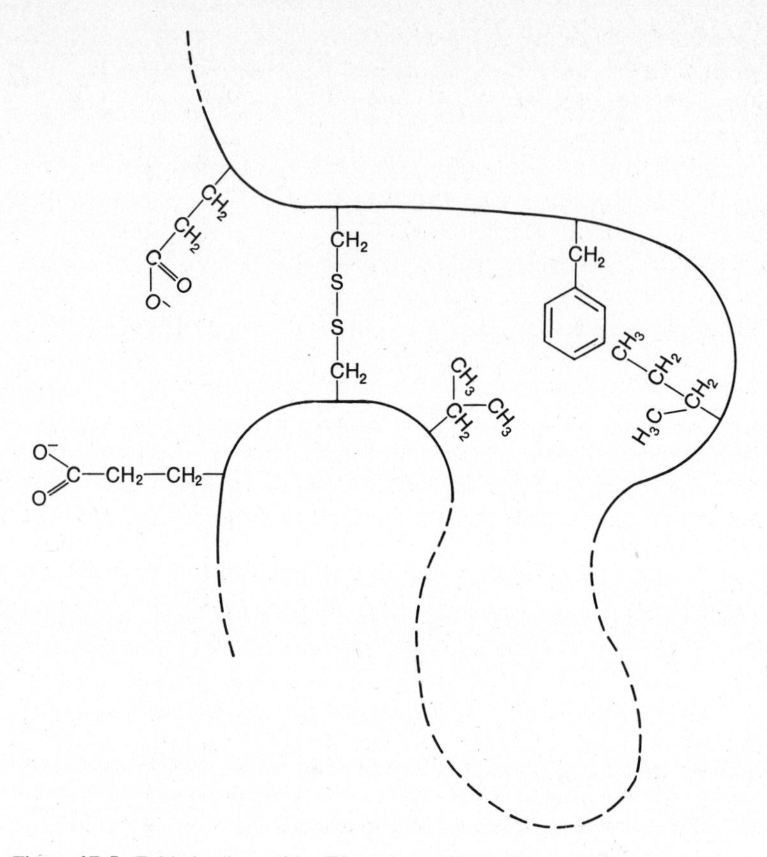

Figure 17-5. Folded polypeptide. The polypeptide backbone is shown as a heavy black line; the side chains of the amino acids extend from the backbone. For simplicity, the diagram is shown in only two dimensions and the side chains are enlarged. One part of the folding is caused by the interactions of several uncharged amino acids. In another region, a disulfide linkage holds two parts of the chain together. In a third region, the parts of the chain are forced apart by the repulsion of two negatively charged amino acids.

The preceding discussion implies that the shape ultimately taken by a polypeptide will depend on the placement of amino acids within the chain. If two cysteine residues are next to one another in the primary structure, they can interact to form a disulfide linkage. However, what if they are separated by another amino acid? In this case, it would be necessary for the polypeptide to bend in order for the two to come close enough to interact and for the intervening amino acid to be displaced to one side. In fact, this process would be almost impossible to achieve, for the peptide bonds holding the chain together are not that flexible. In reality, it would be necessary for the two cysteines to be rather far apart in order to get enough flexibility in the chain to permit them to be brought close together. Exactly the same thing is true for any of the other possible interactions, which suggests that the placement of the various amino acids in the chain is far from random. The placement mechanism must be one that ensures that each amino acid is put in the right place in the chain such that it can contribute to the internal bonding that gives the protein its final form.

Is there any justification for the hypothesis that the placement of the amino acids in the primary structure is the determinant of the final form of the protein, that the shape of the molecule is solely the result of the spontaneous interactions of the various residues in the polypeptide chain? In fact, there is. It is possible to cause enzyme molecules to unfold. This step causes an immediate loss of enzymatic activity. However, if the proper conditions are restored, the protein will spontaneously refold into the same configuration it had before the treatment, and the enzymatic activity returns to normal. There appears to be no need of "little mem" to fold up the polypeptide into its unique shape, nor is it necessary to invoke a multitude of enzymes for that purpose. Once the primary structure has been determined, the rest follows spontaneously. This fact is of great significance, of course, since it suggests that genes could control enzymatic activity by controlling the placement of amino acids in the polypeptide chain.

The same generalizations that have just been made about the three-dimensional configuration of a single polypeptide apply to more complex proteins as well. As stated earlier, not all proteins are single polypeptide chains. In many cases, two or more chains may associate to give rise to the functional protein. The forces holding the chains together are the same forces that operate within the chain: the interaction of the various amino acid side chains. In many cases, disulfide linkages are formed that cross link the molecule rather permanently. In other cases, no such covalent linkages are utilized, and the chains are held together only by relatively weak interactions. In such cases, treatment with mild reagents can be used to separate the chains in the laboratory. However, under the conditions of the cell, the summation of the large number of these interactions is sufficient to give the molecule stability. Consequently, we treat the associated polypeptide chains as a single molecular unit.

The varieties of such associations are rather large. For example, the enzyme *alkaline phosphatase* has two identical chains. Insulin also has two chains, but each of the chains is different. Human hemoglobin consists of four chains, two of the α type and two of the β type. One of the most extreme cases, *ferritin*, a protein of 480,000 molecular weight, is composed of 20 identical smaller polypeptides. As with the primary structure of proteins, there appears to be no regularity of pattern in the multichain proteins. How many chains are present and what types of chains associate appear to be simply a matter of the primary structure of the chains themselves, which provides the right configuration such that the various chains can fit together physically and form a large enough number of cross linkages to provide the necessary stability. In this context, it is worth noting that there appears to be an upper limit to the size of a single polypeptide. The largest so far encountered has 400 amino acids, but generally the individual chains have no more than 200. Very large proteins are invariably made by the interaction of several polypeptide chains.

Finally, we should mention that two or more proteins (not polypeptides) may associate spontaneously to form a macromolecular complex. The formation of virus particles occurs in this way, as does the association of certain multienzyme complexes of the cell. Frequently such aggregates involve the participation of molecules other than protein to stabilize the interactions, the two most common being nucleic acid (in the formation of virus particles and even whole chromosomes) or lipids (as is characteristic of the membranous organelles of the cell, such as the mitochondria and plastids). The evidence to date suggests that the associations are spontaneous. And although much more evidence is needed in this area before final conclusions can be drawn, the data are highly suggestive of the idea that most of cell structure results from the spontaneous aggregation of large molecules, with proteins acting as the principal source of functional organization. In this light, our original proposition that whatever controls the primary structure of proteins controls the living condition seems very close to the truth, even though at first it appears much too sweeping a generalization.

The Protein "Surface"

Figure 17-6 presents a drawing of the three-dimensional structure of myoglobin. The function of this protein is to carry oxygen in muscle. It is a single polypeptide that has attached to it the heme group that is the functional unit of the molecule (shown in black). (The oxygen molecule is bound to an iron atom that is part of the heme group.)

The highly convoluted structure of this polypeptide is evident. The chain folds back on itself many times to form a compact mass. In many ways, the structure is analogous to that of a sponge. Much of the volume is free space, into which water and small ions can penetrate. Still, it has a definite shape. Just as different species of sponges have distinctly different forms, so different proteins have distinctly different shapes. And so it is possible to talk about the surface of a protein in a real sense, even though it may be highly indented and very porous. The ability of two polypeptide chains to fit together will depend on the existence of spaces on the surfaces of the two that permit them to fit together, much like pieces in a jigsaw puzzle. Similarly, areas of the protein that are to be used for special reactive purposes must be carried on the surface of the molecule, as is the heme group in myoglobin, in a region where the molecule(s) with which the protein is to react can gain access to the reactive site.

Proteins present an added dimension when we talk about their surface, something that makes the sponge and jigsaw analogies inaccurate. The interaction between two polypeptides or a polypeptide and some smaller molecule involves the formation of chemical bonds. It is not enough that the general shape be right to permit the two "pieces" to come together. If the spaces do not provide possibilities for bonding between the two, nothing will come of it. For example, if the cavity of one piece is lined with negatively charged amino acid side chains, it will be impossible to insert a negatively charged piece into it, no matter how good the geometrical fit

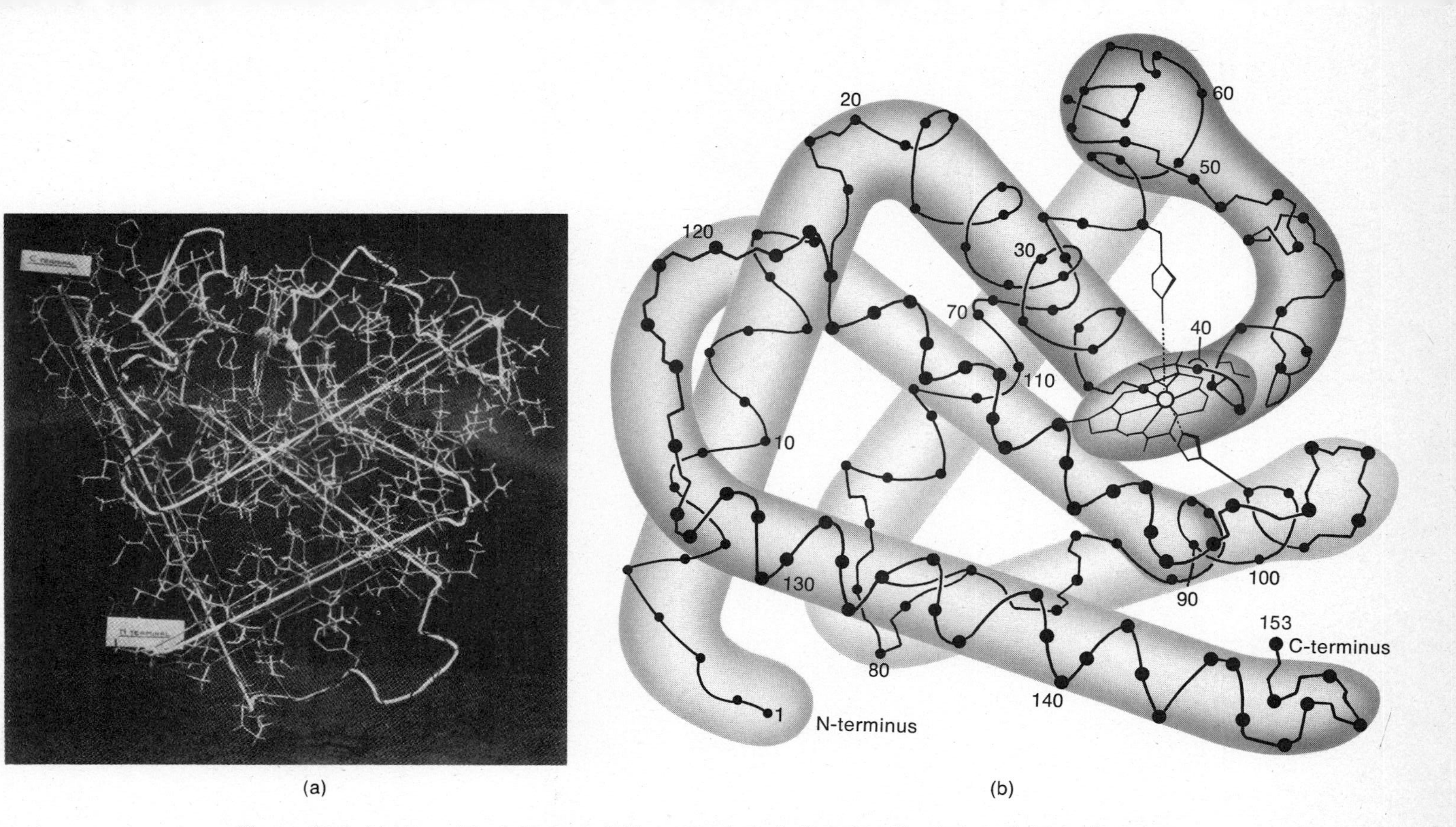

Figure 17-6. (a) A model of the myoglobin molecule including the side chains of the various amino acids. The course of the main chain is indicated by a white cord woven through the model. The iron atom is indicated by a grey sphere. Note that much, but not all, of the space within the folded molecule is occupied by the side-chains. (From Kendrew, J. C., H. C. Watson, B. E. Strandberg, R. E. Dickerson, D. C. Phillips, and V. C. Shore. 1961. Nature *190:* 666.) (b) An outline of the conformation of the myoglobin peptide backbone. The numbers represent amino acid residues, starting with the N-terminal position. (From Sidney A. Bernhard. *The Structure and Function of Enzymes.* © 1968 W. A. Benjamin, Inc. Menlo Park, California.)

may be. Similarly, if the pieces do not form at least a few bonds, the interaction between the two is likely to be too unstable to last for any appreciable time. Consequently, we must think not only of the general shape of the protein in describing its *functional* surface but also of the three-dimensional placement of various reactive groups on that surface. It is this ability to place reactive groups in exactly the right positions that determines the ability of protein to undergo reactions with molecules of one particular configuration but not with others. This property of specificity of reaction is essential to the function of enzymes, to which we now turn.

Enzymes

As indicated in Chapter 15, enzymes are proteins that act as biological catalysts. As such, they can do no more than increase the rate of reactions within the cell. However, the enzyme has one important property as a catalyst. It is specific for the

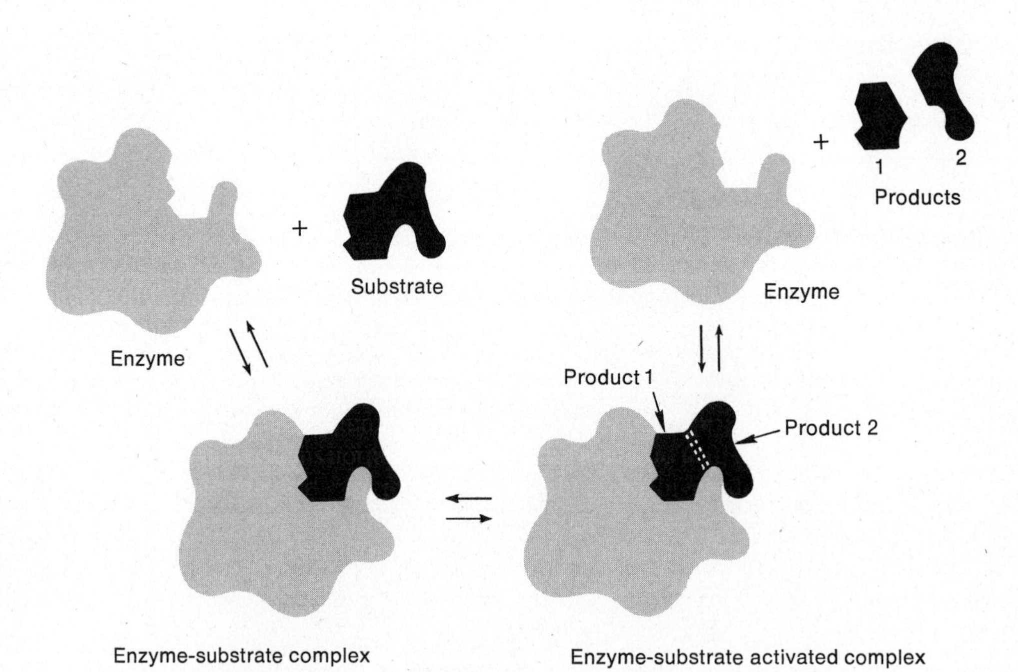

Figure 17-7. Reaction of enzyme with substrate. The enzyme has a configuration that permits it to accept only one molecular type. Once the substrate has attached to the enzyme, to form an enzyme-substrate complex, the enzyme can accelerate the reaction in which the substrate is converted to product. The reaction pictured involves a single substrate molecule ($A \rightleftharpoons B + C$), but two (or more) molecules may be used ($A + B \rightleftharpoons C + D$). (After McElroy, W. D. 1964. *Cellular Physiology and Biochemistry*. 2nd ed. Reprinted by permission of Prentice-Hall, Inc., Englewood Cliffs, New Jersey.)

reaction that it catalyzes. In some cases, this specificity may be absolute, such that it will react with one type of molecule and one only (capable even of distinguishing optical isomers). In other cases, the specificity may be broader, such that a wide variety of molecules will be accepted by the enzyme, provided only that they have a certain region in common. For example, alkaline phosphatase (*ase* is the suffix denoting an enzyme) will catalyze the removal of phosphate groups from any of a large number of molecules (at alkaline pH). In all cases, however, we are dealing with a catalyst that apparently "recognizes" a certain three-dimensional configuration and that will react with that configuration only.

For simplicity, the molecule on which the enzyme acts initially is called the *substrate* and the molecule that is produced as a result of the activity is called the *product.* Thus in the reaction A $\rightarrow$ B, A would be the substrate and B the product. Similarly, the enzyme is named in terms of its substrate rather than its product. For example, alcohol dehydrogenase is an enzyme that removes hydrogens from alcohol, and a protease is an enzyme that breaks down protein.

The fact that enzymes show specificity with regard to substrate, accepting certain substrates and rejecting all others, can best be explained on the basis of the surface configuration of the protein molecule. Figure 17-7 shows the substrate and enzyme fitting together like pieces in a jigsaw puzzle. Obviously this is an oversimplification, since it attempts to do in two dimensions what must happen in three and neglects bond formation. On the other hand, the available data suggest that the degree of matching between the configuration of the part of the enzyme that accepts the substrate (the active site) and the substrate molecule itself is high indeed.

One of the most important pieces of evidence for the configurational hypothesis of enzyme specificity comes from the inhibition of enzyme activity by compounds that "look" like the normal substrate but are sufficiently different to preclude the reaction from being completed. For example, malonic acid is a potent inhibitor of the enzyme that normally catalyzes the oxidation of the very similar succinic acid (Figure 17-8a). In this case, the malonic acid is sufficiently similar to the normal substrate to be able to enter the active site and initiate binding; however, it is too small to complete it. We can propose that normally the succinic acid is bound to the active site through its two carboxyl groups and at a third site where the two hydrogens are removed from the carbons between the two carboxyl groups. The malonic acid lacks one of these carbons, and, consequently, no reduction can be effected. The normal result of adding malonic acid to this system is a decrease in the rate of conversion of succinic acid to the oxidized product, fumaric acid. Thus the malonic acid acts as an inhibitor of the reaction, presumably by occupying the active site of a certain number of enzyme molecules and reducing the effective concentration of active enzyme in the system (Figure 17-8b).

On the other hand, if the concentration of succinic acid is raised to very high levels, relative to the malonic acid, normal enzymatic activity is restored. That is, the inhibition caused by malonic acid can be reversed by the normal substrate of the enzyme. This situation shows two things. In the first place, it indicates that the combination of the enzyme with either the substrate or the inhibitor is a reversible

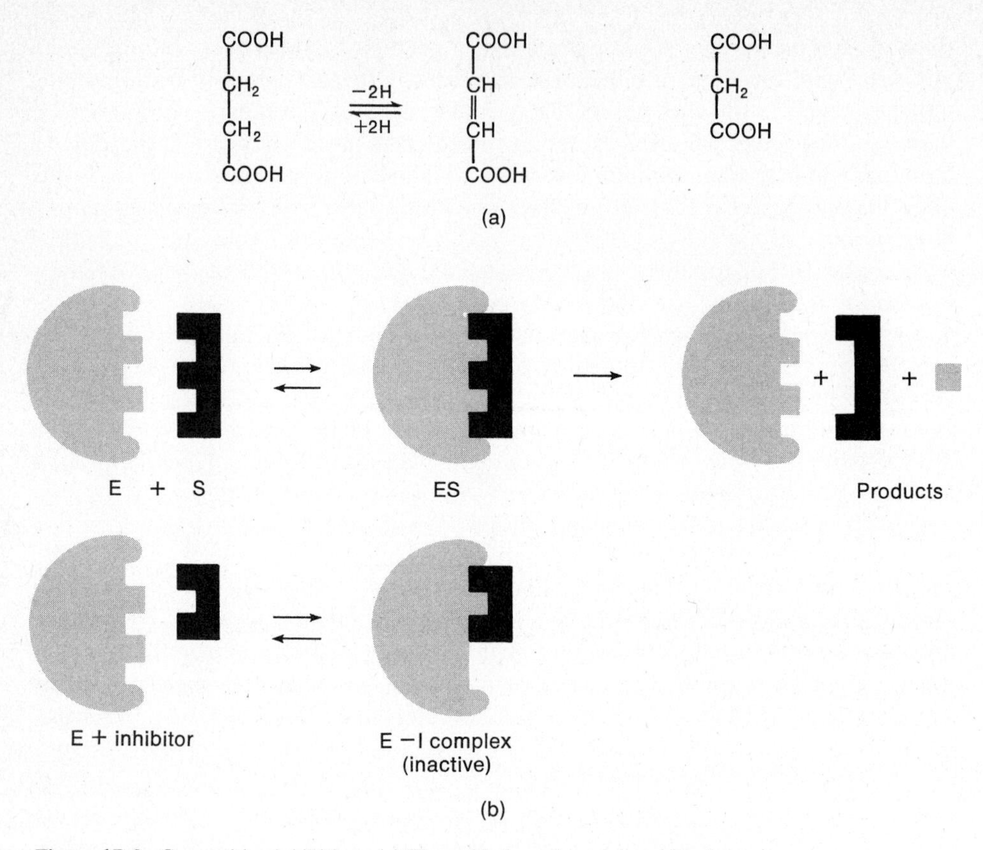

Figure 17-8. Competitive inhibition. (a) The oxidation of succinic acid to fumaric acid is catalyzed by the enzyme succinic dehydrogenase. Malonic acid, which bears a strong resemblance to succinic acid, is a strong inhibitor of the reaction. (b) A diagrammatic representation of competitive inhibition by molecules that closely resemble the normal substrate of an enzyme. (After McElroy, W.D. 1964. *Cellular Physiology and Biochemistry*. 2nd ed. Reprinted by permission of Prentice-Hall, Inc., Englewood Cliffs, New Jersey.)

process. In this case, the enzyme reacts with the malonic acid, holds it for a short period of time, and then releases it, at which point it may pick up either another molecule of malonic acid or one of succinic acid. If there are many more succinic acid molecules than malonic acid molecules, the active sites of the enzymes are soon occupied by the succinic acid and the reaction proceeds at its maximum rate (inhibition is removed). It also suggests that, like a true catalyst, the enzyme molecule itself is not damaged in any way in the interaction between enzyme and inhibitor or enzyme and substrate.

If a molecule that does not resemble or that is considerably larger than succinic acid is added to the system, no such competitive inhibition is observed. Thus the active site is discriminating in what it will accept. It binds only specific three-dimensional configurations of atoms, which suggests, once again, that the placement of the reactive groups on the surface of the active site has a specific pattern.

For our purposes, it is unnecessary to consider the variety of reactions that can occur between substrates and enzymes. However, it should be evident from the preceding discussion of the various levels of structure of proteins that the bonding possibilities provided by the side chains of the various amino acid constituents of the polypeptide chains are numerous indeed and are quite satisfactory to account for the specific acceptance of any particular kind of organic molecule. An enzyme whose normal substrate contains a negatively charged group might well have a positively charged lysine side chain in its active site, and so forth. When we consider the fact that the protein contains these specific kinds of bonding possibilities and through its tertiary configuration can construct a region in which the proper kinds of binding posts are placed in the proper relationship to one another in three-dimensional space, it becomes clear that the protein is the logical choice for the role of enzyme. No other kind of molecule can offer this specificity of function.

It is important to emphasize that an enzyme only *accelerates* a reaction. Any reaction that is enzymatically catalyzed could occur spontaneously. The function of an enzyme is to increase the rate of reaction. However, the rate of the uncatalyzed reaction in many cases would be exceedingly low, so low, in fact, that needed compounds often would not be synthesized within the lifetime of the cell. Life itself is a rapid process. A bacterium completes its life cycle in a matter of minutes. Even the human being passes from a fertilized egg to senility in only three score years and ten. Compared to most spontaneous processes, such as the weathering away of rocks, a living system can be seen to be carrying out its reactions in almost magical times. It does so only because each step in a reaction sequence within the cell has an enzyme, which ensures that the rate of conversion of substrate to product in each reaction is as rapid as possible. Enzymes speed up reaction rates by 10^3 to 10^6; and since each reaction is carried out rapidly, the entire network of reactions that constitute the cellular metabolism goes on at great speed. Once again, we can see the importance of proteins to life.

Finally, it is worth emphasizing that the enzyme diagrammed in Figure 17-7 is not a solid structure like a lump of clay but is rather a structure that is made by the folding up of a single polypeptide chain (or several polypeptide chains) into a glob. When we speak of the proper placement of the reactive amino acid side chains in the active site, there is no reason to presume that these amino acid side chains are in any way close to one another in the *primary* structure. In fact, they may be far apart and may only appear together at the active site through the regular folding of the polypeptide chain. Figure 17-9 makes this point in a simple manner. It is important for the reader to understand this fact, since it is on this basis that we assert that it is ultimately the placement of amino acids within the primary structure that determines the enzymatic activity of a protein. Should some accident occur to alter one of these amino acids, the enzyme would lose its activity. Should the enzymatic protein be unfolded experimentally, the enzymatic activity of the protein will be lost,

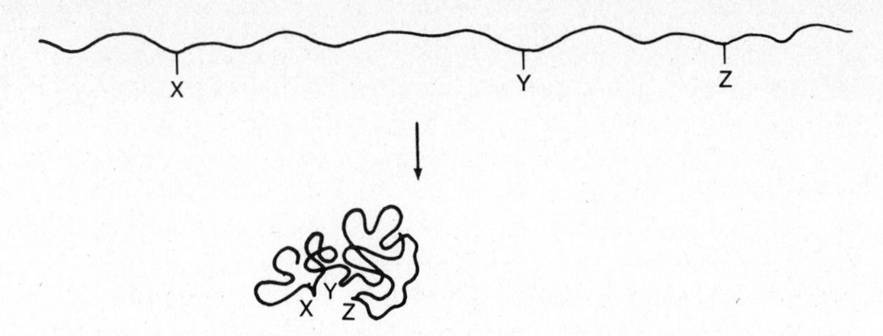

Figure 17-9. Formation of the active site. In this diagram the polypeptide chain containing the three amino acids whose side chains will contribute the three important binding posts of the reactive region of the enzyme (the active site) is shown at the top in an extended condition. The three reactive amino acids are at some distance from one another. At the bottom, the same polypeptide is shown in a folded configuration, which brings the three reactive side chains close together. In the folded molecule the reactive groups are brought into the proper spatial relationship to react with a particular substrate and only with that substrate.

even though all the amino acids remain unchanged. Activity depends both on certain reactive groups being present and on certain spatial relationships between these groups being preserved. Both factors are determined by the primary structure of the protein. With this fact in mind, it is possible to project what the function of genes must be. Clearly, if the one gene–one enzyme hypothesis is correct, then it is probable that genes can somehow ensure the proper placement of the amino acids within the polypeptide chain at the time of its synthesis.

APPENDIX

THE SECONDARY AND TERTIARY STRUCTURE OF PROTEINS

Beyond the primary, four levels of structure can be used to describe proteins. Two of these levels, the secondary and tertiary, are common to all proteins. The other two, quarternary and quinary, are used in discussions of rather complex proteins and protein aggregates. As described in the body of the text, proteins may consist of more than one polypeptide. When they do, we say that they exhibit quarternary structure. Similarly, proteins may aggregate to form multiprotein complexes. In this case, we speak of quinary structure. Since the forces involved in these last two special cases are the same as those responsible for the secondary and tertiary structure, we will not discuss them in more detail. Rather, in this appendix, we will present a brief account of forces that maintain (and generate) the three-dimensional structure of a single polypeptide. This material is presented for those readers who have an aquaintance with chemistry and who would like a somewhat more complete understanding of the material presented in this chapter.

Secondary Structure

From our point of view, the secondary structure is the least interesting. Briefly stated, it describes the fact that some proteins can assume a rodlike, helical* configuration. We will discuss this fact here to make three points about it. (1) The discovery of the helical structure of proteins was the first indication that biologically important macromolecules could assume a complex but regular configuration; they need not be either straight chains or random coils. (2) The forces responsible for the generation of the helix are the hydrogen bonds formed between different regions of the peptide backbone. (3) The helix characteristic of proteins is a hollow rod, which leaves the various side chains of the amino acids free, extending from the surface of the rod. This last feature is what makes possible the tertiary configuration of proteins, and we will discuss its consequences in that section. Here we will expand on the first two points, for they tell us a great deal about all macromolecules.

The first indication of a regularity of structure came from X-ray diffraction studies. X-ray diffraction is a rather complicated technique by which atoms can be located in three-dimensional space. We can summarize the process briefly by saying that when atoms are regularly arranged, they can act in the same way that lines on a grating act to diffract light. Since the space between atoms is very small, very short wavelengths must be used (hence X-rays). By observing the diffraction pattern generated when X-rays are passed through a collection of molecules, it is possible to determine exactly the location of each of the various atoms in the molecule, since, like the lines on a diffraction grating, the relative separation of the different atoms will have strong effects on the pattern produced. The technique of X-ray diffraction has been used since 1912 to establish the structure of crystals, in which the spacing of the atoms is regular and repetitious. In the 1930s the first attempts were made to apply this technique to the structure of the macromolecules of cells, particularly to proteins. Initially they met with little success, for it is difficult in most cases to obtain crystals of these molecules. However, it was highly suggestive that when *fibrous* proteins were examined, relatively clear diffraction patterns were observed. This fact could only mean that in the fibrous proteins the atoms were held relatively rigidly, in what we might call a semicrystalline condition.

The problem was to explain how a regular pattern of the spacing of atoms could be achieved with molecules whose internal constituents (amino acids) were so different. There was no evidence for the repetition of different kinds of amino acids. Therefore it was proposed that the regularity resulted from some higher order of structure than the primary sequence. In 1951 Pauling and Corey published a

*Note that a helix is *not* a spiral. A helix is a regularly coiled structure, the coils of which all have the same diameter.

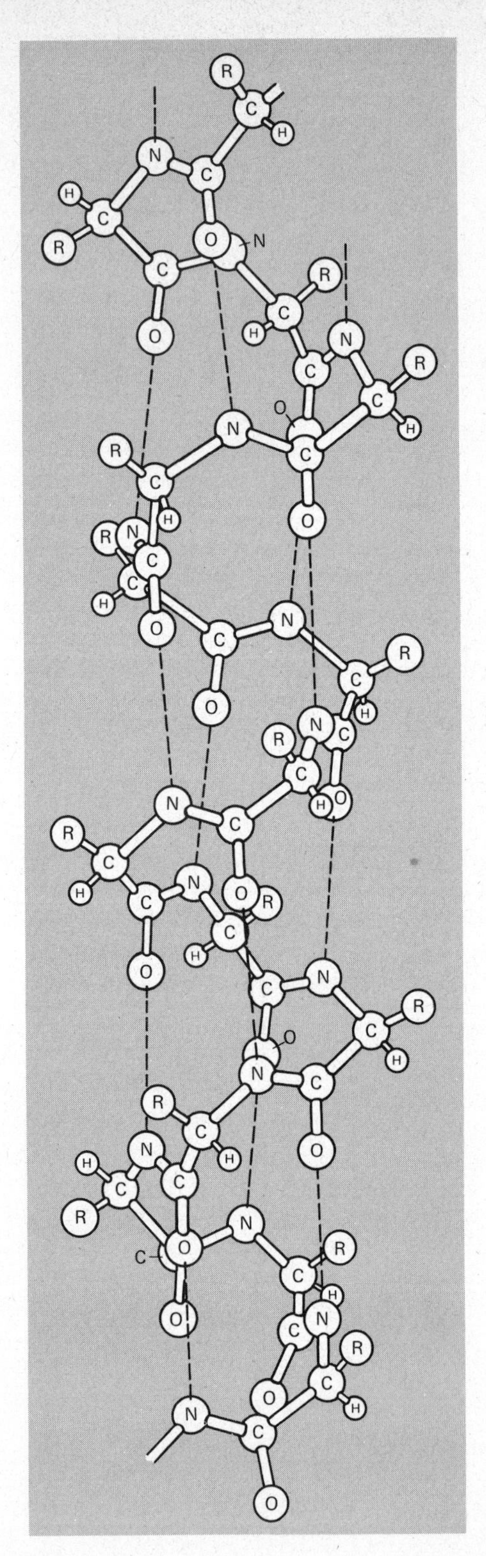

Figure 17-10. The α-helix of a protein. The helix, which constitutes the secondary structure of a protein, is formed by the coiled chain of amino acid units. The peptide backbone of the polypeptide forms a repeating sequence of carbon (C), oxygen (O), hydrogen (H), and nitrogen (N) atoms. The R stands for the side chain that distinguishes one amino acid from another. The configuration of the helix is maintained by hydrogen bonds (broken lines). The hydrogen atom that participates in each of these bonds is not shown. (After Pauling, L., and R. B. Corey. 1954. *Fortschr. Chem. Org. Naturstoffe* **11**:180.)

paper that described this secondary structure of proteins in terms of a helical configuration of the peptide backbone. This description was the result of a series of brilliant experiments elucidating the three-dimensional configuration of small peptides and then extrapolating from those results to the larger structure of the protein itself. Of the several different helical structures that they could postulate, one, the α-helix, gave the best fit to their data. On the basis of this finding they projected the diagram of protein structure that is presented in Figure 17-10.

The emphasis in the diagram is on the chain of peptide linkages that makes up the backbone of the protein. The side chains of the amino acid are represented by the designation R. At this level of structure, as in the primary, we are able to ignore the side chain of the amino acid. Note that the force holding the protein in the helical configuration is the hydrogen bonding between the unoccupied oxygen of one peptide linkage and the hydrogen carried on the nitrogen of another linkage farther along the chain. Hydrogen bonds are different from ordinary bonds in that there is no direct linkage between the two groups that are held in position relative to one another. Essentially, the hydrogen is shared between the negative oxygen and the nitrogen, and this sharing provides a sufficient force to keep the two groups fixed in space. Each such bond is relatively weak; it has only a small fraction of the strength of the usual covalent bonds that link atoms together in organic molecules. Nevertheless, since the peptide linkages are numerous, the possibility of forming hydrogen bonds is great, and the summation of all these H-bonds may be a considerable force.

The bending of the polypeptide backbone is restricted by the covalent bonds linking the various atoms together. These bonds cannot be deformed very much. Nevertheless, all molecules "strive" to obtain the minimum energy condition, and this minimum energy level is characterized by the maximization of bonding. (Merely remember that to put energy into a molecule breaks bonds, and it follows that making of bonds goes on with the release of energy.) Consequently, the most stable condition (the minimal energy state) is achieved when the maximum number of hydrogen bonds is formed. Apparently the α-helical configuration is the one that permits a maximum number of hydrogen bonds with minimal deformation of the covalent bonds holding together the atoms of the polypeptide chain. The coiling permits reactive groups to come close enough together to form hydrogen bonds. Since such bonds can form only between the oxygen atom of one peptide linkage and the nitrogen of another linkage approximately three amino acids farther along the chain, and since hydrogen bonds have a definite length, the molecules are held in a rather regular array, which accounts for the regularity observed in the X-ray diffraction patterns. In other words, the α-helix gives the polypeptide its semicrystalline structure.

Tertiary Structure

The third point we made about the secondary structure of proteins was that in the α-helix the side chains of the amino acids (designated R) extend from the coiled structure. In this position they are free to interact with other molecules or with one another should they perchance come close enough together. To understand the significance of this statement, it is necessary to consider the side chains of the amino acids in a less-abstract manner.

Review the structure of the 20 amino acids presented in Figure 17-4. They have been grouped together according to the similarities of their side chains. For example, some (glutamic and aspartic acids) have side chains carrying a negative charge, whereas lysine has a positive charge. Should a lysine residue come into contact with glutamic or aspartic acid residue, they would tend to stay together through the electrostatic interaction of the positive and negative charges. Such bonds can be relatively strong, although much weaker than covalent linkages. Cysteine carries a sulfhydryl group (SH) on its side chain. Should two such residues come close to one another in the presence of a oxidant, a disulfide bond will be formed (—S—S—), linking the two molecules together covalently. Such a bond is very strong, although it can be broken by reduction more easily than could the covalent linkages of the peptide bond or those holding two carbon atoms together. Finally, there are groups of amino acids that have nonpolar side chains, for example, leucine and valine. In fact, you will notice that this group represents the largest single group of amino acids. These nonpolar portions of the molecule have their electrons evenly distributed so that there are no charged regions. For this reason, they do not attract other charged molecules, including water. (Water is a dipole, having a negatively charged oxygen at one "end" and two positively charged hydrogens at the other.) Such groups are commonly referred to as hydrophobic, which is something of a misnomer. The truth is that uncharged groups of this type have a net attraction for one another through forces that are generally called van der Waals interactions. Consequently, when such groups are given a choice of similar molecules or water molecules, they "prefer" their own kind, since bonding can occur that will keep them in the minimal energy condition. It is only in this sense that they "dislike" water, and they only resist its presence when it tries to force its way into their otherwise stable associations.

The foregoing discussion suggests that four kinds of bonds can be generated between the various side chains of the amino acids in the polypeptide. Of these, only one is of the relatively strong, covalent type (the disulfide linkage involving cysteine). The others are weaker. In decreasing order of relative strength, they are electrostatic interactions, hydrogen bonds, and van der Waals associations. However, the low strength of these individual bonds can be misleading. While it is true that a small amount of thermal energy will cause paraffin (composed of large hydrocarbon molecules) to melt, it is also true that a block of paraffin can be a relatively rigid structure that requires considerable mechanical force to break at room temperature. The van der Waals associations holding the molecules together in a block of paraffin are weak individually but relatively strong when taken altogether. The same thing is true for protein molecules. When several different hydrophobic side chains come together, they will resist separation. When there are many such centers of interaction, the stability given to the molecule (or molecular aggregate) by such association may be considerable.

As the foregoing material suggests, once a polypeptide has been generated, it can undergo a number of possible reactions. We have already discussed the generation of the α-helix, a secondary level of structure. However, the formation of the α-helix is spontaneous only as long as the formation of the hydrogen bonds does not interfere with the formation of other bonds. Let us suppose that, in one region, a positive and negative group come close enough together to interact and that, in order to complete the formation of the α-helix, the association between these two "R" groups would have to be broken. If it were simply a matter of whether a single hydrogen bond should be formed or this electrostatic association be kept, it is clear that the hydrogen bond would not form, since more energy would be expended to separate a positive and negative charge than would be gained by the formation of the hydrogen bond. If, however, that single electrostatic interaction is preventing the formation of very many hydrogen bonds, the situation may well be reversed.

The energy level of a molecule is a property of the molecule as a whole, and the net result of all spontaneous interactions will be to place the molecule at its minimum energy state. This may be a complicated problem in detail, but, in general, it should be easy enough to see that the configuration taken on by the molecule will depend on the manner in which all the various bonding possibilities interact with one another. Those regions in which the side chains can interact strongly with one another will distort the backbone such that there is little or no α-helical character to the protein. For convenience, we say that in this case the tertiary structure of the protein displaces its secondary structure. Globular proteins have a high degree of tertiary structure and a low degree of secondary structure. Fibrillar proteins have a high degree of secondary structure and a low degree of tertiary structure. The functional proteins of the cell (enzymes, contractile proteins) tend to have a globular form. Many structural proteins (silk, keratin, and so on) are fibrillar.

We are now in a position to say what it is that determines either the globular or the fibrillar nature of a protein. It must be the direct result of its primary structure, because the primary structure specifies exactly which amino acids are present and the relative spacing of those amino acids within the chain. If two of the side chains that could interact with one another are so far apart that they cannot, then their interaction is of no consequence. If we construct a protein with amino acids having very small side chains or with amino acids whose side chains tend not to interact with one another, then secondary structure will predominate. Silk is such a protein. On the other hand, should the primary structure be such that a large number of amino acids with strongly reactive side chains are present, tertiary structure will predominate. Should it be particularly desirable to keep the chain extended maximally, possibly even beyond that obtainable with secondary structure, the introduction of a large number of similarly charged amino acids will effect that. Under these circumstances the placement of cysteine residues in the polypeptide is of great importance, since the formation of the disulfide linkages, the strongest of all, will have a profound effect on the structure of the polypeptide. In each case, the final form of the polypeptide depends entirely on the placement of the amino acids in the primary structure, and it is not necessary to invoke other forces for the higher levels of structure characteristic of protein.

REFERENCES

BERNHARD, S. 1968. *Enzymes: Structure and Function.* W. A. Benjamin, Inc., New York. (A very good treatment of the subject that will provide more detail on protein structure as well as enzyme function.)

HAYNES, R. H., and P. C. HANAWALT (Eds.). 1968. *The Molecular Basis of Life.* W. H. Freeman and Co., San Francisco. (A collection of papers originally published in *Scientific American.* There are several fine papers on proteins in the first section. The paper by Phillips gives a very good description of the relation of structure to enzyme function.)

KENDREW, J. C. 1961. The three-dimensional structure of a protein molecule. *Sci. Amer.* **205**:96–110. (An excellent paper that describes the use of X-ray diffraction in quite comprehensible terms.)

PAULING, L., and R. HAYWARD. 1964. *The Architecture of Molecules.* W. H. Freeman and Co., San Francisco. (A pretty book that makes sense of molecular structure and bonding in simple terms.)

WATSON, J. D. 1970. *Molecular Biology of the Gene.* 2nd ed. W. A. Benjamin, Inc., New York. (The early chapters of this book give a nice summary of metabolism and the role of enzymes. Chapter 4, "The Importance of Weak Chemical Interactions," is especially recommended for the reader who needs to brush up on the various types of bonds available in biologically important molecules.)

QUESTIONS AND PROBLEMS

17-1. Why is it necessary to understand something about the structure of proteins before we can understand modern genetics?

17-2. Define the following: amino acid, dipeptide, polypeptide.

17-3. What constitutes the primary structure of proteins? What are the "higher" levels of structure?

17-4. Are all proteins single polypeptides? Explain.

17-5. In general terms, define the forces that hold proteins in a folded configuration. Are they the same forces that hold two polypeptides together?

17-6. Suppose that you isolated two polypeptides, each from a different enzyme, and mixed the two together. Would they be likely to form a new protein made up of the two chains? Why?

17-7. Define the following: enzyme, substrate, active site.

17-8. In what way does an enzyme differ from many inorganic catalysts?

17-9. Of what are enzymes composed?

17-10. If two enzymes have very different functions, what would you predict about their primary structures?

17-11. Why do cells have so many different enzymes?

17-12. Tryptophane synthetase is a large enzyme (its molecular weight = 159,000) required for the synthesis of the amino acid tryptophane. The enzyme contains two different polypeptides, A and B. The A chain has a molecular weight of 29,500, and the B chain has a molecular weight of 49,500. (The weights are rounded off.) Assuming that, on the average, each amino acid has a molecular weight of 110, how many amino acids are present in the A chain? In the B chain? How many A and B chains are present in the functional enzyme?

17-13. On the basis of the one gene–one enzyme hypothesis, we predict that a mutation in a gene will remove enzyme function. Certain mutations, however, produce enzymes that have normal activity at one temperature but are not active at a higher temperature. Can you give an explanation for such temperature-sensitive mutants?

18 / What is the gene?

"Walter, explain DNA just once more and I promise I won't ask you again."

As soon as it became clear that genes might function by directing the synthesis of proteins, two important questions remained to be answered: What is the mechanism of protein synthesis, and what molecules in the cell have the necessary properties for directing that mechanism? Intensive investigation of both problems proceeded concurrently and interdependently, so that it is not always possible to separate the two lines of investigation. However, initially, they were separate, and for a number of years they proceeded more or less parallel to one another. We will treat the questions in this parallel manner, beginning with the question of what the gene is chemically and leaving the question of how it effects protein synthesis to later chapters.

Nucleic Acids

Even in Mendel's time it was presumed that the nucleus was the seat of heredity. Consequently some of the early biochemists turned their attention to the composition of nuclei. Among them, the most outstanding was Friedrich Miescher, who did an analysis of pus cells and salmon sperm, both of which he knew had nuclei that were very large relative to the rest of the cells. By 1868, three years after Mendel's first publication, he had identified a new cellular component that contained a much higher fraction of phosphorus than the other well-known cell components (proteins, carbohydrates, and lipids). He presumed, at least partially correctly, that this new component was largely contained within the nucleus and gave it the name nuclein, later to become *nucleic acid.*

By 1900 it was established that nucleic acids were present in all tested plants and animals and that they contained four principal materials: two kinds of nitrogen-containing base, *purines* and *pyrimidines*; a five-carbon sugar; and phosphoric acid. In the 1920s it was established that, in fact, there are two kinds of nucleic acid, each having many components in common but differing in certain important ways. These two are designated *RNA* (ribonucleic acid) and *DNA* (deoxyribonucleic acid). As their names suggest, the primary difference between the two is the five-carbon sugar (pentose). RNA contains *ribose*, and DNA contains *deoxyribose* (Figure 18-1). The difference between the two is that the carbon occupying position 2 of deoxyribose lacks oxygen; otherwise the two molecules are identical. A second difference between RNA and DNA is that the pyrimidine bases of the two are different (see below).

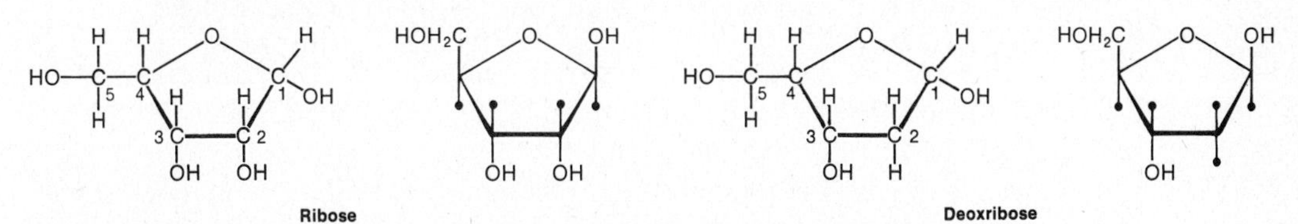

Figure 18-1. Ribose and deoxyribose. The first and third parts of the figure give the structural formulae for the two five-carbon sugars found in the nucleic acids. The second and fourth parts show the more commonly used, abbreviated versions.

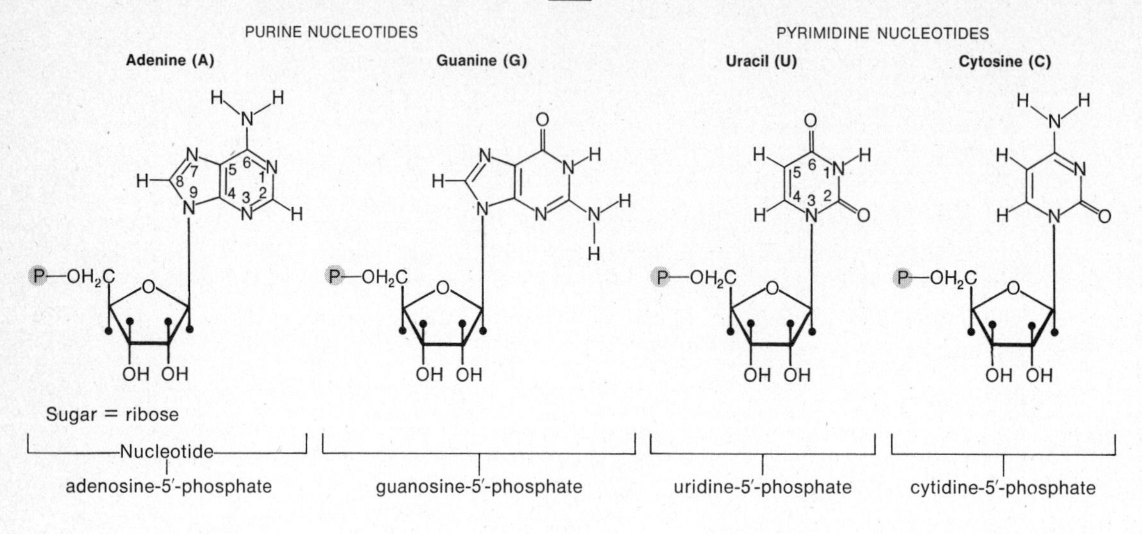

DNA

Adenine (A) Guanine (G) Thymine (T) Cytosine (C)

$P-OH_2C$ OH

Sugar = deoxyribose

Deoxyadenosine-5′ -phosphate

Deoxyguanosine-5′ -phosphate

Deoxythymidine-5′ -phosphate

Deoxycytidine-5′ -phosphate

Figure 18-2. Nucleotide building blocks of the nucleic acids. Eight different nucleotides are shown, four for RNA (upper) and four for DNA (lower). On the left are the purine-containing nucleotides, and those on the right are pyrimidine-containing The P symbolizes a phosphate group $^-O-\overset{\overset{O}{\|}}{\underset{\underset{O^-}{|}}{P}}-O^-$ attached to the 5′ carbon of the sugar. The capital letters following each name are used as abbreviations for the type of nucleotide in this text.

In the 1930s it was demonstrated that nucleic acids are relatively large polymers, the monomer of which is a *nucleotide*, consisting of a nitrogenous base (purine or pyrimidine), a sugar molecule, and a phosphate group. Figure 18-2 presents structural formulas for the eight major components of the nucleic acids, four for RNA and four for DNA. It can be seen that the difference between the nucleotides is relatively small—the absence of one oxygen atom on each sugar in the DNA and the presence in DNA of the pyrimidine base *thymine* rather than the *uracil* that is characteristic of RNA. It should be clear from these rather small differences why it took so long to determine that there are two kinds of nucleic acid and to analyze those differences. Nevertheless, it is of critical importance that these molecules are always manufactured in this way. Thymine is not a usual constituent of RNA, nor is uracil usual in DNA.

The joining together of the nucleotides to form the nucleic acid polymer occurs by the removal of water in the reaction between the terminal phosphate group of one nucleotide and the hydroxyl (OH) group on the number three (3′) position of the sugar.* The nucleotides then are joined by a phosphate diester linkage (Figure 18-3). Once again it is interesting that although the bond could be formed by reaction with the 2′ carbon of the sugar molecule in RNA, it does not happen that way. In both RNA and DNA the linkage is made between the 5′ carbon of one sugar and the 3′ carbon of the other sugar. Apparently the enzymes responsible for catalyzing this reaction are sufficiently specific to guarantee that the linkage will always be of the 5′–3′ type. This is of some importance in the discussion of the nucleic acids. These molecules are polymers in the same sense that the proteins are; that is, they are polymers of sugar-phosphate linkages. The backbone is composed of a series of phosphate diester bonds. At one end of the molecule there will be an unreacted phosphate group on the 5′ carbon of the sugar, and at the other end of the molecule there will be an unreacted OH group carried on the 3′ carbon of the sugar. Consequently, we can differentiate between the two ends of the molecule, the 5′-end and the 3′-end. (It should be remembered that, in the proteins, one end of the chain has a free amino group [NH_3^+] and the other end a free carboxyl group [COO^-].) In both these polymers, then, we can talk about a *polarity* existing in the molecule, since each has two "poles": 5′-end and 3′-end or N-terminal and C-terminal.

Like the proteins, the backbone of the polymer is uniform, but the side chains are different. For this reason, they are generally called *polynucleotides*, indicating a repeating series of nucleotides, each of which may be somewhat different. In each type of nucleic acid, four different bases are available for reaction with one another or with other molecules. Clearly nucleic acids do not present the variety of side chains to be found in the proteins; nevertheless, there is some variability in their structure, which, one would project, could lead to higher levels of structure, as is found in the proteins. We will consider these higher levels when discussing the role of the various types of nucleic acids in succeeding sections.

*The numbering begins with the carbon attached to the purine or pyrimidine and ends with the carbon carrying the phosphate group. Since there are two ring systems to be numbered, the sugar and the N-containing base, a distinction must be made. The "prime numbers" (3′, 5′ . . .) are used to designate the sugar residue.

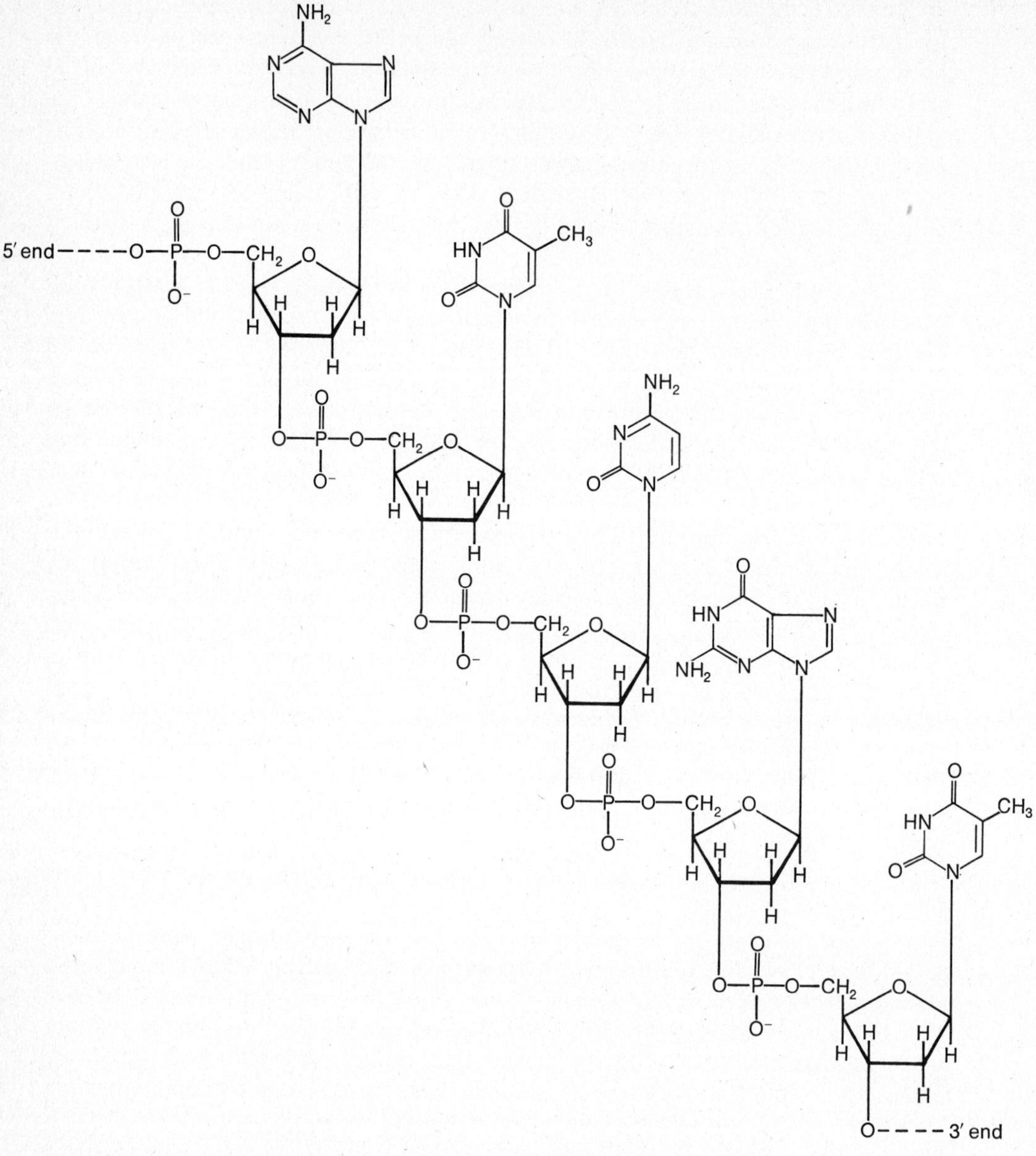

Figure 18-3. A polynucleotide chain. The diagram indicates that a polynucleotide is constructed by the formation of a phosphate diester linkage between the 5′ carbon of one sugar and the 3′ carbon of its neighboring nucleotide. The chain has one free phosphate group on the 5′ carbon, the 5′ end, and one free 3′ OH on the other terminal sugar, the 3′ end. Interactions between different chains or within the chains are effected by bonding between the free purine and pyrimidine bases.

The Composition of Chromosomes

By the 1940s it was firmly established that the genes were carried on the chromosomes, and so it was of critical importance to determine the composition of chromosomes. In general, indirect methods had to be used, depending on the absorption of specific wavelengths of light by chromosomes in fixed cell preparation (particularly the very large chromosomes of certain insects) or by interpretations based on the chemistry of certain staining reactions. Such indirect evidence suggested that chromosomes were composed almost exclusively of DNA and protein. In 1949 whole chromosomes were successfully isolated in quantity from calf thymus cells, and it was possible to do a direct chemical analysis. The results confirmed the supposition that chromosomes are largely composed of DNA and protein. Small quantities of RNA were present, but the concentration was sufficiently low to suggest that this RNA could not be the genetic material. Furthermore, by this time it had been demonstrated that although rapidly metabolizing cells contain far more RNA than DNA, the great bulk of the RNA is found in the cytoplasm, not in the nucleus. Consequently, few people even considered RNA as a possible genetic substance.

The proteins that were isolated from chromosomes appeared to be primarily of a special class—*histones.* They are relatively small proteins rich in the amino acids lysine and arginine; consequently, they have a net positive charge (see Figure 17-5). There are other proteins as well, but as in the case of RNA, they are present in too small a quantity to be seriously considered as likely candidates for the gene. It appeared, therefore, that the gene must be DNA, a histonelike protein, or a combination of the two, a nucleoprotein. Let us now consider each of these three classes of compounds to see which is the most likely candidate.

Indirect Evidence for DNA

Transformation

The first suggestion DNA might be the gene came from the earliest work that indicated nucleic acids were associated with the nucleus and the later work that showed DNA to be a major component of chromosomal structure. This suggestion was supported by the finding that the amount of DNA per chromosome is constant in the cells of different tissues of the same organism. However, the critical piece of evidence that implicated DNA was the discovery of bacterial transformation (Chapter 13). Recall from that previous discussion that a substance can be extracted from heat-killed cells that will transform the hereditary properties of certain strains of bacteria. When it was demonstrated in 1944 that the active principal of the extract was DNA rather than fragments of cell wall material, which it had been presumed to be, DNA became a prime candidate for the hereditary material. Unfortunately, there are two defects in this experiment. In the first place, in the purification of DNA it is impossible to free it completely of protein; there is always some residual protein present. To avoid this difficulty, the purified extract was treated with DNase, an enzyme that specifically attacks DNA, degrading it to individual nucleotides. When this was done, all transforming activity of the preparation was lost. On the other hand, if the extract was treated with trypsin, an enzyme that specifically degrades protein, there was no loss of transforming ability. Clearly this is highly suggestive evidence that DNA carries the hereditary material. Nevertheless, it can be argued that trypsin-resistant proteins could still be present and be acting in the transformation process. Furthermore, the fact that DNA is responsible for transformation does not necessarily indicate that it is, in fact, the gene. Rather, the gene could be a protein that is somehow modified by interaction with this specific DNA, or it could be a nucleoprotein (with the DNA carrying part of the genetic information but not all of it). Thus although the experiment is highly suggestive that DNA is the gene, it was not widely accepted as critical proof during the 1940s and early 1950s.

Bacterial Viruses and Transduction

A second line of evidence supporting the concept that DNA is the gene came from the study of transduction (Chapter 13). It should be remembered that in this process the bacteriophage attaches to its host bacterium and, after digesting away a small part of the cell wall to which it is attached, "injects" its nucleic acid into the cell, a process that occasionally results in the carrying of genetic material from one strain of bacteria to another. Regardless of whether bacterial genes are being transferred in this process, it is clear that the life cycle of the bacteriophage depends on the reconstitution of many new phage within the bacterial cell, in response to the injected nucleic acid. For this reason alone, it is reasonable to presume that the injected nucleic acid carries all the hereditary information required for the synthesis of new bacteriophages. Certainly it is not unreasonable to talk about hereditary information even for such a simple system as a virus. Each specific kind of bacteriophage has its own particular shape and is composed of its own specific proteins. Furthermore, each strain of phage has its own "behavior" pattern. That is, it will attach to certain host cells but not to others. To the extent that we talk about genes controlling the synthesis of proteins as a primary function, it is reasonable to talk about the control of the synthesis of coat protein, tail protein, and attachment protein as a normal genetic process, and so it is reasonable to talk about the transfer of hereditary material by the injected phage nucleic acid.

In 1952 an experiment was designed to test this general hypothesis. Bacteriophages were grown in a medium containing radioactive isotopes of phosphorus and sulfur (^{32}P and ^{35}S). Since sulfur is found almost exclusively in proteins, and since the great bulk of phosphate is found in nucleic acids, these two radioactive markers labeled the molecules specifically. All the coat protein would contain ^{35}S, and all the DNA would contain ^{32}P. (Since the energy of the emitted beta rays of each is different, it is possible to distinguish between the two isotopes by using proper radiation-counting devices.) Such labeled bacteriophages were mixed with unlabeled cells and allowed to remain in contact sufficiently long to ensure that the phage had absorbed onto the bacteria and injected their DNA, but not so long that

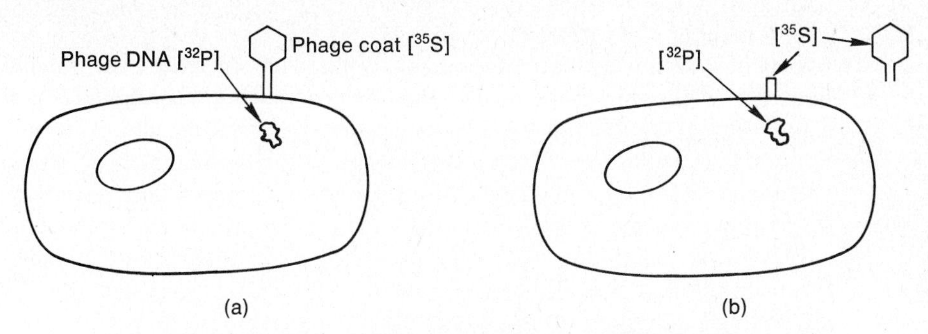

Figure 18-4. Transmittal of DNA in bacteriophage infection. The protein coat of the phage was labeled with ^{35}S and the internal DNA was labeled with ^{32}P. After infection, the cells were subjected to the shearing force of a Waring blender, which detached the phage coats ("ghosts") from the bacteria. Very little labeled protein was found in the bacterial cells. A small amount remains on the cell wall, the stump of the phage tail piece. It appears that the information for the synthesis of new phage is contained in the DNA alone.

they were able to produce new phage by replication (Figure 18-4a). The shearing force of a Waring blender was used to detach the phage "ghosts" (the head and tail protein remaining after DNA injection) from the bacterial cells, and the two were separated from one another by centrifugation, the bacteria being removed as a pellet and the much less dense phage ghost remaining in the supernatant. Analysis of the two components indicated that all the ^{32}P was in the bacterial cells and that most of the ^{35}S was in the supernatant. However, 20 % of the radioactive sulfur was found with the bacteria. Most of this latter ^{35}S could be accounted for by small fragments of the tail of the bacteriophage that were still attached to the bacterial cell wall (Figure 18-4b). The shearing forces break the phage through its tail, leaving a "stump" attached. Nevertheless, there was still 7 % of the ^{35}S in the bacterium (not too surprising, since we now know that there is some protein inside the bacteriophage head). Consequently, once again, the evidence is highly suggestive that the genetic information is transferred by the DNA. However, there is some residual protein, and it is always possible that, even though it appears to be too little to carry enough information, it might be the genetic material or that the gene requires the presence of both DNA and protein.

Tobacco Mosaic Virus

The studies with tobacco mosaic virus (TMV) are some of the most interesting, for initially they appeared to give strong support for the hereditary role of protein. This virus attacks tobacco plants, causing splotches on the leaves, regions where cells have been destroyed in the course of viral replication. (In this regard, TMV is similar to bacteriophage.) In the mid-1930s it was thought that TMV was made exclusively of protein. Assays of the particles showed little else; furthermore, the viruses can be crystallized and kept on the shelf for long periods of time without loss of function, properties the virus shared with a number of proteins. Consequently, it appeared that at least the hereditary information required for proper self-replication could be contained in protein. However, when it was shown that the virus particles contained 5 % RNA, the claim for a self-replicating pure protein had to be abandoned.

In the 1950s experiments showed that it is possible to dissociate the protein and nucleic acid components of TMV and then let them spontaneously reassociate to form infectious particles. This finding was of great general importance, for it suggested the possibility of the spontaneous aggregation of subunit molecules to form a unit that was biologically active. Moreover, it opened the door to an interesting experiment, using man-made plant viruses.

As a test of the role of nucleic acid and protein in hereditary, the protein and nucleic acid of both TMV and Holmes-Ribgrass (HR) virus were dissociated and separated. The nucleic acid from TMV was then mixed with the protein from the HR virus, and a "hybrid" virus was generated by the spontaneous aggregation of these particles. Similarly, the nucleic acid from the HR virus was mixed with the protein from TMV virus and another "hybrid" formed. These two hybrid viruses were then placed on their host cells, which they successfully infected, and on which they replicated to form a new virus. Significantly, the virus produced from the hybrid containing TMV nucleic acid (and HR protein) produced TMV viruses only. These progeny viruses had both TMV RNA and TMV protein. Similarly, the hybrid virus containing HR nucleic acid produced HR progeny only. Thus it seems clear that in this case the nucleic acid alone determines the kind of progeny produced. The protein is inert, both in the transmission of hereditary information and in the modification of the expression of the information contained in the nucleic acid. (On the other hand, the bare nucleic acid is not infective.)

Clearly, in this case, we have an example in which it appears that the nucleic acid is the gene. On the other hand, the nucleic acid of these viruses is RNA; and, for the reasons previously outlined, no one was ready to accept the role of RNA in determining the genetic information of higher organisms. In fact, the whole question of whether plant viruses of the TMV plant type are alive confused the issue. It could too easily be argued that TMV is a special case and so dismissed.

Mutation Studies

One of the most important properties of the gene is that it mutates. Although normally these changes are rare events, the process can be accelerated in a number of ways, including the exposure of cells to the action of ultraviolet radiation. This fact provides a tool for the analysis of which chemicals were being changed, since ultraviolet is absorbed specifically by certain molecules at certain wavelengths (unlike X rays, whose energy is so great that they are absorbed by all organic molecules equally well). Fortunately, proteins and nucleic acids have distinctly different absorption spectra. Put simply, nucleic acids absorb strongly at wavelength 2600 Å, and proteins absorb maximally, but much less strongly, at 2800 Å.* Consequently, it is possible to expose cells to radiation at

*The symbol Å stands for the angstrom, a unit of length equal to 10^{-8} centimeters. Most covalent bonds are about 1 to 2 Å long.

2600 Å or 2800 Å and determine the increase in mutation rates resulting from this exposure. In this way, it was demonstrated that only the 2600 Å radiation was effective, which suggested once again that protein was not the genetic material and that nucleic acid was. Unfortunately, however, experiments of this type are never as clear-cut as might be hoped. They had to be performed with microorganisms, since the external covering of the germ cells of higher organisms prevents the entrance of ultraviolet light. In addition, there was always the possibility that the energy absorbed by the DNA was transferred to the gene, which might be either protein or nucleoprotein. Although this possibility seems unlikely, it cannot be ruled out altogether.

Objections

When the evidence from studies of transformation, transduction, virus replication, and mutation was taken all together, the genetic role of nucleic acid was hard to resist. But it was not conclusive. A great number of people, perhaps the majority, felt that the nucleic acids did not provide the variety of structure necessary to carry all the information required for the construction of an entire higher organism. The nucleic acids contain only four different nucleotides, the differences being exclusively in the nitrogenous bases. The number of permutations of these four seemed rather limited. This point of view was reinforced by the findings that, in many organisms, all the bases appeared in equal concentrations. Consequently, it was proposed that the nucleic acids were composed of repeating sequences of the four types—the tetranucleotide hypothesis. Using the first letter of the name of each base to symbolize the nucleotide, the structure of a typical nucleic acid would be AGCTAGC-TAGCTAGCTAGCT . . . continuing ad nauseum. Such a structure would permit relatively few different kinds of molecules, between 1 and 256 (4^4), too few for the variety of proteins whose synthesis was ascribed to the genes. The proteins, composed of 20 different subunits, are potentially far more various, and so they were the logical candidate. But what many people overlooked was the fact that not all proteins are found in the nucleus. Only those few associated with the chromosomes could be the genetic material.

Chromosomal Proteins

If it could be argued that the DNA was too monotonous a molecule to be the gene, there were also serious defects with proteins. Perhaps the most important problem was that the basis of biological action of proteins was understood solely in terms of enzymatic activity. While it must be true that the joining together of the various amino acids in the polypeptide chain is catalyzed by enzymes, the use of these enzymes to specify the relative positioning of the various amino acids is illogical. All the evidence indicates that each protein has a unique sequence of amino acids and that each enzyme has a specific reaction that it catalyzes. To invoke enzymatic activity as a means of determining the primary structure of proteins leaves us with a serious problem: What makes the enzyme that makes enzymes? What is required is an enzymatic molecule that can make itself while making other proteins—that is, a self-replicating protein. In the absence of a self-replicating molecule of any kind, it was difficult to accept any particular molecule as being genetic.

Looking back from our present knowledge, there are other important objections. One of the most important is that chromosomal proteins are rather homogeneous. Proteins make up approximately one-half the mass of the chromosome, but the majority are *histones*. Histones are small proteins (10,000 to 20,000 MW) that are positively charged. Presumably they associate with the negatively charged DNA by electrostatic interaction. The histones are divided into three major classes: arginine-rich, moderately lysine-rich, and very lysine-rich (see Figure 17-5). Each of these groups may have several subdivisions; however, there appear to be no more than ten different molecules. In addition to the histones, there is a special group of even smaller proteins, the *protamines,* having a molecular weight of 3000 to 5000. They are very basic proteins, two-thirds of their structure being arginine, with proline and alanine making up most of the rest. It is not surprising that, with such a limited amino acid constitution, there are not very many different protamines. It is probable that there are only three distinctly different molecules. Besides their lack of variation, the protamines are never found in quantity in most organisms; however, they are readily isolated from the sperm of fish and certain birds, which suggests that their role is special rather than general. Certain other nonhistone proteins are associated with chromosomes. Unfortunately, we know little about them, but they are present in too low a concentration to be seriously considered to be the gene. If the gene were protein, it would almost have to be histone. However, with their lack of variety, the histones cannot account for the transfer of hereditary information from one generation to the next.

Today it is clear that the chromosomal proteins are an unlikely candidate for the gene, but it was not so clear in the late 1940s and 1950s when this question was being asked. The structure of the histones was unknown, and the methods for analyzing protein primary structure were just being developed. In fact, at that point it was presumed that we knew more about the structure of DNA than of protein; and since the structure of DNA did not appear to permit it to be the gene, it was reasonable to invoke that *unknown* class of molecules, the proteins, as the most likely candidates. It is not at all unusual for human beings to attribute to things whose properties are not understood functions that are also not understood. Although it is probably difficult to do so now, it is worth trying to recapture the attitudes of those scientists, particularly biochemists, who were convinced, quite reasonably, that proteins were the only class of molecules having sufficient variability in their structure to encode all the information that genes must carry. It is amusing that in the late 1940s biologists could be divided into three groups: those who "irrationally" believed that DNA was the gene on the basis of transformation studies, and so on, even though the molecule apparently did not have the necessary properties to be the gene; those who quite rationally believed that only the proteins could be the gene;

and those who ducked the issue by asserting that the gene must be a combination of both. We must now consider how the irrational point of view prevailed.

The Double Helix

The first important discovery was made possible by the development of better techniques for the isolation of the nucleotide components of DNA molecules and their identification (particularly paper chromotography). Once it was possible to separate the various nucleotides cleanly and rapidly and to determine accurately the amount of material present, it became evident that DNA was not all that monotonous. In 1950 it was demonstrated quite clearly that the bases in DNA were not present in exactly equal proportion. It certainly was true that they were present in proportions that *approximated* 1 : 1 : 1 : 1, but they were sufficiently far off so that deviation could not be accounted for on the basis of experimental error. For example, the DNA from calf thymus nuclei contained 28 % adenine, 28 % thymine, 24 % guanine, and 20 % cytosine. Furthermore, these proportions did not hold for all DNAs; that is, DNA from different sources had different percentage compositions of the four nitrogenous bases. Nevertheless, from individual to individual within any one species, the composition remained the same. This correlation between species difference and DNA difference would be expected if DNA were the gene, and it lent strength to the proposition of the molecular geneticist that DNA was the important molecule to study.

Study of a variety of different DNAs led to the conclusion that, in all the nucleic acids examined, the molar ratio of total purines to total pyrimidines was not far from 1 : 1, and the same was true for the ratio of adenine to thymine and guanine to cytosine. As we shall see, this was a most important discovery, but its significance was not apparent at the time. Unfortunately, there was no context for the discovery. There was no way to show how the DNA molecule could have the three essential properties of the gene: self-replication, mutation, and carrier of the "code" that specified the placement of amino acids within the polypeptide chains. The last factor was the most critical. If there are only four nucleotides, how could they specify the position of twenty-odd amino acids? Even though the tetranucleotide hypothesis was incorrect, an approximate equivalence of adenine to thymine and guanine to cytosine still exists. Superficially, this fact appears to reduce variability within the molecule. The only answer to the question could come from a knowledge of the exact structure of DNA, a knowledge that was unavailable.

The first attempt to obtain detailed information about the three-dimensional structure of DNA was made in the early 1940s, using the technique of X-ray diffraction (see the appendix to Chapter 17). This was no easy task, but, on other grounds, it could be proposed that it was a long molecule whose nitrogenous-base-containing side chains were stacked close to one another. Since the bases are flat molecules, it was easy to envisage a "stack" of these bases separated by very small distances indeed. The diffraction pictures agreed reasonably well with this model. However, since it is not possible to get crystals of DNA, pictures were quite irregular, and exact analysis did not seem possible at the time.

The advent of World War II delayed research. Afterward several groups of scientists resumed the X-ray crystallographic approach. A major advance was the development of a technique whereby preparations of highly oriented DNA molecules could be obtained. Although far short of the crystal structure, it did improve the diffraction pictures remarkably, and a much more detailed analysis of the three-dimensional spacing of the atoms was possible. Perhaps the most important finding was that the distance between the nucleotides was very regular and very small (3.4 Å). Certain other regularities in the picture, however, could not be so easily explained.

The solution came in 1953, when James Watson and Francis Crick published their model of DNA. Using their own data and the data obtained from Wilkins' laboratory, they were able to build a model of the structure of DNA that satisfied all the restrictions coming from both chemical analysis and X-ray pictures. Analysis of

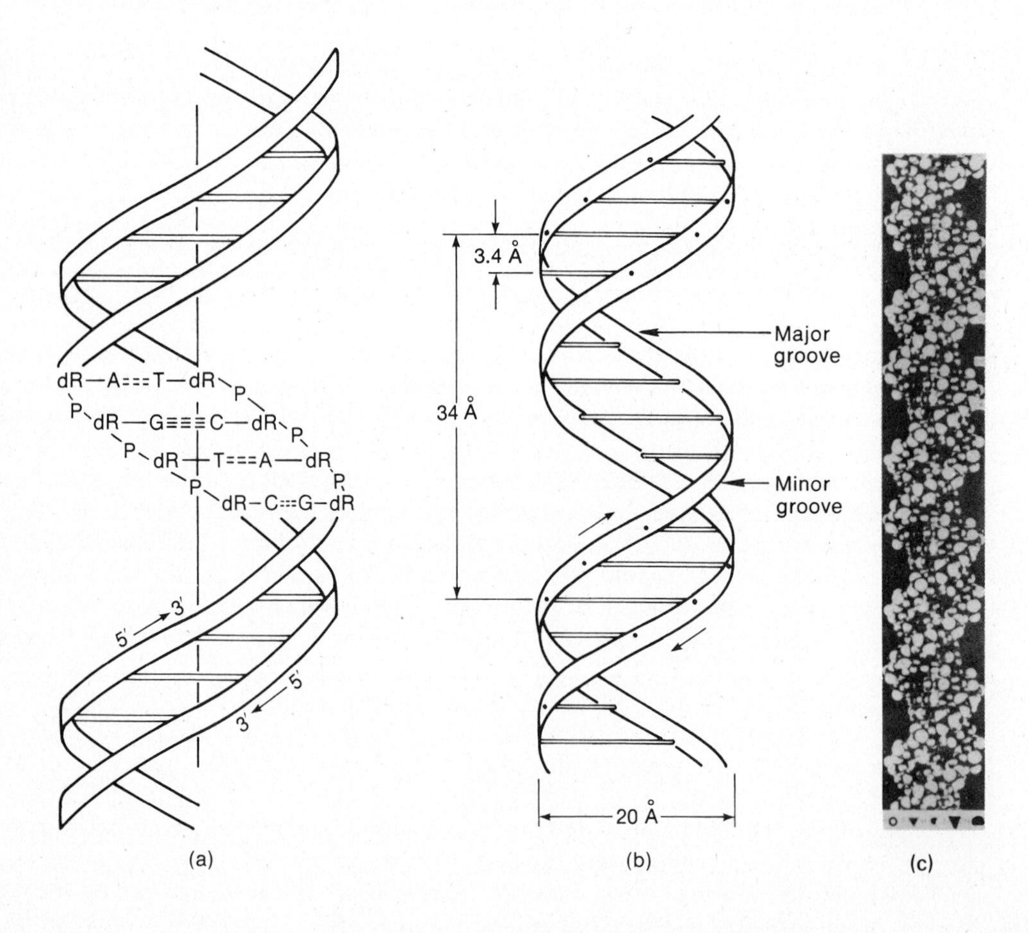

the pictures indicated that the polynucleotide chain of DNA was a regular helix having a diameter of 20 Å. Furthermore, the helix makes one complete turn every 34 Å. The internucleotide distance was known to be 3.4 Å, which implied that there were ten nucleotides within each gyre. This point turned out to be of great interest, since a DNA molecule with that structure would have too low a density. However, the correct density could be achieved if there were two chains (one chain was too low and three chains too high). Thus any model had to account for the facts that there were two chains and that these two chains, wound together in some way, had a uniform cross-sectional diameter of 20 Å. Moreover, it was clear that if DNA were the gene, the sequence of nucleotides within the DNA molecule had to be variable; that is, a large number of DNA molecules was necessary if each is to carry a specific bit of information that somehow directs the placement of amino acids in a particular peptide chain.

The key to the problem came from the realization that, as suggested by the chemical analysis of DNA, the purines and pyrimidines in the DNA molecule were complementary. It occurred to Watson and Crick that the dimensions of a purine and a pyrimidine base were such that they would fit nicely into a cylinder whose cross-sectional diameter was 20 Å. If such were the case, one chain of nucleotides would be joined to the other chain by interaction between the purines and the pyrimidines. In addition, such a structure would explain the equivalence of purines and pyrimidines, and, more specifically, a 1 : 1 ratio of A : T and G : C, without making it necessary to have a redundant molecule. That is, the sequence of nucleotides in one chain could be anything. However, since the other chain would be exactly complementary to it for every purine in one chain there would be a pyrimidine in the other and vice versa), the molar ratios of purines and pyrimidines would come close to 1 : 1, and an exact fit could be obtained, *provided that A always paired with T and G always paired with C.* Using this restriction, it was possible to propose the now famous *double helix model* of DNA (Figure 18-5).

By considering the diagram of DNA structure proposed by Watson and Crick, it can be seen that, taking a single chain of the molecule, any order of bases is possible. It would make no difference if the sequence in one chain were ATCGGGCAATG or GCTTAGCGAA or any other sequence. The restriction imposed by the model is that the second chain of each molecule must be complemen-

---

Figure 18-5. The double helix. (a) A schematic diagram of the DNA molecule showing the base pairing and the helical structure. The polarity of the ester linkages of the two chains is shown by the arrows. (b) The two chains are held together by hydrogen bonding between the complementary base pairs. Approximately 10 base pairs lie between two exactly overlying regions of the molecule. Note that there are two grooves in the helix (compare to c). (c) Space filling model of the DNA double helix prepared by Dr. M. F. H. Wilkins. The major and minor grooves are more evident. The purine and pyrimidine bases are exposed in the major groove.

tary to the first chain. Wherever an A appears in one, a T must appear in the neighboring chain. Wherever a G appears in one, a C must appear in the neighboring chain.

It is worth comparing the α-helix of proteins (Figure 17-10) with the double helix of DNA (Figure 18-5). The protein has a hollow core with the side chains extending to the outside and available for interaction with other molecules or among themselves. On the other hand, the double helix of DNA has a solid core composed of the paired bases. What would be a side chain in a protein is internally located in the DNA molecule. Consequently, cross reaction with other molecules and internal folding of DNA are greatly limited. The exposed groups are chiefly the phosphates that join the different nucleotides together. Thus the DNA molecule exposes a uniform set of negative charges to its environment but cannot possibly give rise to the complicated tertiary level of structure characteristic of proteins. The double helix forces it always to be a rigid rod.

Base Pairing

One of the critical features of the double helix model is that there can be a specific kind of base pairing, such that A will pair with T and G with C. Figure 18-6 shows the hydrogen bonds formed by these bases. A and T pair with the formation of two hydrogen bonds, but G and C pair with the formation of three. Thus the GC combination is significantly more stable than the AT combination. In terms of specific pairing, we may say that the stability of GC is such that G will always preferentially pair with C and C always preferentially with G. To a great extent, this situation excludes other combinations.* Moreover, there is the restriction produced by the double helix itself. The internal core is too small for two purines to pair and too large for effective hydrogen bonding between two pyrimidines. Taken all together, it is reasonable to propose that specific pairing of A with T and G with C in the double helix results both from the geometry of that structure and from energetic considerations of maximizing hydrogen bonding.

Polarity

We have already noted that each polynucleotide chain has a distinct polarity, with a 5′- and 3′-end. In order to achieve the proper fit of the base pairs within the double helix so that the maximum number of hydrogen bonds can be formed, the two chains must run in opposite directions; that is, starting from the "top" of a molecule, one chain would be running in the direction of 5′ → 3′ and the other chain would run in the direction of 3′ → 5′. Figures 18-5(a) and 18-7 diagram this fact. Although it may seem a trivial

*It should be evident from the figure that pairing between C and T will bring two oxygen atoms next to one another in such a way that repulsion of the negatively charged atoms will cause instability. Similarly, C and A do not present a good match, particularly when we consider that their attachment to the sugar-phosphate backbone of the polynucleotide chain prevents free motion in just any direction.

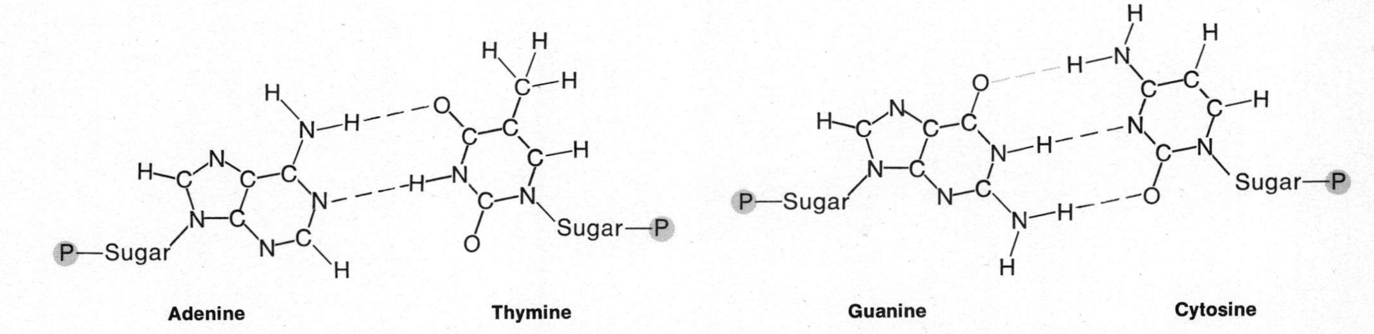

Figure 18-6. The base pairs of DNA. The AT pair has two hydrogen bonds; the GC pair has three. This difference is sufficient to account for the high degree of specificity in pairing that gives the DNA molecule its structure.

piece of information, it is important in later consideration of gene activity, and the student should take the time to make sure that he understands what is meant by the polarity of DNA structure.

Replication of DNA

The significance of the uncovering of the double helix structure of DNA cannot be overestimated. It provided a molecule that could account for all the known properties of the gene. It can account for the specific ordering of amino acids in the polypeptide chain by assuming that the sequence of bases in any one chain is a "code" for the ordering of amino acids in a particular polypeptide. Since the bases can be thought of as letters in the code (ATGC), any arrangement of these four letters and any groupings of these different arrangements could be used to specify particular sets of instructions to the protein-synthesizing machinery. Considering the great size of DNA molecules, a tremendous amount of information can be stored within any single molecule. Furthermore, mutation can be explained by the assumption that a chance chemical event can alter one of the bases such that there would be a change in the sequence and thus a change in the instructions to the protein-synthesizing machinery. But, most importantly, the model also provides the possibility for self-replication of DNA, which is a primary requirement for the gene.

Figure 18-7 presents a model for the replication of DNA. Here the two chains are seen as separating, which exposes the bases that are normally in the interior of the molecule. Other free nucleotides that surround DNA in the medium are now free to pair with the bases, so that each adenine (A) specifically binds with a thymine (T) and each guanine (G) pairs specifically with a cytosine (C). Ignoring for the moment the mechanisms by which the chains separate and by which the newly attached free nucleotides are "zipped up" to form the new double helix, it is clear that, *in principle*, the specific pairing of A with T and G with C provides a mechanism whereby each DNA molecule can duplicate to provide two *exact* copies of the original parent molecule. The double helix has the potential to be a self-replicating structure, and it was this factor above all that convinced geneticists that the model must be correct. Not only did it fit parameters established by the use of physical-chemical techniques, it also had the necessary properties to be the gene. Taken together with the evidence of transformation, phage reproduction, and so forth, it became clear that DNA was the long-sought master molecule of the cell, and the study of proteins as genes could be put aside.

DNA Polymerase

Two things are necessary to establish that the DNA molecule can replicate itself. The first is a mechanism for the replication, and the second is a demonstration that replication of this type does occur. The mechanism was discovered first. In the mid-1950s an enzyme was discovered that can make DNA—*DNA polymerase*. This enzyme has certain special properties. First, it requires all four nucleotides (A, T, G, C), which must be present

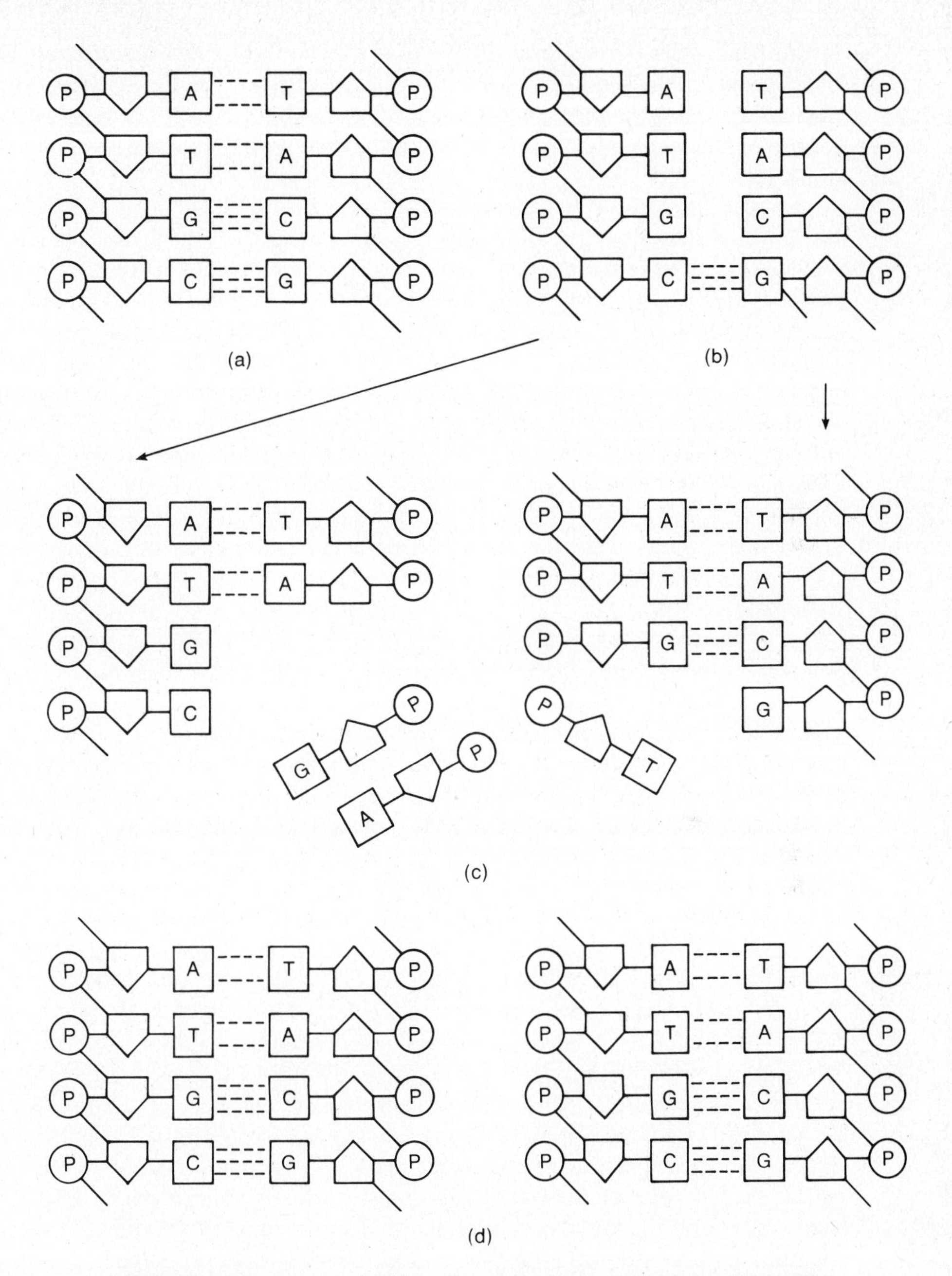

Figure 18-7. Replication of DNA. The replication mechanism by which DNA might duplicate itself is diagrammed. The two DNA chains (a) unwind and separate (b). The two complementary chains begin to attach DNA precursor subunits from the medium (c). When the proper bases are joined, two new helices will build up (d). The letters A, T, G, and C represent the purine and pyrimidine bases.

in the triphosphorylated form (ATP, GTP, CTP, TTP). The requirement for the triphosphates is a requirement for an energy source. All the triphosphorylated nucleotides are at a high energy level, and this high energy form facilitates the formation of the phosphate diester bonds that establish the primary structure of the molecule.

Another interesting feature of the enzyme is that it will make DNA from the four nucleotides only after a lag period of several hours. During this lag period no DNA can be detected, but afterward there is a steady increase in polymer. Such a lag period is unusual for most enzymes, and, furthermore, the lag period is related to the degree of purity. That is, highly purified preparations show a long lag period; crude extracts do not. This observation suggested something was present in the crude reaction mixtures that primed the reaction in some way. It soon became evident that the required substance was DNA itself. As long as some molecules of DNA were present, the reaction went with no lag period, even with the most highly purified enzymes. Thus it would appear that DNA is involved directly in the process of DNA formation, which is precisely the prediction that is made by a double helix that can self-replicate. The suggestion therefore arose that DNA was acting as a template for its own synthesis and guiding the DNA polymerase in the formation of new DNA molecules.

If the foregoing assumption of a template role of DNA is correct, it would follow that the DNA made by the polymerase should be identical to the DNA that is used as a primer in the system. Although it is impossible to do a complete analysis of DNA molecules, it is possible to characterize them. Each DNA molecule has a different molar ratio of purines and pyrimidines. If we were to analyze for all the A and T present in a system and add those two together (they should be 1 : 1 with reference to each other) and then divide that number by all the G and all the C added together, we should be able to get a ratio that is descriptive of the DNA molecule itself. In fact, when this process is done for different DNAs, it is seen that they are markedly different. The A + T/G + C ratio for the bacterium *E. coli* is 0.97; for calf thymus it is 1.25; for T_2 phage it is 1.92. Clearly each DNA is different from the other.

Using the base ratio to characterize different DNA molecules, it is possible to "feed" different DNAs into the system and determine whether a specific kind of DNA is manufactured by the polymerase. The data for such experiments are summarized in Table 18-1. It can be seen that there is a remarkable resemblance between the product DNA and the template DNA provided initially. The last line of data in the table represents a special case. It will be remembered that the enzyme can act in the absence of a primer, although it takes awhile for it to begin working. Apparently the lag period is related to the time required for the random putting together of nucleotides to establish an initial chain, which then acts as a template and produces a uniform product. When no template is present, it is possible to construct a molecule that is made up of only two bases, as long as these bases are complementary (A and T or G and C). Consequently, it is possible to make an AT copolymer, a molecule consisting only of A and T. When this synthetic DNA was fed to the reaction mixture, the DNA formed contained only A and T even though G and C were included in the reaction mixture. Thus the enzyme, which is capable of hooking bases together at random in the absence of a primer, will follow the instructions of the DNA molecule exactly when such a molecule is present. Apparently it preferentially attaches to existing DNA molecules and then "reads" them, presumably from one end to the other, constructing a new DNA double helix that is exactly the same as the parent molecule.

Table 18-1. Purine and Pyrimidine Composition of Enzymatically Synthesized DNA[a,b]

DNA		NUMBER OF ANALYSES	A	T	G	C	(A + T)/(G + C)	(A + G)/(T + C)
M. phlei	primer	3	0.65	0.66	1.35	1.34	0.49(0.48–0.49)	1.01(0.98–1.04)
	product	3	0.66	0.80	1.17	1.34	0.59(0.57–0.63)	0.85(0.78–0.88)
A. aerogenes	primer	1	0.90	0.90	1.10	1.10	0.82	1.00
	product	3	1.02	1.00	0.97	1.01	1.03(0.96–1.13)	0.99(0.95–1.01)
E. coli	primer	2	1.00	0.97	0.98	1.05	0.97(0.96–0.99)	0.98(0.97–0.99)
	product	2	1.04	1.00	0.97	0.98	1.02(0.96–1.07)	1.01(0.96–1.06)
Calf thymus	primer	2	1.14	1.05	0.90	0.85	1.25(1.24–1.26)	1.05(1.03–1.08)
	product	6	1.19	1.19	0.81	0.83	1.46(1.22–1.67)	0.99(0.82–1.04)
T_2	primer	2	1.31	1.32	0.67	0.70	1.92(1.86–1.97)	0.98(0.95–1.01)
	product	2	1.33	1.29	0.69	0.70	1.90(1.82–1.98)	1.01(1.01–1.03)
"Synthetic A-T Copolymer"	. . .	1	1.99	1.93	<0.05	<0.05	$>40.$	1.05

FROM KORNBERG, A. 1959. IN ONCLEY, J. L. (ED.). *BIOPHYSICAL SCIENCE.* JOHN WILEY AND SONS, INC., NEW YORK. REPRINTED BY PERMISSION OF JOHN WILEY & SONS, INC.

[a] A, T, G, and C refer, respectively, to adenine, thymine, guanine, and cytosine, except that C in the case of T_2 phage primer refers to hydroxymethylcytosine.
[b] The data in parentheses represent the range of values obtained.

Semiconservative Replication

All things considered, the properties of DNA polymerase were satisfactory to remove any mystery about the self-replication of DNA molecules. A mechanism that could do just what was needed had been found, and it was even able to do what was needed in the test tube. However, there were bound to be skeptics, and some pointed out that it would appear to be impossible for the DNA molecule to separate its two chains completely, since they are interlocked. Were they to separate by "unpeeling," the rest of the molecule would have to flip around in the medium in a most unlikely way. Since the molecule did replicate, however, it was reasonable to propose some alternative mechanisms for the formation of DNA to the simple unzipping model that had been initially proposed to explain it (Figure 18-7).

Logically, there are only three possible modes of replication. The first would be that the DNA molecule remains intact, with the enzyme traveling along its surface, reading individual bases in some way and making a new DNA

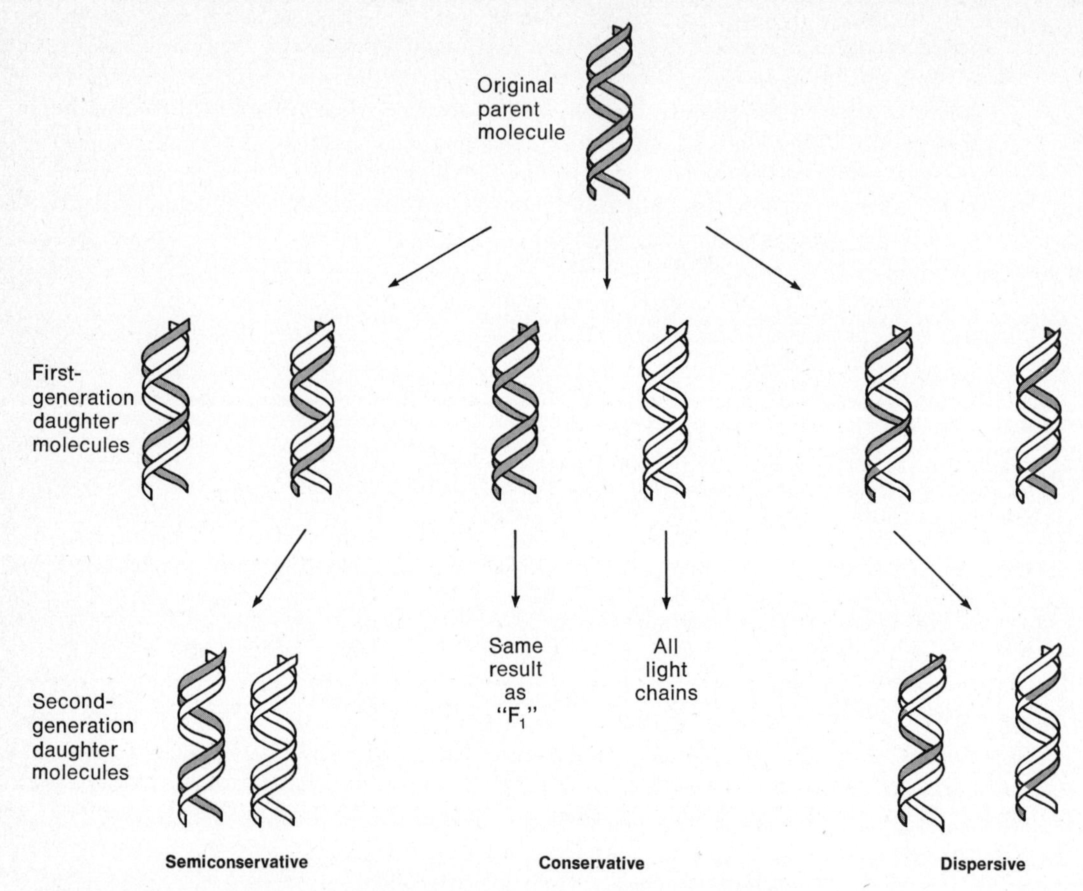

Figure 18-8. Three modes of DNA replication. In *semiconservative* replication, one chain of the original DNA molecule is used to make a new molecule. Thus the progeny molecules are half parental and half newly synthesized. In *conservative* replication the original DNA molecule remains intact, acting as a template to guide the synthesis of other molecules. In the first generation there will be one parental DNA molecule and one newly synthesized one. In the second generation there will still be one parental molecule, but three newly synthesized ones. In *dispersive* replication the chains break down into smaller units, which replicate semiconservatively and then rejoin. The product molecules will all have some parts that are parental and some that are newly synthesized.

molecule. Since this process preserves the parent DNA molecule, it was given the name *conservative replication* (Figure 18-8). The second mode would be the one already presented, the unzipping of the two strands of the helix, each one giving rise to a new DNA molecule. In this case, one-half the original molecule remains intact, one DNA chain going to one molecule and the other DNA chain going to the other. This process was given the name *semiconservative replication.* Finally, it is possible for the DNA to break at different (presumably special) places along the chain and be replicated as smaller pieces that can then rejoin to form the daughter molecules. The great advantage here, logically, is that it is no longer necessary to unzip a very long molecule; only short fragments of helix are separated prior to being replicated. This model was called, alternatively, *dispersive* or *nonconservative replication.* The problem was to test which of these three mechanisms was used in the living cell.

In 1958 an ingenious experiment was performed that could test the three hypotheses for replication exactly. The bacterium *E. coli* was grown in a medium containing heavy nitrogen (^{15}N). Since the bases of DNA contain a large amount of nitrogen, such cells should contain DNA molecules that are more dense than DNA molecules of cells grown on a normal (^{14}N) medium. Presumably, if exactly the same strains of cells are being used, the two sets of DNA molecules will be the same in base composition and will have the same average size. Each will contain exactly the same number of nitrogen atoms; but in one case the nitrogen atoms will be heavier than in the other case, and thus the density (weight per unit volume) will be higher in the ^{15}N DNA than in the ^{14}N DNA. Since there is a density difference, the two kinds of DNA can be separated from one another in a centrifuge, as long as the centrifuge tube itself contains a medium of varying densities.

To understand the experiment in simple terms, consider the fact that human beings sink rather deep in fresh-water ponds, much less deeply in the salt water of the ocean, and not at all in the Great Salt Lake. In each case, the density of the medium in which the particle is suspended is different. As the density of the medium increases, its ability to support other particles increases. When the particle is less dense than the medium, it will be buoyed up by it completely, and only a considerable force applied to the particle can "push it under the surface." If molecules (or other particles) are placed in a centrifuge tube containing a medium whose density increases from the top of the tube to the bottom and are then exposed to a centrifugal field, the molecules will migrate through the medium until they reach a point at which the density of the medium is sufficient to buoy them up and resist their further movement. If two particles of different densities are present in the centrifuge tube, they will stop at different points along the tube, each particle stopping when it reaches that part of the density gradient that exactly counterbalances the particle's tendency to sink further. This prediction is exactly borne out in practice and is now a widely used method for separating different kinds of molecules by centrifugation. (It is frequently referred to as a *buoyant-density* or *density-gradient* centrifugation.) When DNA extracted from cells grown on heavy nitrogen was placed together with DNA extracted from cells grown on normal nitrogen, in the centrifuge, clean separation with the two kinds of DNA was achieved, each one occupying a distinctly different position in the centrifuge tube (Figure 18-9a).

Having the technical ability to separate heavy and light DNA makes possible a test of the various kinds of replication. Cells of *E. coli* were grown on ^{15}N so that the DNA was uniformly heavy. At this point a synchronously growing culture of the cells was transferred to a medium containing only ^{14}N and allowed to grow for one cell division. (In this case, the DNA will have replicated once.) What is

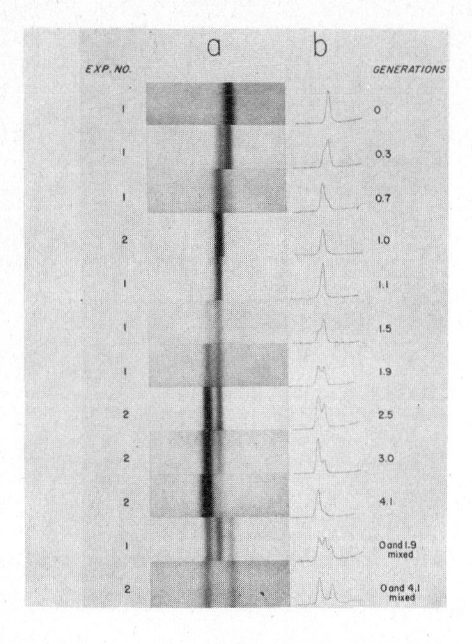

Figure 18-9. Test of semiconservative replication. (a) A diagram of the relative positions of three different types of DNA molecules in the density-gradient centrifuge tube (see text). The molecules composed of ^{14}N alone are the lightest and come to rest closest to the top. The heavy molecules (^{15}N) travel the farthest. Hybrid molecules are intermediate. (b) The results of the experiment. (1) Ultraviolet absorption photographs showing DNA bands resulting from density-gradient centrifugation of lysates of bacteria sampled at different times after the addition of an excess of ^{14}N substrates to a culture previously labeled with ^{15}N. The density of the solvent increases to the right. Regions of equal density occupy the same horizontal position on each photograph. (2) Microdensometer tracings of the DNA bands shown in the adjacent photographs. The deflection of the tracing pen is directly proportional to the concentration of the DNA. The degree of labeling of a species of DNA corresponds to the relative position of its bands between those of fully labeled and unlabeled DNA shown in the lowermost frame as a reference. (From Meselson, M., and F. W. Stahl. 1958. *Proc. Natl. Acad. Sci. (U.S.A.).* **44**:671.)

the prediction that each of the three models of replication makes with regard to the product DNA? Reference to Figure 18-8 will show that conservative replication predicts an equal quantity of heavy and light DNA if only one cycle of replication has occurred. On the other hand, semiconservative replication predicts that a single class of DNA molecules will be present and that these molecules will be of a density intermediate between the parental (all heavy) and the daughter (all light) predicted by conservative replication. The intermediate density results from the fact that one chain of the molecule is parental and thus composed exclusively of ^{15}N and the other chain is newly synthesized and composed exclusively of ^{14}N. Dispersive replication makes exactly the same prediction as does semiconservative replication at the end of the first division. Therefore these two cannot be distinguished from one another in this first test. The result was a single band of density intermediate between all heavy and all light, and so conservative replication could be eliminated.

Distinguishing between semiconservative and dispersive replication requires the growth of the cells for another generation. The prediction for semiconservative replication is clear. One-half the daughter DNAs of the second round of replication will contain nothing but ^{14}N and will be of the all-light density. The other half will contain the remaining heavy chain of the original parent DNA and a newly synthesized light DNA and will thus be of intermediate density. Consequently, one-half the DNA will be intermediate and one-half will be light if semiconservative replication is used. On the other hand, in dispersive replication, the prediction is not quite as exact, since the amount of heavy and light material present in each chain of a particular double helix would depend on how the chains reassembled themselves. However, one point is clear. Because of a mixture of both heavy and light in each chain, it would be impossible to achieve a completely light DNA in the second round of replication, and we would predict a variety of DNAs scattering in density around the intermediate position but never achieving the totally light DNA condition. The results were consistent with the semiconservative model of replication; one-half the DNA was light and the other half intermediate in density.

Denaturation and Renaturation of DNA

The foregoing experiments indicate that the original hypothesis for the replication of DNA is correct. The chains can separate from one another even though, in the double helical form, they would appear to be interlocked. Apparently their unfolding is relatively simple. All that is required is an energy input that will break the hydrogen bonds that hold the base pairs together. In 1959 it was demonstrated that heating isolated DNA molecules would permit chain separation. Figure 18-10(a) diagrams what would be expected when DNA is denatured by heat treatment. The hydrogen bonds holding the two chains together are broken, and the chains can move apart from one another. If the medium is cooled rapidly, new hydrogen bonds will form; but since the sister chains are separated from one another, it is more probable that the purine and pyrimidine bases of a single chain will bond with one another than that they will bond with other chains. Consequently, each single chain assumes a configuration of its own, on a more or less random basis, a configuration generally called a *random coil.* As would be expected, random coil structures have properties that differ from the DNA double helix. For example, many of the base pairs are now external to the backbone and are free to react with certain reagents in ways they could never react while "buried" in the core of the double helix. Of the several changes, however, the most easily measured is a change in the ultraviolet absorption of DNA. For well-understood but rather complicated reasons, the absorption of ultraviolet by DNA molecules is much greater in a random coil than in the DNA double helix. Consequently, the dissociation of the chains, or "melting," goes on with a marked increase in absorption. Figure 18-10(b) presents some of the original data, in which it can be seen that the absorption remains relatively constant until a critical temperature is reached, after which there is a dramatic increase in absorption until it reaches a maximum, presumably indicating the completion of the transition from double helix to random coil structure.

It is interesting to note that different DNAs have different "melting" temperatures. This point may seem surprising at first, but it emphasizes the critical fact that the DNAs are different from one another. In those cases in which there is a high amount of G and C, we would expect the DNA double helix to be more stable, since three hydrogen bonds are formed per pair rather than the two that are characteristic of A and T. The lower stability of the AT base pair is shown dramatically in the melting of the synthetic AT copolymer, which melts well below the temperatures of the four DNAs from various organisms. These data indicate, then, that the coils can separate readily and that the stability of the double helix depends on the base composition of the molecule, which is precisely what the original Watson–Crick model would predict.

In 1960 the remarkable discovery was made that it is possible to renature melted DNA. If DNA is denatured by high temperature and then cooled very slowly (by going from the melting temperature to 60°C over a period of about $1\frac{1}{2}$ hours), random coil formation does not occur. Rather, molecules having properties of

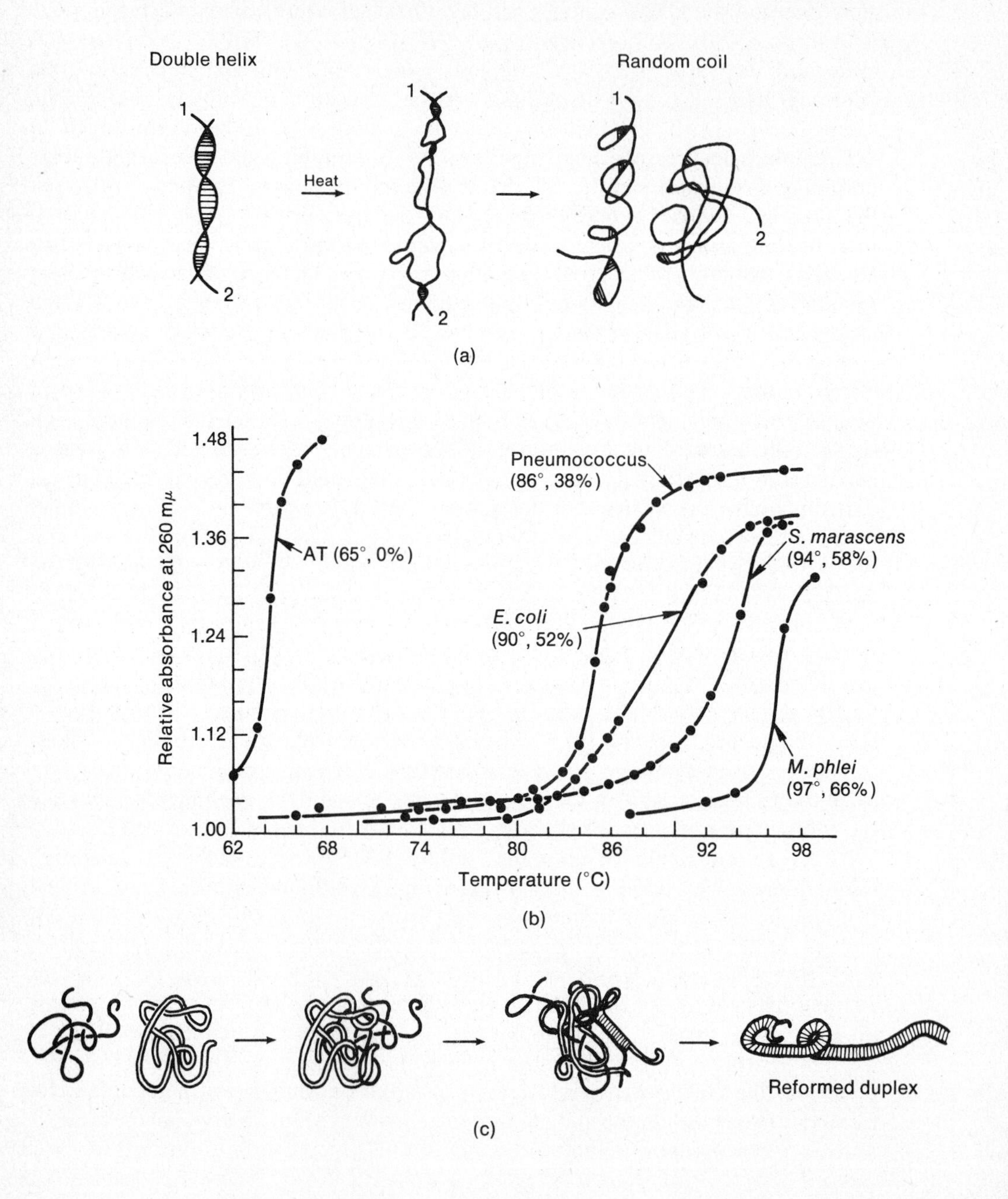

double helices are formed. As Figure 18-10(c) suggests, if the molecules are given sufficient time to interact with one another during the cooling process, there will be a tendency to maximize correct base pairing, since doing so will maximize the total number of hydrogen bonds formed and thus finally bring the molecule to its most stable state. At the early stages all the associations will be relatively unstable (transitory). Only those in which a high degree of homology exists between the two chains will remain together at the higher temperatures. The result is to permit the different chains to "identify" one another such that the proper (complementary) chains are in contact. As the temperature drops, more and more bonds will form; and at this point the two polynucleotide chains shift along one another's surface to maximize the number of hydrogen bonds formed, ultimately restoring a structure that is essentially identical to the original double helix.

The amount of identity in renatured (annealed) DNA is remarkable. This fact can be tested in a variety of ways. One of the most common is the restoration of the ability of renatured DNA to transform bacteria. Other tests of the specificity of the process include attempts to renature DNA molecules by using only a single type of strand, attempts that invariably fail even though there must be a certain amount of base pairing between these identical polynucleotides. Similarly, the amount of double helix formation is much lower when polynucleotide chains of two different organisms are mixed in an annealing process. As expected, the larger the DNA molecule studied, the lower the recovery of complete renaturation. In a spontaneous process, the smaller the molecule, the greater the chance of all bonds being reformed properly.

These experiments indicate that most of the predictions of the double helical model of DNA are correct. In particular, separation of the chains of the double helix can be effected with a relatively small energy input. Obviously the cell does not use local heating to cause the DNA to melt. But the unzipping of the chains in the formation of two daughter double helices should be relatively easy, for there is a clear mechanism for paying the energy debt. In the first place, for each set of hydrogen bonds broken in separating an A from a T, two sets of hydrogen bonds are reformed, since each A and T will pick up a new T and A, and so forth. In a situation where twice as many bonds are formed as broken, we would expect the spontaneous

---

Figure 18-10. Melting and renaturing of DNA. (a) A diagram showing the dissociation of the two chains of a double helix (melting) caused by heating. When cooled rapidly, the separate chains form random coil structures. (b) Melting of DNA from four different organisms and of an artificially prepared AT copolymer. The increase in absorbance is caused by random coil formation. It can be seen that the AT copolymer, which lacks the stronger GC bonding, is the most sensitive to heat treatment. The number in parenthesis gives the melting temperature and the relative GC content of the DNA. (c) A diagram of the renaturation of DNA. The two complementary strands are shown in different shades. If the temperature is kept high enough, tight random coils are not developed; and if cooling proceeds slowly, the chains have an opportunity to come into contact. New bonds can form between the two complementary chains and a new helix is established. (Part b after Marmur, J., and P. Doty. 1959. *Nature* **183**:1427.)

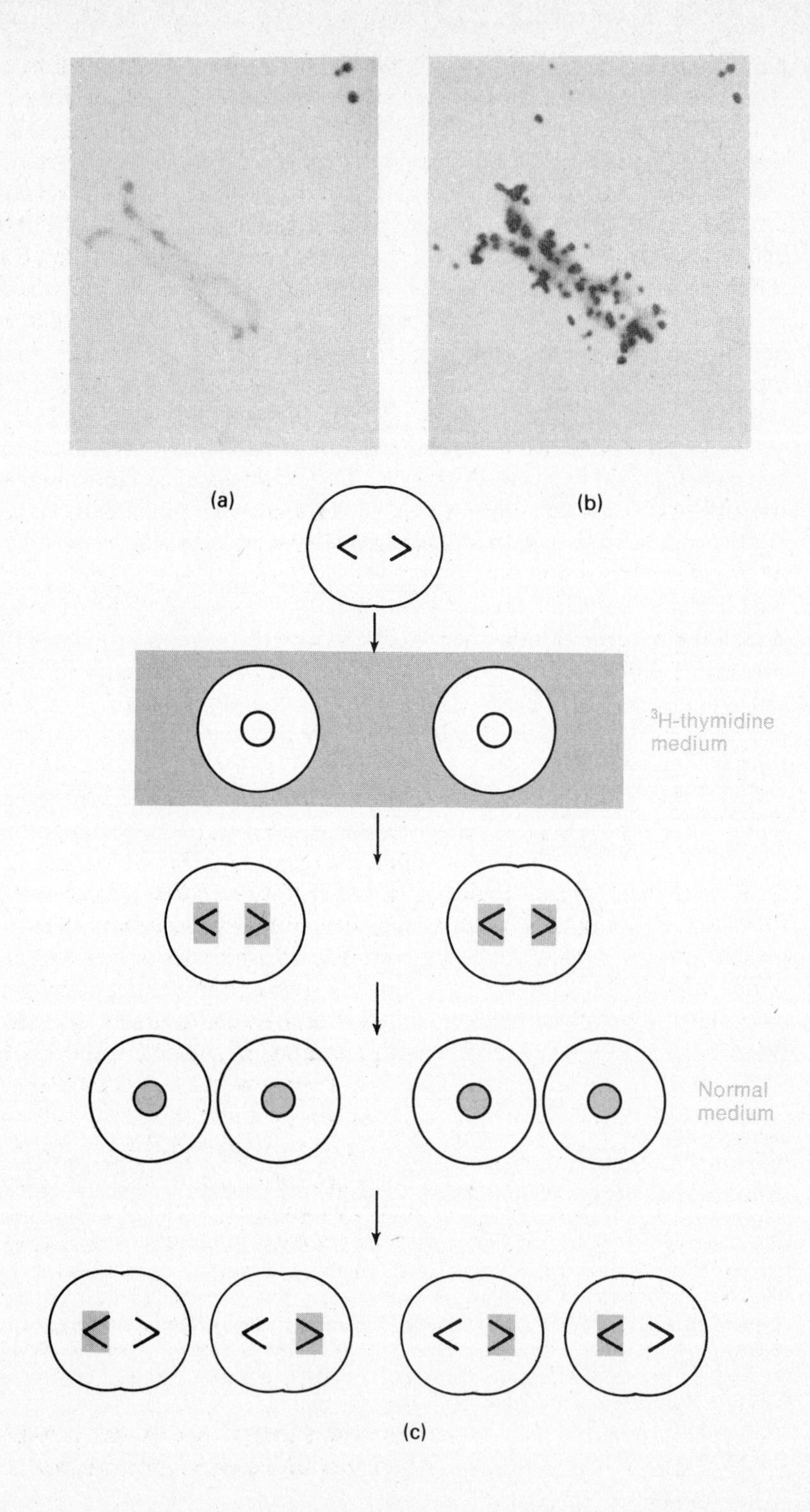
(a)
(b)
³H-thymidine medium
Normal medium
(c)

disassociation of the chains for the formation process. Furthermore, since the new chains are being synthesized from the triphosphorylated forms of the nucleotides, the formation of the backbone itself releases energy, so that the overall process results in a net lowering of the energy state of the entire system and thus is spontaneous. Looking back, the argument that semiconservative replication could not work appears strange. Nevertheless, it was important at the time, for it provoked the experiments that in the end explained how the replication of DNA did, in fact, occur within living cells.

Replication of Chromosomes

Higher Organisms

It is interesting that, prior to any of the experiments on the replication of DNA, it had already been shown that chromosomes of higher organisms duplicate themselves semiconservatively. In 1957 the distribution of radioactively labeled DNA in chromosomes was used in an experiment similar to that already outlined for the test of semiconservative replication of DNA itself. In this case, the roots of the broad bean *Vicia faba* were grown in a solution containing the nucleotide thymidine (T) that had been labeled with the radioactive isotope of hydrogen, tritium (^3H). We will use the common abbreviation ^3HT. Since the ^3HT is incorporated only into DNA, all the radioactivity appears in the DNA of the cell.

The presence of the radioactive isotope can be detected by *autoradiography*. The beta radiation emitted by ^3H will darken a photographic film emulsion. Slides of the rapidly dividing root tips are prepared in the traditional way and are then coated with a photographic emulsion sensitive to beta rays. The decay of the radioactive elements in the cell will darken the emulsion immediately over the isotope. Wherever there is label, a black spot will appear in the emulsion. It may take days or weeks before the exposure is sufficient to be seen, but since the cells are fixed, time is no problem. Once sufficient time has passed to be certain of the exposure, slides can be removed and exposed in the usual way so as to take a picture of the cells themselves with ordinary light. This procedure results in a double

---

Figure 18-11. Chromosome replication. (a) Autoradiographs of replicated chromosomes. The chromosomes in these photomicrographs are in cells of broad bean seedlings (*Vicia faba*). The seedlings were grown in a medium containing thymidine labeled with ^3H (tritium). When the cells synthesized new chromosome material, the DNA contained labeled thymine. The photomicrograph (a) shows a single bean chromosome, and (b) shows a layer of photographic emulsion above the same chromosome. Each dark spot in the emulsion was made by a particle emitted in the decay of a tritium atom. (c) Diagram of the experiment demonstrating semiconservative replication of chromosomes. The cells are prepared as in (a) and are uniformly labeled. The black Vs represent chromosomes. The shading represents the presence of radioactive tracer (^3HT). After exposure to ^3HT, all chromosomes are labeled. (One-half of each is newly synthesized.) After a passage through an unlabeled medium, only one-half of the chromosomes are labeled. (Parts a and b from Taylor, J. H., P. S. Woods and W. L. Hughes. 1957. *Proc. Natl. Acad. Sci. (U.S.A.)* **43**:122.)

exposure on the film, in which the photographic image of the cells is superimposed on the tracks made in the film by the radioactive decay. If, for example, all the radioactivity were in the DNA, we would expect to find it in the chromosomes and thus only part of the photographic image containing chromosomes should also contain decay tracks. Allowing for the fact that the beta rays can sometimes travel relatively large distances and that not all the tritium will be absorbed in the DNA at the time of exposure, this predicted image is precisely what is seen (Figure 18-11a).

Once it is established that the chromosome can be labeled specifically with a substance known to be incorporated only in DNA, an experiment can be carried out in which the logic is the same as that used to establish semiconservative replication of bacterial DNA. Cells are grown in the presence of ^{3}HT and uniformly labeled in their chromosomes. The cells are then switched to an unlabeled medium and allowed to grow for one division cycle, fixed, and autoradiographs are made. As would be predicted on the basis of semiconservative replication, both halves of the metaphase chromosome are labeled. Each chromatid therefore contains old DNA as well as new. When the cells are allowed to go through two cycles of replication in the absence of label, however, one-half the metaphase chromosomes at that second division are unlabeled. Again, this observation follows semiconservative replication. Presumably one chromatid is the product of the original labeled DNA and the other is the product of DNA synthesized in the absence of label just prior to the first division (Figure 18-11b). It thus appears that the semiconservative mode of replication is common to both DNA molecules and whole chromosomes, which strengthens the proposition that it is the DNA that is the continuous part of the hereditary material.

Bacteria

As noted earlier, the bacterial chromosome, in many cases at least, appears to be circular (Chapter 12). The preceding models of DNA replication applied to linear DNA molecules. The replication of a circular molecule is a much more difficult problem topologically. Where does replication begin? How do the chains separate when we are dealing with the unraveling of a circle that has no free ends to rotate in space?

The experiment confirming semiconservative replication in bacteria suggests that there is only a single point of starting and stopping of replication of the chromosome of *E. coli.* In most cases, the entire genome must be synthesized semiconservatively before a second round has begun in any other part. Otherwise the results would not have been as clean-cut as they were in the experiment. The densities would have been altered by that fraction of second-round replication that occurred before completion of the first round. That is, some parts of the DNA molecules would have had only ^{14}N atoms, whereas most of the strands contained one-half ^{14}N and one-half ^{15}N. (That this mechanism is a little too simple to be used as a general model of DNA replication is discussed in the next section.)

A number of investigations have been made to determine how many points of replication actually occur within the *E. coli* genome. It now seems reasonably well established that there is only one. Of these experiments, perhaps the easiest to understand is the one reported by Cairns in 1963. This work was addressed to the problem that the molecular weights of the DNA molecules isolated in the tests of semiconservative replication turned out to be entirely too low to account for the entire bacterial DNA content. It was presumed, therefore, that the DNA isolated was, in fact, fragments of the bacterial chromosome. Consequently, it was desirable to isolate an intact chromosome from *E. coli.* Only very delicate methods make it possible to isolate this giant DNA molecule without breaking it. Once the technique was perfected, *E. coli* DNA was labeled with ^{3}HT, allowing the cells to go through slightly less than two complete generation times (1.*n* generation times). The cells were then isolated, gently lysed, and the DNA collected on special filters. These filters were next overlaid with a photographic emulsion and stored for 2 months. The autoradiograph thus produced is shown in Figure 18-12(a). It can be seen that, as postulated, the DNA molecule is a closed circle and that the replication of this molecule produces two circles that will be ultimately interlocked at one point only.

Neither the analysis of the labeling of the chromosome nor the relationship of that labeling pattern to the replication of the giant DNA molecule that composes the bacterial chromosome is easy. Figure 18-12(b) presents a simplified analysis prepared by Stent (1971). The interpretation assumes that replication of the grandparent chromosome had begun at P_1 and proceeded through P_2 and P_3 *prior* to the addition of the ^{3}H label. The time required to replicate the sector P_2 to P_3 is $\frac{15}{100}$ of the time required for the replication of the entire chromosome. Therefore the second picture in the diagram is labeled 0.15*g*.

After the ^{3}H-thymine is added, replication continues around the circle from P_3 to P_1. Since replication had occurred all the way around to P_3 before that time, only a small portion of each of the daughter chromosomes, the region between P_3 and P_1, carries the radioactive isotope. Since the replication is semiconservative, one strand of each of the daughter chromosomes will be labeled. This situation is shown in picture 0.3*g*. The result is two circles, locked at position P_1, which can detach as the final step of replication. One of each of the circular chromosomes will go to each of the daughter cells at the time of cell division. As soon as the two chromosomes have detached, another cycle of replication can begin.

The second round of replication occurs in the presence of ^{3}HT. The next picture shown (1.0*g*) is that in which the replication (always beginning at P_1) has proceeded to P_2. One strand of each of the potential daughter chromosomes is labeled between P_1 and P_2. Since replication has not yet passed point P_2, there is no label in the region P_2 to P_3. However, since the region P_3 to P_2 was labeled in the first round of replication, some label is already present. When the second round of replication is complete (1.3*g*), two daughter chromosomes will be produced—one that is completely labeled in one strand only (upper) and one that is completely labeled in one strand and doubly labeled only in the region P_3 to P_1.

We presume that the chromosome shown in Figure 18-12(a) is derived from one like that pictured in the lower part of 1.3*g*. That chromosome has undergone a third round of replication as far as P_2. In this case, labeling again began at P_1 and gave rise to one chromosome completely labeled in the region P_1–P_2 (B) and its sister

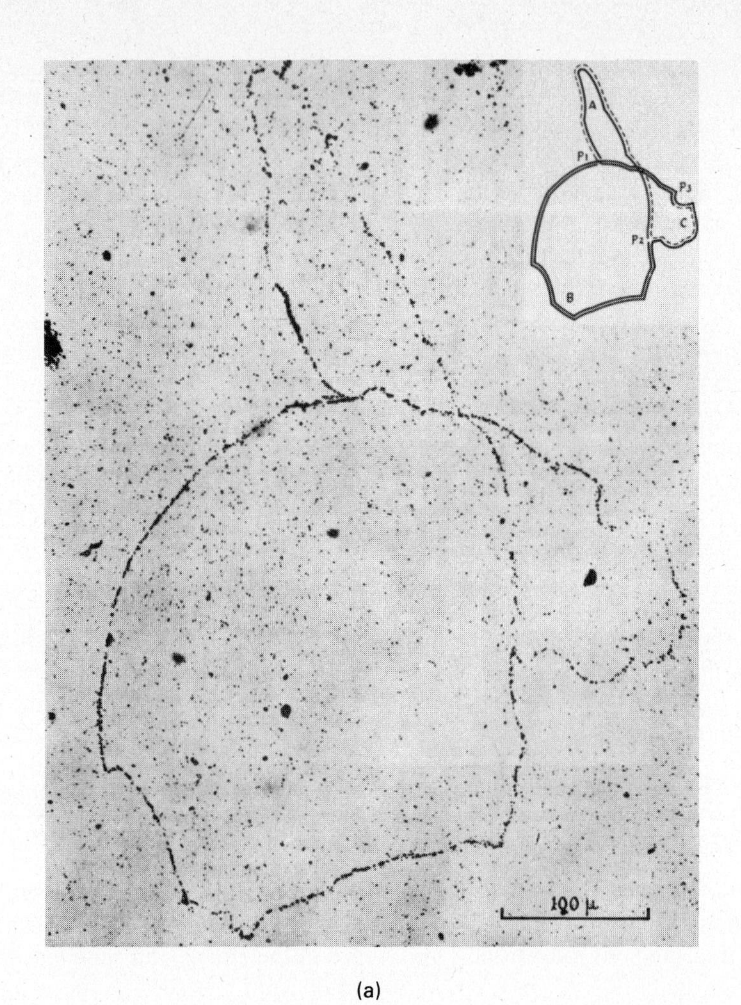
A
P1
P3
C
P2
B
100 μ

(a)

A
P1
C
P3
0 g
P2
B
P1
P3
0.15 g
P2
³H-thymine
added here
0.3 g
A
P1
P3
1.0 g
C
P2
B
1.3 g
A
P1
C
P3
2.0 g
P2
B

(b)

chromosome, which has only one strand labeled in the same region (A). In region C, P_3–P_1 is already doubly labeled and P_2–P_3 is singly labeled, since replication had not passed the point of P_2 when the DNA was isolated for study.

The foregoing interpretation is entirely consistent with the photograph obtained, and it suggests that, in fact, the giant, circular DNA molecule of *E. coli* is perfectly capable of replicating from one point only, going from that point around the circle until two daughter circles are produced. Other experimental evidence supports this view. Genetic mapping studies indicate that *E. coli* has a circular chromosome (Chapter 12). On the basis of the data presented, that circular chromosome corresponds to a circular DNA molecule, and apparently the circularity poses no special problems for the replication of the molecule in a purely semiconservative fashion.

Before concluding, we should mention the separation of the chromosomes of bacteria. In these cells there is no formal division process—no spindle, no centromere, no mitosis in the strict sense. The question must be raised, then, as to how the daughter chromosomes can be distributed in an orderly manner between the two daughter cells. The answer did not come until 1965 when it was observed that the replicating DNA molecule is attached to the bacterial cell membrane. The attachment is at the point of replication itself, the rest of the circle "feeding through" the replication center. Once replication is complete, the replication centers are free to separate from one another. Presumably this process occurs through the growth of

---

Figure 18-12. Replication of the bacterial chromosome. (a) Autoradiograph of the intact chromosome of an *E. coli* bacterium that has been allowed to incorporate ^{3}H-thymine into its DNA for not quite two generations. The lines of dark grains were produced by electrons emitted by the decaying ^{3}H-thymine atoms in the DNA molecule. (The scattered grains were produced by background radiation.) The inset diagram shows the portions of doubly and singly stranded labeling that can be inferred to be present in the three sectors A (P_1 to P_2), and C (P_1 to P_3 and P_3 to P_2) of the branched chromosome on the basis of the grain density. (b) Diagrammatic history of the *E. coli* chromosome seen in the autoradiograph. The double lines represent the two DNA polynucleotide strands of the chromosomal double helix; the broken lines indicate unlabeled strands, and the solid lines, labeled strands. At 0g, exactly two generation periods before extraction of the chromosome, the entirely unlabeled DNA of the grandparental bacterium was at the same stage of two-thirds completed replication as its granddaughter nucleus (2.0g) and (a) above. Let P_1 represent the point at which replication started and P_2 the position of the replicating fork. A short while later (0.15g) the replicating fork had moved to P_3. It was at this point that ^{3}H-thymine was added, and from then on all newly synthesized DNA strands contain the ^{3}H label. One-sixth of a generation later (0.3g) replication of the chromosome was complete, since the replicating fork had reached P_1. The two daughter chromosomes contained half-labeled double helices in the region from P_3 to P_1, but were otherwise entirely unlabeled. At 1.0g we diagram one of the two daughter chromosomes in which the replicating fork had reached P_2. It is two-thirds of a generation later. At this stage, sectors A and B, and sector C from P_3 to P_1 contained half-labeled doubled helices; sector C from P_2 to P_3 was still entirely unlabeled. One-third of a generation later (1.3g) the replicating fork was once more at P_1. The two granddaughter chromosomes contained fully labeled double helices from P_3 to P_1 and half-labeled double helices in the rest. At 2.0g we diagram one of the two granddaughter chromosomes in which the replicating fork has reached P_2 two-thirds of a generation later. Such a chromosome would have a labeling pattern precisely like that actually observed in the autoradiograph (a). (Autoradiograph from Cairns, J. 1963. *Cold Spring Harbor Symp. Quant. Biol.* **28**:43; inset and interpretation (b) after Stent, G. S. 1971. *Molecular Genetics.* W. H. Freeman and Co., San Francisco.)

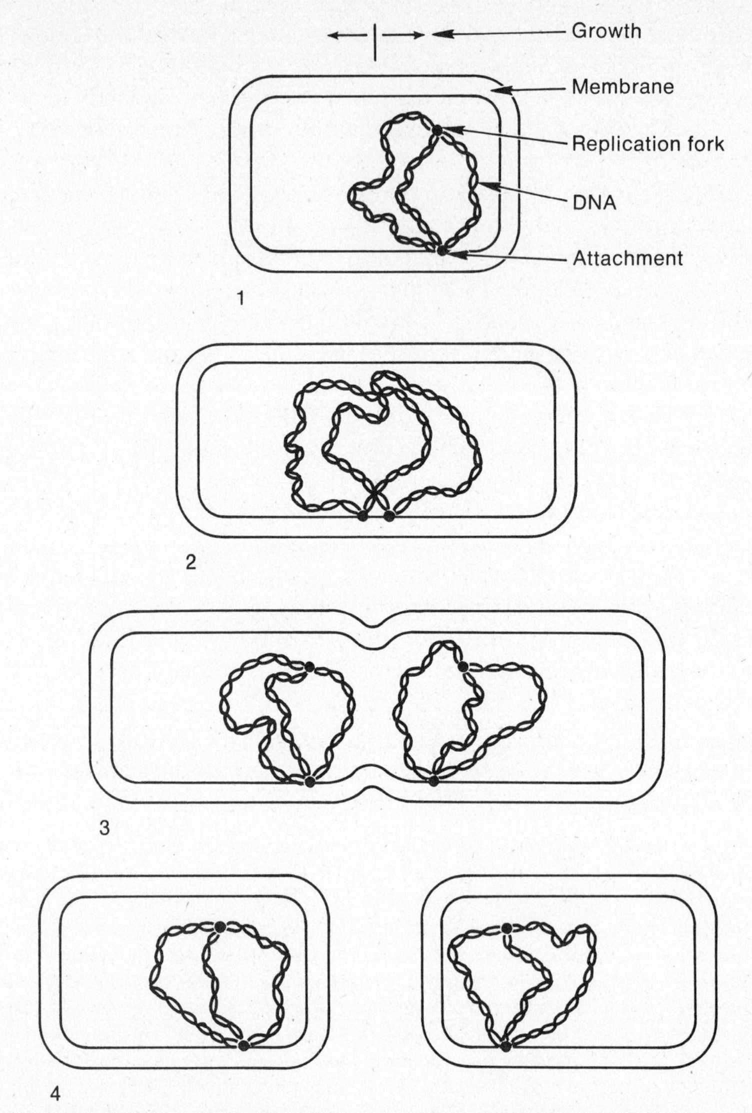

Figure 18-13. Mechanism of distribution of the bacterial chromosome. (1) A daughter bacterium has just been born. It contains a partially replicated circular DNA molecule attached to the bacterial membrane at the site of replication of the circular structure. Replication started at the replication fork in the parent cell. (2) The bacterium is midway through the generation period. Replication of the circle has just been completed, and the two replica circles have separate points of attachment. The sector of the bacterial membrane that has been formed concurrently is lightly stippled. (3) The bacterium is nearing end of generation period. Both replica circles are already partially replicated. Their points of attachment have moved apart because of growth of the sector of the bacterial membrane shown in light gray. (4) The bacterium has divided in two between the points of attachment of the two partially replicated genomes.

the membrane, which pushes the two replication centers apart, as indicated in Figure 18-13. As the cell membrane grows, the two chromosomes are separated from one another until, finally, a new septum develops that will divide the two daughter cells without damaging either of the chromosomes. Thus the growth of the membrane rather than the formation of a mitotic apparatus is utilized. It is a far more economical way, and although it may seem primitive, it is perfectly satisfactory as long as a single chromosome is involved and as long as the center of replication is attached to the membrane. This last factor is important, for otherwise premature separation of the cells might occur, thereby leading to abortion of at least one of the cells. In order to achieve synchrony of DNA replication and cell division, it is essential that the cell have control of the replication process.

It should be noted that the diagram of bacterial cell division given in Figure 18-13 implies that replication of the DNA has begun for a second round before the separation of the daughter cells. Available evidence suggests that such is the case (and explains more rationally the results discussed above). Apparently the replication of DNA in *E. coli* is essentially continuous and the second round has begun before complete separation of the daughter cells. This point is of minor interest to us at the moment, but it is worth pointing out that it is distinctly different from what happens in the cells of higher organisms, in which the replication of DNA is limited to a single part of the cell cycle.

The Mechanism of DNA Replication

The preceding discussion has presented a general model for replication of DNA. It provides an understanding of the general process. However, while accurate, it is nonetheless superficial. A number of difficulties are glossed over for simplicity of presentation. Consequently, we shall now turn to a brief description of some of the more recent findings about the replication process. The student can then begin to see how complex is the in vivo mechanism.

It is somewhat of a paradox that the DNA polymerase used to demonstrate that DNA could act as a template for its own replication (pages 300–303) is not the enzyme normally used by cells to replicate DNA molecules. It has been shown that mutant strains of bacteria lacking this enzyme can still replicate their DNA. This was an interesting finding, for it suggested, first, that a different DNA polymerase was used for normal replication and, secondly, that it should be possible to obtain mutants that lack the "true" replicating enzyme. A series of investigations in search of such mutants resulted, and it has now been shown that there are three distinctly different DNA polymerases in most cells, called DNA polymerase I, DNA polymerase II, and DNA polymerase III. Mutants that lack DNA polymerase III are incapable of replicating their DNA, and so it appears that this is the enzyme that is normally used in the in vivo replication process. The other two polymerases have special functions that will be discussed later (Chapter 23).

During the search for the replicating enzyme a number of other replication-deficient mutants were discovered, but these mutational events do not affect the replicating enzyme. Rather, they control other parts of the replication process. One group of genes (probably three different loci) controls production of *initiation proteins.* Without these proteins, the double helix cannot open up to provide the *replication fork* (Figure 18-13) produced by the separation of the two chains. Careful analysis indicates that there is a special initiation region (or regions) to which the initiation proteins attach, and this region serves to begin the replication process. In some organisms there is only one such initiation region; in others there are multiple sites of initiation.

Yet another replication-deficient mutant is at a locus controlling an enzyme that makes "nicks" in the phosphate backbone of one strand of the DNA double helix. This *nickase* has the function of relieving the tension in the double helix as it unravels; since only one strand is nicked, the DNA molecule remains essentially intact, the paired bases holding the two chains together. On the other hand, the nick permits one chain to rotate around the other as the replication fork moves along the molecule and thus permits easy chain separation during replication. In some organisms there may be only a few sites of nicking; these sites act as more or less permanent swivels for the unraveling of the DNA molecule. In other cases, multiple nicks are induced, and such nicks are far more transitory in duration, the nicks being made and then repaired relatively rapidly. In both cases, however, the nicking does not appear to be random; therefore there must be certain sites that the nickase recognizes as proper for the introduction of nicks. Mutants lacking the nickase enzyme are incapable of unraveling their DNA, and thus replication is blocked.

A complementary function to the nickase is that of the *ligase.* This enzyme repairs the cleaved phosphate linkages induced by the nickase (or by other events) and so reestablishes the intact DNA molecule. Mutant strains lacking this enzyme produce replicated but fragmented DNAs. As expected, in many cases one-half a DNA is intact (the unnicked chain), and the other half consists of relatively short fragments of replicated DNA (in those cases in which there is a relatively large number of nicks in the molecule).

Figure 18-14 is a diagram of the replication process in higher organisms that incorporates the facts discussed above as well as some other models for the replication process. In Figure 18-14(a) the DNA molecule can be seen to be open over a short stretch, producing a structure known as the *replication bubble* or *replication eye.* Presumably the initiation proteins are attached to the DNA molecule and have moved along it, opening up the two chains to make replication possible. Such bubbles can be seen in electron micrographs of replicating DNA, and the incorporation of radioactive markers into the structures indicates that, in fact, they are regions of replication. At first only a few replication bubbles are seen in any one region of the DNA. Later, others appear in different regions. Replication occurs via bidirectional growth of the bubbles. Consequently, when later bubbles appear, the initial bubbles have grown considerably. Eventually the bubbles merge to give a completely replicated molecule. Figure 18-14(b) indicates that

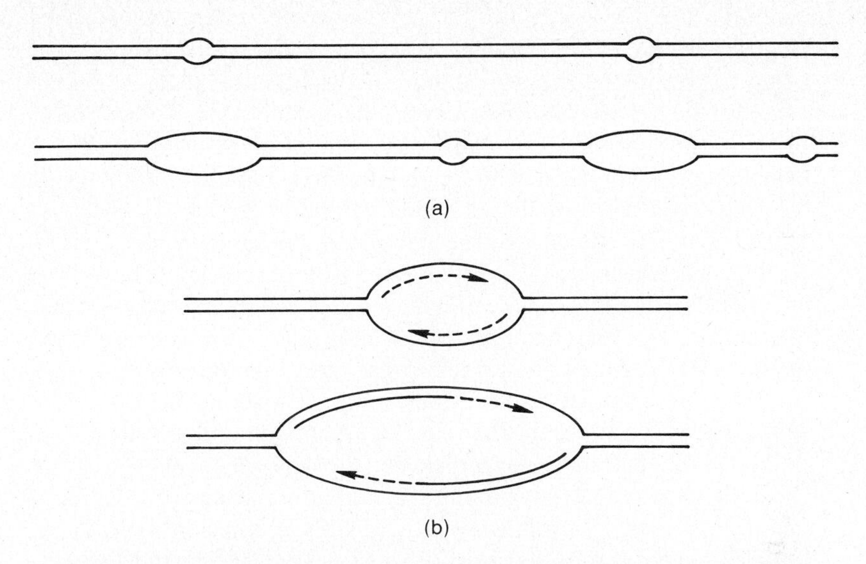

Figure 18-14. Replication of DNA. (a) The DNA molecule has been opened up and the two chains separated. Initiation proteins, which attach at one place in the chromosome, move away from that place either in one direction or in both directions, generating two replication forks, and the "bubble" between them. Replication of the DNA occurs in the bubble. In the upper diagram, two sites of initiation are shown. Beneath, the diagram shows two new sites that originated later and the two original sites now enlarged by the continued replication process. (b) In an enlargement of the bubble in (a), the newly formed strand is indicated by the dotted lines; the arrows indicate the polarity of replication. The process is discontinuous. The ligase enzyme joins the shorter replicated fragments together into a long, single molecule, as indicated in the lower diagram. The possible significance of the discontinuous replication is discussed in the text.

replication is polar; that is, one or more enzyme molecules move along each strand of the DNA in the 5′-3′ direction. This is true of all the known DNA polymerases—they cannot replicate the molecule in the 3′-5′ direction—which means that at least two molecules of DNA polymerase are required in any region that is replicated (one for each strand) or that one polymerase molecule replicates one strand in the bubble and then proceeds in the reverse direction along the other strand replicating it.

Perhaps the most curious feature of DNA replication in eukaryotes is the fact that it is discontinuous (Figure 18-14(b)). That is, even in the replication bubble the replication process goes on in short segments that are linked together later into the longer strands of the replicated DNA molecule. The enzyme responsible for attachment of the various segments into the completed daughter strand is the ligase. At first this method of synthesis seems strange. Nevertheless, it must be recognized that the DNA normally exists as a double *helix*, not as two straight rods. (To this extent, Figure 18-14 is an oversimplification.) Just as it is necessary to introduce breaks into one of the chains in order to permit the DNA to unravel, it is similarly necessary to introduce breaks in the daughter chain as it is replicated in order to permit the proper rotation of the bases relative to one another so as to establish the helical character of the replicated DNA molecule. The ligase presumably joins the short replicated strands together only after the helical configuration is established. All of which leads to an interesting paradox. Although the experimental evidence clearly establishes semiconservative replication as the means by which DNA is synthesized, yet it is true that certain aspects of the dispersive (nonconservative) mechanism are, in fact, employed to achieve this end. The problem of uncoiling and recoiling the DNA molecule during replication is sufficiently great that breaks are required in at least one strand of the molecule. This is precisely the point that was made by the original critics of semiconservative replication, and they deserve credit for their farsightedness. Nevertheless, as an overall picture of the process, the semiconservative mechanism is correct. Although one strand may be nicked at one point, the other strand is never nicked in the same place (or even close by). And thus the replication process, beginning at an initiation site and proceeding along the DNA molecule to some terminus (most probably a collision with another replication bubble growing in the opposite direction), never results in the dispersion of the DNA into small fragments of double helix. Present evidence suggests that this blending of semiconservative and dispersive models of replication has resolved the essential problems of DNA replication. It is a far more complicated process than the early section of this chapter suggests, and the details of the process are still the subject of intense investigation. Still, for the geneticist, it is sufficient (and safe) to assume that the DNA molecule and the chromosome are replicated semiconservatively.

The replication process may be either unidirectional or bidirectional. That is, from the initiation point, replication of DNA may proceed in one direction only, or it may move away from the initiation point in both directions with two growing replication forks. Bidirectional replication appears to be the case in the majority of organisms studied. However, there are a significant number of unidirectional mechanisms. In a circular chromosome, a unidirectional mechanism will eventually grow all the way around the circle until it reaches the initiation point. In a linear chromosome, however, unidirectional replication implies the necessity of multiple initiation points along the chromosome in order to achieve complete replication of the DNA molecule. From the data available to us now, it appears that most chromosomes are replicated with several to many initiation sites, and it is relatively rare that a single initiation site is used.

General Comments

The discussion in this chapter has centered around the problem: What is the gene? The evidence presented here should make it clear that the only candidate is DNA. Where it can be tested, isolated DNA clearly has a genetic function. Its structure is such that the most important properties of the gene can be explained. Most importantly, the property of self-replication, which is one of the unique properties of living systems, is the exclusive property of DNA. Under the circumstances it would be hard to withhold the title of gene from DNA. However, one part of the puzzle remains unresolved. If a gene functions through the control of protein synthesis, how is that control exerted? In principle, we can see how DNA might direct the mechanism; but until that mechanism is completely elucidated, and the DNA is shown to be the controlling factor, we cannot be unequivocally convinced of the hypothesis. Consequently, we must now turn to the study of protein synthesis.

REFERENCES

HAYNES, R. H., and P. C. HANAWALT. (Eds.). 1968. *The Molecular Basis of Life.* W. H. Freeman and Co., San Francisco. (This volume has several good articles, originally published in *Scientific American.* The articles especially recommended as supplements to the material covered in this chapter are *The structure of the hereditary material,* by Crick, *The bacterial chromosome,* by Cairns, and *The duplication of chromosomes,* by Taylor.)

STENT, G. S. 1971. *Molecular Genetics.* W. H. Freeman and Co., San Francisco. (This somewhat advanced text has a very good treatment of the structure and replication of DNA. Chapters 8 and 9 are recommended.)

WATSON, J. D. 1969. *The Double Helix.* Mentor Books, New York. (An autobiographical account of the discovery of the structure of DNA. It is a rather biased but entertaining and often enlightening story of the events surrounding an important scientific achievement.)

Original Papers and Advanced Reviews

CHARGAFF, E. 1950. Chemical specificity of the nucleic acids and mechanism of their enzymatic degradation. *Experientia* **6**:201–209. (The paper that established the 1 : 1 ratio of A to T and G to C.)

FRANKEL-CONRAT, H. 1956. The role of the nucleic acid in the reconstitution of active tobacco mosaic virus. *J. Amer. Chem. Soc.* **78**:882–883. (This paper shows that only the nucleic acid is important in heredity.)

HERSHEY, A. D., and M. CHASE. 1952. Independent functions of viral proteins and nucleic acid in growth of bacteriophage. *J. Gen. Physiol.* **36**:39–56. (The paper showing that only viral DNA is injected at the time of bacteriophage infection.)

JACOB, F., S. BRENNER, and F. CUZIN. 1966. On the regulation of DNA replication in bacteria. Cold Spring Harbor Symp. *Quant. Biology.* **28**:329–348. (A review of the evidence leading to our present understanding of how the bacterial chromosome is replicated and separated at division.)

KORNBERG, A. 1960. Biologic synthesis of deoxyribonucleic acid. *Science* **131**:1503–1508. (A full account of the properties of DNA polymerase by the man who discovered it.)

MARMUR, J., R. ROWND, and C. L. SCHILDKRAUT. 1963. Denaturation and renaturation of deoxyribonucleic acid. *Prog. Nucleic Acid Research.* **2**:231–300. (The melting and annealing of DNA are described in some detail.)

MESELSON, M., and F. W. STAHL. 1958. The replication of DNA in *E. coli. Proc. Natl. Acad. Sci. (U.S.A.)* **44**:671–682. (The important experiment that proved semiconservative replication.)

MIRSKY, A. E., and H. RIS. 1947. The chemical composition of isolated chromosomes. *J. Gen. Physiol.* **31**:7–18. (The paper reporting the first study of isolated whole chromosomes.)

WATSON, J. D., and F. H. C. CRICK. 1953. A structure for deoxyribosenucleic acid. *Nature* **171**:737–738. (The classic paper that first presented the double helix model of DNA. The two papers following this one in the volume, by Wilkins and his associates, present the crystallographic data on which the model is based.)

QUESTIONS AND PROBLEMS

18-1. Define the following terms:

DNA	RNA
purine	pyrimidine
nucleotide	base
polynucleotide	polarity
histone	adenine
guanine	thymine
cytosine	

18-2. Summarize the five major arguments for DNA as the gene. Why were these arguments not completely convincing prior to the formulation of the double helix model for DNA?

18-3. What are the essential features of the double helix model of DNA?

18-4. In what ways did the double helix model explain the essential properties of the gene?

18-5. If one chain of a DNA molecule has the base order ATTGACGT . . . , what is the base order of its complementary chain?

18-6. Explain the forces that enable A to pair with T and G with C. Why do not A and C or G and T pair?

18-7. Of the bases AGCTU, which would you expect to find in DNA? In RNA?

18-8. What is the role of DNA polymerase III in the normal replication of DNA?

18-9. State the distinction between conservative, semiconservative, and nonconservative (dispersive) replication.

18-10. What is the critical evidence that supports the semiconservative model of DNA replication?

18-11. What is the evidence that supports a semiconservative model for the replication of whole chromosomes? What does this suggest about chromosome structure in higher organisms?

18-12. If the chromosome is a single DNA molecule, what restrictions does its polarity place on the production of inversions?

18-13. What is the function of the replication center in bacteria?

18-14. In general terms, explain denaturation and renaturation of DNA.

18-15. If you denature two different DNAs and then mix the separated chains together in such a way that they can renature, would you expect to find hybrid molecules or only the original types? Explain your answer.

18-16. If you did find hybrid molecules (18-15), what would that suggest about the two original DNA molecules?

18-17. If you discover that two relatively unrelated groups of organisms have the same AT/GC ratios, what would you conclude?

19 / Genetic control of protein synthesis

Throughout the preceding chapters the assumption was made that the gene exerts its control over heredity through the control of protein synthesis. This particular point of view has been held for a long time by a wide variety of biologists. However, prior to the work that established the one gene–one enzyme hypothesis, much of what was said about gene action was highly speculative. Early studies showed that the great bulk of protein synthesis (if not all of it) is carried on in the cytoplasm. At the same time it was clearly shown that the genes resided on chromosomes in the nucleus. It was necessary to prove, therefore, that the nucleus had a direct control over cytoplasmic events.*

*The material discussed here focuses on the point at which the fields of genetics, cytology, biochemistry, and developmental biology converge. Obviously the mechanism of protein synthesis is important to all these disciplines. Different teams of investigators in each of these areas are pursuing research on the mechanism of protein synthesis. Recent studies in molecular genetics that united these different disciplines were a great achievement, since they unified biology in an important way. On the other hand, in writing about the development of this unification, it is impossible to summarize all the literature. So we would like to caution the reader that we have been very selective in the experiments that we chose to discuss in some detail. In many cases, they have been selected primarily for didactic reasons, although we have attempted always to include those experiments that are recognized as major contributions to the resolution of this problem. The choice, however, is that of a geneticist; a biochemist might make a different one.

The Role of the Nucleus

A great amount of research has been devoted to this problem. Some of the most interesting experiments were those carried out with the green alga Acetabularia. This alga develops as a very large, uninucleate cell, which may be 4 to 6 centimeters in length. The cell is somewhat differentiated. It is attached to rocks or pebbles by projections called rhizoids; the nucleus of the cell resides in a rhizoid. From this basal attachment rises a long, narrow stalk and then, at the top, a terminal cap about 1 centimeter in diameter (Figure 19-1). Two of the species used in these experiments were *A. mediterranea*, which is found in the Mediterranean Sea, and *A. crenulata*, which is commonly found in the West Indies. The Mediterranean species has a cap that is divided into approximately 80 compartments, each of which is rounded on the outer edge. The West Indian species, however, has fewer segments to the cap (approximately 30), and the tips of each of these sections are pointed (hence its name). As Figure 19-1 indicates, they are distinctly different organisms in their overall appearance.

In the normal growth of these plants, the stalk develops from the rhizoid end and continues to elongate until finally the cap is produced. Immature plants can be grafted together by removing the tips of each of the stalks and fusing the two. This process produces a "hybrid" individual that grows a new stalk and cap (Figure 19-1). Interestingly, the cap produced showed some of the features of each species. That the intermediate appearance is not merely an artifact of the grafting procedure can be shown by grafting together two individuals of the same species. When this is done, a stalk and cap characteristic of the species are produced. Is the intermediate cap of the graft hybrid a result of the presence of two different nuclei or two different cytoplasms?

The answer is complex. If the growing stalk from an immature plant is removed and grafted onto the rhizoid region of another plant that has been decapitated (Figure 19-2), the new "hybrid" will have the nucleus from the plant providing the rhizoid end, which is designated the "host." When interspecific grafts of this type were made, the cap usually resembled that of the host species; thus it appeared that the nucleus controlled the event that led to cap production in Acetabularia. Whatever nucleus was present determined the type of cap formed. Unfortunately, the results were not completely unambiguous, since the cap formation was at least partially dependent on the amount of stalk grafted onto the nucleus. When a large piece of stalk was grafted onto a small host rhizoid, a cap of intermediate appearance was formed; whereas when a very small stalk was grafted onto the rhizoid, the cap resembled that of the rhizoid host. The most reasonable interpretation of these results is that cap formation proceeds with materials that are present (or formed) in the stalk and that the formation of these materials is ultimately under nuclear control. If a stalk has already developed, a fair amount of this material will be available during cap formation whether or not the original nucleus is present; and thus the cap will be formed from a mixture of materials that were formed in the stalk under the control of the original nucleus and those formed in the stalk under the control of the host nucleus.

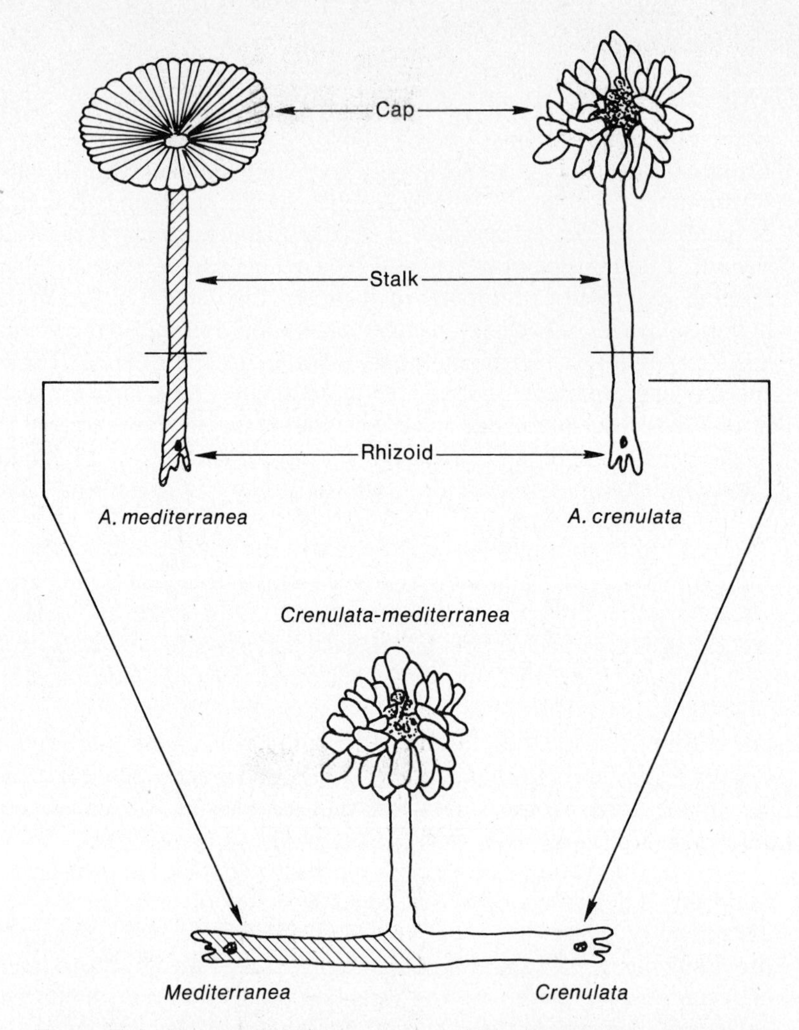

Figure 19-1. Acetabularia. The upper part of the diagram shows the essential differences between the caps of the two species most commonly used in experiments. Acetabularia is a unicellular green alga; differences in shading are for diagrammatic purposes only. The lower part shows the result of fusing two decapitated plants. The cap formed is intermediate between the two species.

Since the nucleus of a plant lies in the rhizoid, it is relatively easy to prepare anucleate plants, either by removal of the rhizoid containing the nucleus or by taking a piece of immature stalk and letting it regenerate both rhizoid and cap. Interestingly, the plant can live for a relatively long period of time (several months) in the absence of a nucleus, and the formation of the cap goes on perfectly well. (Rhizoid

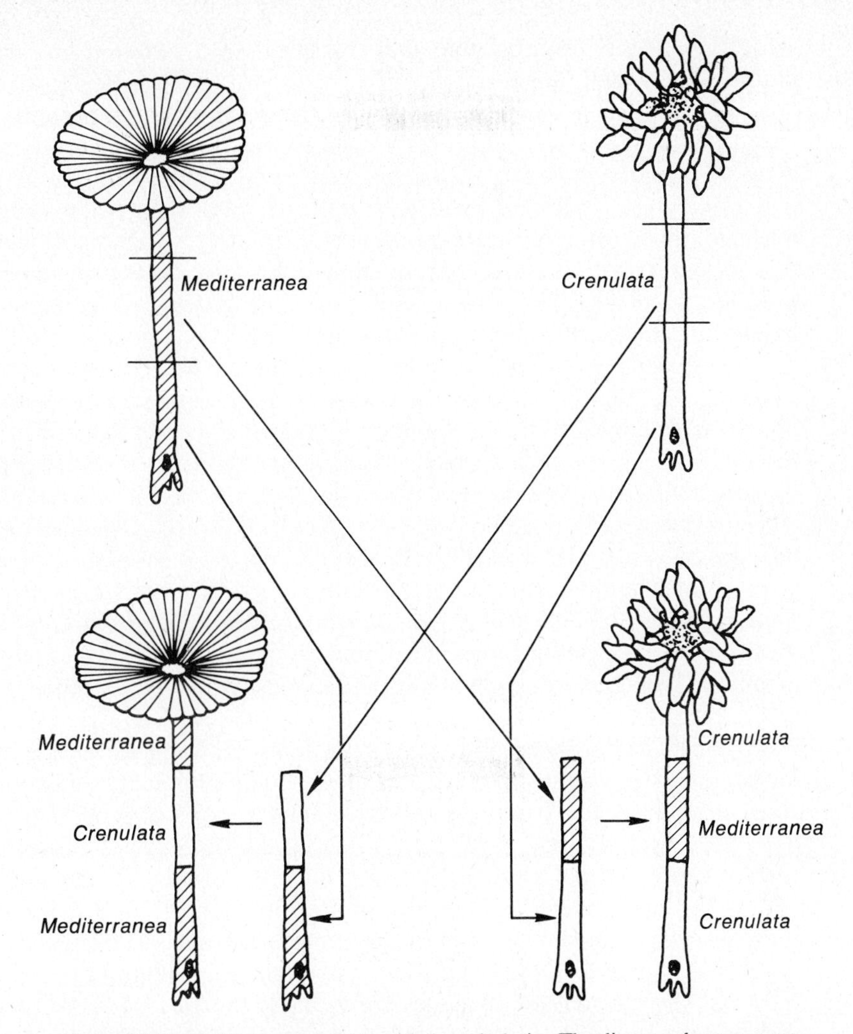

Figure 19-2. Grafting experiments with Acetabularia. The diagram shows two sets of experiments in which a piece of stalk of one species is grafted onto the rhizoid (containing the nucleus) of another species. The cap formed resembles most strongly that normally produced by the nucleus-containing host. (From Swanson, C. P. 1957. *Cytology and Cytogenetics.* Reprinted by permission of Prentice-Hall, Inc., Englewood Cliffs, New Jersey. After Hammerling, J. 1953. *Internat. Rev. Cytol.* **2**:475.)

production, however, is often incomplete.) In these experiments the kind of cap produced is characteristic of the species as long as a sufficiently large portion of the stalk is used to make the new anucleate plant. When small portions of the stalk are used, the morphogenetic capacity is greatly reduced, and the plants are frequently incomplete. Thus it would appear that the nucleus provides the cytoplasm with something that is essential and that, for want of a better term, we can call a morphogenetic determinant.

As long as a sufficient quantity of the determinant is present, the stalk is capable of forming a cap. However, if such anucleate organisms are decapitated and allowed to regenerate a cap, the morphogenetic determinants are soon used up; and it is rare for an anucleate plant to regenerate more than twice, even though its nucleate sibling can regenerate essentially indefinitely. Apparently, in the absence of a nucleus, there is no further production of the determinant; and once the store is used up, the plant is incapable of further morphogenesis. Once again the experiments indicate that the nucleus provides an essential component to the cytoplasm.

The obvious question raised by these experiments concerns the nature of the determinant. What is made by the nucleus that is responsible for determining the normal morphogenesis of Acetabularia? Over the years it was supposed that the nucleus was the site of a variety of functions. At the turn of the century it was suggested that the nucleus was the main center for cellular oxidation. But the evidence was never very good, and by 1925 E. B. Wilson had changed the hypothesis to suggest that the nucleus was the site for enzyme synthesis or accumulation. Considering the later finding of the relationship of genes to enzymes, this hypothesis was quite prophetic. On the other hand, there has never been very good evidence for the localization of enzymatic function within the nucleus; most enzymes are found in the cytoplasm, either free floating or attached to the various membranous systems of the cell, such as the mitochondria and plastids. In 1941 it was suggested that the nucleus was the main center for the formation of proteins. This suggestion was based largely on the observation that the nucleolus is extremely prominent and contains a high concentration of RNA in those cells that are carrying out protein synthesis. This hypothesis was modified in 1952 to state that the nucleus was concerned with the replacement of certain materials in the cell that are necessary for metabolic function.

The replacement hypothesis was based on a number of experiments similar to those discussed above. It is possible to remove the nucleus from certain cells and have the cell survive. Different cells survive for different lengths of time. Sometimes they will survive only briefly; sometimes, as is the case with Acetabularia, they survive for long periods of time. But in no case will the cell survive forever in the absence of a nucleus. Consequently, it was reasonable to propose that the nucleus provides something that the cell uses for maintenance purposes.

The maintenance hypothesis has been borne out by a variety of experiments, some of the most extensive of which were done in the 1950s in a study of protein synthesis in Acetabularia. They demonstrated that although protein synthesis could continue in the cytoplasm after the nucleus had been removed, it could not be maintained indefinitely. *Prolonged* protein synthesis—that is, beyond 2 or 3 weeks—is not maintained. Similarly, the maintenance of RNA synthesis was demonstrated to be dependent upon the presence of the nucleus. After 2 weeks RNA is synthesized much more poorly in anucleate cells than in nucleate cells. Essentially similar results were obtained with amoebae. When a cell is found that has a relatively small amount of cytoplasm surrounding the nucleus, it is possible to cut the cell such that a nucleate and an anucleate portion are obtained. Both portions heal and the cells live; but in the anucleate fragments RNA synthesis drops remarkably. By 10 days the RNA content is down to 40 % of normal.

Perhaps the most interesting experiments are those in which amoebae were grown in the presence of food particles containing ^{32}P. This labeled the nucleic acids of the amoeba with the radioactive isotope. The nuclei were removed from such radioactive amoebae and transplanted into normal, unlabeled amoebae and anucleate fragments of amoebae. After a time the cells were fixed and autoradiographs (see Chapter 18) were prepared. The pictures showed clearly that the cytoplasm of the amoeba carrying a radioactive nucleus becomes radioactive after 12 or more hours. Furthermore, it could be shown that all the autoradiographically detectable ^{32}P was in the form of RNA. It appears that the labeled RNA was formed in the nucleus and subsequently passed out into the cytoplasm. What is perhaps most interesting is the finding that, in binucleate amoebae in which one nucleus is radioactive and other is not, there is no transfer of radioactivity into the unlabeled nucleus. That is, the transfer appeared to be a one-way process from the nucleus to the cytoplasm.

In both the amoeba and in Acetabularia, removal of the nucleus leads to a loss of protein-synthesizing ability. In each case, there seems to be a parallel between the production of RNA and the production of protein; and this parallelism was too much to be ignored. Taking the evidence as a whole, it was proposed that, in the normal course of events, DNA made RNA, and it was this RNA that, when transferred to the cytoplasm, directed the synthesis of protein. This hypothesis would clearly explain the results obtained with Acetabularia. The amount of RNA present in the cytoplasm at the time of enucleation would be sufficient to permit regenerative capacity for one or two times; but since it is used up in the formation of protein, the anucleate cell ultimately runs out and regeneration must stop. Similarly, in the graft hybrid experiments in which two different forms are put together, two different kinds of RNA would be sent to the cytoplasm of the stalk, each presumably making its own type of protein. The interaction of these different kinds of proteins would produce an intermediate cap. The data were far from complete, but with a leap of the imagination it was possible to suggest that RNA was somehow intermediate between the gene and the protein-synthesizing mechanism.

The Role of RNA in Protein Synthesis

We must now turn to a discussion of RNA. Rather than deal with this subject historically or logically, we will merely describe the three major classes of RNA, propose a model for protein synthesis, and then describe some of the experimental tests of the model.

Ribosomal RNA

One of the most important discoveries in unraveling the mechanism of protein synthesis was the finding that a large amount of the RNA of the cytoplasm is contained in small particles (originally called microsomes but now more correctly called *ribosomes*). In the early 1950s it was demonstrated that these particles were the site of the great bulk of protein synthesis. The ribosomes are composed exclusively of RNA and protein. They are complex structures consisting of a larger and a smaller subunit, which associate during protein synthesis to form the functional, large ribosomes, having a molecular weight of about 70 S* in bacteria (and other prokaryotes) and 80 S in organisms having a true nucleus (eukaryotes). Corresponding to the two parts of the ribosome are two ribosomal RNAs (*r*RNA). One has a sedimentation value of 16 to 18 S and the other of 23 to 28 S, depending on the source of the material, the lower weight being typical of bacteria and the higher one of higher organisms. Each of these *r*RNAs combines with specific proteins to produce the two ribosomal particles that associate in a 1 : 1 ratio to give the active 70 S ribosome. In addition, there is a 5 S *r*RNA that associates with the 23 to 28 S molecule to produce one subribosomal unit.

In general, *r*RNA appears to be a singly stranded molecule that is highly flexible, with a random coil structure. There is good evidence for the existence of some regions of a rather ordered helical nature; apparently certain regions have stacks of AU and GC pairs that are hydrogen bonded to give the more highly organized secondary structure.

Ribosomal RNA constitutes 50 to 80 % of the total cellular RNA, and it is by far the largest of the cytoplasmic RNAs. Its upper molecular weight limit of approximately 10^6, however, is obviously smaller than the usual values of DNA. Because of its large size, it has been difficult to do an extensive analysis of the structure of the *r*RNA, but for our purposes that is unnecessary. It is only necessary to understand that because it is very large and found exclusively within the particles, *r*RNA is relatively easy to isolate as a separate fraction of RNA that can be characterized easily.

Transfer RNA

In 1957 another class of cytoplasmic RNA was reported. This RNA is now called transfer RNA (*t*RNA). Since it is too small to be precipitated by centrifugation at 100,000 g (where g = gravity), it is sometimes referred to as "soluble" RNA. Between 10 and 20 % of the total cellular RNA falls in this category. Transfer RNA has a sedimentation value of 4 S, corresponding to a molecular weight of about 10^4. The

*The molecular weight of the very large molecules of the cell (proteins and nucleic acids) and of cellular particles is generally obtained from calculations based on the velocity of movement of the particle in a centrifugal field. Since the numbers are frequently large, it is simpler to use the designation of the number of "sedimentation units"—8 S, 70 S, and so on. The relationship of molecular weight to sedimentation is not linear, and these designations can be confusing. They are *not* additive. For example, the 70 S ribosome is composed of a 30 S particle and a 50 S particle (30 + 50 = 70). However, the larger the number, the greater the molecular weight, and that is sufficient for our purposes.

molecules contain approximately 70 to 80 nucleotides and are the smallest of the known RNAs.

Transfer RNA combines with amino acids in the presence of an enzyme and an energy source (ATP) to form a *t*RNA-amino acid complex. Investigation of this reaction has shown that there is at least one specific *t*RNA for each of the 20 amino acids encoded into protein, as well as a specific enzyme that attaches each amino acid to its particular type of *t*RNA. The *t*RNA-amino acid complex is required for protein synthesis. Free amino acids are not directly incorporated into protein; only the *t*RNA-amino acid form will work.

There are several interesting features about *t*RNA molecules. They are singly stranded, and all have the same terminal groupings, A—C—C at one end and G at the other end. The amino acid is always attached to the A end, and thus the specificity of the different *t*RNAs must reside in the intervening portion. Furthermore, it is interesting to note that although single-stranded molecules, they show a high degree of helical structure. They are the only class of RNA to show regular X-ray diffraction patterns and the only ones whose base composition suggests a high degree of complementarity. It has been determined that extensive internal folding takes place in the molecules, a schematic representation of which is given in Figure 19-3(a). A number of *t*RNAs have been analyzed for their nucleotide sequence, the first of which was the alanine-*t*RNA from yeast. Its primary sequence and the most probable secondary configuration, the "cloverleaf" form, are shown in Figure 19-3(b). Note that the free G of one end pairs with the C of the other end, leaving the A of that end exposed as the amino acid acceptor.

Messenger RNA and the Central Dogma

With the discovery of *r*RNA and *t*RNA two essential components of the protein-synthesizing machinery had been found, both of which involved cytoplasmic RNA. This evidence was sufficient to establish firmly what has been called the *central dogma*—namely, that DNA (the gene) makes RNA, which moves from the nucleus to the cytoplasm, where it guides the synthesis of protein. The intermediate between the gene and protein synthesis is the RNA molecule (or molecules).

Since the placement of amino acids in the polypeptide chain (the primary structure) determines all other levels of structure and the function of proteins, it is necessary for the RNA in some way to control the specific placement of amino acids in the polypeptide chain at the time of synthesis. To give a specific example of the problem that we face, consider the major protein of egg white, egg albumen, which has a molecular weight of 40,000 and contains 288 amino acids. Let us assume that this protein is made by random collisions of amino acids. If such were the case, something in excess of 10^{300} different kinds of polypeptide chains could be formed (each of which would have a molecular weight of 40,000 and would contain 288 amino acids). Nevertheless, an animal as stupid as the chicken is capable of making the egg albumen protein the same way every time. There is, in fact, one and only one type of egg albumen. Clearly, this fact indicates that the protein-synthe-

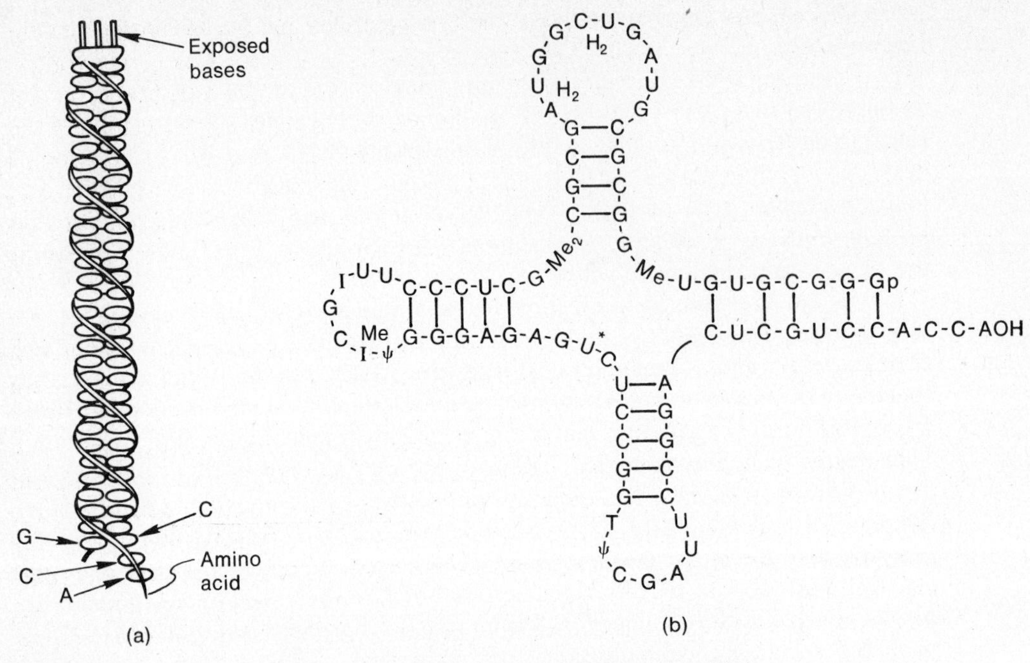

Figure 19-3. Structure of transfer RNA. (a) An idealized diagram of a *t*RNA molecule. At one end, three bases are exposed that can bond specifically with bases in the *m*RNA molecule. At the other end, the amino acid is attached to the terminal A nucleotide. The intervening bases are hydrogen-bonded to give the molecule a high degree of double helix structure, even though it is a single chain. (b) The cloverleaf form of *t*RNA. This was the first *t*RNA to have its nucleotide sequence determined. In the cloverleaf form there are three sets of exposed bases. Probably the shape of the different *t*RNAs varies to provide specific recognition possibilities for the enzymes that attach the amino acids. The *t*RNAs contain a number of unusual bases (see Chapter 21).

sizing system must contain a high degree of specificity. What is the basis for that specificity?

Since we know that the gene is a DNA molecule, and we know that DNA is merely a long string of nucleotides, we must presume that somehow the sequence of nucleotides in the DNA is translated into the proper sequence of amino acids in the polypeptide chain. We are now ready to propose that the intermediate in this process is RNA. Thus the ready hypothesis is that an RNA is made by a direct transcription of the DNA; and it is this RNA that is the template on which the polypeptide chain is organized, the definite sequence of nucleotide bases in the RNA being in some way related to a definite sequence of amino acids in the polypeptide. This hypothesis seems a reasonable one in view of the data already discussed; however, as pointed out, the great bulk of the RNA in the cell consists of *t*RNA and *r*RNA. The initial question, therefore, must be: Which one of these could carry the specificity of the genetic message?

It is easy to rule out the *t*RNA. Although one amino acid is attached to each *t*RNA, it is attached in a specific manner. It is impossible to see how a long string of amino acids can assemble themselves together through random collisions of *t*RNA-amino acid complexes. Little is to be gained with this mechanism that could not be achieved by the simple random collision of the amino acids themselves. Therefore it appeared reasonable that the ribosome must somehow contain the specific information. On this basis, it was initially proposed that there was one ribosome for one protein, the ribosomal RNA presumably being made by the gene to determine the sequence of amino acids.

The one ribosome-one protein hypothesis was perfectly reasonable, but it turned out to be incorrect. A number of experiments made it seem improbable. An analysis of the *r*RNA from *E. coli* indicated that the AU/GC ratios were invariant from one strain to another and were very unlike the ratios of DNA (allowing the U of the RNA to stand for the T of the DNA). Clearly, if the RNA template is to carry the same basic sequence of nucleotides as the DNA, the ratios should be similar. Since different strains of *E. coli* make different proteins, the different strains should have different kinds of templates. But since the analysis of *E. coli r*RNA showed it to be invariant, it could not be the template itself. Furthermore, there was some outstanding evidence against this hypothesis. As early as 1948 it had been demonstrated that the infection of *E. coli* by certain bacteriophages (even-numbered T phage) led to the rapid synthesis of certain proteins that were very different from the proteins made by the bacterium prior to the infection. But unfortunately for the hypothesis, there was no net synthesis of RNA after the phage infection. These experiments suggested that protein synthesis could go on in the absence of net RNA synthesis (and hence an absence of ribosomal RNA synthesis) and proved to be a contradiction to the entire scheme.

In 1953, however, an ingenious experiment was done in which *E. coli* cells infected with T_2 phage were exposed to ^{32}P for a brief period of time. RNA was isolated from the cells shortly after the "pulse" of isotopes had been administered, and it was found that one small fraction of the RNA had a relatively high degree of labeling. However, about 40 minutes after infection, the isotopically labeled RNA had disappeared. These observations suggested that although there is no *net* synthesis of RNA during phage growth, there is, apparently, the rapid formation of some small amount of RNA that is very unstable. This interpretation was confirmed when the unstable RNA fraction was isolated. It constitutes a distinct class of RNA, having an intermediate molecular weight (8 S). The molecules are single stranded and show little internal base pairing. Most important was the finding that the base ratio of this unstable RNA produced in response to phage infection is the same as that of the phage DNA (letting the U in the RNA stand for the T in DNA). This is precisely what is necessary if the phage DNA is specifying the synthesis of its own proteins through the formation of an RNA template. This special class of RNA has been given the name messenger RNA (*m*RNA), since its function appears to be the carrying of the sequence-determining message from the gene to the site of protein synthesis in the cytoplasm.

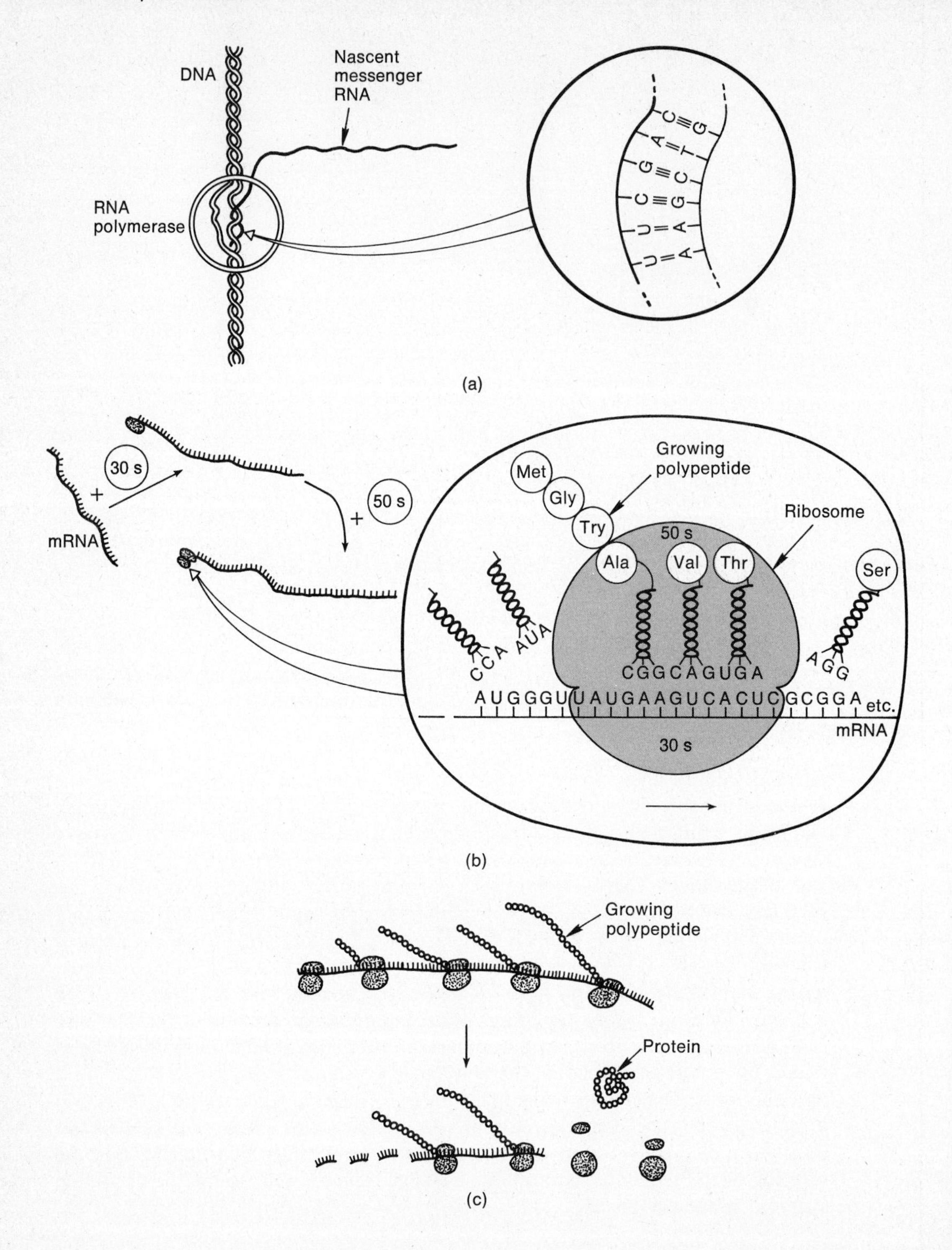
DNA
Nascent messenger RNA
RNA polymerase
(a)
30 s
50 s
mRNA
Met
Gly
Try
Growing polypeptide
Ribosome
50 s
Ala
Val
Thr
Ser
CCA
AUA
CGGCAGUGA
AGG
AUGGGUUAUGAAGUCACUCGCGGA etc.
mRNA
30 s
(b)
Growing polypeptide
Protein
(c)

Protein Synthesis

The discovery of three distinctly different classes of RNA makes it possible to propose a more detailed model for protein synthesis. Presumably, in some way, the DNA is transcribed into an extended *m*RNA molecule, which travels to the cytoplasm, where it attaches to the ribosome. The ribosome acts as the "factory," and to this factory the *t*RNAs bring various amino acids to be placed, each in the right sequence. If this much of the hypothesis be true, then it should be possible to effect cell-free synthesis of protein, using only isolated and purified *t*RNAs, *m*RNA, and ribosomes, a suitable energy source, the necessary activating enzymes, and the 20 amino acids. In fact, such is the case, and as soon as it was demonstrated that cell-free synthesis of protein required nothing other than these components, it was accepted that the hypothesis was close to reality.

Figure 19-4 is a diagrammatic representation of the various stages of protein synthesis, omitting the complication of the many different enzymes required for the steps. At the first stage, the messenger is transcribed from DNA. This transcription process can only be effected by base pairing. As pointed out in the previous chapter, there is a high degree of specificity to base pairing; G preferentially bonds with C and A with T. We merely need to postulate, then, that the RNA is formed as a "mirror image" of one strand of the DNA (remembering to replace T with U in RNA). In this case, for every place that has an A in the DNA there will be a U in the RNA that is transcribed from it; for every place that has a G in the DNA there will be a C in the RNA that is transcribed from it, and so forth. Figure 19-4(a) is a diagram of the generation of *m*RNA from a section of DNA.

When the messenger RNA arrives in the cytoplasm, ribosomes attach to it. There is apparently a specific point of attachment at one end of the *m*RNA, and the ribosome attaches to that end and then moves along the message from one end to the other. Because of the length of the message, more than one ribosome may be attached to the *m*RNA at any one time. In bacteria, the usual number is five. That part of the *m*RNA to which the ribosome is attached (remember that it can cover

---

Figure 19-4. General model for protein synthesis. The overall diagram gives a view of protein synthesis. (a) Initially the messenger (*m*RNA) is transcribed from the DNA of the gene. The RNA polymerase a very large enzyme, is shown lying over a region of chain separation in the DNA. This region of uncoiling might be larger or smaller than indicated. On the right, an enlargement indicates the base pairing between one strand of the unravelled DNA and the newly forming *m*RNA. (b) Once the *m*RNA has been formed, the 30 S subparticle attaches; then the 50 S subparticle attaches to the complex to form a functional 70 S ribosome-*m*RNA complex. The enlargement on the right shows how the *t*RNAs, carrying their amino acids, attach to the *m*RNA within the ribosome. Each donates its amino acid to the growing poypeptide chain and then returns to the cytoplasm to pick up another amino acid of the right type. (c) A diagram of the *m*RNA molecule being "read" by several ribosomes. Each ribosome attaches at one end of the *m*RNA and proceeds forward along the message, attaching each amino acid to the growing polypeptide according to the base pairing between the *m*RNA and the several *t*RNAs. When the ribosome reaches the end of a message, it detaches, separating into its two subunits (30 S and 50 S) and releasing the finished polypeptide. After the passage of several ribosomes, the *m*RNA molecule is degraded.

only a small portion of the *m*RNA at any one time) is held in an extended state, and it is to this extended part of the message that *t*RNA-amino acid complexes attach.

One of the central problems in explaining the formation of protein concerns the way in which an amino acid could recognize a part of a nucleic acid template that specifies its particular position in the polypeptide chain. The only mechanism available with sufficient specificity appears to be base pairing. Thus the formation of the *t*RNA-amino acid has as its primary function the specific attachment of a small piece of RNA that has the potential for reading the *m*RNA and effecting the attachment of a particular amino acid. As indicated in Figure 19-4(b), one part of the *t*RNA is unfolded (compare Figure 19-3), and these unpaired bases are able to combine with a specific part of the message. The diagram presents it as three base pairs specifying the position of any one amino acid.

As each *t*RNA-amino acid complex arrives at the surface of the message, the amino acid it is carrying is transferred to the growing polypeptide chain. As its amino acid is detached from the *t*RNA, the free *t*RNA is ejected from the ribosome-*m*RNA complex, and the ribosome moves a few notches farther along the message. The various proteins of the ribosome provide the enzymatic machinery necessary to effect the formation of the peptide linkage between the different amino acids and the motion of the ribosome along the messenger. The details of this process are beyond the scope of this book, and the interested reader is referred to any of the several very good books on biochemistry or cellular physiology that are presently available.

Since several ribosomes may be attached to the message at any one time, several copies of the specified protein can be made from a single message. However, in the microorganisms at least, the *m*RNA is unstable. It is degraded after the passage of approximately five ribosomes, which generate five polypeptide chains. As indicated in Figure 19-4(c), when the ribosome reaches the end of the message, it detaches and releases the polypeptide chain, which is then free to take up its function in the cytoplasm.

Tests of the Hypothesis

The preceding section has presented a general picture of the mechanism of protein synthesis as it is now understood. It must be clear, however, that in going from the evidence that there are three kinds of RNA to a more detailed mechanism for the synthesis of protein, we took a giant step. Each part of this hypothesis has been verified independently, and before we go on to matters of the most important genetic consequence, it seems well to examine at least some of the evidence on which the hypothesis is based.

One Gene–One Polypeptide

The work leading to the one gene–one enzyme hypothesis clearly established the relationship between a specific gene and a specific protein. However, rather early the hypothesis encountered two kinds of contradictory evidence. The first was the existence of genes that appeared to control more than one enzyme. This is a complex problem, and we will return to it later (Chapters 21 and 24). The other was that certain enzymes appeared to be controlled by more than one gene. This finding posed the greatest threat to the entire hypothesis. Two genes–one enzyme made little sense.

Fortunately, the contradiction was removed by research done on sickle-cell anemia. This hereditary disease of humans is under the control of a single allele. The homozygous sickle condition gives rise to an anemia that, without special treatment, is lethal. The anemia is the result of a defective hemoglobin molecule that has a low oxygen-carrying ability. When the oxygen tension is low, the red cells (which are largely hemoglobin) take on the characteristic sickle appearance that gives the disease its name. In homozygotes, the sickling is followed by rupture of the red cells and a consequent serious anemia. Heterozygotes are apparently normal; however, when put in a low oxygen tension, the red cells collapse into the sickle shape. Apparently the heterozygote has some of the defective protein and is only partially as good as the normal homozygote.*

Consider this fact for a minute, for it tells a great deal about protein synthesis. In diploid individuals, two genes are present for any one protein; but in a heterozygote, one gene is presumably producing a normal protein and the other the sickle-type protein. The heterozygote thus has only half the normal hemoglobin and a lower oxygen-carrying capacity. This prediction was verified in 1949, when it was shown that the hemoglobin from sickle-cell patients could be separated from that of normal patients if the proteins were placed in an electric field. Normal hemoglobin is slightly more negative in its character than is the sickle-cell protein. Heterozygous individuals had both kinds of protein in their blood.

Hemoglobin is a multichain protein consisting of four polypeptide chains, two each of two different types. One type, called α, has 144 amino acids, and the other type, called β, has 146 amino acids. Using A to symbolize a normal adult, we can write the "formula" for normal hemoglobin as $\alpha_2^A\beta_2^A$ and that for the homozygous recessive sickle-cell anemia patient as $\alpha_2^A\beta_2^S$. This formulation indicates that only the β chain is affected; the α chain is completely normal. Analysis of the β chain of sickle-cell hemoglobin revealed that it differed from the normal by only one amino

*We cannot pass by this subject without at least commenting on the prevalence of the sickle-cell trait among black Americans. The heterozygous individual is resistant to the malarial parasite. Apparently the abnormal hemoglobin confers this resistance. The homozygous condition is lethal, but the heterozygous condition is more favorable than the homozygous normal, since the chances are greater of encountering the malarial parasite than they are of encountering very low oxygen tension. For this reason, the incidence of the sickle gene in the malarial regions of Africa is very high. Since many ancestors of black Americans came from this region and since there has not been as extensive outbreeding as there might have been, the black population still retains a relatively high incidence of the trait. In America, where the incidence of malaria is extremely low, the heterozygous individual is not as well off as the homozygous normal, and the occasional homozygous sickle-cell offspring are quite sick. This situation has created the problem for black Americans.

acid, as indicated below:

A valine-histidine^{+}-leucine-threonine-proline-glutamic^{-}-glutamic^{-}-lysine^{+} . . .

S valine-histidine^{+}-leucine-threonine-proline-valine-glutamic^{-}-lysine^{+} . . .

It can be seen that the change is the substitution of an uncharged valine for a negatively charged glutamic acid. Apparently the insertion of an uncharged hydrophobic amino acid for the charged hydrophilic glutamic acid is sufficient to make the hemoglobin molecule abnormal in function.

The preceding analysis illustrates two important points. First, it indicates how two genes could control one functional protein. If that protein is composed of more than one kind of polypeptide chain, two genes would be controlling its synthesis. In this case, one gene controls the synthesis of the α chain and the other that of the β chain. Mutational events taking place in one gene have no effect on the structure of the other polypeptide chain, but they may affect the function of the protein as a whole, since the interaction of the different chains ultimately gives the configuration necessary for function. Therefore we must modify our one gene–one enzyme hypothesis. To be exact, we must make it one gene–one polypeptide chain. Whenever a single polypeptide chain is the functional unit, one gene–one enzyme is equivalent to one gene–one polypeptide chain. But when more than one kind of polypeptide chain is used to make a protein, the two statements are not equivalent. The best statement, the one that is exact at all times, would appear to be one gene–one polypeptide chain.*

The second important observation from this study is that a mutational event in one gene gives rise to a change in a single amino acid within the polypeptide chain. Mutation, therefore, cannot be envisaged as necessarily destroying all the information carried in the gene. Rather, a specific mutation (point mutation) is more likely to affect a very local part of the genetic message, and apparently that part controls the specification of one amino acid, and one only, in the polypeptide. Thus our view of genetic control of protein synthesis must be altered, for this finding suggests that each amino acid is placed in a specific place in the polypeptide chain under the control of the gene responsible for the synthesis of that particular polypeptide. This piece of evidence was essential for the overall model.

Colinearity of the Gene and the Polypeptide

If we are to assume that a particular nucleotide sequence specifies a particular amino acid and that the sequence is read as in Figure 19-4, it follows that the hypothesis predicts an exact correspondence between the position of a mutational event in the genetic message and the position of an amino acid within the polypeptide chain

*To bring this terminology into line with the various levels of organization revealed by the fine-structure analysis of the gene (Chapter 16), we should say one cistron–one polypeptide. In this chapter we use the term gene as if it were equivalent to cistron. That need not always be the case.

coded for by the gene. This prediction has been tested by a long and skillful analysis of the primary structure of tryptophan synthetase in various mutant forms of the bacterium *E. coli.*

The enzyme catalyses the final step in tryptophan synthesis:

$$\text{indole-glycerol-phosphate} + \text{serine} \rightarrow \text{tryptophan} + \text{triose-}\mathbf{p}$$

It has a quaternary structure involving two distinctly different polypeptide chains, designated A and B. The A chain has been completely analyzed (Figure 19-5), and, consequently, it is possible to determine alterations in its primary structure that result from mutational events leading to a loss of tryptophan synthetase activity. As expected, such mutations invariably result in the replacement of one amino acid within the polypeptide chain. Presumably, when amino acid substitutions of this type are sufficiently different, the tertiary and quarternary configurations of the final enzymatic protein are distorted to the extent that it can no longer react with the substrate and/or effect the conversion to product. Those amino acid substitutions that are sufficiently similar to the original might go unnoticed when screening for mutational events by using absence of enzymatic function as the assay. Consequently, the information in Figure 19-5 is based primarily on rather drastic differences (from the enzyme's point of view).

An experiment of this type has a major technical difficulty. In general, when studying biochemical mutations, we look for an absence of enzymatic function. However, when a protein is to be isolated for sequence analysis, it is usual to look for the presence of function. If a mutational event has removed the enzymatic activity, how is the investigator to know that the inactive protein he isolates is, in fact, in all other respects the same as the active form for which the amino acid sequence is already known? (Once a sequence analysis has been done on the inactive form, the correspondence between amino acid patterns will be sufficient to ensure that it is the proper protein. However, to do sequence analysis on a variety of proteins until the right one is found is an elaborate and wasteful procedure.) Therefore before this experiment could be completed, a simple technique was needed to identify the inactive protein that is a defective form of tryptophan synthetase. Antigen-antibody techniques were used. Simply stated, the active form of tryptophan synthetase is injected into a suitable mammal and the antibodies made against the enzyme are collected and purified. Assuming that the inactive tryptophan synthetase is similar in many ways to the active form, it can be argued that any protein material reacting with the tryptophan synthetase antibody is a form of the enzyme. Such material, precipitated by treatment with antibody, is called cross-reacting material (CRM), since it reacts with an antibody for the enzyme but is not itself an active enzyme. By using this technique, it is possible for investigators to isolate protein from mutant cells with relative ease and then analyze the CRM to determine the amino acid sequence. The various substituted amino acids indicated in Figure 19-5 were determined in this way. For example, mutant 446 has a cysteine substituted for

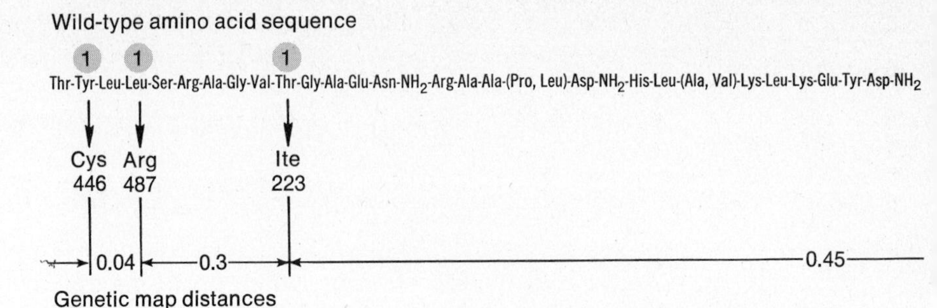

Figure 19-5. Colinearity of a gene and its protein product. The upper line presents the sequence of amino acids in the A chain of tryptophan synthetase isolated from *E. coli.* The lower line gives the locations of the mutational sites in the genetic map, as determined by recombination analysis. The circled amino acids have been substituted as indicated by the arrows, as a result of mutation. The numbers under the substitutions are the designations of the mutants. The numbers over the circles indicate the number of mutational events identified at that particular locus. Note that the position of the amino acid substitutions in the polypeptide chain are exactly the same as the spacing of the mutational sites within the gene. (After Mahler, H. R., and E. H. Cordes. 1971. *Biological Chemistry*, 2nd ed. Harper & Row, Publishers, New York.)

tyrosine, and mutant 187 has a valine substituted for glycine. Each of these substitutions is sufficiently important to remove the normal enzymatic activity.

Once the foregoing analysis had been done, it was possible to test the hypothesis that the placement of amino acids in the polypeptide chain is colinear with the placement of mutated sites within the gene. This test is done by recombination analysis of various mutants. Using this genetic technique, it is possible to construct a map of the mutated sites (Figure 19-5, bottom). The map units are relative, but they can be correlated with the relative placement of the amino acid substitution in the polypeptide chain. The two correspond almost exactly. A mutational event at the right end of the gene causes an amino acid substitution at the right end of the polypeptide chain, and so forth. An essential prediction of the hypothesis has been proved: the colinearity of the gene and its polypeptide product.

Role of Ribosomes and Messenger

Figure 19-4 suggests that the ribosomes are protein-synthesizing factories that will make any polypeptide that the *m*RNA instructs them to make. The experimental basis for this postulate comes from an elegant study of the mechanism in *E. coli.* The cells were grown on a medium containing compounds labeled with ^{13}C and ^{15}N. All materials synthesized by these cells should be "heavy," since they contain heavier than normal isotopes. Ribosomes that are generated in such cells can be shown to be heavy by the technique of density gradient centrifugation, discussed in the previous chapter. The "70 S" ribosomes of such bacteria are, in fact, heavier than the normal 70 S ribosomes of cells grown on nonisotopic medium. When such cells are lysed and their ribosomes isolated by density gradient centrifugation, there are two major peaks of material. One, farther down the tube, containing the 70 S

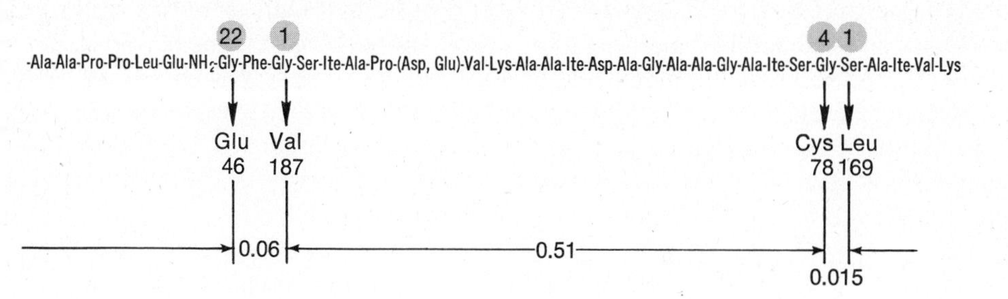

complete ribosomes, and another, much broader peak containing the 30 S and 50 S subribosomal particles. The two regions are clearly distinct; thus it is possible to distinguish between subribosomal particles and functional ribosomes.

Cells that had been grown on heavy medium were transferred to nonisotopic medium and infected with T_2 bacteriophages. After a period of time, the cells were lysed and the ribosomes recovered and separated by centrifugation. All the ribosomes isolated were heavy. Consequently, no new ribosomes had been formed. Had there been synthesis of new ribosomes, they would have been made from the normal isotopes and thus would have appeared in the centrifugation as a lighter band of 70 S ribosomes. The absence of any synthesis of new ribosomes after phage infection indicates that, in fact, the ribosomes are "for hire." They can apparently attach to any messenger and translate that message into protein.

The same system can be pulse-labeled with either ^{14}C-uracil or ^{35}S-cysteine. Both are radioactive and so can be detected by normal counting procedure. Each emits radiation at a very different energy level, which means that they can be distinguished from one another. Pulse-labeling with ^{14}C-U will label only RNA; whereas pulse-labeling with ^{35}S-cysteine will label only protein. When such cells were lysed, once again only the two ribosomal peaks characteristic of cells grown on heavy medium were found. However, any RNA synthesized after phage infection has been labeled with ^{14}C, and all this label appeared at the "heavy" 70 S peak. The new messenger produced by the phage DNA apparently attaches only to the preexisting, functional ribosome (70 S). It is not attached to the subribosomal particles, and it does not "float free" in the medium to any extent. Furthermore, at very brief times after phage infection, all the ^{35}S label was also found at the 70 S peak, thereby indicating that the newly forming protein is associated with the messenger-ribosome complex, as would be the case if the complex were responsible for the synthesis of new protein. Taken altogether, then, these data provide a good fit for the general mechanism outlined earlier.

The preceding experiment certainly suggests that the messenger governs the specificity of the protein formed. However, a direct test in this system is not possible, for too little protein is made to be certain that it is only T_2 protein. A more direct test of this part of the hypothesis was done by isolating, from rabbit reticulocytes, the rather stable messenger RNA that governs the synthesis of rabbit hemoglobin. This protein is large and distinctly different from any proteins found in microorganisms. When the hemoglobin messenger is added to an in vitro system whose ribosomes, *t*RNA, and activation enzymes all have come exclusively from *E. coli*, protein is made, and the protein is rabbit hemoglobin. Clearly, then, the messenger is the all-important part that determines the specificity of protein made. The rest of the machinery is capable of accepting any messenger, apparently no matter how foreign.

Role of tRNA

Another question that can be asked about the hypothesis concerns the role of the *t*RNA. We have postulated that it is responsible for guiding the amino acid to its proper place in the polypeptide chain and that, in a sense, the amino acid itself is inert. This postulate would seem to be the only logical one if base pairing is to be used as the mechanism for specifying the position of the amino acids. Fortunately, the proposition can be tested. The amino acids cysteine and alanine are very similar, differing only by the presence of one S atom:

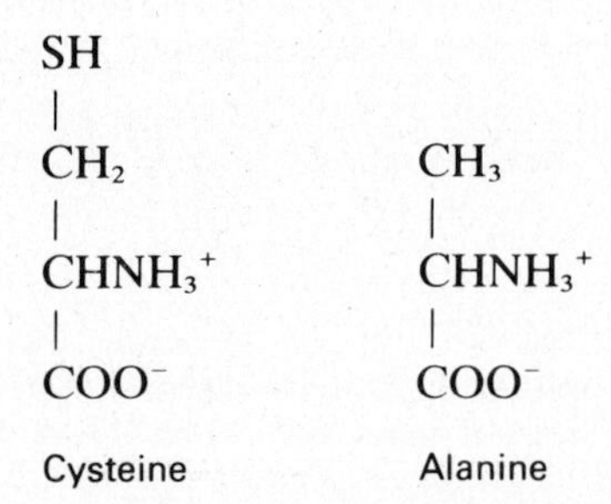

The SH group can be removed from cysteine while it is still attached to its *t*RNA without breaking up the complex. This step converts the amino acid to alanine. Consequently, by a simple chemical trick, it is possible to have alanine attached to *t*RNA that normally carries only cysteine. The question is obvious. What will happen if such a cysteine-type-*t*RNA-alanine complex is fed to the protein-synthesizing system? If the specificity of placement of the amino acids depends solely on the *t*RNA, alanine should appear in the protein wherever cysteine formerly appeared, which turns out to be the case. This experiment provides a nice test of the part of the hypothesis that suggests that the reading of the message is done by the *t*RNA.

Polysomes

Yet another part of the hypothesis is that several ribosomes attach to the *m*RNA and produce several polypeptides. It has been possible to obtain electron micrographs of several ribosomes attached to a single *m*RNA molecule. Figure 19-6 presents such an electron micrograph. A variety of tests can be applied, all of which indicate that the very fine strand joining the ribosomes is, in fact, an RNA molecule. Thus we have direct visual evidence that the general model proposed above is satisfactory. In almost all cases studied to date, this complex of a long messenger RNA molecule with

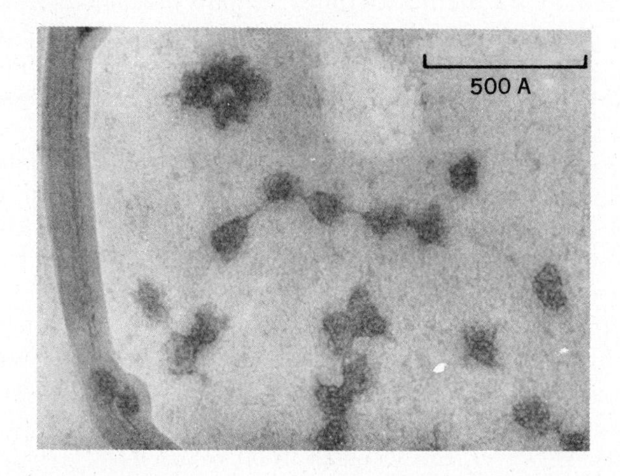

Figure 19-6. Polyribosomes. The electron micrograph shows several polyribosomes. The one in the center is most evident; five ribosomes are attached to a single *m*RNA in this case. The preparation is from rabbit reticulocytes. (From Slayter, H. S., J. R. Warner, A. Rich, and C. E. Hall. 1963. *J. Mol. Biol.* **7**:652.)

several ribosomes, generally called a polysome, appears to be the active protein-synthesizing system of the cytoplasm.

Homology of DNA and mRNA

If the first step diagrammed in Figure 19-4 is correct, two questions must be asked. The first is: Is the *m*RNA, in fact, a "mirror image" of the DNA? That is, does each A in the DNA correspond to a U in the *m*RNA, and so forth? The second question is: Is it true that only one strand of the DNA is utilized in this process? Let us consider them sequentially.

The discovery of an enzyme that can make RNA only in the presence of DNA (*DNA-dependent RNA polymerase*) provided the beginning of an answer to the first question. Such an enzyme would need to be present in cells if we are to propose that the DNA somehow "directs" synthesis of *m*RNA. Furthermore, once such an enzyme has been isolated and purified, it is possible to do the studies in vitro, free from the complications of the entire cell. We need only provide the enzyme with a DNA molecule to act as a template and the four nucleotides (AUGC) in energy-rich form. When they are present, RNAs are made. Omission of the DNA, or any one of the nucleotides, stops the reaction. Obviously the removal of the DNA removes the template that guides the enzyme in the formation of the RNA, as would be expected on the basis of our model. The fact that omission of any one of the nucleotides also stops the reaction is suggestive of the same thing. Since if we were to omit, for example, the U nucleotide every time the enzyme reached an A in the DNA, there would be no U to put in; and since we must presume that the mechanism is exact, reading of the DNA would stop at this point. This latter finding is extremely important, for it suggests that a very exact mechanism prevents substituting C for U or G for A, and so forth. Once again, such a mechanism would be required by the general model.

Even more interesting is an experiment with this polymerizing enzyme that demonstrates directly the complementary nature of the *m*RNA to the DNA. (Before continuing, it is best for the reader to be sure that he is familiar with the structure of the DNA molecule by referring to Figures 18-3 through 18-6). It is possible to synthesize (under special conditions) DNAs that consist only of adenine and thymine nucleotides. If the RNA synthesis is specifically guided by the DNA molecule, the RNA produced should contain only U and A even though G and C are present in the medium during the time of synthesis. Such was found to be the case, and thus the initial proposition that DNA was acting as a template could be accepted. However, this test does not prove that the RNA produced has the same base sequence as that contained within the DNA. To do so, it is necessary to know the order of bases both within the DNA and within the RNA. The base sequence is extremely difficult to establish with native DNAs, but the AT copolymer can be determined by using the technique of nearest-neighbor analysis (Figure 19-7).

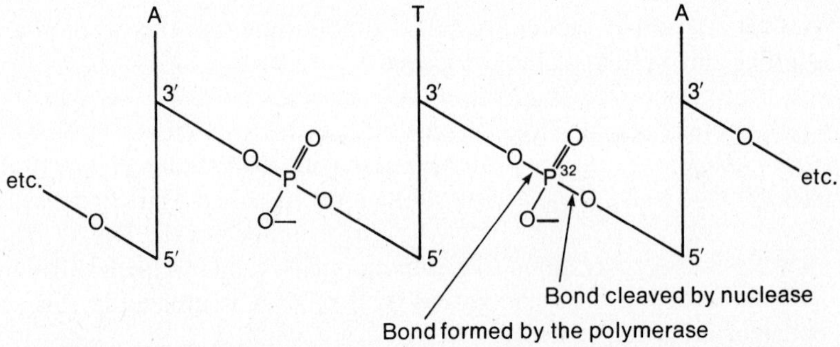

Figure 19-7. Nearest-neighbor analysis. The phosphate joining any two nucleotides was contributed originally by the one bonded at the 5′ position. The polymerase making the nucleic acid joins it to the 3′ position of the next nucleotide. At cleavage of the polymer, the phosphate linkages are broken on the 5′ side. Thus all phosphates are shifted from their original nucleotide to that of their nearest neighbor. If the A nucleotide contained the phosphorus isotope originally it will be attached to the T nucleotide after cleavage.

In nearest-neighbor analysis, the synthesis of the nucleic acid is carried out in a medium containing one type of nucleotide (in this case, assume the A) with ^{32}P carried in the phosphate. If the base sequence is pApTpApTpA and so on, the phosphate linkage joining any A and T will run from the 5′-position of the thymine nucleotide to the 3′-position of the adenine nucleotide. Once such a nucleic acid has been synthesized, it can be digested by a specific enzyme that cleaves the molecule such that the bond is broken only on the 5′ side, releasing nucleotides that are all 3′ phosphates. The result is that the phosphate group has been transferred from its original position (for example, on A) to the nearest-neighbor nucleotide in the molecule (in this case, T). In a simple system, such as the AT copolymer structure postulated above, there would be a transfer of ^{32}P from the A nucleotide to the T nucleotide throughout. Thus if the structure is an alternating set of A and T, the results of nearest-neighbor analysis should show that *all* the radioactive phosphorus has been transferred from A to the T. This is the case. Had there been an appreciable number of As side by side, some ^{32}P would have been associated with released A nucleotides. Similarly, had there been any appreciable number of Ts side by side in the molecule, some unlabeled T nucleotides would also have been found. It can be concluded that the structure of the AT DNA is, in fact, pApT-pApTpA, ad nauseum.

Having established the base sequence of A and T in the copolymer by nearest-neighbor analysis, it is possible to utilize the same technique to establish the base sequence of U and A in the RNA synthesized in the presence of the AT copolymer. Once again it turns out to be an alternating series of Us and As, with the U corresponding to the position of the A in the DNA and the A of the RNA corresponding to the position of the T. Consequently, it is presumed that synthesis of RNA by the polymerase is highly specific and that the "message" is transcribed by base pairing for the synthesis of the RNA molecule just as it is for the replication of DNA (Chapter 18).

The foregoing discussion provides an enzymatic mechanism for transcription, but it does not test the hypothesis that in vivo a specific *m*RNA is transcribed from a specific region of the DNA. Since it was impossible to do a sequence analysis of either DNA or RNA until very recently, there was no way of making a direct test of the hypothesis that the *m*RNA is directly transcribed from DNA. However, it is possible to make indirect tests. Our proposition is that the *m*RNA will be a "mirror image" of a specific region of the DNA. That is, wherever an A appears in the DNA, a U will appear in the *m*RNA. Wherever a C appears in the DNA, a G will appear in the *m*RNA, and so forth. If this proposition is correct, then any specific *m*RNA should be capable of hydrogen bonding with a particular region of the DNA in a highly specific manner. To test this hypothesis, T_2 bacteriophages were prepared in such a way that their DNA was labeled with tritium (^3H). Such labeled phages were then used to infect *E. coli*. These cells were placed in a medium that contained radioactive phosphorous (^{32}P). As indicated, as soon as infection occurs, the bacterial machinery is shut down and only a small amount of RNA is made, an amount that corresponds to the messenger RNA required for the protein synthesis of the bacteriophage. Thus the RNA made in this system will be messenger RNA of the T_2 type and should be labeled with ^{32}P. Any pre-existing *m*RNA will be unlabeled. Similarly, the phage DNA will be labeled with ^3H and the bacterial DNA will be unlabeled.

If the infection is allowed to proceed for only a short period of time, relatively little phage *m*RNA will be produced, and the phage DNA will have no chance to replicate (and thus dilute the label). In this case, any nucleic acid labeled with ^{32}P represents recently synthesized messenger and any unlabeled DNA represents bacterial DNA. When the cells are lysed and the various nucleic acids separated, it is possible to denature the DNAs by heat treatment (Chapter 18). The strands separate and, since they have different nucleotide contents, will have slightly different densities. If such DNAs are subjected to density gradient centrifugation, there will be four "bands" of DNA in the tube, two of which correspond to the two strands of bacterial DNA and two of which correspond to the two strands of phage DNA. The two labeled strands are phage DNA, and the unlabeled strands are bacterial.

Suppose that the heat-treated DNA is mixed with the RNA that is isolated from the system prior to centrifugation. Clearly, any unlabeled RNA is of no consequence to us. Fortunately, however, the various RNAs have very different densities from the singly stranded DNA and are readily separable in the density gradient. What will happen to the labeled phage messenger RNA, however? There are two possibilities. One is that the labeled messenger will be with the rest of the RNAs, as its density would dictate. The other possibility is that the messenger has such a strong homology for a certain region of the DNA that it will hydrogen bond (anneal) with that DNA and form a new RNA-DNA hybrid molecule, which will have a yet different density, although one that will probably correspond more closely to the relatively large DNA fragments than to the various RNAs.

In fact, what happens is that the labeled phage *m*RNA is found associated with the phage DNA peak. There is no binding of the labeled messenger to the bacterial DNA, nor is there any binding between the labeled messenger and other RNAs (as would be expected). If such labeled phage messenger is centrifuged with heat-treated DNAs from other sources, there is little or no hybrid formation, even in cases where DNAs were prepared that had an *overall* base composition almost identical to that of the bacteriophage used. Thus it can be seen that it is not overall base composition that is critical but rather it is the sequential order of nucleotides within the DNA that determines this specific hybridization, a point that is in complete agreement with the proposition that a specific message is transcribed from a specific part of the DNA.

The results of the RNA–DNA hybridization experiments answer our second general question as well. The *m*RNA binds selectively with one strand of the phage DNA rather than with both. So it must be presumed that when the DNA template is acting to impose a specific sequence of nucleotides during the synthesis of RNA, only one strand of the DNA is "read," with the other strand remaining essentially inert in the process. Interestingly, it has been well established that such RNA transcription from DNA will occur in vivo only if the DNA is doubly stranded. It is apparently necessary for the synthetic mechanism that both strands of DNA be present, even though only one strand is read by the RNA polymerase. It has been shown that messenger RNA is produced in response to infection with bacteriophage T_2 with no change in the physical properties of the T_2 DNA. Similarly, extracted bacterial DNA can be used to produce six times its own amount of RNA with no loss in the ability of the DNA to transform other cells (a process that requires doubly stranded DNA). Perhaps the reason is that the double helix is extended, which should make "reading" of the nucleotide sequences easier. Another reason may be to prevent transcription of *m*RNA during the period of DNA replication, a time when the DNA must be single stranded for a while.

Although the extended nature of the double helix is an asset, the amount of base pairing that can occur with the intact helix would seem to be small compared to the amount necessary to get exact transcription of the *m*RNA. It thus seems more likely that the DNA chains are separated locally at the time of RNA synthesis. There is suggestive evidence to support the possibility of local DNA "unzipping." The timing of RNA and DNA synthesis is such that they are separate processes. The synthesis of DNA occurs during a relatively restricted period of the cell's history, whereas RNA synthesis occurs over a long period of time. It is unlikely that the entire DNA is uncoiled during this large part of the cell's cycle, for separation of the DNA chain requires energy, and spontaneous separation presumably occurs only during DNA synthesis. Since the *m*RNA molecule is very small compared to the entire DNA molecule, only very local separation of the chains would be necessary to permit extensive base pairing during synthesis.

Finally, we should point out that a certain problem results from the fact that only one strand of DNA is read. Under these circumstances how does the enzyme choose between the two strands? As a partial but satisfactory answer to this question, we need only point out that, in fact, the two strands of the DNA molecule are different. As was demonstrated in Figure 18-6, each strand has a different polarity, one reading in the 5′-3′ direction and the other in the 3′-5′ direction. All that is required is that the enzyme have a specificity of function such that it reads only one of these two strands. The great specificity of enzymes is sufficient to account for the ability of the polymerase to distinguish these two with ease. On the other hand, the problem is not quite that simple. It has now been shown that, in a number of organisms, although only one strand is read, it need not always be the same strand. That is, one region of the DNA molecule may be read from one strand and another region read from the complementary strand. (Under no circumstances are the two strands of a single region transcribed.) As a result, it is necessary to propose that specific sequences of nucleotides initiate the transcription process. Presumably they are regions of binding that the enzyme can recognize as proper points of attachment. Once the enzyme has attached to a particular strand in a particular region, it will only move in the 5′-3′ direction, and thus only one strand can be read. For this mechanism to be complete, however, there must also be specific detachment sites within each strand, so that only certain stretches of a particular DNA chain are transcribed. This also appears to be the case. As we shall see later when discussing the transcription process in more detail (Chapter 24), this particular enzyme is very complex and quite remarkable in its properties.

Summary

Taken altogether, the various experiments cited in this chapter and a number of others that space forces us to omit have been in agreement with the general model. As an overall mechanism, there seems little doubt that proteins are synthesized in accordance with the steps diagrammed in Figure 19-4. The central dogma seems justified, but we have omitted any test of the nature of the genetic message itself and hence any detailed analysis of how that message is translated into a polypeptide. To that question we will turn in the next chapter.

REFERENCES

BRACHET, J. 1957. *Biochemical Cytology.* Academic Press, New York. (An advanced book that gives an extensive review of the evidence supporting the role of nucleic acids in protein synthesis. See Chapters 6 and 7 in particular.)

HAYNES, R. H., and P. C. HANAWALT. 1958. *The Molecular Basis of Life.* W. H. Freeman and Co., San Francisco. (The most pertinent articles are *The Nucleotide Sequence of a Nucleic Acid,* by Holley, *Hybrid Nucleic Acids,* by Spiegelman, *Polyribosomes,* by Rich, and *Gene Structure and Protein Structure,* by Yanofsky. They provide simple descriptions of much of the research that established the role of nucleic acids in protein synthesis.)

INGRAHAM, V. 1958. How do genes act? *Sci. Amer.* **198**:69–74. (A very good article on the research done with sickle-cell hemoglobin.)

INGRAHAM, V. 1966. *The Biosynthesis of Macromolecules.* W. A. Benjamin, Inc., New York. (A good review of the entire subject. It is accessible to the reader who has had an introduction to chemistry. Chapter 6 presents a review of the material covered in this chapter.)

SPIRIN, A. S., and L. P. GAVRILOVA. 1969. *The Ribosome.* Springer-Verlag, Berlin. (An advanced but excellent account of all features of ribosomes, including their role in protein synthesis.)

Original Papers and Advanced Reviews

BRENNER, S., F. JACOB, and M. MESELSON. 1961. An unstable intermediate carrying information from genes to ribosomes for protein synthesis. *Nature* **190**:576-581. (This is the paper that unequivocally demonstrated the role of *m*RNA and the ribosomes in protein synthesis.)

CRICK, F. H. C. 1958. On protein synthesis. *Symp. Soc. Exptl. Biol.* **12**:138–163. (In this remarkable article the author first proposes the role of *t*RNA. It gives a fine overall view of protein synthesis, much of it hypothetical at the time it was written.)

HURWITZ, J., J. J. FURTH, M. ANDERS, P. J. ARTIZ, and J. T. AUGUST. 1961. The enzymatic incorporation of ribonucleotides into RNA and the role of DNA. *Cold Spring Harbor Symp. Quant. Biol.* **26**:91–100. (This paper presents proof that DNA directs the synthesis of RNA.)

ZAMECNIK, P. C., M. L. STEPHENSON, and L. I. HECHT. 1958. *Proc. Natl. Acad. Sci. U.S.* **44**:73–78. (One of the first demonstrations of the role of *t*RNA in protein synthesis.)

QUESTIONS AND PROBLEMS

19-1. Define the following terms:

*r*RNA	*t*RNA
*m*RNA	70 S ribosome
polysome	transcription
translation	

19-2. As early as 1941 it was shown that cells active in protein synthesis are also rich in RNA. Explain this finding.

19-3. What is the significance of the experiments done with grafting and regeneration of Acetabularia?

19-4. What is the "central dogma" of gene action?

19-5. Make a diagram of protein synthesis that illustrates the central dogma without referring to the diagrams given in the text.

19-6. Outline the evidence that supports the diagram drawn in answer to Problem 19-5.

19-7. Why does the mechanism require *t*RNA as an "adaptor" to place a particular amino acid in its proper locus in the polypeptide?

19-8. State clearly the difference between the one gene–one enzyme hypothesis and the one gene–one polypeptide hypothesis.

19-9. If you were to compare the RNAs of different organisms, which of the three types (*m*RNA, *t*RNA, *r*RNA) would you expect to be the most similar and which the least similar? Explain your answer.

19-10. How does the DNA-dependent RNA polymerase recognize the proper strand of the double helix in transcription?

19-11. Explain how DNA–RNA hybridization studies can be used to establish homologies between the two types of molecules.

19-12. In view of what you now know about gene action, how would you interpret the following terms: recessive, dominant, and intermediate

(Chapter 2); complementary factors (Chapter 4); quantitative inheritance (Chapter 5); multiple alleles and lethals (Chapter 6); and hypomorph (Chapter 15)?

19-13. If you were asked "What do genes do?" what would your answer be? (Compare your answer to that given to Question 15-12.)

19-14. What is the initial product of a cistron? What is its ultimate product?

20/The genetic code

"Now that they've cracked the genetic code, I suppose everyone will want to be Peabodys."

In Chapter 19 we developed a mechanism for protein synthesis that, in its broadest aspects, must be considered correct. However, we have omitted the rather difficult question of exactly how the information contained in a sequence of nucleotides can be converted into a sequence of rather different chemicals—amino acids—in the polypeptide chain. That, somehow or other, the conversion must occur was intuitively obvious as soon as it was clear that the gene was DNA and that a single mutational event within a gene resulted in the replacement of a single amino acid in the polypeptide chain. The question was only: How can the information be coded in the nucleotide sequence?

The Triplet Code

The obvious restriction to the coding problem is that all the "words" in the message must be written with four letters and four letters only—A, T, G, and C—the purine and pyrimidine bases of DNA. (In *m*RNA, they would be A, U, G, and C.) Suppose that we had only a single-letter code. In this case, only four amino acids could be specified, since only four single letters are available—ATGC. Similarly, a two-letter code provides only 16 (4^2) possible combinations, which is not sufficient to control the placement of the 20 amino acids that appear routinely in proteins. On the other hand, a three-letter code provides 64 (4^3) coding possibilities, which is considerably more than necessary. Consequently, although a triplet code seemed the most reasonable on purely intuitive grounds, it was far from a perfect fit. Obviously there could be higher-letter codes (any number), but on the generally valid principle that one should always work with the least-complicated hypothesis, we can assume a three-letter code.

The second problem is that of reading such a code. The DNA molecule is a very long sequence of nucleotides, which could be read either in an overlapping or a nonoverlapping manner. As can be seen in Figure 20-1, reading the code in these two different ways has important consequences. Nonoverlapping reading (in triplet form) will give only four words for the 12 letters shown. All are quite reasonable words, and presumably each one stands for a different amino acid, on the basis of our preceding analysis. With overlapping reading, however, each letter is read three times, and in each case it is part of a different word, thus making it possible to code much more information into the system than would be the case in the nonoverlapping method. Clearly if the assumptions we have made so far are correct, any mutational event (T $\rightarrow$ C) would give rise to three new code words when the reading occurs in an overlapping manner. On the other hand, in the nonoverlapping reading of the code, such a mutational event would change only one code word. From the information available from the study of normal and sickle-cell hemoglobin (Chapter 19), it appears that the code is read in a nonoverlapping manner, since a single mutational event resulted in the substitution of only one amino acid; and thus it is reasonable to assume nonoverlapping reading.

There are certain difficulties with a nonoverlapping reading of the code, the most important of which is whether or not there are "commas" between the individual code words. If there are not, there must be a way of specifying that the code is read from one end to the other in a polarized manner. Otherwise reading might begin at any point, and the code words so derived would depend on the triplet chosen as the beginning. In the absence of commas, any three letters constitute a triplet, and thus the message would change randomly. Unfortunately, it was impossible to treat the data from the hemoglobin study as critical. First, it was not certain that the code was triplet in nature, and, secondly, a triplet code (or quadruplet code) provides more code words than amino acids. Consequently, the code may well be degenerate. That is, an amino acid may be specified by more than one set of letters. If such is the case, it is possible to have a mutational event that does not change three amino acids in a particular polypeptide chain even though the code is read in an overlapping manner, since it is *conceivable* that two of the changes resulting from a single mutational event specified the original amino acid, merely using different code words. Before any further analysis could be done, it was essential to have an experimental test of the nature of the genetic code.

The nature of the reading of the code was first established in 1961 in a brilliant experiment. The analysis was done by using the $\mathbf{r_{II}}$ region of T_4 phage, the same one that had been used in the analysis of the fine structure of the gene (Chapter 16). In studies of mutations induced in this region of T_4, it was found that the agents inducing mutations (mutagens) could be divided into two classes: those that changed the character of one of the bases (for example, $A \rightarrow G$) and those that inserted or deleted an extra base, producing addition/deletion mutations. As would be expected, base-change mutations can back-mutate when treated with the same kind of mutagen (that is, if $A \rightarrow G$, then $G \rightarrow A$). On the other hand, no such back mutation of addition/deletion mutants could be observed when a base-change mutagen was used to increase the rate of back mutation. This finding is reasonable, since it would be difficult for an agent that changes one base to another to have any effect on a missing base. On the other hand, occasional back mutations of +/− mutants could be observed when a mutagen that causes addition/deletion changes was used to accelerate the process.

Figure 20-2 proposes an explanation for revertants caused by addition/deletion mutagens, using a simplified genetic message for clarity. *Assuming* that the code is triplet and read in a nonoverlapping manner, it can be seen that the removal of a single base will change the reading from the point of the deletion (or addition), and thus the entire message will be altered, giving rise to a severely mutant form. Consequently, we would expect mutants induced by agents of this type to be extreme in their phenotype, which is the case. If, however, a second mutation of the opposite type is induced in such a mutant (Figure 20-2c), the reading of the code will be reestablished beyond the place of the second mutation, and only a short region of the message between the two mutations will be altered. Consequently, we would expect such revertants, or "back-mutants," to be more like the normal cell, but not quite normal, and such is also the case. Revertants of this type are called *pseudowild.* Obviously if the +/− mutants are too far apart within the cistron, so much of the function will be changed that it will still be scored as a mutant. Such individuals are discarded in the analysis because they appear to be

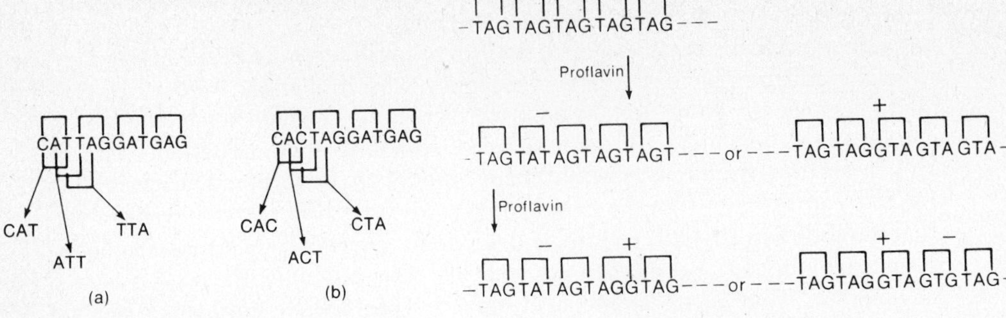

Figure 20-1. Reading the genetic message. (a) Nonoverlapping reading (top) will give fewer "words" than overlapping reading (bottom). (b) A mutation (T→C) results in one word change (CAT→CAC) if the code is nonoverlapping and in three changes if read in an overlapping manner.

Figure 20-2. Mutations induced by proflavins. These mutagens act by deleting or adding a base in the nucleotide chain. Either change will alter the reading of the entire sequence following the place of mutation. "Back mutation" is explained by a change of the opposite type occurring at some distance from the place of the original change which reestablishes the normal reading of the code.

nonrevertants. The analysis can only be done with those addition and deletion mutants that can be combined to generate a pseudowild individual.

The foregoing is an hypothesis to explain the difference in action of the mutagens on this particular organism. However, as is characteristic of good scientific hypotheses, it makes certain predictions that can be tested. The most important is that the revertants to pseudowild-type are *double* mutants rather than back-mutants in the strict sense. Since these are changes that are proposed to be at a certain distance from one another within the gene, it should be possible to separate the two by genetic recombination (Chapter 16). This test was done, and it was demonstrated that there are two separate mutants, which, when combined in a single genetic message, produce a pseudowild organism, although either one alone produce a definite mutant form. (This is a case where two wrongs do make a right!)

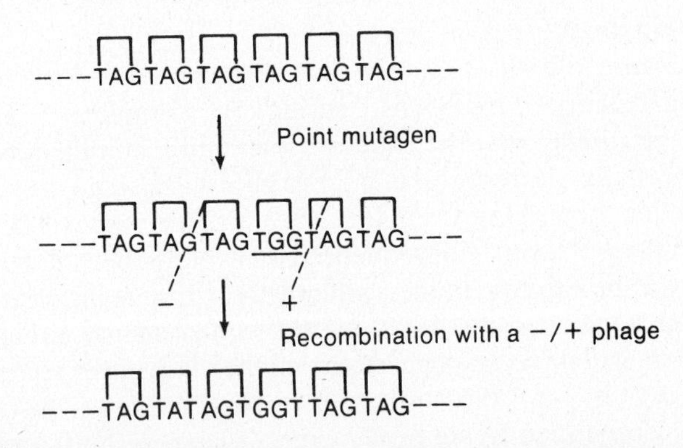

Figure 20-3. Back mutation by use of +/– mutants. A point mutagen was used to induce the initial change of one code word to a nonsense triplet (TGG). Changing the frame of reading by surrounding the nonsense triplet with – and + mutations converts the nonsense to missense and produces a pseudowild phage.

A second test of the addition/deletion hypothesis is possible by using bacteriophages with a previously induced base-change mutation. Generally such mutations do not result in a serious malfunction of the particular gene under study. However, occasional base-change mutants have a complete loss of function similar to that observed with addition/deletion mutants. Presumably this result is caused by the production of a code word that cannot be read by the organism at all (a "nonsense" mutation). As Figure 20-3 suggests, it should be possible to change the nonsense code word by surrounding it with +/− mutants (placed in the genome by recombination with other mutant phage). Assume that the reading of the code occurs from left to right. The removal of a G to the left of the nonsense triplet will change the reading so that the message is altered as follows:

TAGTAGTAGTGGTAGTAG...

↓ −G

TAGTATAGTGGTAGTAG...

The nonsense triplet, TGG, is gone, but the rest of the triplets are changed in the process. However, if an addition of a base is made to the right of the nonsense triplet as well, the reading of the base sequence will be put back on the original track as follows:

TAGTATAGTGGTAGTAG...

↓ +T

TAGTATAGTGGTTAGTAG...

Surrounding the postulated nonsense triplet (TGG) with the +/− mutants removes the nonsense triplet and creates a short, altered region of the gene to produce a pseudowild revertant.

In order to carry out the experiment suggested above, it is only necessary to obtain a doubly mutant (+/−) phage whose addition/deletion events map close to, but on either side of, the presumed nonsense mutant. Recombination between the +/− mutant phage and the nonsense mutant phage will produce the pseudowild revertant illustrated in Figure 20-3. The predicted results were achieved, and the pseudowild revertant recovered as a result of the recombination can be demonstrated to be a *triple* mutant, carrying +/− and the original base-change mutation.

A third test of the hypothesis is that it predicts that two + mutations or two − mutations will not restore functions but a +/− double mutant will. Consider only the +/+ case illustrated below:

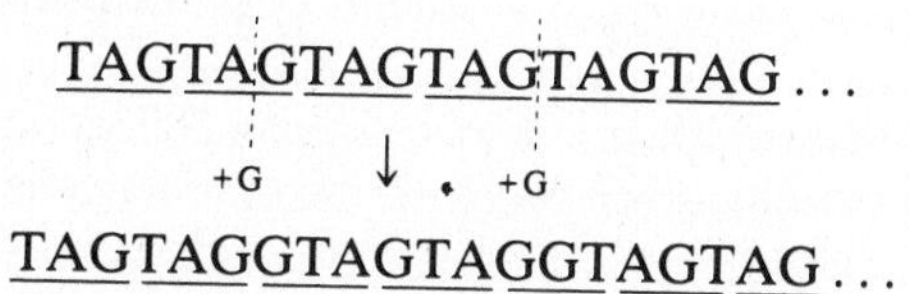

The reading of the code is still completely altered from the point of the initial insertion. In this case, as in the cases cited above, the simplest hypothesis is the one presented originally in Figure 20-2—that the +/− mutants cause a change in the reading of the code—and thus they have been called *reading frame* mutations.

The preceding hypothesis is attractive, but it is important to realize that, initially, the whole set of hypotheses was a construct. Let us pause for a moment to ask: How could we know whether or not we have an addition or a deletion (a+ mutant or a−mutant)? Clearly there is no way that we could with our present techniques. What, then is the meaning of + and −? It is simply this: we designate a loss of function as a − initially and then designate the second mutational event that restores that loss of function as a +. Everything else will be named from these two initial events. Any mutant that cannot restore the function of the original − mutant by recombination must also be −, and, similarly, any mutant that cannot restore the function of the original + mutant must be +. It is that simple. All that is needed is a large number of mutants and a great deal of patience to do all the possible combinations in order to see whether the function is restored. It is certain now from other evidence that, in fact, the addition and deletion of bases is exactly what does occur in this system, but there is still no way of knowing whether the original mutant was an addition or a deletion. We only know that the mutants designated as + are capable of reverting mutants designated − to a pseudowild condition. Of course, it makes no difference if the original − mutation was a deletion or an addition of a base. As long as the + mutants are the opposite of − mutants, the system works perfectly.

The form taken by the reading frame hypothesis involves two central assumptions. The first is that there are no commas in the code; that is, the code is read from one end to the other without interruption. If this were not the case, alterations of the reading by one unit should not have a far-reaching effect on the total genetic message. The assumption implicit in this hypothesis is that the shifting of the reading by one letter (in either direction) causes a change in the reading of the entire message from that point on. Thus we must assume that there is nothing within the cistron proper to put the reading of the code back into the right reading frame. Similarly, the reading of the code must be polar (that is, from one end to the other), since reading at random from any point in the message (in either direction) would give results different from those obtained above. However, it does not answer the critical question of whether the code is triplet.

Fortunately, the question of the triplet nature of the code can be answered with the same system. In T_4, two cistrons govern the $\mathbf{r_{II}}$ function (rate of lysis), and they lie exactly next to one another in the chromosome (Figure 20-4). These two regions are designated A and B. The studies discussed to this point were conducted with the B locus; however, it is equally possible to study mutational events in the A locus that immediately precedes it on the phage chromosome. Interestingly, mutational events of any type in a A locus have no effect on B. Thus although we must say that within any one cistron there are no commas, there must be a "semicolon" that separates the independent clauses of the genetic message. Such a stop-start region (stop reading message one, start reading message two) is logical if we are to assume

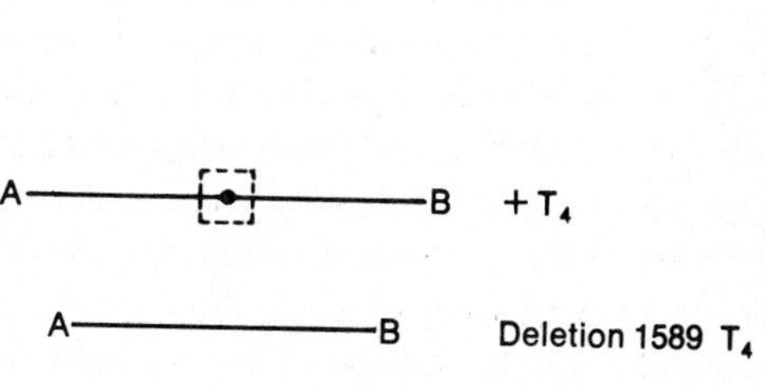

Figure 20-4. Schematic representation of the r_{II} locus in T_4 bacteriophage. The component loci (cistrons) are regions of separate genetic function. In the phage carrying deletion 1589, the region separating the two cistrons has been lost. In the wild type phage, addition/deletion mutations in region A have no effect on B function. In the deleted phage, addition/deletion mutations in A alter B function. The deletion removes the *stop* that separates the two cistrons and reorders reading of the message of the B cistron.

that a number of different functional units must be coded on a single chromosomal unit (in the case of the microorganisms, at least, on a single giant macromolecule). Apparently the reading of the A code cannot cross the semicolon, and thus the reading of B is reordered at that point.

One mutant form T_4 has a deletion of the terminal region of the A locus and the initial region of B. This deletion seems to be so small that the organism still functions rather normally and thus can be used for other mutational studies. In this strain, however, + or − mutations in the A locus alter the B function as well. Therefore the deletion has removed the semicolon and joined the two loci into a single reading unit. This observation does more than confirm our previous hypothesis of reading frame mutations. It now makes possible a test of the magnitude of the code word itself. By accumulating a number of + (or −) mutations in the A region, it is possible to add (or subtract) one code letter at a time by genetic recombination. If the A region has one + mutant, the B region shows no function, as would be expected on the basis of the initial hypothesis, since the entire region will be shifted one letter and misread. If two + mutations are introduced in A, there is still no function in B. If, however, three + mutations are introduced in the A region, the B function is normal. Therefore it would appear that it is only necessary to add three code letters (equal to the addition of one triplet) in order to restore the normal reading frame and hence no change at all in B. (In the A region this would amount to the addition of an additional amino acid to the A polypeptide chain, and so we could hardly expect normal A function.) The test can be extended. If four + are added to the A region, there is, once again, no function in B, since this step is more or less the same as adding a single nucleotide (+). The same is true of five, but with six mutants in the A region, the function of the B region again is normal. It is clear from this study that the genetic code is, in fact, a triplet code, that it is commaless, and that it is read in a polar, but nonoverlapping, manner.

The Amino Acid Code

The experiment with the r_{II} region of T_4 phage gave a great deal of information about the nature of the genetic code. It could not, however, indicate which specific nucleotide triplets specify which particular amino acids. It is a triplet code read in a nonoverlapping manner and apparently lacking any punctuation within a cistron. However, we could not tell whether the code was degenerate—that is, with a number of different triplets specifying one particular amino acid—or whether only certain code words were used and others never appeared. With only 20 amino acids to be specified and perhaps some punctuation to indicate stop and start signals for the reading of specific messages, there is still an excess in a code that provides for 64 different triplets. Unfortunately, we cannot do a chemical analysis of DNA that will reveal the nucleotide sequence of any particular cistron even if we were able to isolate a single cistron in sufficient quantity to make the analysis possible. In the early 1960s no one even attempted such an experiment. There are, however, other ways to approach the problem.

Two developments of modern biochemistry made possible the analysis of the code. The first was an in vitro system for the synthesis of protein. This topic has already been discussed in Chapter 19. The relatively few necessary enzymes, the various *t*RNAs, and the ribosomes can be prepared in a reasonably pure condition from a number of cells. The other necessary materials—amino acids, energy sources, and cofactors—can all be obtained commercially. When this mixture is incubated, no protein synthesis occurs. However, upon the addition of *m*RNA (from any source), protein is produced, and it is produced in proportion to the amount of messenger (or the stability of the messenger) given to the system. As noted, the type of protein produced in vitro is totally dependent on the nature of the *m*RNA added.

The in vitro system provides the experimenter with a method for investigating the relationship between the *m*RNA added and the protein synthesized. Techniques are readily available for the sequential analysis of amino acids in the protein; and if a single kind of protein is formed in such systems, it is relatively easy to determine its amino acid sequence. The problem is to determine the nucleotide sequence of the messenger. Adequate methods for determining such nucleotide sequences, even of the relatively short *m*RNA molecules, are only beginning to be developed today. They were not available in the early 1960s, and so more indirect methods had to be used.

The second technical achievement came from the discovery in the mid-1950s of polynucleotide phosphorylase, an enzyme that could make RNA. Unlike the DNA-dependent RNA-polymerase used to transcribe *m*RNA from DNA (Chapter 19), this enzyme requires no primer or template. When it is given a set of nucleotides in the energy-rich diphosphorylated form (ADP, UDP, TDP, and CDP), it will put the nucleotides together randomly to generate RNA. In this sense, the enzyme is similar to the DNA polymerase used to replicate DNA in vitro (Chapter 18). The most useful feature of polynucleotide phosphorylase for the study of the code is that if it is given a single type of nucleotide, it will make a polymer of that nucleotide. Thus if the enzyme is given ADP only, it will make polyA RNA; or if it is given UDP, it will make polyU, and so forth.

In 1961 these rather strange RNAs were put to use in the study of the code. PolyU was used as a messenger in an in vitro synthesizing system, and it produced a protein that contained only the amino acid phenylalanine. Consequently, the triplet code word for phenylalanine is construed to be UUU. Similarly, polyA gave rise to a protein containing only lysine.

The foregoing results suggested the possibility of "breaking the code" by the use of synthetic messengers. For example, it is possible to make a message containing only U and A. In a random polymer, eight different triplets will be available: UUU, UUA, UAA, UAU, AAA, AAU, AUU, AUA. We would expect, then, that eight different amino acids could be coded in such a message. In fact, the protein formed under these conditions contains only six: phenylalanine, lysine, tyrosine, leucine, isoleucine, and asparagine. Phenylalanine and lysine are expected on the basis of the two code words UUU and AAA. The other amino acids presumably represent different combinations of U and A. The question is: Which combinations represent which amino acids? The answer can be partly determined by varying the relative composition of U and A in the synthetic messenger. For example, if the messenger-forming system contains U and A in the ratio of 3 : 1, the polymer would contain far more UUU than AAA and far more UUA than AAU. Such a messenger produces a protein with a much higher percentage of leucine than of asparagine. Thus we might deduce that there are more Us in the leucine triplet than in the asparagine triplet (see Table 20-1). Obviously such a message should also produce a protein containing far more phenylalanine than lysine, which is the case. By making a series of messengers of different compositions, it is possible to approximate the code—to a high degree of completion.

Unfortunately, the synthetic messenger method lacks the precision that we would most like to have in unraveling the code. It is difficult to distinguish between UUA and AUU by methods that depend solely on the random assortment of nucleotides in a synthetic messenger. Furthermore, the analysis is complicated by the fact that there is degeneracy in the code. For example, a mixed polymer containing only U and C produces a protein containing leucine, among other amino acids, as does a messenger containing only U and G. In fact, small quantities of leucine are incorporated into proteins in the presence of polyU messenger. The simplest interpretation of these data is that the code has considerable degeneracy and that UUX codes for leucine, where X might be any nucleotide. (The alternative to this interpretation is that UUU specified both.) The fact that UUU specifies phenylalanine 30 times more efficiently than leucine lends support to the interpretation that the code is degenerate. However, a final resolution of this problem could not be obtained until a more sophisticated method of determining the triplet sequence was developed.

The final solution to the nature of the triplet code came from the discovery that the various RNA components of the protein-synthesizing system will adhere to one another on a cellulose nitrate filter. Such filters were used initially to isolate ribosomes from microorganisms; the *t*RNAs wash through and leave the ribosomes behind on the filter. However, it was soon discovered that the various *t*RNAs would also adhere to the filter, as long as both ribosomes and *m*RNA were present. In the absence of *m*RNA, the *t*RNAs washed through the filter quite readily. To determine the specificity of such adhesion, ribosomes were prepared on the filter and a polyU messenger was added. Then various *t*RNAs, carrying their specific ^{14}C-labeled amino acids, were passed individually through the filter. As would be expected from the studies discussed, the phenylalanine-*t*RNA was bound exclusively on the filter, and all other amino acids were passed through. Consequently, this system presents a mechanism for the analysis of the code. We can add a specific messenger to the ribosome-filter complex and determine which amino acid-*t*RNA complex is bound to the filter. If we know the character of the message, the nature of the specific reading of the code can be determined.

The essential feature of the filter-binding system is that it is possible to use trinucleotide messengers (UUU, UUA, and so on) rather than complete *m*RNA. When this is done, satisfactory binding of specific amino acid-*t*RNAs can be obtained. Using this "minimessenger" technique, the exact coding sequence was

Table 20-1. The Genetic Code[a]

FIRST LETTER	SECOND LETTER U		C		A		G		THIRD LETTER
U	UUU	Phe	UCU	Ser	UAU	Tyr	UGU	Cys	U
	UUC		UCC		UAC		UGC		C
	UUA	Leu	UCA		UAA	non.	UGA	non.	A
	UUG		UCG		UAG	non.	UGG	Tryp	G
C	CUU		CCU	Pro	CAU	His	CGU	Arg	U
	CUC		CCC		CAC		CGC		C
	CUA		CCA		CAA	GluN	CGA		A
	CUG		CCG		CAG		CGG		G
A	AUU	ileu	ACU	Thr	AAU	AspN	AGU	Ser	U
	AUC		ACC		AAC		AGC		C
	AUA		ACA		AAA	Lys	AGA	Arg	A
	AUG	Met	ACG		AAG		AGG		G
G	GUU	Val	GCU	Ala	GAU	Asp	GGU	Gly	U
	GUC		GCC		GAC		GGC		C
	GUA		GCA		GAA	Glu	GGA		A
	GUG		GCG		GAG		GGG		G

[a] The genetic code, consisting of 64 triplet combinations and their corresponding amino acids, is shown in its most likely version. The importance of the first two letters in each triplet is readily apparent. UAA, UAG, and UGA are "nonsense" mutations that act as signals for terminating polypeptide chains. The code words are written so that the first letter is at the 5′ end of the sequence and the third letter is at the 3′ end.

determined, as given in Table 20-1. The first two code letters seem to be more important in specifying the particular amino acid, although this is not always the case. Apparently three sequences are "nonsense," and they are used as termination points in the message.

Using the filter-binding technique, it is possible to determine a number of important facts about the reading of the code. For example, the deoxy forms of the triplets will not work. Thus we must assume that the 2′—OH group of the ribose is necessary to bind to the ribosomal surface. It is possible to make different kinds of triplet messages: for example, one in which the order of the nucleotides is such that there is a 5′-terminal phosphate (pUpUpU), and one in which the base sequence is reversed such that there is a 3′-terminal phosphate (UpUpUp). When these two forms of the message are used, the one having the 5′-terminal phosphate is greatly preferred. Furthermore, doublet code words are unrecognizable. It would appear that a specific trinucleotide sequence attaches to the ribosome in proper order to be read, thereby supporting the hypothesis that the message is read from one end to the other in a particular order; that is, the message has a specific polarity ($5' \rightarrow 3'$).

Only 20 amino acids are specified by the code. Two other amino acids found in proteins are not included—cystine and hydroxyproline. Cystine is a double amino acid formed from the interaction of two cysteine molecules. The disulfide linkages that are often of great importance in maintaining the tertiary configuration of proteins are the result of such interactions. Therefore, in principle, there is always cystine in the protein molecule, but the placement of the amino acids in the primary structure is done in the form of cysteine, followed by an oxidation that produces the disulfide linkage between the two residues. Similarly, hydroxyproline occurs in a number of proteins (particularly collagen). The hydroxyl (OH) group is added to proline after the protein has been synthesized, and so the triplet specifying proline is also the one that "specifies" hydroxyproline. How many other alterations of amino acids occur after the primary polypeptide has been formed is uncertain.

Finally, we should mention the more recent, very elegant work that has produced a synthetic message. Starting with U and A, the dinucleotide U—A is made. Then by adding two UAs together, we obtain U—A—U—A; and by adding two of these together, and so on, we get a rather large "message"—that is, UAUAUA indefinitely. Using the same chemical technique, it is also possible to combine the doublet UA with the doublet UC to get UAUC, which gives the more interesting short message UAUCUAUCUAUC. When fed to an in vitro system, this latter messenger gives rise to a quadrapeptide composed of tyrosine, leucine, serine, and leucine. These four amino acids correspond exactly to the four triplets generated in the synthetic message, UAU, CUA, UCU, AUC. Thus, in a completely synthetic system, the code words determined by the other techniques appear to work. In time it should be possible to make long messages in the laboratory and test the code in yet another way. More importantly, such synthetic messages should be capable of directing the synthesis of any protein desired. The message, in this case, is directly under the control of the chemist.

Some Consequences of Unraveling the Code

Mutation

The amino acid code has certain important genetic consequences. We can now discuss the nature of mutation in strictly molecular terms. Table 20-2 presents some known mutational events as seen at the polypeptide level. The changes are shown in the *m*RNA for ease of comparison with Table 20-1. It is important to remember, however, that the mutation occurred in the DNA and that the change in *m*RNA is only a reflection of that fact. In hemoglobin, all the substituted amino acids are simple one-letter transitions. These are events that have occurred naturally (that is, are not induced by specific agents in an experimental situation). Consequently, the mechanism of mutation is uncertain and may result from a number of different causes. Some very drastic changes have occurred (the substitution of pyrimidines for purines, for example). We can only presume that the alteration occurs by a two-step process (Figure 20-5), leading to the production of some mutants forms. If these mutant forms survive and yield progeny, they will appear in the population later on as individuals with altered proteins. The mutagen used in the tobacco mosaic virus studies is known to act as an oxidizing agent that in vitro has the specific action of converting C → U and A → G by oxidative deamination (removal of $—NH_2$ to give =O). The substitutions observed all represent single-letter changes in which the appropriate mutational event has occurred. From these and similar studies, it would appear that the molecular basis of mutation is an alteration of the code. (Mutation is discussed in more detail in Chapter 23.)

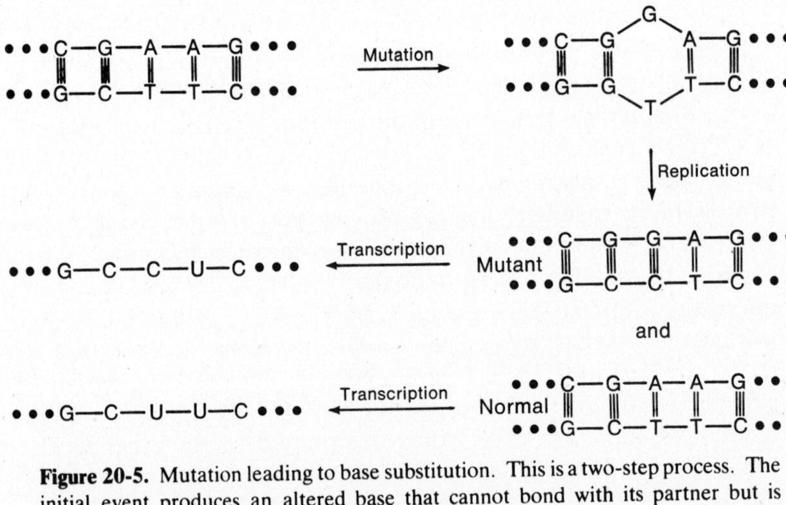

•••C—G—A—A—G•••
•••G—C—T—T—C•••
Mutation →
•••C—G—G—A—G•••
•••G—G—T—T—C•••
↓ Replication
Mutant
•••C—G—G—A—G•••
•••G—C—C—T—C•••
← Transcription
•••G—C—C—U—C•••
and
Normal
•••C—G—A—A—G•••
•••G—C—T—T—C•••
← Transcription
•••G—C—U—U—C•••

Figure 20-5. Mutation leading to base substitution. This is a two-step process. The initial event produces an altered base that cannot bond with its partner but is maintained in place by the rest of the molecule. When replication occurs, the altered base guides the insertion of a suitable partner, and both chains become "mutant" at that point. Such a mechanism would produce progeny one-half of which would be mutant and one-half normal.

Table 20-2. Amino Acid Changes in Two Proteins Resulting from Genetic Mutation

A. Hemoglobin: Spontaneous Mutants

AMINO ACID IN NORMAL HEMOGLOBIN		AMINO ACID IN MUTANT HEMOGLOBIN	
Lysine (AAA)	⟶	Glutamic acid (GAA)	A → G
Glutamic acid (GAA)	⟶	Glutamine (CAA)	G → C
Glycine (GGU)	⟶	Aspartic acid (GAU)	G → A
Histidine (CAU)	⟶	Tyrosine (UAU)	C → U
Asparagine (AAU)	⟶	Lysine (AAA)	U → A
Glutamic acid (GAA)	⟶	Valine (GUA)	A → U
Glutamic acid (GAA)	⟶	Lysine (AAA)	G → A
Glutamic acid (GAA)	⟶	Glycine (GGA)	A → G

B. TMV Coat Protein. Nitrous Acid Induced Mutants

Proline (CCC)	⟶	Serine (UCC)	C → U
Proline (CCC)	⟶	Leucine (CUC)	C → U
Isoleucine (AUU)	⟶	Valine (GUU)	A → G
Isoleucine (AUA)	⟶	Methionine (AUG)	A → G
Leucine (CUU)	⟶	Phenylalanine (UUU)	C → U
Glutamic acid (GAA)	⟶	Glycine (GGA)	A → G
Threonine (ACA)	⟶	Isoleucine (AUA)	C → U
Threonine (ACG)	⟶	Methionine (AUG)	C → U
Serine (UCU)	⟶	Phenylalanine (UUU)	C → U
Serine (UCG)	⟶	Leucine (UUG)	C → U
Aspartic acid (GAC)	⟶	Glycine (GGC)	A → G

FROM, JAMES D. WATSON. 1970. *MOLECULAR BIOLOGY OF THE GENE.* 2ND ED. W. A. BENJAMIN, INC., MENLO PARK, CALIFORNIA.

If, by some chance, the "nonsense" code words UAA, UGA, or UAG are generated by mutation, there can be no reading of the triplet. In these cases, only a polypeptide fragment is produced from the message prior to the chain-terminating mutation. Such drastic mutations are known to be generated by base-change mutagens and to occur relatively rarely, as would be expected on the basis of the low frequency of nonsense triplets. However, by patient testing, we can obtain a series of such mutants that map at different positions within one cistron. Such forms can then be analyzed for the protein produced by use of the antigen-antibody technique to isolate the proper material. When this is done, it is found that the size of the polypeptide fragment produced is directly proportional to the map distance between the beginning of the genetic message and the location of the nonsense triplet. This finding is in agreement with the postulated colinearity of the gene and the polypeptide. Even more important, however, it is an indication that partial polypeptides can be released from the site of protein synthesis, which supports the suggestion that the nonsense triplets have a *stop* function in the normal reading of the *m*RNA. Apparently the ribosomes detach when they reach such a triplet and the partial peptide is released. It follows, then, that the ribosomes cannot reattach to the unread part of the message. Specific "start" regions are the only places on the *m*RNA to which they can attach.

Proof of the Reading Frame Hypothesis

The unraveling of the code made possible a further test of the reading-frame mutation hypothesis. In this hypothesis, it is presumed that the combination of addition and deletion mutational events can restore a proper reading of the code such that a pseudowild protein is produced. In this case, we could talk about one mutational event suppressing the effect of the other. That is, a mutational event at one point within the cistron acts to remove (or to reduce) the effect of a previously occurring mutational event within the same cistron. Therefore it is called *intracistronic suppression.* This hypothesis is consistent with, and useful in, the determination of the nature of the triplet code. However, it remained an untested hypothesis for some years, until a similar experiment was done with the T_4 bacteriophage involving mutants at the locus that controlled the production of the enzyme lysozyme. (This enzyme is required to release the completed bacteriophage from the cell in which it is made.)

Two mutants that were completely defective for the production of active lysozyme were crossed, and they produced, in combination, a phage capable of forming a lysozyme of reduced activity. By genetic recombinational analysis, it was possible to show that both mutants were present in the pseudowild phage and that the genetic system was presumably of the +/− type. In this case, however, the protein produced was a specific enzyme, and so it could be isolated and analyzed. (The protein made by the $\mathbf{r_{II}}$ locus is still unknown.) Upon analysis, it was possible to show that the pseudowild protein differed from the wild-type enzyme over a stretch of five amino acids, as shown in Figure 20-6. Only a short segment of the polypeptide involving the altered region is shown. The figure also presents the mechanism by

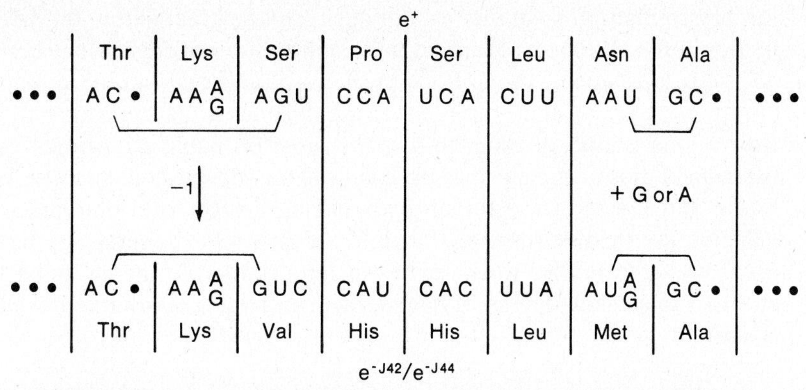

Figure 20-6. Intragenic suppression. A pseudowild lysozyme is produced by recombination between two completely defective mutants. The pseudowild organism produces an enzyme 50% as effective as the wild-type. Genetically the pseudowild organism can be shown to be doubly mutant. In e^{-J42}, a deletion of one base has altered the entire reading of the code, so e^{-J44}, is completely defective. When mutant e^{-J44} is integrated to form the double mutant, its additional base restores the code to the proper reading frame. The altered amino acids of the pseudowild polypeptide reduce the enzymatic activity. (After Terzaghi, E., Y. Okada, G. Streisinger, J. Emrich, M. Inouye, and A. Tsugita. 1966. *Proc. Natl. Acad. Sci.* (*U.S.A.*), **56**:500.)

which the wild-type protein is converted into the doubly mutant pseudowild form, which involves an initial deletion somewhere in the code specifying threonine, lysine, or serine and a later addition in the region specifying either asparagine or alanine.

The analysis of the lysozyme protein exactly fits the interpretation placed on reading frame mutations in general—namely, that the deletion altered the frame of reading by one nucleotide such that all succeeding nucleotides were altered to code for different amino acids in the polypeptide chain. If this misreading were to continue indefinitely, the entire sequence, from the point of deletion, would be altered, and the protein produced could hardly take the proper tertiary configuration required for enzymatic activity. However, the addition of a base slightly "downstream" in the chain corrects the reading of the code and restores the normal sequence of the amino acids. Presumably the altered amino acids between the two mutational events are not of sufficient importance to remove the activity completely; they only lower it by 50 %. In other cases, the combinations of different code letters would produce a variety of enzymatic activities ranging from 0 to 100 %.

In addition to giving proof for the overall reading frame hypothesis, the preceding experiment makes another important point. Within the region of substituted amino acids, leucine occupies the same position both in the wild-type and in the mutant form. This is obviously a result of the degeneracy of the code—that is, both CUU and UUA specify leucine (Table 20-1)—and thus an alteration of the code may not always lead to a difference in the polypeptide. It should be noted that leucine is one of the amino acids showing the highest degree of code degeneracy. Since the tertiary configuration of the protein depends to a great extent on the interaction of hydrophobic side chains of amino acids, a high order of degeneracy of the fatty amino acids (such as leucine) would obviously make an organism more highly tolerant to mutational events. It is possible that the degeneracy of the code has been adapted, over a long period of time, to suit the most probable mutational events and the function of those amino acids most essential for enzymes. It is an interesting fact that serine and leucine, two amino acids having a high level of representation in all proteins, are the two showing the highest degree of degeneracy in the code. On the other hand, histidine, which plays an important role in the active site of many enzymes, has a low degree of degeneracy, and thus the argument is not completely consistent.

The Code in Vivo

It is now possible to consider the coding problem in reverse. Knowing the amino acid sequence, we might attempt to determine the genetic code for a particular DNA sequence in a particular organism. This attempt has been undertaken in *E. coli* with the tryptophan synthetase system described in Chapter 19. It should be recalled that the A chain of the enzyme is known completely and that a variety of mutations have been obtained, each of which can be located on the genetic map and correlated with a particular amino acid substitution within the polypeptide chain (Figure 19-5). Let us consider only one case, the glycine for which a number of different amino acid substitutions have been obtained and which gives rise initially to mutants 23 and 46 (arginine and glutamic acid). Figure 20-7 shows a series of amino acid replacements that have been determined for this one residue. The original (wild-type) glycine can be mutated either to arginine or to glutamic acid in a single step. From the arginine mutant, three other types can be obtained (threonine, serine, and glycine). The problem is to utilize the code words for the various amino acids to work back to the code in the DNA molecule. In the following analysis we will use the triplets found in the *m*RNA rather than those projected for DNA, since doing so will make it easier to follow the code words given in Table 20-1. To convert these triplets to DNA words is a simple matter (AUC = TAG).

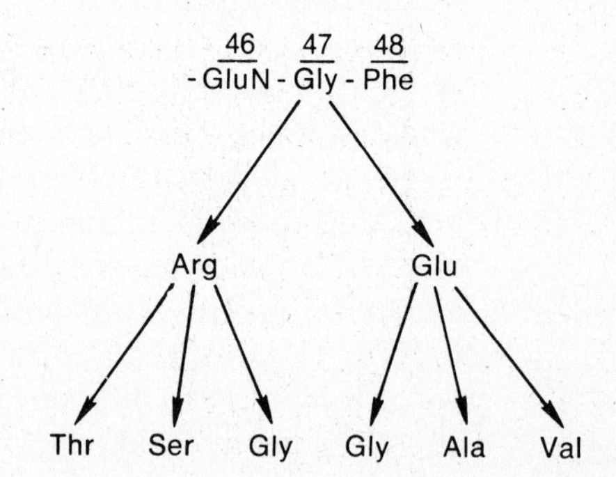

Figure 20-7. Mutations at position 47 in the A chain of tryptophan synthetase. Two lines of mutants are obtained, giving seven different amino acids at this position. The glu mutant never gives rise to ser or thr; the arg mutant never gives rise to ala or val, indicating that they differ from one another by at least two code letters.

Four possible *m*RNA triplets could stand for the original glycine: GGA, GGG, GGU, or GGC. Since the mutations that convert glycine to arginine or glutamic acid can be induced by point mutagens, it is presumed initially that the mutant forms differ from the original glycine by only a single code letter. Glutamic acid can be coded in only two possible ways: GAA or GAG. Since both triplets differ from GGU and GGC in two ways, we make the simplifying assumption that the original glycine was coded either as GGA or GGG. We can test this assumption by crossing the original glycine strain with the glutamic acid mutant. The diploid so obtained could be written alternatively as

$$\underset{\text{(a)}}{\frac{\mathrm{GGA}}{\mathrm{GAA}}} \quad \text{or} \quad \underset{\text{(b)}}{\frac{\mathrm{GGG}}{\mathrm{GAG}}} \qquad \underset{\text{(c)}}{\frac{\mathrm{GGU}}{\mathrm{GAA}}} \quad \text{or} \quad \underset{\text{(d)}}{\frac{\mathrm{GGC}}{\mathrm{GAG}}}$$

(The other possible diploids will give the same results and are omitted for simplicity.) When new haploid cells are produced, some of them will be recombinants. By examining the possible recombinants from the preceding crosses, it is evident that (a) and (b) cannot give *any* new types, whereas (c) and (d) can give two new types each (GGU/GAA → GGA, GAU, and so on). Two will be glycine and thus undetectable, but the other two will be an aspartic acid substitution. Therefore analysis of the progeny to detect any such recombinants will determine whether the glutamic code differs from the original glycine code by more than one letter. (If all three letters are different, six different recombinants are possible.)

Turning to the arginine mutant, GGA could give rise to this in a one-step manner, by conversion either to the code word CGA or AGA. Similarly, GGG could give rise to arginine by mutation to the code words CGG or AGG. We can eliminate the code words CGA and CGG by examining the secondary mutants produced from the arginine-substituted mutant. Note that threonine can be produced by only those code words beginning AC–. One-step mutation could give rise to such code words only from the triplets AGA and AGG. Similarly, it would be impossible to obtain serine as a one-step mutant from either CGA or CGG. As above, the validity of the single-letter change can be verified by recombination analysis.

We have projected two possible formulations for the glutamic acid and arginine substitution, each of which uses code words that differ by two letters:

$$\underset{\text{(a)}}{\frac{\mathrm{GAA}}{\mathrm{AGA}} \;\; \begin{matrix}\text{glu}\\ \text{arg}\end{matrix}} \quad \text{or} \quad \underset{\text{(b)}}{\frac{\mathrm{GAG}}{\mathrm{AGG}} \;\; \begin{matrix}\text{glu}\\ \text{arg}\end{matrix}}$$

Diploids formed from these mutants should yield two new types, one of which will be a glycine revertant (GGA or GGG) and the other a lysine mutant (AAA or

AAG). The results obtained are consistent with this initial hypothesis. Similarly, the fact that the glutamic mutant can give rise to alanine and valine substitution but not serine or threonine (Figure 20-7) supports the two-letter difference. Unfortunately, the degeneracy of the code makes a clear choice between the two possible initial code words for glycine impossible. It should be clear, however, that this method can ultimately provide such a distinction when all the possible mutant forms are obtained. Although it is a laborious process to obtain all the necessary mutants, isolate, and analyze the proteins for differences, the end result should be a determination of the actual code used in the gene.

An interesting genetic sidelight to emerge from the preceding experiment is the fact that the predicted recombinants actually are obtained. This means that genetic recombination in *E. coli* (and probably in all other organisms) involves the exchange of nucleotides in the DNA chain. The method of genetic recombination must be envisaged as the ability to trade single bases between two DNA molecules lying in close proximity. Obviously such recombination involves the exchange of both chains of the DNA molecule; otherwise a mutational event would always occur as a result of genetic recombination. How this exchange is accomplished remains one of the unsolved problems of biology, but we will present some possible models in Chapter 23.

Codons and Anticodons

The general model for protein synthesis developed in Chapter 19 requires that the coded message (*m*RNA) be translated into a sequence of amino acids (polypeptide) as the final stage. The code words are triplets and are given the name *codons.* They are translated by base pairing with the *t*RNAs that carry the amino acids. The exposed triplet on the *t*RNA molecule that "reads" the code in the message must be complementary to the codon, and therefore it has been given the name *anticodon.* The codon for the amino acid methionine is AUG. Its anticodon in the corresponding *t*RNA must be UAC. Furthermore, the *m*RNA is read in a polar manner, starting at the 5′-end and ending at the 3′-end. Thus the codon must be written $^{5'}$AUG$^{3'}$. (The reverse, GUA, is one of the codons for valine!) Polar reading of the message ensures that the proper amino acid is put in the proper place. Since two nucleic acid chains can only pair properly when they run in antiparallel directions (Chapter 18), the anticodon must have the opposite polarity. Thus the anticodon for methionine should be written $^{3'}$UAC$^{5'}$. In the discussion that follows it is important to remember that the first letter in the codon is at the 5′-end but the first letter in the anticodon is 3′.

The amino acid code consists of 64 codons, three of which are nonsense triplets. Therefore there are 61 codons representing 20 amino acids. Are there 61 different *t*RNAs, each with its unique anticodon? The best evidence says no. There are fewer than 61, which implies that the anticodon of some *t*RNAs can "read" more than one codon. This was a dilemma. How could the base-pairing mechanism be inexact? A second problem arose from the discovery that the anticodon of certain *t*RNAs contained the unusual purine inosine (see Chapter 21 for the origin of such bases in *t*RNA). In 1966 the "wobble hypothesis" was advanced to explain these problems. It assumes that the third letter in the anticodon is not fixed as rigidly as the other two. Its ability to shift about in space permits it to pair with more than one base, and thus one anticodon can pair with more than one codon. For example, the anticodon $^{3'}CAU^{5'}$ could pair with both $^{5'}GUA^{3'}$ and $^{5'}GUG^{3'}$. However, it could not bond with GUU nor GUC, since U—U and U—C bonding is not feasible.

Inspection of the amino acid code in Table 20-1 will demonstrate a nice feature of the wobble hypothesis. Note that the redundancy in the code is not random. For the most part, the two or more codons standing for a particular amino acid all have the same first two letters (the exceptions are leucine, serine, and arginine). For example, all four codons beginning GU stand for valine, and, similarly, all four beginning GC stand for alanine. Furthermore, in the cases where only two codons are used to specify an amino acid, the third letters of the pair are either two purines or two pyrimidines, never one of each. For example, UUU and UUC specify a phenylalanine, but UUA and UUG specify leucine. Similarly, CAU and CAC specify histidine, but CAA and CAG specify glutamine. The arrangement of code letters begins to make some sense. The wobble hypothesis suggests that, in the case of phenylalanine, both UUU and UUC might be read by the single anticodon AAA; and, for glutamine, the codons CAA and CAG might be read by the single anticodon GUU.

The preceding material should indicate that the wobble hypothesis fits well with the known amino acid code. More importantly, it permits certain testable predictions to be made. Take the case of leucine. Six codons are used to specify this amino acid: UUA, UUG, CUU, CUC, CUA, and CUG. What is the minimum number of *t*RNAs that can be used to translate these six? Obviously there must be at least two, since UU- and CU- are distinct in the first two positions. However, there is no base that can pair will all four bases; therefore CUU, CUC, CUA, and CUG cannot be read by the same anticodon. It follows that the minimum number is three: one that is presumably AAU to read the UUA and UUG, one that is GAA to read the CUU and CUC, and one that is GAU to read the CUA and CUG. The discovery that, in fact, there are three leucine-*t*RNAs was an important test of the hypothesis.

Let us return now to the matter of the unusual base inosine. If allowed to wobble, this base could pair with three other bases (U, C, and A). Its pairing possibilities are diagrammed in Figure 20-8. The presence of such a base in the third position of an anticodon would permit a single *t*RNA to read three codons, but is there such a case? In fact, there is. Isoleucine is the only amino acid having three codons: AUU, AUC, and AUA. The terminal letters of these codons (U, C, and A) can all pair with inosine. It is interesting that the one that could not, AUG, specifies a different amino acid—namely, methionine. This is another achievement for the hypothesis.

Good as it is, the wobble hypothesis is still tentative. A complete three-dimensional model of the various *t*RNAs is needed before we can verify its primary assumption that the third base in the anticodon is more free to "wobble" than the first two. One such model has been presented, and it is in agreement with the

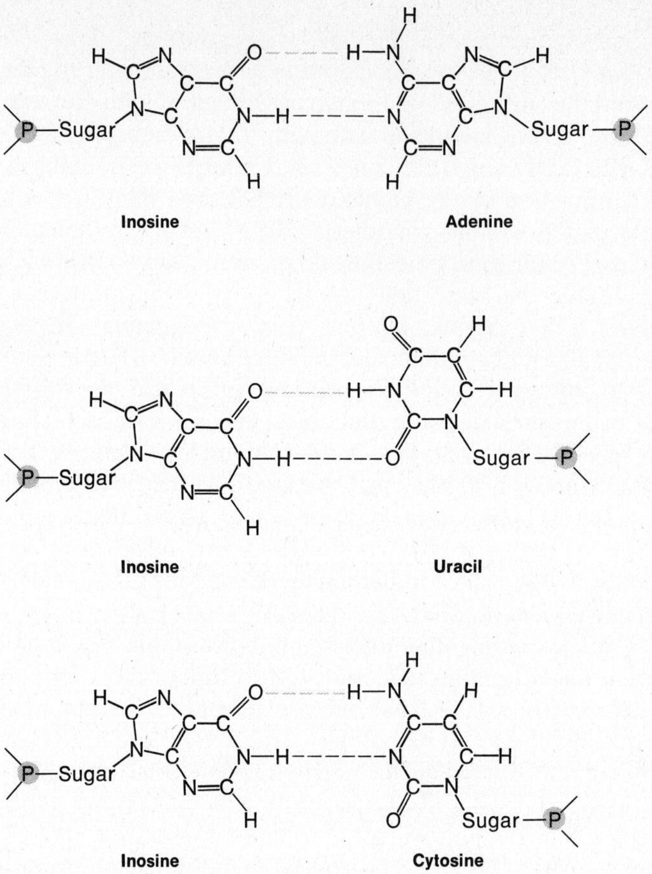

Inosine Adenine

Inosine Uracil

Inosine Cytosine

Figure 20-8. Codon–anticodon pairing of inosine. Inosine is an unusual base occurring in *t*RNA. It can pair with three bases of the *m*RNA (A, U, C) provided only that the base can rotate somewhat on its sugar-phosphate linkage. This "wobble" in the anticodon region of *t*RNA permits one *t*RNA to read several codons of the *m*RNA.

hypothesis. However, several cases are needed before we can safely conclude that this example is not unique.

REFERENCES

HAYNES, R. H., and P. C. HANAWALT. 1968. *The Molecular Basis of Life.* W. H. Freeman and Co., San Francisco. (This collection contains three good, general articles on the code, two by Crick and one by Nirenberg. The article on *Antibodies and the Genetic Code* by Gorini is also pertinent.)

FRISCH, L. (Ed.). 1966. The genetic code. *Cold Spring Harbor Symp. Quant. Biol.* **31**. (This volume is a compendium of original and review papers covering all aspects of the subject. There is a very good introductory article by Crick. For the advanced student.)

YCAS, M. 1969. *The Biological Code.* John Wiley & Sons, New York. (This is rather too comprehensive for the undergraduate reader, but it is the most complete work available.)

Original Papers

CRICK, F. H. C. 1966. Codon-anticodon pairing: the wobble hypothesis *J. Mol. Biol.* **19**:548–555. (This paper was the first to present the wobble hypothesis.)

CRICK, F. H. C., L. BARNETT, S. BRENNER, and R. J. WATTS-TOBIN. 1961. General nature of the genetic code for proteins. *Nature* **192**:1227–1232. (The classic paper that proved the genetic code to the triplet and read in a polarized manner without punctuation.)

NIHIMURA, S., D. S. JONES, and H. G. KHORANA. 1965. The *in vitro* synthesis of a copolypeptide containing two amino acids in alternating sequence dependent upon a DNA-like polymer containing two nucleotides in alternating sequence. *J. Mol. Biol.* **13**:302–324. (This paper demonstrates the triplet code using synthetic nucleic acid.)

NIRENBERG, M. W., and J. H. MATTHAEI. 1961. The dependence of cell-free protein synthesis in *E. coli* upon naturally occurring or synthetic polyribonucleotides. *Proc. Natl. Acad. Sci. U.S.* **47**:1588–1602. (The first paper in the series that "cracked" the code.)

STREISINGER, G., Y. OKADA, J. ENERICH, J. NEWTON, A. TSUGITA, E. TERIZAGHI, and M. INOUYE. 1966. Frameshift mutations and the genetic code. *Cold Spring Harbor Symp. Quant. Biol.* **31**:77–84. (A good account of the experiments that demonstrated the correctness of the reading frame hypothesis.)

QUESTIONS AND PROBLEMS

20-1. Define the following terms:

genetic code
codon
degenerate code
pseudowild-type
overlapping code
anticodon
reading frame

20-2. Explain how addition/deletion mutants were used to show that the genetic code is a triplet code, that is commaless, and that is read in a polar manner.

20-3. What is the evidence that the genetic code is nonoverlapping?

20-4. How was the filter-binding (minimessenger) technique used to crack the code?

20-5. Given the sequence AAATACTGGACG . . . in the DNA from which a message is transcribed, what is the corresponding sequence of amino acids in the polypeptide? What is the *m*RNA sequence? What *t*RNA anticodons are required?

20-6. What is the role of the triplets UAA, UGA, and UAG?

20-7. How would you now explain the term "point" mutation?

20-8. What is a frameshift mutation? What is a nonsense mutation? Why are these two types of mutations often more drastic in their effects than other mutations?

20-9. Frequently, mutations do not completely remove the function of an enzyme. Such mutants are called "leaky." How would you explain leaky mutants?

20-10. What is the advantage of a degenerate code?

20-11. What is the smallest unit of crossing over? How does it explain the results of Benzer's fine-structure analysis (Chapter 16)?

20-12. What is the gene?

20-13. Explain the "wobble" hypothesis of *t*RNA action.

20-14. On the basis of the wobble hypothesis, what is the minimum number of different *t*RNAs required to read the *m*RNA code?

20-15. It has been estimated that the "molecular weight" of the entire genome of *Drosophila melanogaster* is 1.2×10^{11}, which, in turn, indicates about 2×10^8 nucleotide pairs. Assuming that the average polypeptide contains 100 amino acids, how many cistrons could the Drosophila DNA encode?

21/Genetic problems related to RNA synthesis

Synthesis of Ribosomal RNA

As indicated in Chapter 19, the functional ribosomes that translate the *m*RNA into the amino acid sequence of the polypeptide are complex particles made up of both *r*RNA and protein. There are two large *r*RNAs—one having a sedimentation value of 16 to 18 S and the other 23 to 28 S—and a small *r*RNA, 5 S. The values vary, depending on the source of the material, the lower values being typical of bacteria and the higher ones of higher organisms. Each of the larger ribosomal RNAs combines with protein to produce a subribosomal particle, and these particles, in turn, associate in a 1 : 1 ratio to give rise to the active 70–80 S ribosome. The 5 S RNA is included in the particle containing the 23–28 S RNA.

The association of the various RNAs and ribosomal proteins is a completely spontaneous process. The 30 S particles can be dissociated and the *r*RNA and proteins separated. These isolated components can then be mixed together, and they will aggregate to give 30 S particles spontaneously. These particles are completely functional. However, if proteins from 50 S particles are mixed with the *r*RNA from 30 S particles, there is no aggregation, nor will 30 S proteins from rat liver aggregate with 30 S *r*RNA from bacteria. Apparently the interaction between the *r*RNA and the proteins forming the several ribosomal particles is selective, more so than the similar self-aggregation process by which viruses are formed within cells.

Analysis of the ribosomal proteins has only just begun. So far the ribosomes of *E. coli* have been the most extensively studied. The 30 S particle yields 19 different proteins, each present as a single copy. The 50 S particle has 55 protein molecules, but it is not yet certain that they are all different. Clearly there is a great deal of diversity in the protein content of these very small structures, but their exact role in protein synthesis is uncertain. They must supply the enzymatic activity necessary for polypeptide synthesis, but there are too many different kinds of proteins for those functions alone. As yet we cannot be completely certain about the mode of synthesis of the ribosomal proteins. However, there is little reason to doubt that their formation differs in any way from other proteins. This poses the dilemma of how the initial ribosome was ever formed, but that problem is outside the scope of this book. Once ribosomes exist, they can certainly be used for the synthesis of proteins from which other ribosomes will be formed. In cells of higher organisms, it appears that the ribosomal proteins are made in the cytoplasm and transferred to the nucleus, but the formation of ribosomes themselves occurs within the nucleus. The messenger RNA molecules made within the nucleus appear to have ribosomes already associated with them prior to their migration to the cytoplasm.

The fact that *r*RNA synthesis is under the control of a specific locus within the chromosome was first demonstrated in studies of anucleolar mutants of Xenopus, the South African clawed toad. The mutation is recessive. In the homozygous mutant, no nucleoli are present and the embryo dies early in development (at the time of gastrulation; see Chapter 25). Analysis of these aborted embryos indicates that they are incapable of forming ribosomes and that their death is the result of an inability to carry on the rapid protein synthesis required at this stage of development. Until this stage the embryos survive with the ribosomes that had been previously manufactured during the development of the egg. (It should be realized that the homozygous, anucleolate individual is produced by crossing two individuals heterozygous for the character, and thus the egg contains ribosomes made prior to meiosis.)

Work with anucleolar mutants confirmed what had long been suspected—that the nucleolus has as its special function the manufacture of ribosomes. The rather high RNA and protein content of the nucleolus is consistent with this finding, and its normal cycle of growth throughout interphase with the discharge of its contents during the early stages of division is also consistent with its role in ribosome manufacture. Presumably the region of the chromosome to which the nucleus is attached, the *nucleolar organizer*, is a region containing the genetic information that in some way governs the synthesis of ribosomes. The function of the genes results in the production of a large quantity of ribosomal precursor material. At each division, this material is released in quantity for the manufacture of ribosomes, to provide the protein required for the continued life of the cell during the next interphase. As good as the preceding model is, however, it must be pointed out that it is not completely satisfactory. There must be exchange between the nucleolus and the cytoplasm at all stages, since protein synthesis continues rapidly in many cells that do not divide and in which the nucleolar cycle is not seen. This and other evidence that confirms exchange between the nucleolus and other parts of the cell suggest that two processes occur within the sac. One is the accumulation of materials for ribosomal synthesis, and the other is the leakage of these materials to the cytoplasm to maintain the steady state of the cell.

The precursor material coded for in the nucleolar organizer region is ribosomal RNA. This fact has been most directly demonstrated by experiments with Drosophila. By the mid-1960s it was recognized that the gene controlling the *bobbed* phenotype (very short bristles) was located at, or at least very close to, the nucleolar organizer on the X chromosome, and this fact suggested that the bobbed phenotype might result from a deficiency in the ability of the nucleolar organizer to manufacture ribosomal material. If the bobbed locus and the nucleolar organizer are one and the same, the mutant phenotype bobbed would presumably result from the inability of the organizer to function properly. In the absence of a critical number of ribosomes, bristle formation, involving extensive protein synthesis, would be defective and the bristles would be short. It should be noted that this case is different from the anucleolar mutants of Xenopus, described above, which are lethal. The presumption here is that the bobbed phenotype results from a reduction in the *amount* of gene product. It is not altogether absent; if it were, the homozygous bobbed would be a lethal gene. The hypothesis, then, is straightforward. The mutant is defective in that it has a decreased output of ribosomes. The question is how to measure the nucleolar function in some quantitative way.

In Chapter 19 we discussed the tests of the complementarity of *m*RNA to regions of the DNA. In those tests specific binding between RNA molecules and at least one strand of the DNA was used as an assay procedure. This technique has been applied to the problem of *r*RNA synthesis on the relatively safe presumption that *r*RNA is transcribed directly from some region of the DNA (presumably in the region of the nucleolar organizer). There should be some fractions of the DNA that bind *r*RNA selectively. This was found to be the case, both in Drosophila and in other organisms. The interesting feature, however, is that the amount of DNA that showed specific binding of *r*RNA was much higher than necessary to account for the production of all the ribosomal RNA. In Drosophila, there are approximately several hundred times as much DNA as needed to code for the ribosomal RNA, which suggests that the genes coding for ribosomal RNA synthesis are present in multiple copies, presumably as a way of producing large numbers of ribosomes in a relatively short period of time.

Proceeding on the assumption of genetic redundancy, it is now possible to make a test of the general hypothesis put forward for the bobbed phenotype. The bobbed mutant would be one in which a certain fraction of the genes coding for *r*RNA synthesis were mutant (or missing), and thus the total amount of *r*RNA would be reduced. If true, the amount of DNA that would specifically bind with *r*RNA would be lower in a fly that was heterozygous for bobbed than in one that was homozygous wild-type. And, similarly, a fly that was homozygous for bobbed should have less binding of *r*RNA than one that was heterozygous. This turned out to be the case. Apparently the bobbed mutation is one in which some part of the genetic material coding for *r*RNA is deleted. The organism can survive the deletion because the multiple copies still present are producing sufficient *r*RNA for survival but not enough to give the normal bristle formation If, however, a third bobbed gene is added (see Chapter 15), the added copies given by the third gene, even with its deleted parts, should be sufficient to produce enough *r*RNA to return the phenotype of the fly to normal. Titrations of *r*RNA with such mutant flies confirm that such is the case.

The foregoing somewhat complex set of experiments demonstrates two things. First, the nucleolar organizer region of the chromosome contains the DNA from which *r*RNA is transcribed. Secondly, that region is highly redundant; that is, many copies of the same DNA sequence are present within a single nuclear organizer. Studies with other organisms indicate that this is always true. Both the nucleolar organizer of higher organisms and the corresponding region of bacterial chromosomes appear to be complex regions of DNA containing many copies of the same gene that produces *r*RNA.

The nucleolar organizer produces *r*RNA, but we have already indicated that there are three kinds of *r*RNA. Only two are produced by the nucleolar organizer. The natural presumption would be that there are two different cistrons controlling *r*RNA synthesis. That is not the case, however. It turns out that there is only one cistron from which a single, very large RNA is transcribed. This molecule is broken down in a series of steps to give rise to the functional *r*RNAs. Figure 21-1 shows the scheme by which these two ribosomal RNAs are produced. The mode of production is complex, but it ensures that the major kinds of the *r*RNA will be produced in exactly equal quantities. If there were separate cistrons for the 18 S piece and the 28 S piece, the time required for transcription of the two would be different, and more 18 S would be produced than 28 S. The process of uniting them in a single cistron and then breaking the rather large precursor RNA molecule into fragments, ensures that these RNAs will be present in equal quantities at all times.

In the 1970s the region of synthesis of the third component of *r*RNA, the 5 S fraction was located. This was done by using the property of specific binding of RNA to the cistron from which it is transcribed. The 5 S *r*RNA is prepared with ^{3}H, and it is used to anneal with the chromosomal DNA. The chromosomes are then examined by autoradiography to see where the radioactivity is located. The 5 S *r*RNA will bind only to the proper cistron, and therefore the radioactivity will be present only at the location of that cistron in the chromosome. Surprisingly, many 5 S cistrons are revealed by this method, at least one and probably two per chromosome. Furthermore, these cistrons appear to be localized at the ends of the chromosomes (the telomeres). The presence of multiple copies of the 5 S cistron is consistent with the

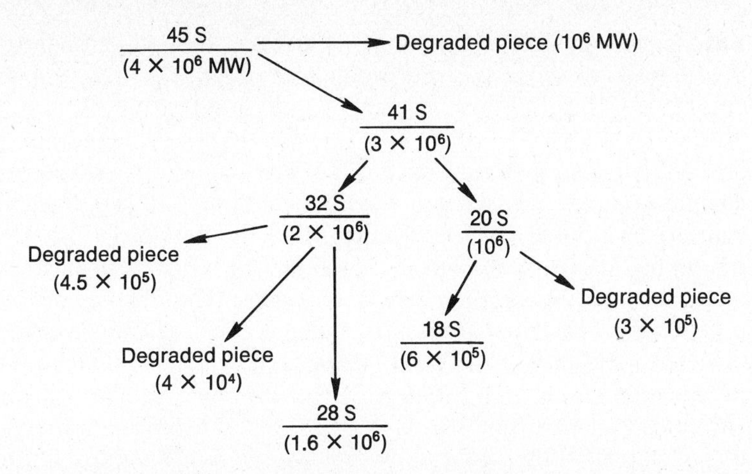

Figure 21-1. Production of ribosomal RNA. The *r*RNA is transcribed as a single large RNA (45 S), which is subsequently cleaved into the two major classes of *r*RNA. In the process a substantial part of the parent molecule is discarded. The mechanism insures that these types of *r*RNA are produced in equal quantity and simultaneously.

redundant nature of the other *r*RNA cistrons. Presumably the redundancy permits rapid synthesis of the ribosomes. However, the question of how the 5 S cistrons came to be so widely scattered, and always at the ends of the chromosomes, remains unanswered.

The mode of production of two RNAs from the 45 S precursor molecule shown in Figure 21-1 indicates that there is no direct correspondence between the sedimentation coefficient (S value) and molecular weight. The separation of the different fractions in a centrifuge depends on the density of the molecule rather than on its weight; the two are only indirectly related. As changes in the shape of the molecule occur (more or less coiling and so on), there will be changes in the density without consequent changes in the internal composition of the molecule. For this reason, when the large 45 S nucleolar RNA is broken down into two components, the sedimentation coefficients do not add up (32 S plus 20 S = 41 S). In the conversion of 41 S to 32 S and 20 S, there is apparently no loss of material, only a change in the configuration of the molecules with a corresponding change in density. It is probably for the same reason that the *r*RNAs from different organisms do not have exactly the same sedimentation coefficients. In general, the *r*RNAs of microorganisms are somewhat less dense than those of mammalian cells (the figures used for Figure 21-1).

Synthesis of Transfer RNA

As indicated in Chapter 19, the transfer RNAs are the smallest type of RNA, and there is at least one, and in most cases more than one, *t*RNA for each amino acid. In addition to their small size and the high degree of base pairing within the molecule that the *t*RNAs show, they are peculiar in that they contain a number of abnormal bases (see Figure 19-3). If we are to presume that *t*RNAs are transcribed from regions of the DNA, it is difficult to see how the normal base-pairing mechanism can be used for the transcription of RNA that contains such anomalous bases. This dilemma was resolved when it was discovered that, in fact, the abnormal bases are made from normal bases *after* the *t*RNA is produced. The precursor *t*RNA is transcribed as any other RNA, containing the usual four bases (AUGC). These bases are later modified by specific enzymes. The purpose of this modification is probably twofold. First, it is essential that certain specific configurations be adopted by *t*RNA molecules so that certain triplets remain unpaired for anticodon reading of the message and so that the specific enzyme that attaches a particular *t*RNA to its amino acid can recognize the *t*RNA molecule. The greater variety of bases found in *t*RNA may assist this recognition process. The second purpose for modification of the bases may be more important. The *t*RNA molecules are utilized over and over again in the cytoplasm, but the cytoplasm contains various enzymes that degrade RNA. Modification of the bases probably protects the *t*RNA against such degradation. (Ribosomal RNA is presumably protected by being "buried" in the accessory proteins of the ribosome.) Regardless of the reason for the modification of the bases, there is no difficulty in envisioning the transcription of DNA into precursor *t*RNA as the initial step.

It is possible to "titrate" DNA with the various *t*RNAs to see whether they will bond specifically with certain regions of the DNA. Although this process is complicated somewhat by the abnormal bases, there is sufficient specificity of pairing left so that it can be shown that a particular *t*RNA bonds with a particular part of the DNA. In this context, it is interesting to note that in *E. coli* far less of the DNA is given over to the manufacture of *t*RNA than to the manufacture of *r*RNA. Approximately 0.2 % of the total DNA of *E. coli* codes for ribosomal RNA, whereas only 0.02 % codes for *t*RNA. Because of the great difference in size between *r*RNA and *t*RNA, such data alone do not permit us to decide whether the *t*RNA sites are redundant. However, there is indirect evidence to suggest that they probably are, although at a much lower level of redundancy than that characteristic of *r*RNA. We will consider this indirect evidence in the next section.

Suppressor Mutations

In the previous chapter we considered two kinds of intragenic suppressor: the case in which the addition of a base can suppress the deletion of one previous to it in the reading frame and the case in which +/− mutations can alter the reading of a nonsense triplet such that it makes either sense or relatively minor missense. Either event can convert a mutated gene to the pseudowild condition. Strictly speaking, however, when geneticists talk of suppressor mutations, they are referring to intergenic (not intragenic) events. Here a mutation at one genetic locus can suppress the effects of a mutation at an entirely different locus, even one that is on a different chromosome.

One of the most interesting sets of suppressor mutants are the *amber** mutations of the T_4 bacteriophage. Amber mutants cannot grow on most of the strains of *E. coli* on which wild-type T_4 normally grows. However, certain strains of *E. coli* will permit their growth (permissive strains). With careful investigation it was discovered that there were many amber mutations. Some were defective in the initiation of DNA synthesis, others were defective in the production of phage coat proteins, and so forth. Thus the amber mutant was, in fact, a whole group of mutants, all of which had in common the ability to grow on permissive strains of *E. coli* but not on others. (Even so, it should be noted that their growth on these strains was never as good as the wild-type T_4.) The question raised by this observation is clearly: How is it possible for some strains of *E. coli* to restore the function of mutated gene in the bacteriophage DNA?

The question was answered shortly after it was realized that nonsense mutations would terminate translation of *m*RNA by ribosomes. If a messenger contains one of the three nonsense triplets (UAA, UAG, UGA), the ribosome will read the message until it comes to that nonsense triplet; then it detaches, releasing an unfinished polypeptide chain. The curious feature of the amber mutation is that, in certain strains of *E. coli,* these universal nonsense triplets apparently are not chain terminating. Careful study of the various *E. coli* strains indicated that three well-defined and separate genes suppress the UAG "stop" codon. One acts to insert serine in the position that would otherwise be chain terminating. Another gene inserts tyrosine, and the third one inserts glutamine. The simplest interpretation of these data is that the suppressor genes produce *t*RNAs that can read the nonsense triplet. We would presume that, in the case of the suppressor substituting tyrosine for the nonsense codon, a mutational event has occurred such that the anticodon of the tyrosine *t*RNA, instead of being AUG, as is normal, now reads AUC, an anticodon that would read the nonsense triplet UAG perfectly well. This interpretation suggests, then, that the permissive character of the strain is the result of a mutational event at a locus controlling *t*RNA synthesis. The substitution of a single base at the proper position changes the reading of the code such that the nonsense triplet is now read by this particular species of *t*RNA. In another case, serine is inserted because the mutational event occurred at a gene specifying the production of serine *t*RNA, and so forth.† It is, of course, no accident that serine, tyrosine, and glutamine are the substituted amino acids. They are the ones whose *t*RNA-anticodons can be changed to read UAG with only one base change.

That the preceding interpretation is reasonable was first clearly demonstrated by adding isolated *t*RNA from a suppressor strain to a system that would otherwise

*The word amber is in no way descriptive of the phenotype of the mutant. In fact, it is the translation of Bernstein, the name of the original discoverer, into English. This is a classic example of a scientific in-joke: can you imagine the *Bernstein mutant* becoming part of the language of science?

†Of course, the insertion of an amino acid at the position of a nonsense codon is effective only if that insertion produces a functional protein. If the inserted amino acid is of the wrong kind, the phenotype will still be mutant, and the altered *t*RNA gene will not be classified as a suppressor.

have been nonpermissive; and when this *t*RNA was present, suppression of the nonsense codon was achieved.

Since the altered *t*RNA is all that is necessary for suppression, it seems reasonable to propose that this type of suppression is caused by mutations in genes controlling the synthesis of *t*RNA. Several other lines of evidence support the hypothesis. The first is that any particular suppressor mutation can only suppress a limited number of mutational events at other loci. As expected, an altered *t*RNA molecule cannot make substitutions for every altered triplet. If the suppressor gene were one that permitted serine to be substituted for the nonsense triplet UAG, it would be difficult to imagine a mechanism whereby that suppressor gene could be used to suppress either a missense triplet or some other nonsense triplet, such as UAA. The second case is the converse; that is, different suppressor mutations suppress different classes of mutational events at other loci. The suppressors are specific, as they would be if each suppressor mutation were effected by the alteration of only one particular class of *t*RNAs. Finally, and most importantly, cells containing suppressor mutations are less effective at reading the normal genetic message. If such a suppressor gene is introduced into an otherwise wild-type cell (by genetic recombination), the amount of wild-type protein produced is significantly reduced. This result is to be expected if a certain class of *t*RNAs has been altered such that they now have available no specific code word to be read in the normal genetic message. This last case demonstrates that what is a suppressor gene in one situation is merely a mutant (thus nonfunctional) gene controlling *t*RNA synthesis in another.

One of the most interesting features of such intergenic suppression systems is that the effects are never complete. The activity of the suppressor gene may be sufficient to produce an adequate level of enzymatic protein for the continued survival of the cell, but the amount of protein made is never equal to that produced in a wild-type cell. (At most, it is about 50 %.) Similarly, when suppressor genes are put into wild-type cells, there is never a complete loss of protein synthesis. The simplest interpretation of these observations is that suppressor genes affect only a certain percentage of the total *t*RNA of a particular type. There must be at least two, and possibly many, spatially distinct genetic loci for the production of each specific type of *t*RNA. If there were two, the suppressor mutation would probably involve only one of these loci, and so both wild-type *t*RNA and suppressor *t*RNA would be made concurrently. In the mutant cell, suppressor *t*RNA can read the message to the extent that a satisfactory level of enzyme can be maintained, and thus this suppressor gene has done its job. The wild-type *t*RNA is used to read nonmutant regions of the genome. Similarly, in a wild-type cell, the message can be read satisfactorily by the presence of the wild-type *t*RNA, and thus a suppressor gene does not result in the death of a normal cell. In evolutionary terms, the selective advantage of such a system should be obvious.

In general, the suppression of the UAG and UGA nonsense triplets is about 50 %. But the suppression of the UAA triplet is relatively poor. This latter fact suggests the possibility that the normal chain-terminating triplet of a message is

UAA. Acquisition of *t*RNAs that read UAA well would result in a disruption of the reading process and such *t*RNAs would themselves be selected against. Only inefficiently read UAAs would survive. This possibility may be true, but some studies have indicated that, in fact, chain termination may involve more than one nonsense triplet. If so, this mechanism would have the distinct advantage of preventing suppressor mutations from interfering with the vital function of chain termination.

On the basis of the general postulate that suppressor genes result from mutational events in loci coding for *t*RNA, it is reasonable to propose that a wide variety of such events occurs and that in some cases they may suppress missense codons as well as nonsense codons. The carefully studied tryptophan synthetase system described in Chapter 19, when we considered tests for the colinearity of the cistron and the polypeptide, supports this proposition. In this system, it is possible to know exactly which amino acid is substituted at exactly which place. In the case of the mutational event that replaces one of the glycines in the chain with arginine, suppressor mutations cause the insertion of glycine in the mutant form that normally would have arginine. The level of suppression is relatively low. Thus we must assume that only a small fraction of the *t*RNA is of the modified form. Once again the suppressing *t*RNA can be isolated, and it appears to be almost identical with the wild-type glycine-*t*RNA.

Taken as a whole, the evidence from intergenic suppression has established that there are at least two and probably more sites of *t*RNA synthesis for each of the *t*RNAs. When this fact is taken together with the fact that there are several different *t*RNAs for some amino acids, it is reasonable to propose that probably quite a few loci in the DNA specify the various *t*RNAs. Any mutational events that replace a large fraction of a particular kind of *t*RNA with a mutant form will undoubtedly be lethal, but those in which only a small fraction of *t*RNA is altered can survive; and, under appropriate circumstances, they may be favorable mutational events, since they provide a suppressor gene action for mutations that have occurred at other loci.

Genetic Redundancy and Gene Amplification

The fact that there are several to many copies of the genes from which *t*RNA and *r*RNA are transcribed is an example of *genetic redundancy*. Just how extensive this situation is in higher organisms is not yet certain, but there appear to be a number of cases in which genes are present in more than one copy. The redundancy of these particular loci is presumably related to the need for rapid protein synthesis. The more copies of a cistron in the genome, the more *m*RNA can be produced per unit time. Both *t*RNA and the ribosomes are relatively stable, but they do not last forever. If they were not replaced, the rate of protein synthesis might be limited by their availability rather than by the rate of synthesis of *m*RNA. It appears that the latter course is the one most frequently used, and organisms have developed redundancy as a means of ensuring a rapid replacement of these units when required.

The best-investigated case is *r*RNA synthesis. In all such instances, redundancy seems to range between a hundred and several thousandfold. These assays are made easier by the fact that it is a relatively simple matter to extract *r*RNA from cells and that all *r*RNAs are very much the same. It is much more difficult to use annealing studies (measuring the complementarity of DNA and RNA) with *m*RNA, since it would be necessary to isolate a specific messenger in pure form *and* in quantity in order to do the annealing study that would tell us how many copies of a particular cistron are present. The most interesting studies have involved the bands of the polytene chromosomes and the chromomeres of certain other chromosomes. Traditionally these structures have been considered the seats of individual genes. When it was discovered that there was far too much DNA to code for a single genetic function, there was speculation about the possible redundancy of these loci. The studies made in recent years have indicated that it is most probable that, in fact, the bands and chromomeres are redundant loci. These findings suggest that the genetic material in the higher organisms may have a rather different organization from the genetic material of the bacteria and viruses; the latter seem to have a low level of redundancy of most genetic functions. The data are not sufficiently extensive to permit us to generalize about the level of redundancy in the higher organisms, but we do know that redundancy does occur for many genes and that it occurs at a lower level than is necessary for the ribosome synthesis.

In the cases where genetic redundancy is not enough to supply the needs of the organism, a more elaborate mechanism is used, *gene amplification.* Perhaps the most elaborate case is that demonstrated in the oocytes of certain amphibians. The egg cell develops while it is "locked" in the prophase of meiosis. In this stage, many genes are very actively producing the materials that the egg will need in its early development. In fact, so much material is stored in the egg initially that there is relatively little transcription of any of the *r*RNAs from DNA until much later in development (see Chapter 25). Large reserves of ribosomes, transfer RNAs, and even quantities of stored *m*RNA are present within these cells by the time the egg is fertilized. In order to produce the large quantity of ribosomes, the nucleolar organizer is replicated in a process that produces a closed circle of DNA, which is cut off from the nucleolar organizer itself in each replication cycle. These small circular DNA molecules, completely detached from the parent chromosome, then organize nucleolar membranes around them. Consequently, the oocyte of the amphibian may contain as many as a thousand nucleoli, each separate from the chromosome to which the nucleolar organizer is normally attached. Each of these nucleoli produces ribosomal precursors. Thus the egg cell is supplied with a vast quantity of ribosomes, which keep it functional without further synthesis until the cell has cleaved and reached a more advanced state of differentiation.

Some very elegant electron micrographs of the isolated DNA molecules of oocyte nucleoli have been obtained. Figure 21-2 presents a portion of such a closed circle, extended so that the organization of the DNA can be seen. In the section presented in the photograph, five separate "genes" are visible. The DNA is represented by the very narrow black thread connecting all of them together. At one end, very short RNA molecules can be seen extending from each side of the DNA;

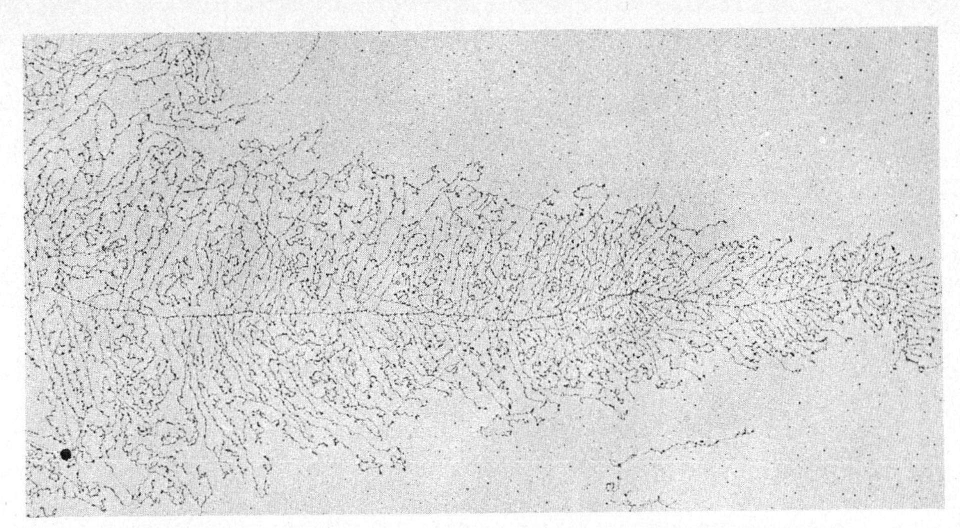

Figure 21-2. Genetic redundancy and gene amplification. The micrograph shows a tandemly duplicated series of genes that code for *r*RNA in the Triturus oocyte. The 45 S RNA chains can be seen growing in each locus from right to left. Note the "spacer" DNA regions between each locus. These do not appear to produce RNA. Presumably the RNA polymerase attaches to a "start" region in each cistron and proceeds along the DNA from left (short chains transcribed so far) to right (long chains almost complete). The picture is one of an isolated circular DNA molecule found in the nucleoli of this organism. It is produced from the "parent" gene in the chromosome. (Micrograph courtesy of O. L. Miller, Jr. and Barbara R. Beatty, Biology Division, Oak Ridge National Laboratory. See also, Miller, O. L. and B. R. Beatty. 1969. *Science*, **164**:955.)

and at the other end, the rather long, presumably almost finished RNA molecules can be seen. These are the 45 S *r*RNA chains. Synthesis begins at the lower left of each gene and proceeds toward the upper right, with the chain increasing in length in that direction. This remarkable electron micrograph shows us a number of things. First, it is a direct visualization of the transcription of RNA from DNA. Furthermore, it shows that the redundant loci are identical in length and general functional morphology. (For example, they are all oriented in the same direction.) In addition, it shows the curious feature that each of the genetically active regions is separated by a short stretch of DNA that is not active. Called "spacer DNA," this region may represent either DNA used for the initiation and control of functions (see Chapter 24) or a region that is not transcribed at all but that is used to separate one gene from another when a large redundant complex is present. Whatever the use of the spacer DNA may turn out to be, there can be no doubt that, for the amphibian oocyte, a large number of loci all coding for exactly the same RNA are developed. These findings suggest the intriguing possibility that gene amplification may be a local event in higher organisms. In certain cases, such as the nucleolar organizer, the locus may always be redundant, but in others it may be made redundant only at the time of special need for protein synthesis. However, until more evidence is forthcoming, we must treat this idea as highly speculative.

Before concluding, other possible advantages of redundancy should be mentioned. One such is protection against mutation. If several loci are coding for the same function, the loss of one may be of little consequence to the organism. The *bobbed* mutation in Drosophila is an example of a relatively minor effect caused by the loss of part of the *r*RNA cistrons. Similarly, the suppressor mutants induced in *t*RNA loci support this idea. The greater the number of loci present for a single function, the more resistant that function will be to mutation.

A somewhat more speculative role of redundancy is its use in the production of new genes. There are advantages and, except for the energy lost in synthesis, no disadvantages to redundancy. A misadventure in the DNA replication process that produced two copies of a gene rather than the normal one copy would probably be preserved. As time progressed, mutational events in one locus could occur with no adverse effects on the organism. (After all, it still has one wild-type gene, as it had originally.) The mutated locus could be carried along indefinitely, accumulating mutational changes until it produced an RNA that had a new role (perhaps as messenger for a new enzyme or as a new *t*RNA). Should this new function be beneficial, the cell would multiply, and the new gene would spread into the population. The idea is attractive, and we know that duplications of regions of the chromosomes do exist (Chapter 13), but it is difficult to test. In any case, it is important to realize that a new function is hard to come by. It would take many mutations in one cistron to produce a completely new enzyme. Consequently, it could be presumed that such a mechanism would operate over a vast period of time.

Heterogeneous Nuclear RNA

It has been known for some time that the nucleus of higher organisms contains a much larger fraction of RNA than occurs in any of the forms found in the cytoplasm. When the 45 S precursor RNA for *r*RNA was discovered, it seemed that the problem of this special class of RNA had been solved. However, it now appears that the 45 S ribosomal precursor is only one of a group of different, large RNAs, and we must offer an explanation for the rest. So far the evidence suggests that the *heterogeneous nuclear RNA* (*hn*RNA) is a class of precursor RNAs, the very large molecules being degraded into smaller messenger RNA molecules. Apparently, the redundant loci produce a single, large, multicistronic *m*RNA precursor, and this precursor is later broken down into the single messenger molecules that are translated in the cytoplasm. Presumably, each *hn*RNA contains many copies of a single type of message, and, consequently, there are many different types of *hn*RNA—hence heterogeneous RNA. As yet we have no hypothesis to explain this rather curious mechanism of *m*RNA production.

Genetic Redundancy and Mendelian Genetics

The preceding discussion has suggested that many genes of higher organisms may be redundant; some scientists believe that the majority may be. But redundancy is difficult to reconcile with the facts of Mendelism. Most genes behave as if they were single units. If most genes are redundant, why do we not observe phenotypes that are always a mixture of mutant and wild-type? Certainly some of the cistrons in a highly redundant locus would be mutant and some would be wild-type. We would only expect to have either all wild-type or all mutant in rare instances. Yet in most cases we can demonstrate that a gene is either one or the other. (We can also show that most mutations are not deficiencies.) Furthermore, studies of genetic recombination indicate that the gene is a unit that is not easily subdivided. If the loci were highly redundant, would we not expect to get mispairing similar to that observed in the *Bar* locus with consequent position effects (Chapter 16)?

Such questions are as unsettling as they are unresolved. A number of suggestions have been made to resolve the contradictions, but none are completely satisfactory and none have been tested. Perhaps the most intriguing suggestion is the *master-slave hypothesis.* In general terms, this hypothesis suggests that one copy of a redundant locus is the master copy and all other cistrons of the locus are checked against the master constantly. Mutations that occur in the slave copies are rectified by a repair mechanism (Chapter 23). When the master is mutated, all of the slaves are "corrected" to the mutant form. Presumably at the time of crossing over, only master copies pair and recombination carries the slaves along as appendages to the master. This is a neat hypothesis, but it is currently under attack from many sources. (Perhaps its most severe limitation is that it is so vague about the mechanism by which the correction process occurs.) We present it here only to put the problem in better focus and to suggest the kind of hypothesis that will be necessary in the future.

Genetic RNA

As the discussion of tobacco mosaic virus (Chapter 18) indicated, there are certain organisms in which RNA rather than DNA appears to act as the hereditary material. Since these organisms are relatively rare, we have largely sidestepped the issue. However, an explanation of the mechanism of hereditary in such organisms has been uncovered, and we are now in a position to resolve what was otherwise a rather sticky business. How can single-stranded RNA make an exact copy of itself?

Most of the RNA viruses that have been studied are relatively simple. They consist of a central core of singly stranded RNA surrounded by a protein coat that generally contains no more than two different kinds of protein. The first rather interesting observation about these phages is that, in a sense, they are manufactured backward. Unlike the bacteriophage, whose DNA replicates and condenses prior to the synthesis of the proteins that will encapsulate it, the RNA viruses produce proteins first and then replicate their hereditary material. Because of the relative simplicity of their structure, this process probably does not cause any serious problems for the spontaneous assembly of the new virus (as it might well for bacteriophage). But it did seem a little illogical until the discovery that one of the proteins synthesized from the infecting RNA was the enzyme, *RNA replicase*, that is responsible for replicating genetic material. In this sense, the life cycle is similar to that of the bacteriophages, which produce a unique DNA polymerase as one of the early functions in the life cycle.

Another aspect of the reverse life cycle of RNA viruses is that the singly stranded RNA molecule that acts as the infective agent can be used as *m*RNA. In DNA systems, it is necessary to transcribe *m*RNA for the production of the necessary proteins. With an RNA virus, the genome itself acts as an *m*RNA. The first stage of the life cycle involves the attachment of ribosomes of the infected cell to the viral RNA, followed by the complete reading of the genetic message so as to produce all the proteins necessary for the synthesis of intact virus particles (Figure 21-3).

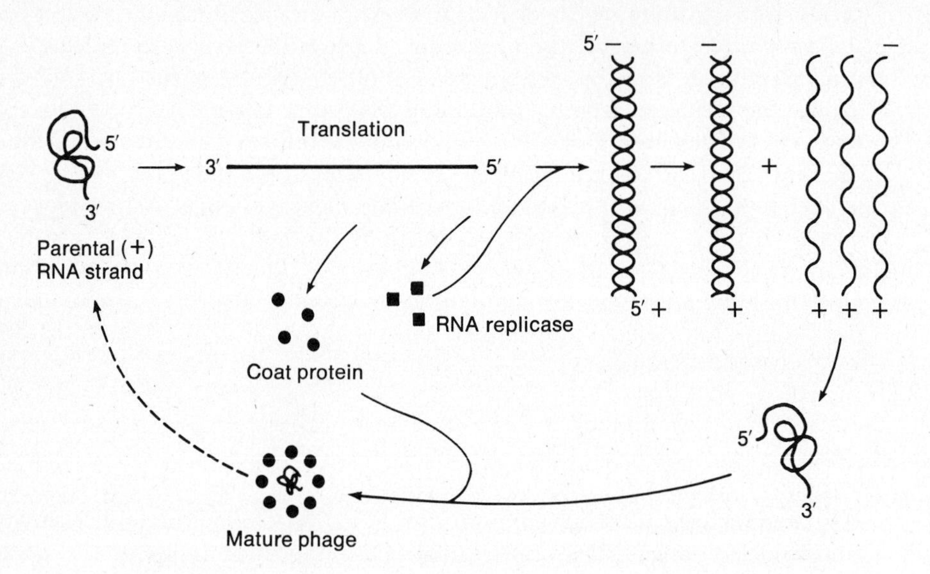

Figure 21-3. Reproduction of an RNA virus. The first event after infection is the translation of the parental (+) strand into protein. One of these, RNA replicase, first facilitates the production of a homologous (−) strand to make a double helix. The − strand is then used to produce a number of + strands that serve as the genetic material of the progeny.

Once RNA replicase has been produced, the genome is replicated. The normal mode of synthesis involves the initial formation of a complementary strand in order to generate a duplex RNA. If the initial infective strand is designated *plus*, the complementary strand then will be designated *minus*. It is the minus strand that is used to generate more infective plus strands. After the duplex has been formed, it appears that only the minus strand is read by the replicase, and a large number of infective plus strands are formed from each duplex. Once a plus strand has been generated, it may undergo encapsulation by the existing proteins to form infective virus. Alternatively, it may be translated by ribosomes, with the consequent production of more encapsulating proteins, more replicase, and, in turn, more infective RNA strands.

The RNA replicase is capable of carrying out all steps in the replication process by itself. Given purified, infective, singly stranded RNA and the four essential nucleotides (ATP, UTP, GTP, CTP), RNA replicase can synthesize infective plus strands of RNA in vitro; nothing else is needed.

So far the study of the RNA viruses indicates that there is only one method for replication in all biological systems—formation of a complementary chain, presumably involving explicit base-pairing arrangement, and the use of the duplex form to generate succeeding progeny molecules. Unlike DNA, the RNA duplex does not make exact copies of itself; rather, one strand is preferentially copied at the expense of the other, and the singly stranded infective agent results. However, this mode of reproduction is not unique to the RNA viruses. At least one DNA phage follows basically the same mode of replication (phage X174), a very small virus that contains singly stranded DNA. The advantage of the singly stranded condition in this phage remains obscure, but it does illustrate that as long as the potential for the replication of the complementary strand exists, one strand is satisfactory. However, beginning with a doubly stranded hereditary molecule is probably an advantage.

Reverse Transcriptase

It is not safe to assume that all RNA-containing viruses are replicated in the same manner. In fact, there may be several different mechanisms, depending upon the type of virus. The most interesting of these alternative mechanisms is that typical of the RNA viruses that cause tumors in mammalian and avian tissues, such as the Rous sarcoma virus. It has been known for some time that these viruses have a rather strange life cycle. After infection of the host cell, the replication of these viruses does not cause cell death but may cause neoplastic transformation. The transformed cells are genetically stable, always producing transformed daughter cells, and for generations they will continue to yield a small amount of the transforming virus. Both the stability of the transformation and the continued production of virus suggest that the viral genome is in some way associated with the host cell genome, but it is difficult to understand how an RNA molecule could have that property.

An explanation of the paradox was offered in 1964, when it was suggested that the RNA of the virus was converted into a DNA *provirus* and that this provirus was first incorporated into the host genome and then transcribed in the usual way to produce progeny viruses. Although this hypothesis was highly speculative at the time, supporting evidence soon appeared. The first finding was that actinomycin and other agents that block transcription of DNA also block the production of the RNA virus by the transformed cells. Then it was shown that a small amount of the DNA from transformed cells was homologous to (would anneal with) viral RNA; whereas DNA from untransformed, but otherwise identical cells lacked that homology. Since both findings are requirements of the hypothesis, it appeared to be more than idle speculation. However, there was no known mechanism by which RNA could produce DNA.

In 1970, two groups of investigators simultaneously reported the isolation of an RNA-dependent DNA polymerase from several different RNA tumor viruses. This enzyme is now generally called *reverse transcriptase.* It has the interesting property of using singly stranded RNA as the template for the production of a doubly stranded DNA. Apparently the DNA produced from the viral genome is incorporated into one of the host's chromosomes, because uninfected cells lack the ability to produce reverse transcriptase, but transformed cells have that ability.

We can now present an overview of the life cycle of the RNA tumor viruses. Initially they are absorbed by the cell, probably by an infolding of the cell membrane that finally discharges its contents into the cytoplasm. The viral proteins are stripped from the RNA, which migrates to the nucleus. The reverse transcriptase, which is bound to the viral RNA, then produces the DNA provirus; and that, in turn, is incorporated into the host genome. If the incorporation does not take place, there is no transformation. Apparently the transcription of the provirus DNA requires its prior integration with the host chromosome. In addition to the requirement for integration, there appear to be other factors that control the transcription of the provirus. The production of new RNA tumor viruses by transformed cells is entirely too sporadic to be explained by continuous transcription. (Some possible mechanisms for control are discussed in Chapter 24.) However, when it occurs, the production of new tumor viruses is by the usual transcription mechanism, producing both viral RNA and proteins, and by the self-assembly of these molecules into the infective agent. When fully assembled, the particles are extruded from the cell, without cell lysis; whereupon they are free to infect other cells

The life cycle of these viruses is unique. It stands somewhere between the lysogenic and lytic cycles common to bacteriophage. Its essential property is to produce a self-replicating entity, the provirus, that can continue to produce progeny. Since the progeny do not destroy the host cells, they can proliferate both by the continued release of viruses and by the division of the transformed cells. From the point of view of the virus, it is a most satisfactory situation.

The isolation of reverse transcriptase has an importance that goes beyond our understanding of the life cycle of the RNA tumor viruses. It provides us with the machinery for the in vitro synthesis of naked genes. It is now possible to isolate *m*RNA and use it as a template to make the corresponding stretch of DNA. The implications of this technique for both the study of individual genes and the potential incorporation of specific loci into organisms ("genetic surgery") are obvious.

REFERENCES: ORIGINAL PAPERS AND ADVANCED REVIEWS

BRENNER, S., A. O. W. STRETTON, and S. KAPLAN. 1965. Genetic code: the "nonsense" triplets for chain termination and their suppression. *Nature* **206**:994–998. (One of the first indications that suppression of mutation could occur by modified *t*RNA.)

BRITTEN, R. J., and D. E. KOHNE. 1968. Repeated sequences in DNA. *Science* **161**:529–540. (A very good review of the problem of genetic redundancy.)

BROWN, D. D., and J. B. GURDON. 1964. Absence of ribosomal RNA synthesis in the anucleolate mutant of *Xenopus laevis*. *Proc. Natl. Acad. Sci. U.S.* **51**:139–146. (The first proof that the nucleolar organizer was the seat of *r*RNA synthesis.)

GAREN, A., S. GAREN, and R. C. WILHELM. 1965. Suppressor genes for nonsense mutations. *J. Mol. Biol.* **14**:167–178. (Another paper on the suppression of amber mutants. In this one, the case is made for three distinct loci coding for a single *t*RNA.)

MILLER, O. L., and B. R. BEATTY. 1969. Portrait of a gene. *J. Cell. Physiol.* **74**: Suppl. 1:225–232. (This paper presents some elegant electron micrographs showing RNA polymerase attached to DNA and transcribing RNA. It also concerns the problem of gene amplification.)

RITOSSA, F. M., K. C. ATWOOD, and S. SPIEGELMAN. 1966. A molecular explanation of the *bobbed* mutants of Drosophila as partial deficiencies of "ribosomal" RNA. *Genetics* **54**:819–834. (A very nice paper showing the nucleolar organizer region of Drosophila.)

SPIEGELMAN, S., I. HARUNA, I. B. HOLLAND, G. BEAUDREAU, and D. MILLS. 1965. The synthesis of a self-propagating and infectious nucleic acid with a purified enzyme. *Proc. Natl. Acad. Sci. U.S.* **54**:919–927. (The first demonstration of the RNA replicase that promotes the replication of viral RNA.)

QUESTIONS AND PROBLEMS

21-1. Define the following terms:

nucleolus | nucleolar organizer
suppressor mutation | amber mutants
redundancy | gene amplification

21-2. Describe the generation of a ribosome.

21-3. How many cistrons are required to code for the three *r*RNA components?

21-4. What is the role of the unusual bases of the *t*RNA molecules? What is the special role of I (inosine) (see Chapter 20)?

21-5. How did the investigation of amber mutants help to elucidate the genetics of *t*RNA synthesis?

21-6. Why do suppressor mutations only partially suppress amber mutants?

21-7. When a mutant "reverts" to wild-type (or pseudowild-type), there can be three different explanations: back mutation at the site of the original change, mutation at another place within the mutant cistron (+/−), or a suppressor mutation at a completely different locus. How would you distinguish between these three possibilities?

21-8. What is the probable role of genetic redundancy?

21-9. When a locus is highly redundant, how is it possible to detect mutational events?

22/Chromosome structure in higher organisms

It is now common to divide all organisms into two groups: the *prokaryotes*, those organisms lacking a true nucleus and other cytoplasmic organelles, and the *eukaryotes*, those organisms having a true nucleus, mitochondria, plastids, and so on. The bacteria, viruses, and blue-green algae are prokaryotes. The rest are eukaryotes.

As the previous discussion of the genetics of bacteria and bacteriophage should have made clear, the prokaryotes have a chromosome composed of a single DNA molecule, frequently circular. Generally this chromosome floats free in the cytoplasm except for one part that is attached to the membrane of the cell. This point of attachment appears to be a replication center, as well as the mechanism for separating two daughter chromosomes at cell division. It is not strictly correct to make the analogy between the attachment point and the centromere of the chromosome of higher organisms, but there are at least some vague similarities. So far as we can tell, the chromosome of the prokaryote is relatively free of proteins; most of the proteins that are attached to it are those associated with its replication, transcription, and control (see Chapter 24). Consequently, it is reasonable to treat the prokaryotic chromosome as if it were a naked DNA molecule.

The chromosomes of higher organisms must be divided into two groups: those that we would call typical of the eukaryotes, having the chromosome structure generally described in Chapter 7, and those that we can term intermediate, the chromosomes of certain algae and protozoa in particular. The structure of this latter group is not well understood, but they seem to fall somewhere between the naked DNA of the prokaryote and the highly organized chromosome of the typical eukaryote. We will omit discussion of these intermediate forms; they are too poorly understood at this time to give us any new insights. What follows is a discussion of the problem of the chromosome of higher organisms, the "chromosomes" of Boveri and Morgan. Consequently, it would be wise for the student to review the terms developed in Chapter 7 before proceeding. Throughout this chapter we assume familiarity with that material.

DNA and Chromosomes

There are three major structural differences between the chromosome of the eukaryotes and that of the prokaryotes: size, presence of a centromere, and presence of large quantities of protein. In addition, most eukaryotes have several to many chromosomes, whereas the well-known prokaryotes have only one. Of these three factors, perhaps the size of the chromosomes is the most striking. It is possible to estimate the total amount of DNA in a normal diploid human nucleus and calculate the total length of double helix that this amount of DNA represents. It turns out to be almost 2 yards! On this same basis, there are roughly 4 centimeters of double helix per chromatid. It is obvious that no chromosome of that length is ever seen; thus we must envisage the chromosome either as being composed of multiple strands of DNA tightly packed together, much like a cable, or as composed of a single double helix that has been folded on itself many times so that it is greatly compacted. For a long time the multiple-strand hypothesis was dominant. This dominance was largely the result of experiments with light microscopy in which it is occasionally possible to see what seem to be subdivisions of chromatids, as well of experiments in which chromosome aberrations are induced by radiation, many of which suggested a multiple-stranded nature to the chromosome. As interesting and unexplained as these observations are, however, the multiple-stranded hypothesis is no longer the dominant view. Since the studies that showed semiconservative replication of the chromosomes (Chapter 18), it has seemed more reasonable to assume that each chromatid is a single double helix. This view is further supported by studies of mutation. Clearly, if the chromosome were a multistranded structure, it would be necessary to assume that each chromatid contained several copies of each gene—something that we might call lateral (side-by-side) redundancy. The mutational event, then, would be masked by the presence of the several other wild-type genes present, but such does not seem to be the case. If it were, Mendelian ratios would not fit chromosomal distribution so well. To avoid this difficulty, it would be necessary to propose that many of the strands were nonfunctional and that only one functioned. Such an hypothesis is too complicated, and certainly until we have more compelling evidence that the chromosome is multistranded, it seems far more reasonable to presume that each chromatid contains a single double helix of DNA.

The next question is whether the DNA is a long, continuous, single molecule stretching from one end of the chromatid through the centromere to the other end of the chromatid or a number of different DNA molecules of shorter length held together by "linkers" of protein. At present, the evidence seems to favor the single DNA molecule hypothesis. It has proved possible to isolate chromosome-sized molecules of DNA from both yeast and Drosophila. These molecules have molecular weights as high as 70×10^{10} in Drosophila. They are, of course, much smaller in the yeasts. The molecular weights of these isolated molecules correspond almost exactly to the amount of DNA present in the chromosome as indicated by other measurement techniques. Careful study of these molecules indicates that they are single, very long DNA molecules, with no protein present as a necessary part of their structure. Furthermore, it has been possible to carefully extract DNA from human white blood cells, using the technique previously described for bacterial chromosomes (Chapter 18). The largest intact molecules have lengths of approximately 2.2 centimeters, which is at least roughly equivalent to the total DNA content of one of the smaller human chromosomes. Today it is reasonable to assume that the failure to isolate longer DNA molecules, of the type that would be necessary to explain the structure of the largest human chromosomes, is largely due to technical difficulties. Single molecules of these dimensions are obviously very fragile; the fact that molecules as long as 2.2 centimeters can be prepared is highly suggestive that the chromosomes contain giant DNA molecules.

There is other, less direct evidence. Digestion of chromosomes with proteolytic enzymes does not result in breakage of the chromosome. Were the various DNA molecules held together by protein linkers, it would seem reasonable to assume that proteolytic enzymes could break up the chromosomal thread. Since they do not, it seems unlikely that any linkers in the chromosome structure are made of protein. On the other hand, treatment of chromosomes with DNase does destroy its integrity, which suggests that it is the DNA that is the backbone of the chromosomal structure.

On the presumption that the backbone of the chromatid is a single DNA double helix, we must assume that the proteins of the chromosome, which make up between 60 and 70% of the dry weight of human chromosomes, surround the double helix. Such a view is consistent with the action of proteolytic enzymes on chromosomes. These enzymes make the chromatids thinner, although they do not break them. Consequently, it appears that the enzymes digest away the proteins that cover the outside of the chromatid. In other words, chromosome structure is similar to that of viral structure—a thread of nucleic acid coated by proteins. In fact, the universality of this type of structure, a feature found as well in viruses and ribosomes, is at least suggestive evidence that this model of chromosome structure is probably correct. As far as we can tell from studies of replication of the chromosome, chromosomal proteins are made at the same time that the DNA replicates; and their association with the new DNA double helices is a spontaneous event, in much the same way that viruses assemble in the cells that they infect. We might even make the analogy that the metaphase chromosome, in its tightly compacted, functionally inert condition, is similar to a giant virus, which, on "infection" of each of the daughter cells, becomes functional and forms a new copy of itself, the rest of the cellular activities being managed only for the preservation and reproduction of the genome. Although merely an analogy it should at least make clear that similar mechanisms are involved in both cases.

Given the preceding general model for the structure of the chromatid, how can we explain the fact that even in the extended condition of interphase the diameter of the finest fibrils found ranges between 200 and 300 Å in diameter? A single double helix is only 20 Å. It would be necessary to surround the DNA with many layers of protein to make a fibril ten times as thick. This problem is compounded by the fact that the amount of DNA that can be measured in stretches of interphase chromosome is far too much for the length. Thus as suggested earlier in general terms, it seems necessary to propose that, even in the relatively extended condition of the interphase chromosome, the DNA is supercoiled rather than being a completely extended double helix. Quantitative estimates of the "packing ratio" (length of DNA double helix calculated from DNA content/length of chromatid) are in agreement with this general proposition. For example, the packing ratio of the giant salivary chromosomes of Drosophila is 50 : 1, and these are relatively extended, interphase chromosomes. In addition, electron micrographs of chromosomes are consistent with the hypothesis. In the regions of interphase chromosomes that can be visualized by present techniques, the DNA is relatively highly compacted, much less so that in metaphase, but still far from being an extended double helix.

At first it might seem extremely difficult to fold up a double helix. Its natural configuration is that of a rod. However, once molecules become excessively long, they can be folded. Such folds may involve local unpairing or even occasional breaks in the phosphate diester backbone of one or the other chain. The folding would be enhanced if proteins, or other molecules, were to interact with two regions of the folded DNA molecule to stabilize their position; here we presume that the bonds formed between proteins and DNA chains are greater in strength than the bonds holding the two chains of the helix together in some small region. In fact, there is considerable evidence for supercoiling of DNA from the prokaryotes. The DNA in the phage head is highly compacted; and as Figure 22-1(a) indicates, the amount of DNA that can be compacted into a very tiny phage head is truly enormous. We know that normally the compaction requires the participation of a special protein that presumably links the various parts of the single DNA molecule in a highly compacted mass. But such supercoiling can be induced with isolated bacteriophage DNA in vitro by alterations in the temperature or in the concentration of ions in the

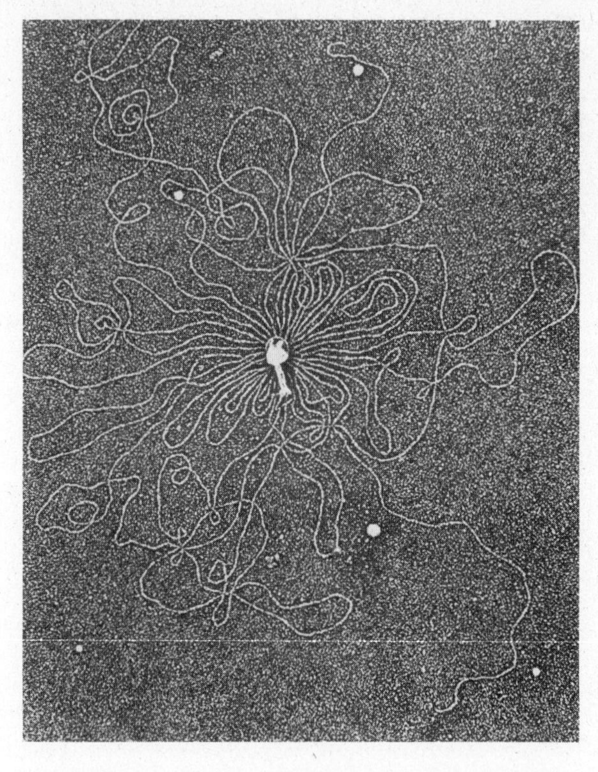

(a)

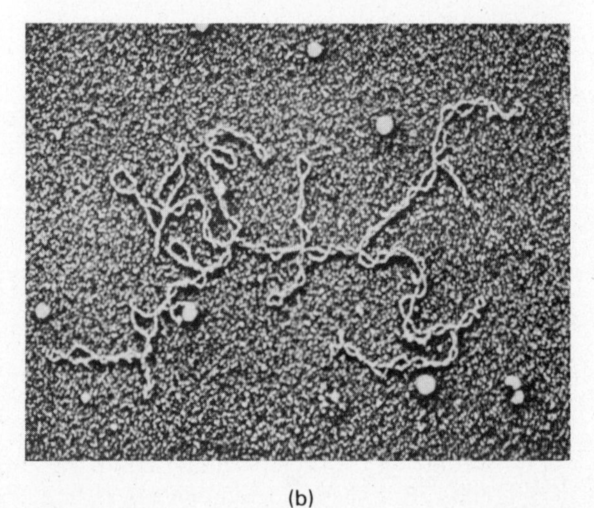

(b)

Figure 22-1. Supercoiled DNA. (a) An electron micrograph of a T_2 phage exploded by osmotic shock. The strands of DNA issuing from the phage head demonstrate how much DNA is compacted into the mature phage. The DNA shown is a single molecule. (From Kleinschmidt, A. K., D. Lang, D. Jacherts and R. K. Zahn. 1962. *Biochim. Biophys. Acta* **61**:857.) (b) The chromosome of λ phage. This chromosome is circular, but it has been highly twisted into supercoils until it resembles a wound-up rubber band. (From Bode, V. C., and L. A. MacHattie. 1968. *J. Mol. Biol.* **32**:673.)

medium. Figure 22-1(b) shows a rather highly coiled, isolated chromosome from bacteriophage λ. It should be noted that the chromosome is fully circular and intact but that is has folded on itself in a variety of ways so that it is already highly compacted. On the basis of such data, we must be willing to believe that the DNA double helix is not a rigid rod once it has passed a certain size and that a particular stretch of chromosome that we happen to view, even in the electron microscope, may well consist of a single DNA molecule that has folded back on itself several times to give the same thickness that would be observed in a multistranded structure. As the micrograph in Figure 22-1(b) shows, it would be relatively simple to form a structure containing four to six DNA strands lying next to one another in a highly compacted manner.

It is important to emphasize at this point that the coiling of the chromosomes that can be seen in the light microscope as they pass from interphase to metaphase is

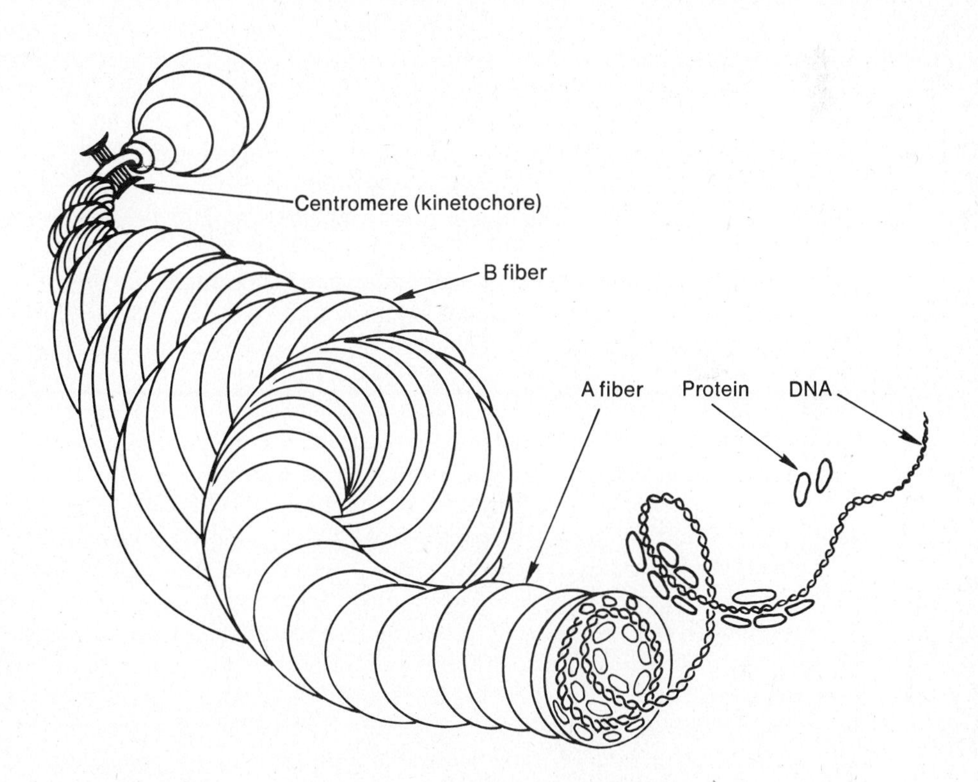

Figure 22-2. A model of the chromosome. The diagram shows a current interpretation of chromosome structure in the eukaryotes. Inside each chromosomal fiber, a single double helix is tightly packed, first by supercoiling to form a small fiber and then by supercoiling again to form a larger fiber. The DNA molecule is held in its supercoiled configuration by histones and other DNA-linked proteins that are shown on each surface.

several orders of magnitude beyond the supercoiling of DNA discussed in the preceding paragraph. What is being proposed here is that the finest filaments of the chromatid are composed of supercoiled DNA and in that sense may be multi-stranded. (Several regions of the DNA molecule are folded back on one another so that a number of strands are lying side by side.) Beyond that level of structure, yet more coiling can occur such that the supercoiled regions can themselves be coiled once or many times until the final dimensions of the metaphase chromosome are achieved.

Figure 22-2 is a diagram of the basic hypothesis for chromosome structure given in the preceding paragraphs. It is obviously diagrammatic, and the role of the proteins in the diagram is far from established. Nevertheless, it does present a clear description of what we mean by supercoiling. It demonstrates two levels of supercoiling of the DNA double helix: first, the coiling of the DNA occasioned by its packing with chromosomal proteins and, secondly, the coiling of that entire nucleoprotein fibril into a much thicker, highly compacted structure. The great advantage of this model is that it does explain the diameter of the fibrils of interphase chromosomes without having to invoke large quantities of proteins, several layers in thickness, coating the DNA, a model that is inconsistent with the ratio of protein to nucleic acid at any stage of the cell cycle. (The ratio ranges from 50 : 50 to 70 : 30.)

The first question raised by the chromosome model presented in Figure 22-2 is how replication occurs. The figure shows a single chromatid, presumably the state of the chromosome during interphase up to the period of the DNA synthesis. Figure 22-3 shows the generation of the duplicated chromosome from a single chromatid. For simplicity, the model begins with an unfolding of the chromatid at each end, with replication proceeding toward the center. A little thought will indicate that this is not an obligatory feature of the process. The unraveling of the DNA chain must first involve a local loss of the chromosomal proteins, followed by the separation of the two daughter strands. Once this process has happened, replication of the DNA could be achieved locally,with new chromosomal proteins then taking up their usual place around the daughter double helices to reestablish the supercoiled condition. Just as the middle part of the figure involves two different sites of replication, many sites of replication could be scattered throughout the chromosome at any one time. It is essential to make this point because the best evidence indicates that DNA replication of chromosomes of eukaryotes does not begin at one end and proceed straight through to the other. Some regions apparently are being replicated while others are still being transcribed. Thus the diagram is an oversimplification, but the essential point is made—the supercoiled structure can be as easily replicated as a naked DNA molecule, presuming only there is some trigger to cause the release of the proteins locally so that the replication process can occur.

Another interesting feature of the model presented in Figure 22-3 is that the region that is holding the chromatids together (i.e., the centromere) is shown as unreplicated. In fact, there is some very suggestive evidence that the DNA of the centromere region replicates very late and that without this replication the chromatids are not free to separate from one another. It is important to emphasize

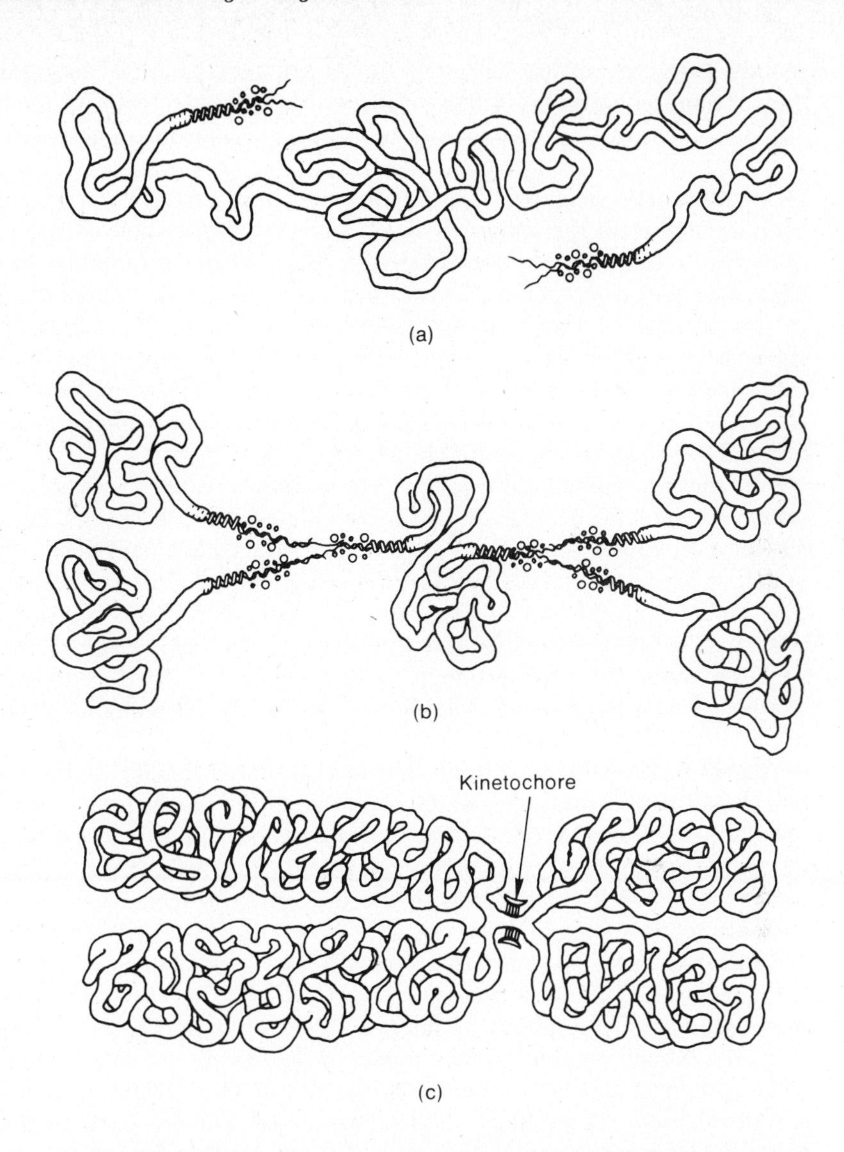

Figure 22-3. Folded fiber model of the chromosome. In (a) each chromatid is a single fiber containing a supercoiled, single DNA double helix (see Figure 22.2). During replication (b), the chromosomal proteins are detached, and the DNA is unwound. After replication, the DNA supercoils are reestablished by spontaneous interaction with the chromosomal proteins. For convenience, the process is shown as moving from the ends toward the center, but it may occur in several places at one time and move in both directions from various centers of initiation. In (c), a metaphase chromosome is shown. Only the centromere region is unreplicated. The supercoiling has produced a very compact chromosome. Note the disc-shaped kinetochores attached to the centromeric DNA. (After DuPraw, E. J. 1965. *Nature* **206**:338.)

that the lower part of the diagram is drawn deliberately and is not there simply for ease of representation. It is there to suggest that a special region of the DNA in the chromosomes of eukaryotes is associated with the centromere and that it is the late replication of this region that accounts for the final separation of the two chromatids at anaphase. It is suggestive that, in a number of cases, a special class of DNA has been isolated from the region of the centromere. It is an unusual DNA consisting only of AT. This unusual polyAT DNA has been isolated in quantity only from a few organisms, particularly crabs. However, there is now evidence that it may be found in the peculiar heterochromatic region that surrounds the centromere of the chromosomes of higher organisms. Such evidence supports the hypothesis that the centromere is actually a region of late replication of the DNA molecule and that the so-called division of the centromere is, in fact, merely the completion of replication of the DNA chain of the parent chromatid.

The preceding discussion of the centromere region makes necessary a statement about the centromere itself. As pointed out in Chapter 7, the centromere has for a long time been called the kinetochore or movement body, since it is the region to which the microtubules of the spindle attach. Electron microscopy now reveals that the centromere is a double structure. It contains two disk-shaped, or coin-shaped, bodies of unknown composition (but probably protein), which, strictly speaking, should be called the kinetochores. These protein bodies are not part of the chromatid in the strict sense. They lie adjacent to the DNA strand, one on each side of the chromosome (Figure 22-3). Each chromatid has its own kinetochore attached to it; thus the metaphase chromosome orients on the spindle by the attachment of one kinetochore to one pole and the other kinetochore to the other pole. The rest of the centromere is composed of the unreplicated strand of DNA, according to the model presented above. The older view that the kinetochore was a single structure that divided is not consistent with the best electron microscope evidence. This evidence tends to support the view that the centromere is more complex than we originally thought. However, the proposal given above is still only an hypothesis. Until there are further tests, the reader must be aware that the problems surrounding the centromere are still open to question.

It should be noted that the model presented in Figure 22-3 implies that the terminal end of the DNA double helix need not lie at the extreme end of the chromatid itself. The folding of the supercoiled structure in the higher levels of coiling can place the terminal ends in almost any position on either side of the centromere. This feature of the model is consistent with certain electron micrographs of a whole metaphase chromosome (Figure 22-4). It is evident that these chromatids are composed of extensively looped strands of a relatively uniform diameter. Since the diagram was made on the basis of electron micrographs of this type, the observation of metaphase chromosomes whose structure can be interpreted as that of a piece of yarn folded many, many times can hardly be considered as proof of the model. It is evident that electron micrographs of this type cannot give the detailed structure that would be necessary to make a final decision about the fine

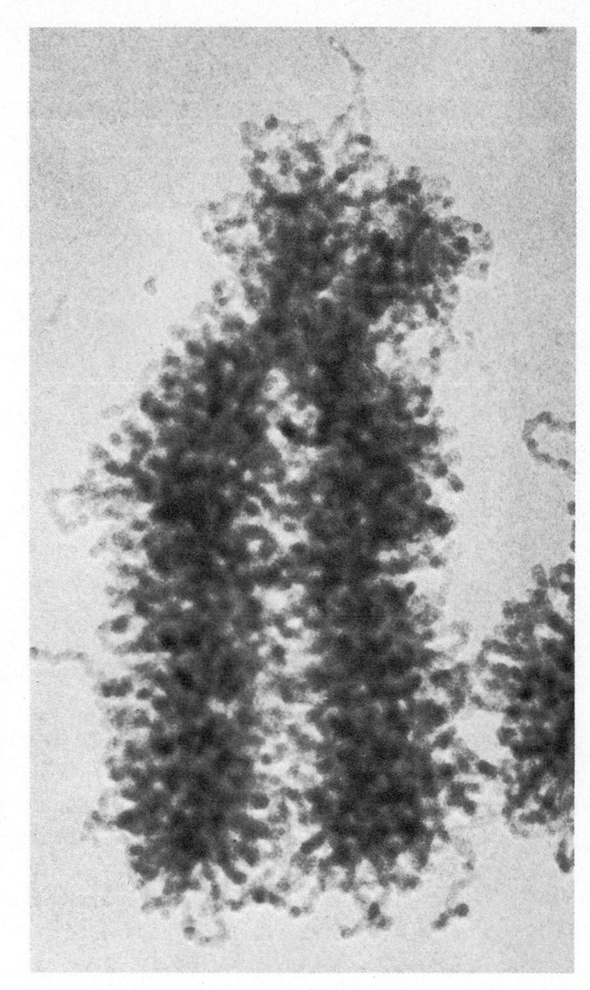

Figure 22-4. The supercoiled chromosome. A single, replicated, human chromosome is shown in this electron micrograph. Compare this to Figure 22.3 to see how the folded fiber model applies. Each chromatid could well be a single, very long, DNA-protein fiber that is folded many times, rather like a hank of yarn. The estimated length of the extended DNA double helix is about 3 cm; the length of the chromatid is only 0.02 cm. The many foldings of the fiber accounts for this great compaction. (From DuPraw, E. J., and G. F. Bahr. 1969. *Acta Cytol.* **13**:188. © 1969 by the International Academy of Cytology.)

structure of the chromosome. But they do indicate that the ends of the chromatids are hardly single-stranded structures. Combining the evidence from such micrographs with the evidence that suggests that the chromosome must be a single DNA double helix requires something at least similar to the diagram given in Figure 22-3. It is important for the reader to recognize, however, that the structure of the chromosome of higher organisms is as yet unresolved. The model presented here is one of many. It is consistent with most of the data, and it requires relatively little that is metaphysical. As such, it is a good hypothesis. On the other hand, it should be approached with caution. It is likely to be some time before we are certain about the exact structure of these chromosomes at the molecular level.

Heterochromatin

Originally the word *heterochromatin* was used to distinguish those regions of chromosomes, or whole chromosomes, that were darkly stained during interphase—that is, regions of the chromosomal material that stained out of phase with the rest of the material (*euchromatin*). As genetic studies progressed, it became apparent that heterochromatic regions carried few if any functional genes, and the emphasis shifted from the staining pattern of the chromosome to its genetic content. Still, it is safe to say that heterochromatic regions are best visualized by staining procedures and that the differences in staining ability imply an underlying difference in organization of chromosomal material. We now know that more darkly staining material is generally more highly concentrated; thus we must presume that the heterochromatic regions of the chromosomes are tightly compacted DNA, probably similar to that found in the phage head or in the metaphase chromosome. In such a highly compacted condition, we cannot expect the DNA to be transcribed, and so the absence of function in heterochromatic regions does not *necessarily* imply a lack of genes. Genetic material might well be present, which, because of the way it is organized, is unable to express itself. In this regard, it is significant that autoradiographic studies indicate little RNA synthesis in the region of the compacted DNA heterochromatin, whereas rapid incorporation is going on in the diffuse euchromatin. The role of the compaction of DNA will become clearer after we consider the problems of the control of gene action in Chapters 24 and 25.

Some evidence has been presented that large regions of the heterochromatin may contain a different kind of DNA. As suggested in the preceding section, the heterochromatic region closely associated with the centromere appears to contain DNA of a unusual type. This DNA contains highly repetitive, short segments with a high A/T content. There is other evidence that a similarly unusual DNA is associated with the nucleolar organizer. These are two regions of chromosomes that always appear to be heterochromatic. Furthermore, it has been possible to demonstrate that certain regions of the DNA appear to be highly redundant, having 10^4 to 10^5 copies of the same gene. These highly redundant regions also seem to be associated primarily with the heterochromatin. Such findings suggest the possibility that heterochromatic regions contain a high proportion of "nonsense" DNA. What the function of such DNA might be is hard to imagine. A case can be made for late-replicating DNA in the centromere as a control factor, but such late-replicating heterochromatin in other regions is not as easy to explain. It is interesting, however, that, in many cytological preparations, a tendency of the heterochromatic regions to clump together even when located on very different chromosomes is observed. This "attractiveness" suggests the possibility that they do contain similar stretches of DNA.

Finally, we must emphasize the need for caution in distinguishing eu- and heterochromatin. In discussing the level of function, the terms may well become confused. To take the most obvious case, consider the banded polytene chromosomes of the salivary glands of the Dipterans (Chapter 13). These chromosomes are truly laterally reduplicated, containing approximately 1024 (2^{10}) chromatids that are tightly synapsed. Consequently, these interphase chromosomes are visible in the light microscope. What is the difference between the darkly stained bands and the much more lightly stained interband region? In fact, there is a marked difference in DNA concentration. The bands contain a great deal of DNA and the interbands much less. Nevertheless, it is important to understand that the amount of DNA in the interband region is still much more than can be accounted for on the basis of a single double helix passing through that region. It is greatly extended compared to the banded region, but it still may be at least partially supercoiled.

The bands themselves appear to be composed of highly supercoiled, tightly compacted DNA. On the basis of what has been said earlier, it is expected that the band regions will show little genetic function in terms of the synthesis of *m*RNA, and such is the case. RNA synthesis goes on only in the regions where the DNA is uncoiled to form relatively swollen or "puffed" structures. Therefore it is reasonable to say that when the DNA is compacted it is inert transcriptionally, but when it is extended it resumes transcription. (This problem is reviewed in more detail in Chapters 24 and 25.) However, it is important to recognize that no one would be willing to call the banded regions of the salivary polytene chromosomes heterochromatin, since they are clearly the seat of a large number of genes, genes that apparently are not functional in the salivary gland but might well be functional in some other tissue.

The foregoing comments should be sufficient to caution the reader about the dangers implicit in the old formula—heterochromatin contains no genes, euchromatin does. What is heterochromatin in one circumstance, in fact, may be a highly compacted region containing many genes that are functional in another cell or under different conditions. The fact that a region of a chromosome stains darkly at interphase cannot be taken as complete proof that it contains no functional genes. Similarly, very small regions of functional DNA may well be surrounded by larger regions that are heterochromatic, tightly compacted, and nonfunctional With our present methods, it would be impossible to detect these regions. Therefore we can never be sure how much genetic activity is associated with a structure that is, by cytological techniques, called heterochromatin. And so we are led to the proposition that there are two kinds of heterochromatin: one that is always transcriptionally inert, such as that proposed for the region of the centromere, and another that is transcriptionally inert when compacted and transcriptionally functional when extended. How extensive this second kind of heterochromatin is remains for future investigations, but the material presented in the next section and in Chapter 25 should make clear that it may well encompass a very large part of the genome of any particular cell in a higher organism.

Lampbrush Chromosomes and Salivary Gland Chromosomes

Some of the most important observations of the chromosomes of higher organisms supporting the general model of chromosome structure presented above were made with the lampbrush chromosomes of amphibia. *Lampbrush chromosomes* can be found in the oocytes of a variety of organisms. The ones in the amphibia are some of the most spectacular, and they have been most extensively studied. These chromosomes occur in the prophase of meiosis. In the development of the egg of many animals, the cell is "arrested" in meiotic prophase. It is during this period, which may last for many days, even months, that much of the material stored in the egg cell is manufactured. It is a period of intense synthetic activity. In most such cases, fertilization of the egg acts as an activating mechanism to allow the completion of meiosis. But until fertilization occurs, the meiotic prophase stage can be considered an important synthetic stage. Figure 22-5 presents a photomicrograph of paired homologs in the lampbrush configuration. The name *lampbrush* came from their generally fuzzy or brushlike appearance. An interpretation of the fine structure of these chromosomes is given in Figure 22-6. As indicated in the diagram, the lateral

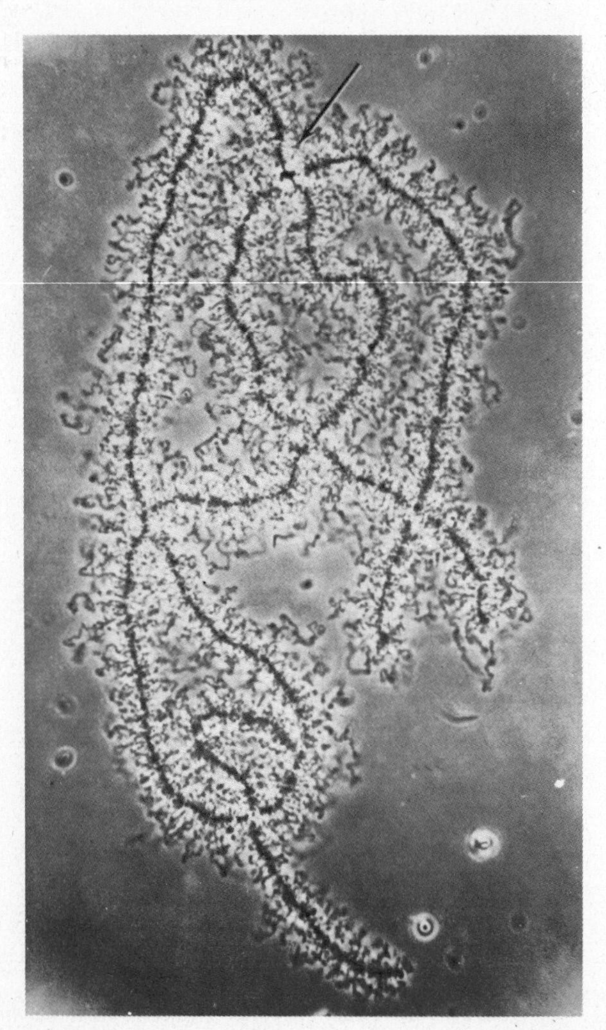

Figure 22-5. Amphibian lampbrush chromosomes. There are two homologous chromosomes united by three chiasmata and by a "clumping" of the centromeres (arrow). The looping out of the material in each homologue produces the fuzzy or "lampbrush" appearance. The loops are interpreted in Figure 22.6. (From Gall, J. 1966. *Methods in Cell Physiology*, vol. 2. D. M. Prescott, ed. Academic Press, New York.)

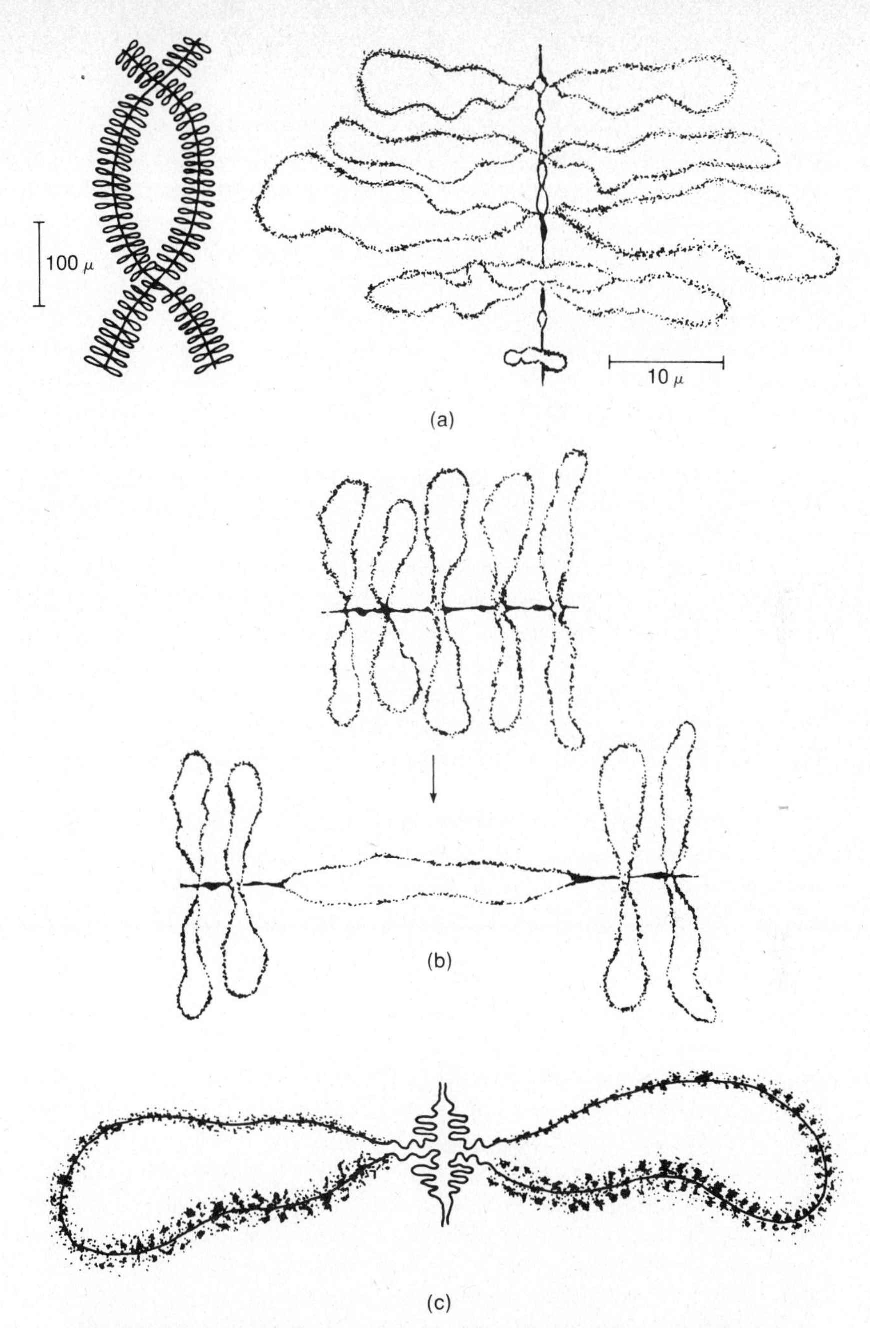

Figure 22-6. Lampbrush chromosomes. (a) A diagram of two meiotic chromosomes joined by two chiasmata. To the right is given a drawing of a region of one chromosome indicating that the loops on each side are identical but that one loop may vary considerably from its neighbors. (b) Stretching the chromosomes reveals that the loops are part of the continuous chromosome structure rather than attached to the main thread. (c) Diagram of the probable structure of a single loop pair, showing RNA chains being transcribed. Compare to Figure 21.2. (After Gall, J. 1955. *Brookhaven Symp. Biol.* **8**:17.)

extensions from the main body of the chromosome are, in fact, loops of material in the highly extended condition. Each of these loops appears to be an unraveling of the very tight structure of the chromomere that is normally characteristic of meiotic stages. This lateral extension of the chromomere is interesting, since the peculiarity of these particular cells is that they are actively synthesizing protein. (In most organisms, the cells in meiotic prophase are not actively synthesizing materials; and even in the male line of the amphibians, there is no such delayed stage and no extension of the chromomeres in this way.) It has thus been suggested that the looped-out portions of the chromomeres are regions of active transcription, the genetic content being read for the production of protein to be stored in the egg cell. The active incorporation of RNA nucleotides into these regions, plus other evidence, indicates that the suggestion is true; and it is to be presumed that the lampbrush chromosomes give us direct evidence of the unpacking of DNA that is required for genetic activity.

One obvious question concerns whether the loops represent only those genes that are active in protein synthesis, the rest of the chromosome (chromomeres and interchromomere filaments) being transcriptionally inert. To a certain extent, the answer must be yes, since all of the incorporation of RNA precursors occurs in the looped regions themselves. On the other hand, some interesting evidence suggests that the synthesis of RNA in the loops is not completely symmetrical. As indicated in Figure 22-6(c), it appears that the transcription of RNA begins at one end of the loop and ends at the other end, which might be the case if the entire loop were transcribed as a single *m*RNA chain. However, the loop is too large and the length of the chains that can be visualized extending from the loops are far too short to have been the product of the transcription of the entire loop as a single message. Therefore it seems much more probable that many *m*RNA molecules are transcribed from a single loop.

A satisfactory interpretation is that the difference in labeling pattern results from the fact that at one end of the loop (for simplicity, we will call it the forward end) synthetic activity has just begun, and at the other end of the loop it has been proceeding for some time. Studies of the incorporation of label with time indicate that this situation does occur and suggest that, in fact, the loops are moving along the chromosome, with the forward end of the chromomere unraveling and the other end coiling up again to reestablish the chromomere. In this manner, there would be a gradual unraveling of the entire chromomere so that the entire unit would be genetically active at different times of the meiotic prophase. Whether the chromomeres themselves are moving along the chromosome or whether this is only a type of sequential "puffing" of the chromomere proper is uncertain, but the latter possibility seems more likely. In the very extended meiotic prophase of these cells, it would be possible to transcribe the entire DNA content of the nucleus, but the evidence so far does not permit us to conclude whether complete transcription occurs or whether only certain regions of certain chromomeres are transcribed.

The finding that looping out of the chromomere is necessary for the transcription of the message agrees with most of the propositions advanced so far. Under normal conditions the DNA is highly compacted, much too tightly compressed to permit transcription, and the activity of particular genetic regions must involve a great extension of the DNA. It is an interesting fact that an addition of certain kinds of chromosomal proteins, particularly histones, causes an immediate contraction of the loops and, with that contraction, a cessation of incorporation of RNA precursors. It thus appears that an interaction between the DNA and chromosomal protein is responsible for the highly condensed condition of the normal chromosome. Presumably it is the freeing of the chromosome from these proteins, by some mechanism, that permits the DNA to extend and thus be transcribed. We will return to this question in Chapters 24 and 25, when we consider genetic control in more detail.

A rather interesting observation has been made with the Y chromosome of Drosophila. As noted, the Y chromosome is heterochromatic and essentially genetically inert. It carries few, if any, alleles of genes on the X chromosome. On the other hand, XO individuals, although perfectly healthy, are sterile. Thus we know that the Y chromosome is the seat of certain genes controlling fertility. It has now been observed that, during meiotic prophase in the male, the Y chromosome assumes a lampbrushlike configuration having at least five paired loops. These loops, apparently, are the genes necessary for the differentiation of functional sperm, the so-called fertility factors. Studies of deleted or translocated Y chromosomes show that five specific regions are associated with the loops and that the loss of any one results in nonfunctional sperm. Thus it appears that, in the normally heterochromatic, transcriptionally inert Y chromosome of Drosophila, the fertility genes are unlocked during the period just prior to spermatogenesis. Once again, this observation suggests that compacted DNA (the Y chromosome contains a great deal of DNA) is inactive and that, in order for activity to be restored, unpacking is required.

Finally, we should note the analogy between the lampbrush configuration and the swelling of the polytene salivary chromosomes of the Diptera. These "puffed" regions are known to be regions of intense genetic activity (see Chapter 25). The development of a puff begins with an initial loss of structure of a tightly banded region, followed by a gradual enlargement of that region to produce the puff. Figure 22-7 is a diagrammatic interpretation of these events. It conceives of the band as a tightly compacted mass (as indicated in our previous discussion) that unravels to give an enormous extension of the DNA chains. These chains loop out in much the same manner as the lampbrush loops to provide an area that can become transcriptionally active.

In the three cases discussed here—the salivary puffs, the heterochromatic Y, and the lampbrush chromosomes of the amphibians—it would appear that compacted DNA is functionally inert and that the extension of that compacted DNA is required for transcription. This observation is in complete agreement with the fact that in all cells the interphase chromosomes are greatly extended. This is the period of maximum genetic activity. The compaction of the chromosomes at metaphase stops transcription. Consequently, we must envisage the structure of the chromo-

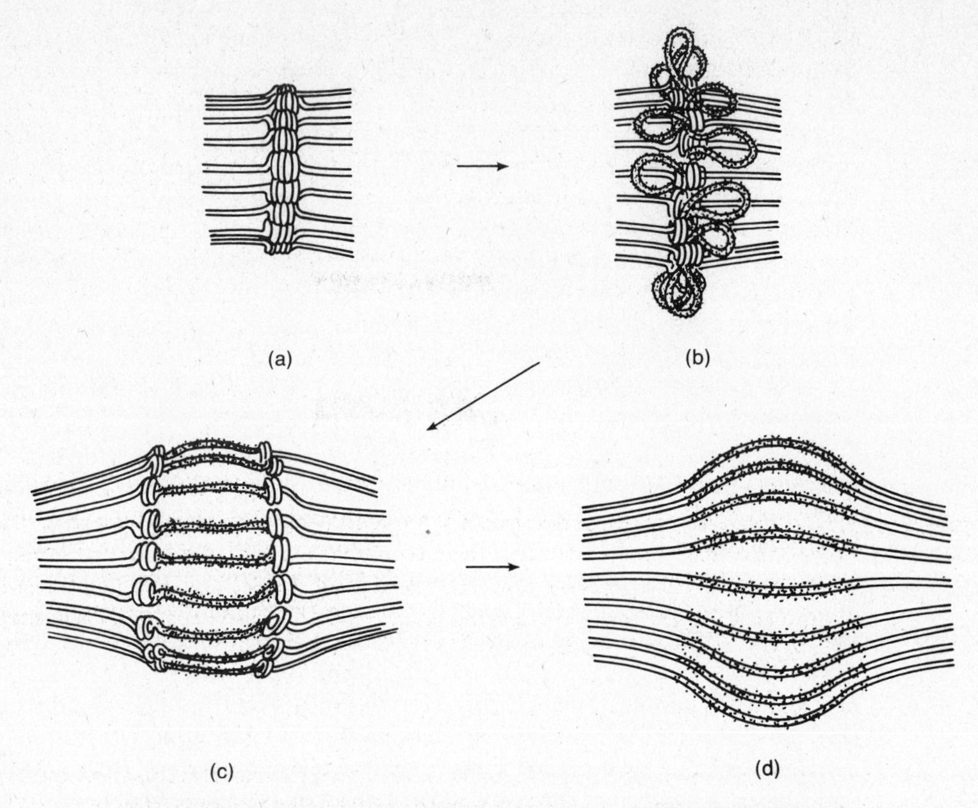

Figure 22-7. Puff formation in polytene chromosomes. (a) The "band" region of a salivary chromosome is interpreted as being a supercoiled region. The apposition of many such supercoiled chromomeres gives rise to the dark staining characteristic of a band. (b) The chromomeres begin to unravel, looping out from the body of the chromosome. (c) The region enlarges greatly as the supercoils relax. (d) The puff is complete. The dark granules on the various strands represent RNA polymerase units transcribing the newly extended DNA into RNA. (After Swift, H. 1962. *The Molecular Control of Cellular Activity*. J. M. Allen, ed. McGraw-Hill Book Co., New York. Used with permission of McGraw-Hill Book Company.)

some of the higher organisms as differentially extended. During certain periods some regions of the DNA will be greatly distended, probably as very fine filaments, much too fine to be seen in the electron micrographs we can presently prepare. But the remainder of the chromosome will be highly compacted DNA sheathed in protein, transcriptionally inactive for the moment but, nevertheless, always potentially capable of genetic function under a different set of conditions.

Excess DNA

The preceding section provides a context for the fact that the amounts of DNA found in higher organisms are greatly in excess of what they appear to need. Man has over

700 times as much DNA as *E. coli.* Even more remarkable, Tradescantia, a garden plant, has ten times more DNA than does man. In general, there appears to be little correlation between phylogenetic position in the evolutionary scheme and DNA content per nucleus. If we plot the *minimum* amount of DNA found in a particular phylum of organisms against the position of that phylum in the phylogenetic scheme, there is a rough correlation. Bacteria have the least and mammals the most. On the other hand, the extreme variation from this pattern observed when we select the data on any other basis makes the relationship seem of little importance. For example, certain amphibia have 25 times as much DNA as man, and it is found that very closely related species may differ in DNA content by a factor of 10. We must assume that there is far more DNA than necessary in most eukaryotes.

One part of this DNA can be accounted for as material of no real importance to the organism, possibly material that has lost its function over a period of time and is merely carried along by the regular events of mitosis and meiosis. For example, the B chromosome of maize is largely heterochromatic and can be gained or lost with little apparent effect on the progeny. Similarly, a large part of the Y chromosome of Drosophila seems to be truly inert. In mice, as well as in other mammals there is a satellite DNA of very large molecular weight (10^8) that is not transcribed into RNA. It appears to be present in a number of places in the genome, scattered amongst the chromosomes, and, as suggested above, may be some special kind of heterochromatin conceivably used for chromosomal pairing or some other function.

On the other hand, it is not reasonable to presume that *E. coli* and man should have the same DNA content, even though the cells of *E. coli* have most of the enzymatic functions routinely required by any cell. It is important to realize that *E. coli* does not make the eye pigment visual purple; nor does it carry the information for the specific proteins that permit the selective attachment of primordial germ cells to the gonad after their long migration; nor does it make the keratin of hair. In fact, adding together all the things that the cells of the mammal can do, it would take a good deal of DNA to supply the code and the control mechanisms. Since each cell of the organism has the complete genome, most of the DNA will be unnecessary (heterochromatic?) to its specific needs. We may assume that large regions of the interphase chromosomes are inert in any particular cell. Only if we were to add the activity of all the cells of the organism together would we expect to get a true picture of the amount of active DNA. One of the important problems that we must unravel is how organisms differentiate such that certain genes are expressed in one cell and a different set of genes are expressed in another. This is a problem we will return to in Chapters 24 and 25.

Another source of the "excess" DNA is genetic redundancy. As indicated in Chapter 21, the region coding for *r*RNA synthesis is multiply reduplicated. There are many copies of the *r*RNA cistron in tandem in that region. The same seems to be the case for a number of other genes. There is now some rather compelling evidence that the chromomeres of meiotic chromosomes and the bands of the salivary chromosomes represent such tandemly duplicated cistrons. We have long known that there is far too much DNA in both chromomeres and bands to represent a single copy of a single gene. Nevertheless, both behave as sites of single genes. Now part of the contradiction seems to be removed by the discovery that a band contains multiple copies of the same gene.

Chromosomal Pairing and Crossing Over

The foregoing model of chromosome structure poses a certain problem for the general concepts of chromosomal pairing during meiosis. The original view of the process was that pairing occurred on a gene for gene basis, each locus in one chromosome pairing with its allele on the homologous chromosome. This view was consistent with the cytological observations that could be made with the light microscope. In fact, the chromosomes do seem to line up chromomere for chromomere during meiosis. Furthermore, in that extremely useful case, the giant salivary chromosomes of Drosophila, which show somatic synapsis, the two chromosomes apparently line up band for band.

If we are to take the model of chromosome structure as a folded, supercoiled, single DNA double helix as correct, we must assume that even in the most intimate pairing that we can see, in the giant polytene chromosomes, a considerable amount of DNA is unpaired. To understand this situation, consider two door springs tightly coiled and lying side by side. Obviously the amount of material that can come into contact with the two is rather restricted. Some parts of each loop will touch, but the majority of the loops cannot. Of course, this model is too simplified, since it is possible to intercalate the two door springs and thus bring much more of the material into close pairing (simply push the two door springs together). But the model we have proposed for the chromosome in the preceding sections suggests that no such simple helical pattern is the rule. The greatly convoluted structure that is required to compact the DNA to the extent that it must be in a polytene band or in the chromomeres of a meiotic prophase chromosome (packing ratios as high as 50 can be observed in these cases, indicating that 50 units of DNA are compressed into the space normally occupied by one unit of extended double helix) makes it difficult to conceive of close pairing gene by gene, much less nucleotide by nucleotide, along the entire length of the chromosome. It is far more probable that chromosomes synapse so that certain regions are in contact and the remainder of the DNA threads do not pair. Figure 22-8 is a diagram of the imagined way that chromosomes synapse at meiosis. For clarity, even this figure shows a chromosome that is entirely too extended. Nevertheless, the point should be clear that we must modify our view of the nature of the synaptic mechanism.

The model of chromosomal pairing presented in Figure 22-8 places a new significance on the *synaptinemal complex.* In Chapter 7 we merely asserted that this structure can be seen between synapsed homologs. We should now elaborate on this point somewhat. The composition of the core material of the complex is not known in detail, but it is most probably protein.

In general, the synaptinemal complex appears to form after the homologs have paired. However, in certain instances, it has been shown that pairing and

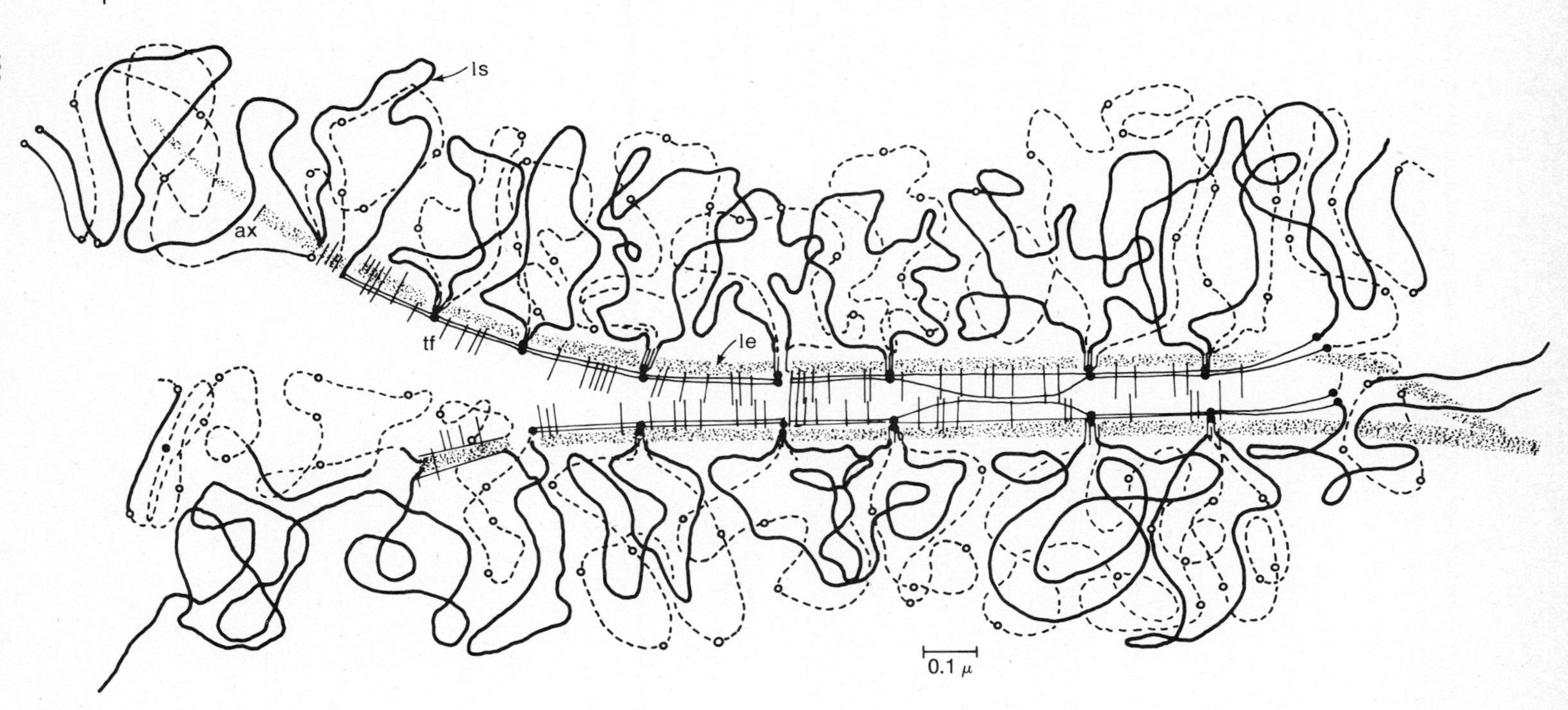

Figure 22-8. The synaptinemal complex. To the left are two unpaired homologues in which the axial cores (*ax*) are being formed. Two alternative modes of formation are shown: the upper core is synthesized *de novo* and the chromosomal strands become associated with it; the lower core is formed from preexisting histones. The loops are 100–200 Å DNA-protein lateral strands (*ls*) of the chromatin. The 600 Å transverse filaments (*tf*)of the two cores meet and interdigitate, locking the lateral elements (*le*) at a distance of 1000 Å from each other. In the center of the complex is shown a hypothetical arrangement of strands at an exchange. To the right is shown the dissociation of genetic material from the complex at the end of pachytene in meiosis. (After Moens, P. B. 1968. *Chromosoma* **23**:418.)

complex formation are simultaneous events. It is unlikely that the complex is the active factor in the synaptic process. There is no synaptinemal complex in somatic synapsis common in the Diptera, and it is lacking even in meiosis in those cases where there is no crossing over. These facts have led to the reasonable postulate that the function of the complex is to facilitate genetic recombination during meiotic prophase. This hypothesis is supported by the observation that agents that reduce crossing over also reduce complex formation. It is also supported by observations of organisms with highly localized chiasma formation. In these forms, there is also a localization of the synaptinemal complex exactly in the regions of chiasma formation.

The fact that strands of DNA from each chromosome penetrate the complex and join in the center is consistent with the hypothesis and has suggested further that the complex stabilizes the regions of close pairing until exchange of homologous parts has been completed. This is all highly suggestive, but it is far from proof. Much more evidence will have to be accumulated about the complex before a definite function can be ascribed to it. Nevertheless, the present evidence does not support the contention that the complex causes chromosomal pairing. It is far more probable that there are special pairing regions in the chromosome that account for the synapsis of homologs during meiosis.

It is one thing to propose pairing regions in chromosomes and another to tell what these regions are. Entirely too many possibilities exist. They could be minute regions of special heterochromatin (untranscribed regions), or they could be particular kinds of nucleoprotein complexes. There is a certain attractiveness to the heterochromatin hypothesis, for it is well known that heterochromatic regions can cause nonhomologous pairing of chromosomes. And since there are certain notable cases in which synapsis is not exact (occasional observations of nonhomologous pairing of one chromosome with regions of itself and so on), a certain general lack of specificity might be invoked. On other hand, the synaptic mechanism is too uniform to be based on large regions of redundant DNA with each region being the same. If certain regions of heterochromatin are, in fact, used as pairing loci, it is probable that they differ from one another in their nucleotide composition. It is important to recognize that not much is required to give the needed specificity in regions where close approximation of the DNA chains can be effected. A stretch of ten nucleotides is more than sufficient to provide the variety of unique pairing sites that would be required for complete synapsis of any genome. (With four bases and a ten-nucleotide sequence, 4^{10}, or in excess of 10^6, different combinations are available.) Obviously other ways of having exact pairing regions exist. However, base pairing is very exact, and it seems likely that the DNA contains within itself the guides to the pairing during synapsis.

It is important to note that certain mutations prevent proper synapsis. These mutants could either be in genes that control the formation of synaptinemal core material or in the "pairing loci" themselves. There is indirect evidence for both. In many cases, failure of the synaptinemal complex seems the most reasonable explanation. On the other hand, in Drosophila, it is possible to produce local synaptic failure in cases of somatic synapsis where normally no synaptinemal complex is found. In this case, there appear to be discrete "pairing loci" scattered along the chromosome whose loss prevents satisfactory pairing. Unfortunately, the biochemical analysis of these mutational events has not proceeded as fast as it should, and we do not know what such mutants lack. Presumably, when that analysis is complete, we will be in a better position to say what the pairing regions may be like. Until that time we will have to consider the pairing hypotheses as speculative and somewhat controversial.

The proposed model of chromosome structure must be reconciled to what we know about genetic crossing over. In the next chapter we will deal with what is known of the molecular basis of recombination. For the moment, let us consider only the special problems relating to the chromosomal structure of higher organisms. It is firmly established that crossing over involves the exchange of parts of homologous chromosomes (Chapter 11). Two very distinct models for this event have been proposed. The first is breakage and exchange, and the second is a mechanism that was originally called copy choice. If we were to relate the first to the present model of chromosome structure, breakage and exchange would involve the breakage of both strands of the DNA double helix of one chromatid at precisely the same locus that the two strands are broken in the homologous chromatid. This process would be followed by the reciprocal exchange of the two parts and their reattachment in precisely the right manner. If this is the mechanism of recombination, it must be precise; otherwise the region of the gene in which the exchange took place would no longer have the same reading sequence in its nucleotides and every recombinational event would be a mutation.

Is breakage and exchange a mechanism compatible with the structure of the chromosome proposed earlier in this chapter? The answer is yes, if parts can be exchanged in an exact way. We have proposed that the supercoiling of DNA that is responsible for the visible structure of the chromosome, including chromomeres and bands, results from an interaction of specific regions of the DNA with specific chromosomal proteins. Switching of one part of the DNA double helix to another chromosome, even if the switching is made of naked DNA molecules, will still result in a transfer of the general morphology of the chromosome. The DNA, finding itself in a new chromosome, will still organize its own structure on the basis of its intrinsic nucleotide sequence.

The same thing is true for a copy choice mechanism. If genetic recombination occurred at the time of DNA replication, from some misreading such that for one stretch of DNA one molecule is read and for another stretch the homologous molecule is read, the result would still be the exchange of parts of the DNA molecule. With that exchange must go the exchange of chromosomal morphology. This view simply states that in order to make what we know of chromosome structure compatible with what we know of crossing over, the fundamental organization of the chromosome must primarily depend on the nucleotide sequence within the DNA. This restriction is far from being unreasonable, and thus there is no inconsistency in the interpretation.

On the other hand, what we have seen about the nature of meiotic pairing does raise questions about the time of genetic recombination. There is a fairly good correspondence between the cytological chromosome map for gene placement and that obtained by mapping through crossing over. This fact is not inconsistent with the general model given above, provided only that the synapsis of chromosomal material *at the time of crossing over* is exact. Conversely, the model presented in Figure 22-8 does suggest that extensive regions of the chromosome might never cross over, which seems highly unlikely. No two genes with really different functions have been found to map at exactly the same place, as would be expected if some sets of genes never synapsed and, therefore, never recombined. Of course, it is possible that there are long stretches of inert DNA between genes. Failure of synapsis in these regions would not affect our genetic analysis, since no exchange of genetic markers could occur here in any case. But somehow this seems rather farfetched.

A more reasonable hypothesis is that once initial pairing homologs are established, the paired chromosomes can undergo a more intimate association in some regions. Pairing in these regions would be nucleotide by nucleotide and would account for the recombinations *within* genes that is the basis for fine-structure analysis. All that is needed is to assume that the supercoiling of the DNA is a variable process, so that the folding of the chains would bring one set of loci close together in one case and another set close together in another case. In one case, crossing over in one region would be detected; in the other case, we would detect crossing over in a different region. Variations in crossing over frequency, then, would be due both to the distance between two genes and to the probability of synapsis of certain regions. This latter part of the hypothesis has the advantage of explaining the discrepancies between the genetic and cytological maps, and it also might explain why so many environmental agents affect crossing over. If the supercoiling of the DNA is a rather labile process, changes in temperature, exposure to radiation, and so forth, could well shift the pattern of synapsed regions and thus alter the frequency of crossing over.

We must point out that the model presented above says little about the time in the cell cycle at which crossing over occurs. There is no requirement that it occur in meiotic prophase. The interphase chromosome is a supercoil that would have all the features we have discussed as characteristic of the synapsed homologs. It is only necessary to propose that some form of synapsis can occur prior to prophase. Consequently, it would be reasonable to propose that what we see as chiasmata are visible results of recombinational events that have taken place sometime before. In this case, the synapsis of homologs at meiosis would merely represent the attraction of specific pairing loci to a synaptinemal complex to provide a mechanism for chromosome reduction and would have little relation to crossing over. However, as we shall see when considering recombination in the next chapter, it is probable that crossing over *does* occur in the prophase of meiosis after the chromosomes have synapsed.

REFERENCES

COLD SPRING HARBOR SYMP. QUANT. BIOL. **38**. 1974. Chromosome structure and function. (A collection of review articles giving the most recent data on chromosome structure. The last article is a good general review of the volume and should be read first. For the advanced student.)

DUPRAW, E. J. 1970. *DNA and Chromosomes.* Holt, Rinehart and Winston, Inc., New York. (A good compilation of data and hypotheses. It presents the basis for the folded-fiber model.)

SWANSON, C. P. 1957. *Cytology and Cytogenetics.* Prentice-Hall, Englewood Cliffs, N.J. (Excellent as a review of some of the older literature. Chapters 5 and 11 are most pertinent.)

SWANSON, C. P., T. MERZ, and W. J. YOUNG. 1967. *Cytogenetics.* Prentice-Hall, Englewood Cliffs, N.J. (A very good, simple account of chromosome structure. See particularly Chapters 2 and 7.)

Original Papers and Advanced Reviews

BEADLE, G. W. 1928. Genetical and cytological studies of Mendelian asynapsis in *Zea mays. Cornell Univ. Agri. Expt. Station Memoir* **129**:1–23. (The first paper to demonstrate that the pairing of homologous chromosomes is under genetic control.)

CALLAN, H. G. 1963. The nature of lampbrush chromosomes. *Internat. Rev. Cytol.* **15**:1–34. (A review of the structure and function of lampbrush chromosomes.)

COOPER, K. W. 1964. Meiotic connective elements not involving chiasmata. *Proc. Natl. Acad. Sci. U.S.* **52**:1248–1255. (This paper presents evidence for the existence of specific pairing sites in Drosophila chromosomes.)

HESS, O. 1968. The function of lampbrush loops formed by the Y chromosome of *Drosophila hydei* in spermatocyte nuclei. *Molec. Gen. Genetics* **103**:58–71. (This paper indicates that when compressed, heterochromatic, chromosomal material is relaxed, it becomes functional.)

MOENS, P. B. 1973. Mechanisms of chromosome synapsis at meiotic prophase. *Internat. Rev. Cytol.* **35**:117–134. (A review article that provides a recent account of work on the synaptinemal complex.)

SWANSON, C. P. 1947. X-ray and ultraviolet studies on pollen tube chromosomes. II. The quadripartite structure of the prophase chromosomes of Tradescantia. *Proc. Natl. Acad. Sci. U.S.* **33**:229–232. (One of the papers whose findings are hard to reconcile with the single DNA molecule model of the chromosome.)

QUESTIONS AND PROBLEMS

22-1. Define the following terms:

prokaryote	eukaryote
supercoiling	heterochromatin
polytene chromosome	lampbrush chromosome
chromomere	

22-2. What is the evidence that the backbone of the chromosome is a single DNA molecule?

22-3. What does the "packing ratio" suggest about the state of DNA in the chromosome?

22-4. Why do we think that heterochromatic regions are highly compacted DNA?

22-5. What is the probable reason for the "swelling" of the loops in lampbrush chromosomes? Is this "swelling" found in other chromosomes?

22-6. Careful studies of the DNA content of lampbrush chromosomes indicate that there is a single double helix in the loop regions and two helices in the regions between the loops. How do you explain this finding?

22-7. Two closely related species of toads have the same chromosome number but their DNA content differs by about 40 %. How do you explain this situation?

22-8. What is the role of the "excess" DNA found in many organisms?

22-9. What is the role of the synaptinemal complex in the pairing of homologs?

22-10. Why do we think that gene-by-gene pairing occurs, even though the coiled state of the chromosome would seem to make that process very difficult?

23 / Mutation and recombination

We have delayed a discussion of mutation and recombination until now because it is better to have an understanding of the structure of the chromosome of higher organisms before considering the processes in detail. Both processes are complicated at the molecular level, and although a great deal is known about mutation, there are still some important questions. It is safe to say that recombination is not really understood. Nevertheless, what is known of the processes suggests that they have some mechanisms in common, and, for this reason, we have chosen to discuss them together. What follows is a rather brief discussion of mutation, emphasizing those mechanisms that are also essential to modern hypotheses of the mechanism of recombination. There are a number of detailed treatises on the mutation process itself, which has produced perhaps the greatest body of literature in genetics. Some are referred to at the end of the chapter, and the reader who is interested in a detailed analysis of the problem should consult those books.

A General Consideration of Mutation

Classical studies of mutation were discussed in Chapter 14, and a sketchy interpretation of the molecular events leading to mutation was presented in Chapter 20. At this point we would like to expand on the general subject in order to make the picture more complete and more unified. Let us begin by asking what is really measured when we attempt to determine mutation frequency. It should be clear that even the more refined terminology given to us by the discovery of biochemical mutation is not in and of itself sufficiently precise. We can treat microorganisms with a mutagen and then select all those organisms mutant in respect to the synthesis of the amino acid histidine. However, histidine is produced by a set of sequentially related reactions, as follows:

$$A \rightarrow B \rightarrow C \rightarrow D \rightarrow E \rightarrow F \rightarrow \text{histidine}$$

Each reaction is catalyzed by a specific enzyme. The one gene–one enzyme hypothesis means that there must be a number of different genes whose normal activity is necessary for the final production of histidine. A mutational event in any one of these genes will produce the mutant phenotype "histidineless." To put it briefly, phenotypes can be elusive as to cause. In the example just given, a mutation in any one of six different genes could produce the same phenotype. Conversely, in a complex, multicellular organism, a single mutational event that interferes with the production of a vital hormone might cause a variety of abnormalities in the phenotype of the organism, since the hormone is required for a variety of different developmental processes. If we wish to determine the mutation frequency for a particular gene, it is essential that we define the genetic function as narrowly as possible. To that extent, it would seem that in most cases the one cistron–one polypeptide hypothesis comes as close to functional unity as possible; and thus in order to study mutation frequency, it would be necessary to work with a cistron, whose end product is a single polypeptide chain.

The measurement of mutation frequency is more complex than the preceding paragraph indicates. For example, within any one cistron there are a large number of mutable sites (mutons). That is precisely what fine-structure analysis has shown us. For the simplest case, we assume that a change of any one nucleotide is sufficient to change the original triplet and thus alter the genetic message. The alteration would be expressed as the substitution of one amino acid for another in the primary structure of the polypeptide chain. (This is exactly the case in the mutation leading to sickle-cell anemia described in Chapter 19.) But certainly this interpretation is far too simple. First the code is degenerate, and so it is possible that the change of a single nucleotide would not result in any change in the primary structure of the polypeptide. For example, should the DNA triplet AAT be converted to AAC, the amino acid leucine would still appear at the proper place in the polypeptide chain. This fact means that we cannot determine the actual base-change frequency without sequencing every DNA molecule. While such a task is probably technically feasible (in the long term), we would hesitate to undertake it. The value of knowing the "true" mutation frequency, as compared to some estimate that must be directly proportional to the true mutation frequency, is far from evident.

There is another way in which we tend to underestimate the mutation frequency. It is not true that the substitution of one amino acid for another in the primary structure of a polypeptide need always result in the loss of function. After all, the tertiary configuration of the polypeptide depends on the interaction of the side chains of the various amino acids. A number of these amino acids are hydrophobic in character and stabilize the structure by making van der Waals associations. Four of the most important are the amino acids phenylalanine, leucine, isoleucine, and valine. Reference to Table 20-1 should indicate that the code words for these four amino acids are similar. A change in a single letter in a codon will quite frequently substitute a similar kind of amino acid. To take the easiest case, the transformation of an A to a G in the first letter of any of four codons would substitute valine for isoleucine, a substitution that is likely to have little effect on the functioning of the enzyme unless it occurred precisely at the very small region of the polypeptide involved in the formation of the active site. At all other regions such a substitution would probably go unnoticed, since the properties of the two side chains are almost identical. We must therefore conclude that a wide variety of amino acid substitutions are possible without our knowing about them. Whenever the function of the enzyme is normal, the growth of the cell will be normal, and no mutational event will have been detected. We call these *neutral mutations.* Consequently, because of the degeneracy of the code and the similar chemical properties of many of the amino acids, we must be constantly underestimating the true mutation frequency.

Considering the preceding statements as a whole, it should be noted that most studies of mutation frequency have two contradictory tendencies. The first is to overestimate the mutation frequency by grouping a large number of mutational possibilities under a single heading as, for example, "histidineless." In this case, many genes are mutating, and the mutation frequency is thereby increased. The opposite tendency is to underscore the mutation frequency when refining one's terms and looking only for the function of a particular polypeptide chain. It is important for the reader to recognize that reported mutation frequencies are always in error. As a rule of thumb, however, it is reasonable to assume (until proven otherwise) that a relatively high mutation frequency (10^{-5}) is probably the result of classifying a particular mutant phenotype under a single mutational heading when a large number of genes contribute to that phenotype. This is undoubtedly the case for the mutation frequency of the human blood disease hemophilia. The blood-clotting mechanism can fail in a number of different ways. Therefore a number of different kinds of mutational events can undoubtedly lead to a very general, somewhat imprecise category of "bleeder." As medical knowledge of the clotting mechanism has grown, so has our understanding of the relatively high occurrence of spontaneous mutations to this disorder. The converse is seen when someone claims that a particular kind of molecule is "resistant" to the action of a mutagen. Frequently, the estimates of mutation are gross underestimates of the actual molecular transformations. The induction of a true mutational event at the level of alterations of the genetic code probably can be achieved rather easily. Certainly there is more than enough energy for a mutational event to occur *every time* an X-ray photon is absorbed by a DNA molecule. Whether such a mutational event will produce a phenotypic change, however, will depend on the nucleotide changed and the position of that nucleotide within the entire genetic message.

The preceding discussion should make clear that only a fine distinction can be made between different genes with regard to their true mutation frequency. Still, that distinction is real. Presumably a cistron coding for a very long polypeptide must have more mutational sites than a cistron coding for a very short polypeptide. Thus the mutation frequencies of the two would differ measurably, even if the chance of substituting one nucleotide for another were equal at every point on the two cistrons. Furthermore, the results of fine-structure analysis of the $\mathbf{r_{II}}$ locus in T_4 bacteriophages (Chapter 16) indicate that not all mutons may be equally mutable. There could be several reasons. Let us consider only three. The first we have already considered—that is, how do you score mutations? An alteration that gives no phenotypic variation will be missed; the ones that appear to be "resistant" to mutation may be cases where the most probable mutational events will lead to no change in the amino acid coded for or to no substantial change in the amino acid sequence. Another possibility, however, is that certain of the purine and pyrimidine bases are more easily altered than others. Particularly when using chemical mutagens, there might be selectivity of the mutagens for certain particular units of the gene. If the transformation A → G is easier to effect chemically than the reverse, there would be more mutational events at A mutons than at G mutons. In subsequent pages we will discuss the chemical basis of mutagenesis briefly, but it is safe to say now that this particular question is not yet resolved. Finally, there is the possibility that certain kinds of mutational events are repaired by the cell more frequently than other kinds. The whole question of the repair of damaged DNA will be considered later in this section. Once again, we are not in a position to draw unequivocal conclusions from the data.

Back Mutation

All the data indicate that the frequency of back mutation is lower than the frequency of forward mutation. Although this idea may seem strange, a little thought will show that it is exactly what would be expected. There are many ways in which a wild-type gene can mutate to give a loss of function. There are few ways that a mutant gene can mutate to restore the wild-type phenotype. Obviously there is only a single way to restore the gene to its original condition. The chance of that particular event happening is much lower than the chance of the forward event happening, since the forward event could happen at any spot in the cistron and the reverse mutation must occur at only one spot.

In a different sense, back mutation can be achieved without reversion exactly to the original condition. It is possible for a mutational event at yet another locus within the same cistron to result in the introduction of a new amino acid that will alter the tertiary configuration of the mutant polypeptide chain in such a way that function is restored. There is evidence that such mutational events do occur. We have already discussed a similar case (Chapter 20) in which addition/deletion mutations can be restored by a second mutational event "downstream" from the original mutant. In both cases, two mutations within a single cistron are better than a single mutational event, and both can be considered kinds of back mutations. However, it is important to emphasize that, in both cases, the restoration of activity is rarely complete. Generally only a pseudowild condition is achieved. As pointed out in Chapter 21, such events can be considered suppressor mutations, in that one mutational event suppresses the effect of another. But since they are within one cistron, they are a special class of suppressor mutations. The effect is frequently referred to as *intragenic* or *intracistronic suppression.*

There is a third form of reversion to wild-type and that is through *intergenic suppression.* We discussed these suppressors in Chapter 21 in the form of modified *t*RNAs capable of reading nonsense or missense mutations. The extent to which other forms of suppression of a mutation in one gene by the action of another exist is not known. But, once again, such suppression rarely restores the mutant form to a truly normal condition.

The essential point remains that even though reversions to wild-type can be accomplished in several ways, the frequency would still be expected to be much lower than the frequency of forward mutation. The observation that it is, in no way contradicts general mutation theory. Furthermore, the discussion of back mutation should have clarified the basis for another observation—that is, that mutation is almost always adverse and rarely beneficial. The wild-type gene has been selected over a long period of time to produce a polypeptide that has the necessary function. The chances of altering that function adversely are great; the chances of inducing a mutation that inserts an amino acid in the polypeptide chain in such a way that tertiary configuration of the protein is improved (the enzyme becomes a better catalyst) are very small indeed.

Conditional Lethals

The suppressable amber mutations described in Chapter 21 were originally classified as conditional lethals. Under normal circumstances the bacteriophage could not reproduce satisfactorily; thus the mutations were lethal. On the other hand, when the phages were in the proper cytoplasm, they were able to reproduce. Consequently, the lethality of the mutation depends on the conditions in which the phage finds itself. We now understand the basis for this type of conditional lethal. The condition is solely that the proper forms of *t*RNA be present in the cytoplasm. Presumably any mutation that can be suppressed by some other genes could be classified as a conditional mutant.

Another set of conditional lethals cannot be explained on the basis of the action of suppressor genes—the *temperature-sensitive* mutations. For example, T-even phages will normally produce progeny equally well in *E. coli* grown at 42°C and at 25°C (although the rates of formation are different). Certain mutants can be obtained, however, that are incapable of producing progeny at 25°C. They are classified as temperature-sensitive mutants and are conditional lethals, since at one temperature they are lethal and at another they are not. As with the amber mutations, it was early recognized that temperature sensitivity was not a single mutational event. Rather, a large number of mutations at almost every part of the phage genome could be found to be temperature sensitive. There were temperature-sensitive mutations for DNA replication, some for head protein, some for tail protein, some for lysozyme, and so forth.

The explanation for temperature-sensitive mutations is relatively simple. It depends on what we know about the higher level of structures of proteins. As pointed out in Chapter 17, proteins depend on a large number of relatively weak bonds to hold them in the proper configuration. Enzymatic activity is dependent on that proper configuration, and anything that distorts the tertiary structure of an enzyme will remove enzymatic activity. With this point in mind, it should not be too surprising to find mutational events that produce temperature-sensitive enzymes. Such enzymes would have alterations in the primary structure of the polypeptide that interfere with the normal bonding pattern. At one temperature, the interference is not sufficient to cause distortion of the molecule. At that temperature the enzyme is active and the organism appears to be nonmutant. At another temperature, however, deleterious bonding patterns are established; the configuration of the protein is distorted and enzymatic activity is lost. At this temperature, the organism appears to be mutant. This hypothesis is easily tested. It is only necessary to grow the organism at the permissible temperature and isolate the enzyme and then show that, when the temperature is changed in vitro, there is a loss of enzymatic activity. In this way, it has been possible to illustrate that, for the majority of temperature-sensitive mutants studied, the effect is entirely on the higher levels of structure of the protein.

Temperature-sensitive mutants are interesting in two ways. First, they demonstrate the general proposition that we have made constantly—once the primary structure of the protein is established, all other levels of structure generate spontaneously. Any alteration in the primary structure will later affect the final form that the protein takes. When such alterations are drastic, it is easy to detect the mutational event, since the loss of function is unambiguous. However, a large number of substitutions of amino acids in the polypeptide chain do not result in drastic changes. In general, these substitutions would go undetected were it not for the fact that they do at least slightly alter the bonding pattern at the higher level of structure. By growing organisms at different temperatures, it is possible to uncover such minor defects in proteins and thus uncover a whole class of mutational events that otherwise would be lost to study.

Secondly, temperature-sensitive mutants have been extremely useful in the study of certain kinds of cellular processes. Suppose that you wished to obtain mutants for the true DNA replicating enzyme. If a mutation is induced in a cell such that the enzyme is inactive, the cell will not be able to replicate its DNA. Consequently, it will never divide, and so we will never detect the mutational event! On the other hand, a temperature-sensitive mutant for this enzyme would grow at one temperature but not at another. We could thus obtain colonies of mutant cells at one temperature and then switch these cells to the nonpermissive temperature to assay for the defect. In this way, temperature-sensitive mutants have proven invaluable. It is rather interesting that this technique was developed in the very early 1950s, before any clear understanding of the cause of temperature sensitivity was available. It was not until almost a decade later that the explanation for temperature-sensitive mutants was universally accepted and that the importance of the initial studies to the one gene–one enzyme hypothesis was understood.

Variability of Gene Expression

The foregoing discussion of suppressor mutations and conditional lethals should demonstrate that a mutant is not always a mutant. It is important for the reader to grasp this fact, since it is this kind of complication that makes it so difficult to study some of the critically important metabolic defects in the heredity of higher organisms. In one genetic background, a particular mutational event will express itself fully and behave as a simple recessive mutation. In another genetic background, the same mutational event will be expressed only partially or possibly not at all. In the absence of purebred lines, it will be essentially impossible to show that the failure of the mutation to be expressed in one individual is the result of a mutational event at another locus (or some other conditional factor). In the study of the inheritance of certain kinds of disease in human beings, it is common to have hints that a single recessive ultimately is responsible, but the variability of expression of genes, the lack of "inbred stock," the wide range of nutritional and other environmental conditions may make it nearly impossible to establish the hereditary basis of the disease unequivocally.

When considering the possibilities for the modification of gene expression, we can see most clearly why Gregor Mendel took such pains to delimit the characters with which he worked. Only by choosing those genes whose expression is unambiguous in succeeding generations was it possible to establish the laws of heredity. We must now reverse the process and understand that variability of expression is not necessarily an indication of lack of hereditary control.

Polar Mutations

We conclude the general consideration of mutation with a brief description of polar nonsense mutations. These mutations were first observed as cases in which a nonsense mutation in one cistron prevented the production of an enzyme (or enzymes) controlled by a completely separate cistron. For example, in *E. coli* three enzymes responsible for the utilization of galactose (a sugar) are under the control of three separate genes (**gal E**, **gal T**, and **gal K**) that are adjacent to one another in the chromosome. It was found that a single mutation, mapping in the **gal E** cistron, could abolish production of all three enzymes. In time a number of such mutations were found, and all were in the **gal E** gene.

Let us assume that the normal transcription and translation processes begin at **gal E** and proceed through **gal T** to **gal K**. A certain class of mutation in **gal E** abolishes **gal T** and **gal K** functions. The mutation somehow interferes with the normal translation process, but it does so in a polar manner. The **gal E** mutation has no effect on the genes immediately preceding it in the genome; it only affects those adjacent genes that immediately follow it. Consequently, the effect is polar in character, and hence the name.

Many of these mutants have been discovered, and we now understand their mode of action. In general, they are found only where a polycistronic *m*RNA is formed. *Polycistronic m*RNAs contain the messages for several proteins. They are more common in prokaryotes, but they can be found in eukaryotes as well, at least in the fungi. Quite frequently a block of adjacent genes, all of whose enzymes control a single metabolic pathway, is transcribed into a single polycistronic message. This process occurs for the galactose enzymes of *E. coli.* Polarity mutants do not interfere with the transcription process. A normal, polycistronic message is made, but it is not translated.

Polar mutations are nonsense mutations. When ribosomes reach a nonsense codon, they detach from the *m*RNA, thus releasing the unfinished polypeptide. Presumably they could reattach at the next cistron (which should have a normal "start" or attachment site. However, in certain cases, they do not. The reason is that the *m*RNA is folded up on itself in a random coil (Figure 23-1). As the ribosomes

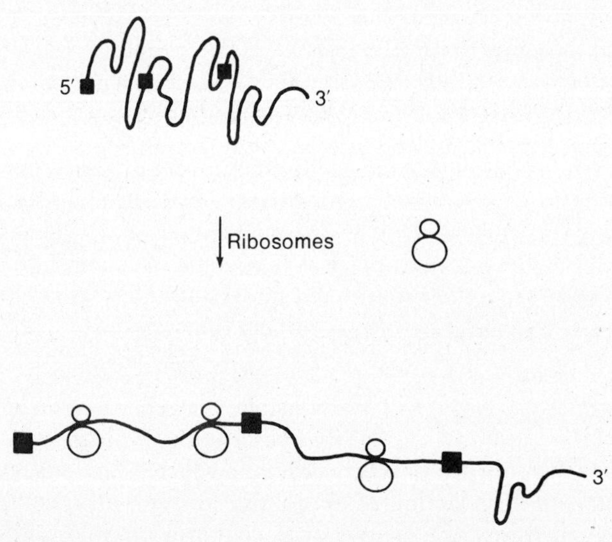

Figure 23-1. Messenger extension by translation. The upper diagram shows an *m*RNA molecule in a folded form. The ribosome binding sites (■) are mostly inaccessible. After ribosomes attach and move along the message, the *m*RNA is unfolded. If the ribosomes detach at a nonsense codon only part way down the molecule, the remaining attachment sites will stay "buried," and this part of the *m*RNA will never be translated. This is the basis for polarity mutations.

proceed along the *m*RNA, they extend the *m*RNA, breaking the hydrogen bonds that hold the random coil together. It is this extension of the *m*RNA that permits translation of the message by base pairing with the various *t*RNAs. Without the extension of the *m*RNA, however, certain "start" regions will be buried in the folded molecule, and no ribosomes will be able to attach to them, as the figure indicates. Consequently, a nonsense mutation in the first cistron of a polycistronic message will cause the ribosomes to detach and prevent the extension of the *m*RNA that is required for its continued translation.

One feature of polar mutants is particularly significant. A nonsense mutation will act as a polar mutation only when it is some distance from the end of the cistron. If it is close to the end of a cistron, the ribosome has already unfolded the *m*RNA where the next "start" is placed. Consequently, other ribosomes can attach to the "start" of the next gene, and all subsequent cistrons can be translated. This feature of polarity mutants, then, is in good agreement with the hypothesis that they exert their effect by detaching the ribosome from the *m*RNA.

The Molecular Basis of Mutation

Spontaneous Mutation

The preceding discussion should have demonstrated that all forms of gene mutation can be explained as either a change of one base into another or as the addition or deletion of one base (or possibly more than one). We must now consider the mechanism by which such changes can occur. The first item that needs to be resolved is how truly spontaneous mutation can occur. In the original paper on the double helix configuration of DNA, Watson and Crick suggested a possible mechanism for mutation. The two parts of the helix are joined together by highly specific hydrogen bonding between A and T, G and C. However, if we compare the structure of the bases given in Figure 18-6 with that given in Figure 23-2, it can be seen that the normal base pairings are not the only ones that can be achieved. Because the purines and pyrimidines are ring structures containing several double bonds, it is possible to get forms (tautomers) of each base in which the distribution of electrons is slightly different. The forms given in Figure 18-6 are the usual forms under cellular conditions, but a certain number of less usual forms of the bases will always be present in the cell. The ones shown in Figure 23-2 are rare forms, but the fact that adenine can take the *imino* configuration means that occasionally an adenine will pair with a cytosine during replication of DNA, and, similarly, the rare *enol* form of thymine can pair with guanine. In either case, a misreplication of the DNA strand will occur, which, at the next cycle of replication, will be expressed as a mutational event (compare to Figure 20-5). Such changes can be only of one type, cases in which the purine of one strand of the DNA molecule is replaced by a different purine (A → G; G → A) or the pyrimidine is replaced by another pyrimidine (C → T; T → C). Such changes are given the name *transitions*.

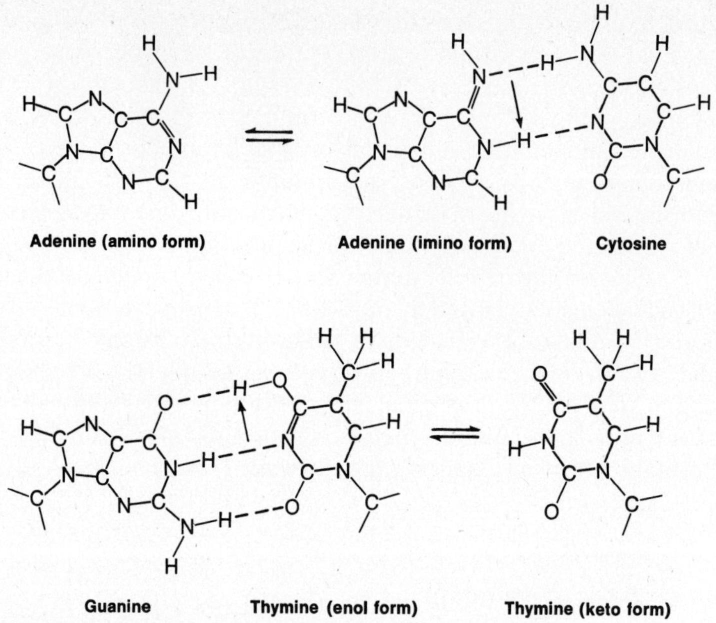

Adenine (amino form) Adenine (imino form) Cytosine

Guanine Thymine (enol form) Thymine (keto form)

Figure 23-2. Tautomers of adenine and thymine. Both A and T have two naturally occurring forms. The imino form of adenine and the enol form of thymine occur rarely under physiological conditions. When they do occur, they can base pair in unusual ways, leading to mutation.

The foregoing explanation is perfectly satisfactory for spontaneous mutation, since a very low frequency of mispairing will occur, which is completely consistent with the low mutation rates observed spontaneously. Furthermore, such an hypothesis makes a clear prediction, which can be tested in one case at least. The prediction is that mutations occur primarily during the process of replication and not while the DNA is in an unreplicated condition. The viruses provide us with a system that can test the hypothesis. Viruses can be kept in an inert condition for long periods of time (up to many years) and then added to cells in which they can replicate. It is possible to test for the presence of mutations at the time of initial infection and after one, two, three, or more complete replicative cycles. When this is done, it can be demonstrated that indeed the frequency of mutation in inert particles is extremely low (so low that for a long time they went completely undetected). On the other hand, as the number of cycles of replication increases, the number of mutants available also increases. Thus it appears that at each replication cycle more mutants are produced, a fact highly consistent with the hypothesis that the mutational event occurs during the period of replication of the DNA. All things considered, then, we can assume that spontaneous mutation is the result of the chemistry of the DNA molecule, and, fortunately for all living things, the rate of spontaneous tautomeric shifts is very low.

Induced Mutation

As pointed out earlier, there are numerous mutagens that increase the frequency of mutation. Since the variety of agents is very great, it would take an entire book to summarize all the special properties of mutagens. A number of such books are available, and the reader is referred to them for a complete description. Here and in the next few paragraphs we would like to discuss only a few well-established cases of mutagenesis. The best way to approach this problem is not historically. The earliest known mutagen, X rays, and the first known chemical mutagen, nitrogen mustard (an antipersonnel "gas" developed for World War I), are so nonspecific in their actions that they tell us relatively little about the mutational process itself. It was not until the mid-1950s that rather specific mutagens were developed, and we should consider the action of a few of them first before returning to the classical mutagenic agents.

The pyrimidine analog 5-bromouracil was the first successful specific mutagen. As Figure 23-3 indicates, this compound bears a marked resemblance to thymine and, in fact, is incorporated into DNA in place of thymine. This fact in and of itself is not sufficient to account for mutation, however, since, for the most part, 5-bromouracil and thymine are functionally identical. Rather, mutagenicity of the substance results from the fact that the relatively rare enol form of the base is much more common in 5-bromouracil. Consequently, there is a far greater chance for copy errors to occur, and thus the mutation frequency is substantially increased. Clearly 5-bromouracil is a specific mutagen to the extent that it involves replacement of T as the initial step. Therefore the mutational events will occur in DNA molecules only at those places where T occurs, and certain kinds of transitions are more common than others. A number of other purine and pyrimidine analogs can act as mutagens. (A particularly effective one is 2-aminopurine.) In all cases, the

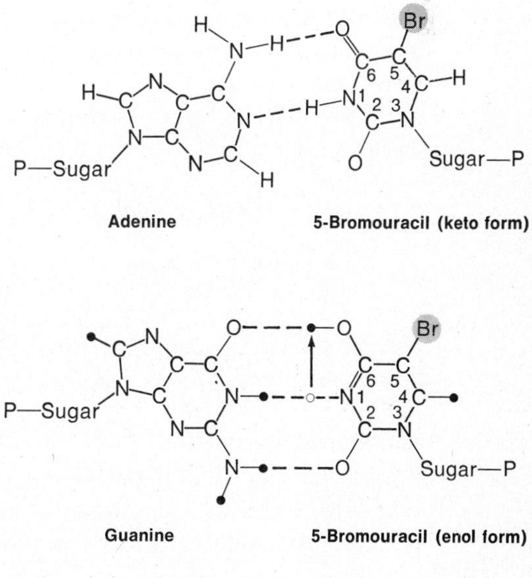

Figure 23-3. Bromouracil action. The mutagen 5-bromouracil is pictured in its two forms. In the keto form, it is incorporated in place of thymine in the DNA. In its enol form, it mispairs with guanine, producing a mutation.

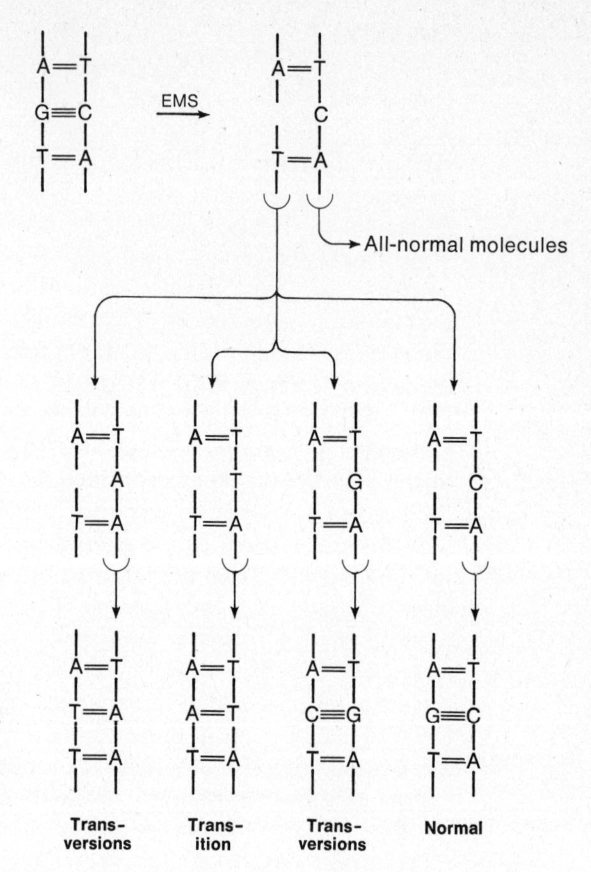

Figure 23-4. Alkylating agents. EMS is an alkylating agent that acts to remove a guanine from intact DNA. At the next replication, any base may be inserted across from the gap. In three cases out of four there will be a mutation.

increased mutation frequency is achieved by increasing copy error during DNA replication. In common with spontaneous mutations, base analogs induce only transitions.

A totally different kind of specific mutagenesis can be obtained with certain *alkylating agents.* Two of the most effective are shown below:

$CH_3CH_2SO_3CH_3$	$CH_3CH_2SO_3CH_2CH_3$
Ethyl methane sulfonate (EMS)	Ethyl ethane sulfonate (EES)

It can be seen that each has an ethyl group that can be donated to the nitrogen in the 7-position of the purine base (thus *alkylating* the nitrogen). The result of this chemical reaction is the hydrolysis of the base-sugar linkage, which frees the bases from the backbone of the DNA molecule. At the time of replication, the alkylated base has been lost, and the one chain will have a "gap" where the base had formerly been (Figure 23-4). Consequently, the enzyme that is replicating the DNA may insert almost anything across from the gap. In certain cases, the insertion will be exactly correct, and no primary mutational event will occur, although it might at some subsequent replication, since the gap may still be present in one chain. On the other hand, a totally incorrect base may be substituted (since there are no restrictions as to pairing at this point), and thus a mutational event must inevitably occur. As a result, mutations of the transition type may be induced but so may those of the *transversion* type, cases in which a purine may be replaced by a pyrimidine or vice versa (G → C, G → T, and so on). With this mechanism of action, it is hardly surprising that both these alkylating agents are extremely potent mutagens. Although EMS can react with any purine, it appears that the N_7 of guanine is the most reactive group. Consequently, EMS, is relatively specific as a mutagen affecting code words having G.*

A third highly specific mutagenic agent is *nitrous acid* (HNO_2). This mutagen specifically deaminates cytosine at the 6-position and thus converts cytosine to uracil (Figure 23-5). Since uracil and adenine can pair perfectly well, and they pair much better than uracil and guanine, essentially the transition is from a C to a T. Wherever the uracil occurs, adenine will be inserted in the complementary chain at the time of

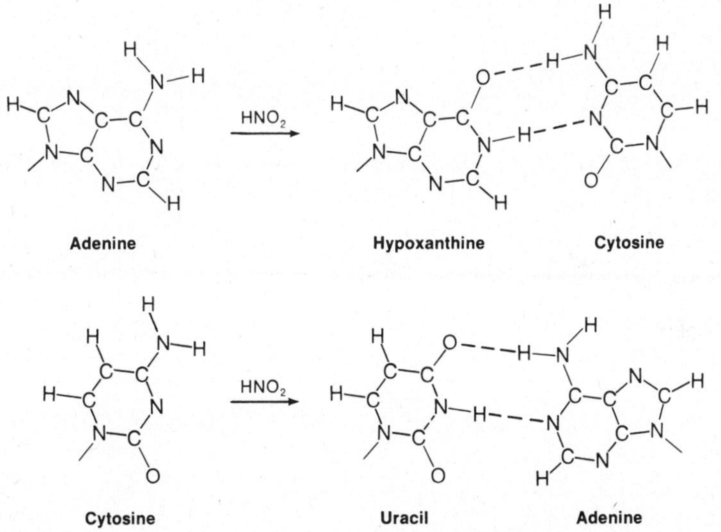

Figure 23-5. Nitrous acid action. HNO_2 acts by deaminating purines and pyrimidines. The resultant deaminated form may lead to a new base pairing association and thus to mutation. Thymine has no amino group and is not affected. Guanine is deaminated to xanthine, which still pairs with cytosine; hence there is no mutation detected. Consequently, HNO_2 produces changes only with A and C and is a "specific" mutagen.

*The alkylation of the N_7 nitrogen of purine is not the only action of alkylating agents. EMS can also attack the oxygen residue of guanine, and its high mutagenicity of guanine bases may well depend on that fact as well as on its action in the N_7 position. In either case, it is probably the filling of a gap in the chain that results in the mutational event.

replication. During the next replication the A will pick up a T in its complementary chain and an effective C → T transition occurs overall. The action of nitrous acid may be the most specific of all known mutagens.

The preceding discussion has shown three different ways in which mutation can occur: increase of copy error, insertion of a base to "fill a gap," and specific conversion of one base to another. It is probable that X rays can cause all these steps to occur but in a totally nonspecific manner. The amount of energy available when an X-ray photon is absorbed is sufficiently great to break any bond in the absorbing molecule. Therefore we must presume that gaps are produced and that one base can be changed (probably by oxidation) to another. But X rays can also cause breaks of the DNA molecule in the backbone, and when breaks occur, whole regions of DNA may be eliminated. It is quite probable that such events lead to lethal mutations through reading frame shifts and so forth. In fact, considering the drastic effects of bombarding DNA molecules with X-ray photons, one wonders how any ordinary mutations are induced. We will return to this problem a little later.

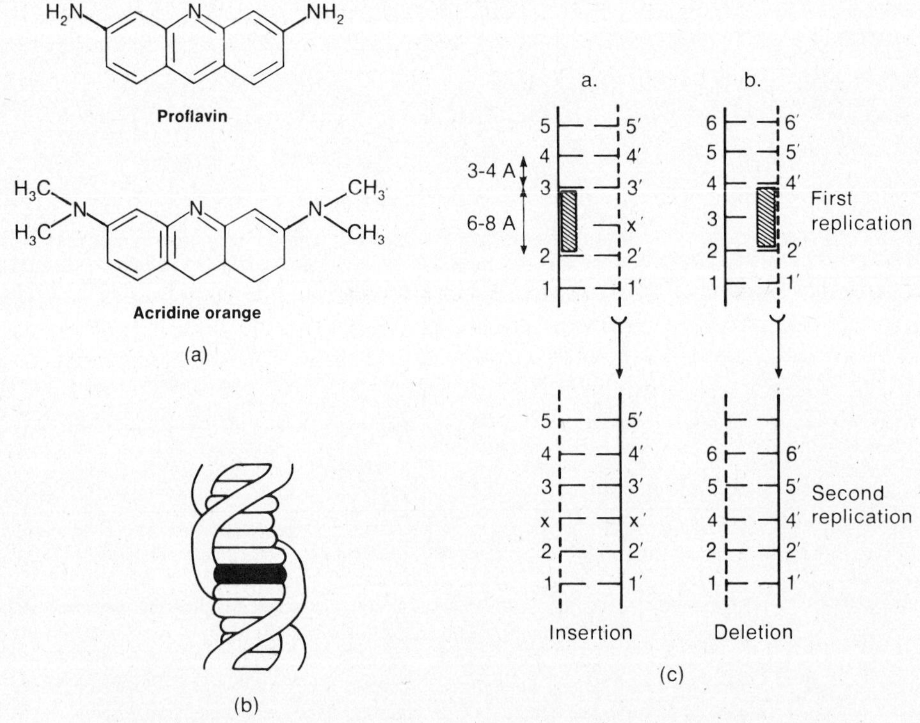

Figure 23-6. The action of acridine dyes. (a) Two acridine mutagens. (b) The method of intercalation of the mutagen between the bases of a DNA chain. (After Lerman, L. S. 1963. *Proc. Natl. Acad. Sci.* (*U.S.A.*) **49**:94.) (c) A diagram of the way in which these mutagens might act to cause either an addition or a deletion of a base. (After Hayes, W. 1968. *The Genetics of Bacteria and Their Viruses.* 2nd ed. John Wiley & Sons, New York.)

Before concluding, let us briefly consider the *proflavin* mutagens, which were previously discussed when we considered the reading frame mutants used to establish the triplet nature of the code (Chapter 20). As shown in Figure 23-6(a) the proflavins strongly resemble the purine bases. Like the purines, they are flat and many of their properties are similar. This similarity is the basis of their mutagenic action. They can slip in between various bases in the DNA molecule even though they are not actually incorporated into the molecule itself (Figure 23-6b). Insertion of these molecules at the time of replication can occur in two ways (Figure 23-6c). First, the molecule may be intercalated between two existing bases on the template strand. The replicating enzyme will then insert a base (presumably at random) in the new chain to "pair" with it. At the next round of replication, that new chain will produce a DNA that has one additional base. Alternatively, the proflavin may be inserted in the new chain as it is being synthesized. This step may "screen" the base in the template strand so that no base can pair with it. The new chain that results will lack one base and at the next round of replication will produce a deficient DNA.

This discussion has demonstrated the mechanism of addition/deletion mutations. We must add it to our preceding list as a fourth method of mutation. In some ways, the mechanism is similar to that in which a gap is produced by the alkylating agents. But it is quite different, since in the latter case a point mutation occurs through random insertion of a base to replace the gap. Here an entirely new base may be added or one of the old bases may be lost. Since the special properties of such mutations were discussed extensively in Chapter 20, no further discussion seems necessary here.

Fractional Mutations

The mechanism of both induced and spontaneous mutations indicates that at the time of the initial mutational event the DNA molecule affected is half mutant, half wild-type. Separation of the chains and replication of each, therefore, should produce one totally mutant molecule and one wild-type molecule. In this sense, the initial progeny of the cell in which the mutation first occurred should be one-half mutant and one-half wild-type. The original cell, then, is only fractionally mutant (a *mutational heterozygote*). A number of tests have been made in microorganisms of this logical consequence of the postulated mechanism of genetic mutation, and all are consistent with the hypothesis. In fact, in certain cleverly designed experiments, it is possible to demonstrate that the parent cell contains DNA molecules that are half mutant and half wild-type. The general proposition about the nature of mutation seems to be confirmed. However, it is difficult to run such tests in multicellular organisms, and the evidence from these forms is contradictory.

In the mid-1960s it was pointed out that Drosophila should be a good organism in which to test the mutation mechanism. It is possible to irradiate mature sperm to induce mutations. Sperm are haploid, and each chromosome presumably consists of a single DNA molecule. Since a mutational event occurs in the mature sperm, the mutant DNA molecule is presumably half mutant and half wild-type.

When it enters the egg cell, it should replicate and produce one totally mutant molecule and one that is totally wild-type. Consequently, all mutations induced in Drosophila sperm should be fractionals. If the cross is designed such that the progeny flies do not cover up a mutational event, it should be possible to examine the progeny and find that some parts of the fly are wild-type and others are mutant. Such tests have been made, and, in fact, some fractional mutations were found. However, they were few in number, and it thus appears that generally the sperm behaves as if it is either totally mutant or totally wild-type. Unfortunately, we cannot be certain of the meaning of this experiment. For technical reasons, X rays rather than chemical mutagens were used, and so the conclusion may well be that the effect of X rays is rather different from that of mutagenic chemicals. On the other hand, it may be true that in some way chromosomes contributed to the zygote from the sperm are repaired after entry into the egg (see below) and that fractional mutations are not the rule in higher forms. Further tests will be necessary before we can be certain of the meaning of these findings.

Repair of DNA

The discovery that cells have enzymes that can repair damaged DNA was made somewhat by accident. An investigator who had exposed bacteria to ultraviolet radiation inadvertently exposed some of the cells to strong visible light. He discovered that there was much more growth in those cultures that had been exposed to the visible light than in those that had been treated with UV radiation alone. Careful study of this process indicated that, in fact, exposure to visible light activates a DNA-repair enzyme that removes the lethal effect of UV radiation.

When cells are exposed to UV in the region of 2600 Å, they are rapidly killed. This lethal effect is, in part, the result of the linking together by covalent bonds of adjacent thymine residues to form thymine dimers (Figure 23-7). Dimer formation distorts normal structure of the DNA double helix, and it is apparently this distortion that the repair enzyme "recognizes." It excises the region of the dimer and rebuilds the DNA molecule, restoring it to its original condition.

Since this discovery of photoreactivation, investigations have shown that there is another repair enzyme in all cells that does not require light for activation. Study of this somewhat more complicated system indicates that the initial event in the process is the excision of the thymine dimer. The excision process apparently involves the removal of a number of bases in the same strand of the DNA in which the dimer originally resided (Figure 23-8a and b). After the damaged region has been removed, a second enzyme restores the proper bases, using the undamaged chain as its template (Figure 23-8c). Finally, the newly inserted bases are joined to the intact strand, and the molecule is completely repaired. This sequence of events requires three separate enzymes: an exonuclease to remove the damaged region, a DNA polymerase to repair the damage, and a ligase to make the final junction. All three enzymes have been found, and they seem to be all that is necessary to effect DNA repair. The major differences between the photodependent repair system and the dark repair system is that the former is a very large enzyme that carries out all three

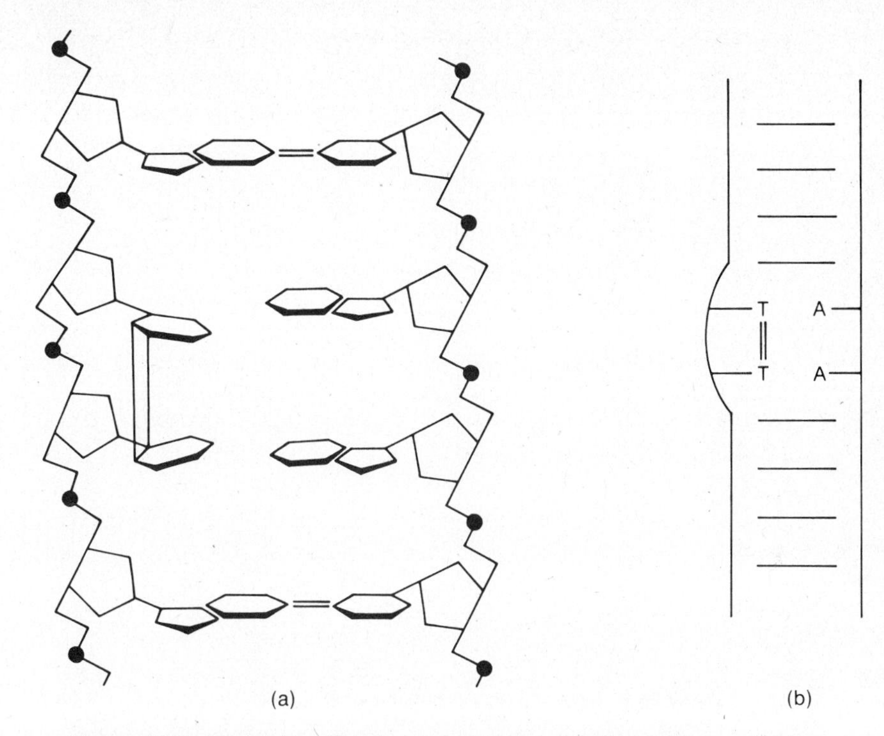

Figure 23-7. Ultraviolet-induced thymine dimers. (a) Two adjacent thymine residues have been joined together to form a dimer. This crosslinking is induced by exposure to ultraviolet radiation. (b) A diagram of the distortion of the DNA molecule caused by the dimer.

functions by itself (and requires light for activation). The more usual dark repair system is apparently a three-component system.

Discovery of the repair system in bacteria explained the observation that some strains of bacteria are far more sensitive to ultraviolet radiation than others. Apparently the UV-sensitive strains are mutants that lack the DNA polymerase. Without the enzyme, they are unable to repair the damaged DNA and they die. An interesting sidelight of this finding is that the missing enzyme seems in all ways exactly the same as the DNA polymerase that was originally presumed to be the enzyme normally used to replicate DNA (Chapter 18). By selecting strains that are very UV sensitive, it is possible to isolate cells that apparently are totally lacking in the DNA polymerase. Nevertheless, these cells are capable of replicating DNA and dividing normally. Apparently the DNA polymerase that we have been studying for so many years is a repair enzyme and the normal replication of DNA goes on with the use of another polymerase. The replicating enzyme is DNA polymerase III and the repair enzyme is DNA polymerase I (Chapter 18).

The existence of repair enzymes raises the question of what proportion of mutational events is removed by repair processes. From what is known so far, it

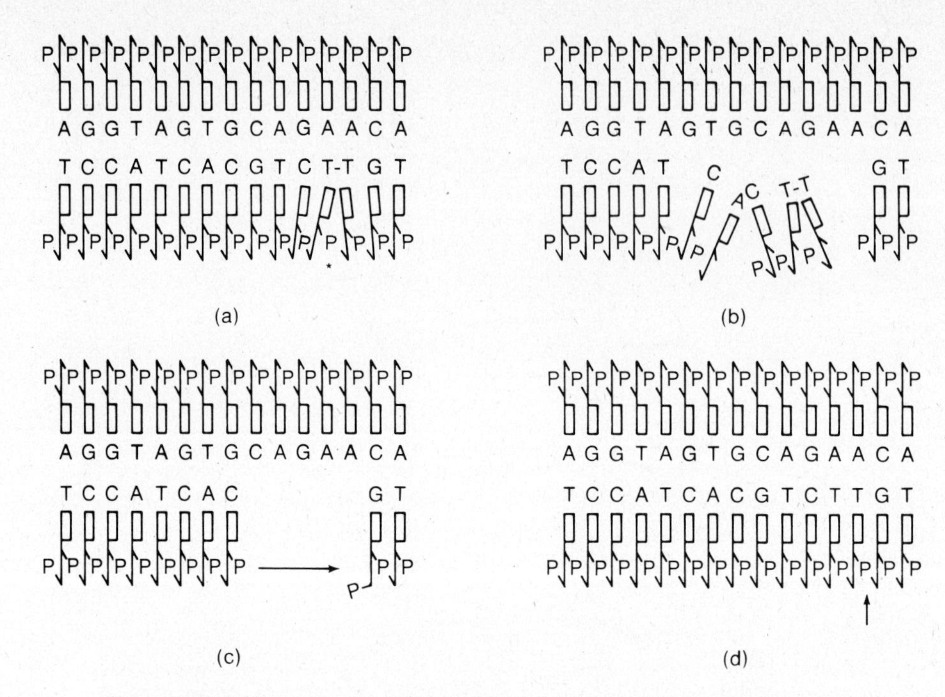

Figure 23-8. DNA repair. (a) A thymine dimer has been induced in the DNA. The repair enzyme complex can detect the distortion in the molecule. (b) The exonuclease function of the repair complex removes the distorted region. (c) The polymerase function of the complex uses the intact strand to synthesize a new complementary chain. (d) The ligase function of the complex makes the final bond attaching the new strand to the older part of the molecule.

appears that the repair systems recognize only changes in the general shape of the DNA molecule. They are incapable of sensing mispairings of the type induced by tautomerization and so forth. It thus appears that the repair system is present to prevent the rather drastic changes in DNA that lead to immediate cell death and, possibly, to reduce the total amount of mutation in the cell. But it is not a system that can reduce mutation to zero. In fact, if there were no mutation, there would be no variability; and, without variability, evolution would be impossible (see Chapter 29). That repair of DNA is an important process, however, is justified from the information indicating that all cells appear to have some kind of repair mechanism.

Recombination

Breakage and Exchange or Copy Choice?

The previous discussion of recombination (Chapters 10 to 12) presented the basis for our belief that the process of crossing over that leads to genetic recombination within linkage groups is effected by breakage and exchange of homologous chromosomes. The exchange of parts must be an exactly reciprocal event. This mechanism is conceptually simple as long as we believe that the genes are arranged on the chromosome like beads on a string. Then breakage and exchange can be envisioned to occur in the strands between the beads, and the beads themselves would be exchanged whole and intact. But once it was demonstrated that crossing over could occur within a cistron, the breakage and exchange model of crossing over came under attack. Certainly, crossing over within the cistron would need to involve an *extremely* exact breakage and exchange process. As studies with proflavin mutagens had already indicated (Chapter 20), the addition or deletion of a single base would totally destroy the genetic message of a cistron. Therefore the recombination process must involve breakage of the two DNA molecules in exactly the same place and the joining of the two strands to be exchanged in exactly the right way (having the same polarity and so on). If not, recombination would lead to mutation, and there is no correlation between the two processes. For too long it had been presumed that chromosomes were broken more or less as a twig is broken. Failure of the breakage/exchange hypothesis to present a mechanism that would ensure the exactness required for the intracistronic recombination made scientists somewhat uneasy about the mechanism.

Perhaps the most important evidence against the breakage and exchange mechanism was the discovery that bacteriophages show a high degree of nonreciprocal recombination. Traditionally, genetic recombination is a totally reciprocal event. In Neurospora, the reciprocal nature could easily be demonstrated and was well documented; and those cases in which nonreciprocal recombination had been discovered were generally explained as some aberration of the normal process. But the new data with bacteriophages suggested that, in this case at least, nonreciprocal recombination was the expected event. If the breakage and exchange model was the mechanism for recombination. it was necessary to answer the question: What happens to the other half of the material produced by the recombination process? Since no one had an answer (or has, for that matter) to this question, it seemed reasonable to propose that, at least in the bacteriophage, there was a different mechanism of genetic recombination. While breakage and exchange might be considered the normal mechanism for recombination in eukaryotes, another mechanism was the preferred one in prokaryotes. Presumably each organism could have both mechanisms (which would explain why rare cases of nonreciprocal recombination could be found in eukaryotes). The difference between the two groups of organisms would be which mode of recombination dominated.

It was generally assumed that copy choice was the mechanism by which recombination occurred in the prokaryotes. As Figure 23-9 indicates, the modern version of copy choice is somewhat different from the original hypothesis (see Chapter 11). It is presumed in this case that during the process of DNA replication the two different but homologous DNAs can be used as a template for the formation of a totally new DNA molecule that has been produced partially from one molecule and partially from the other. Such a mechanism could easily be nonreciprocal, and it is a good explanation for such events. However, it has an intrinsic difficulty of its

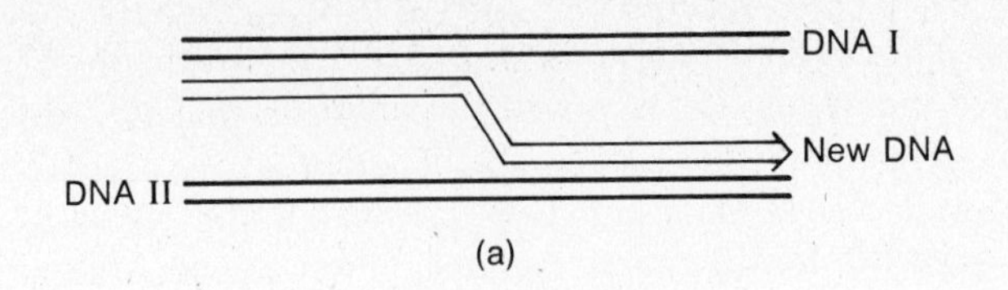

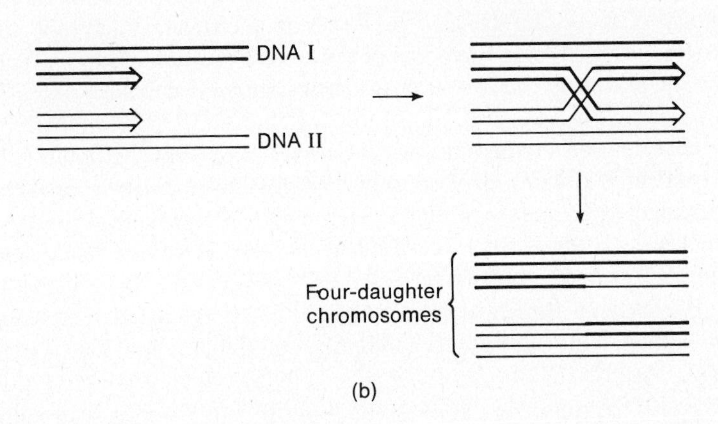

Figure 23-9. Copy choice. (a) A diagram of a nonreciprocal recombinant produced by copy choice. (b) A diagram of the same mechanism producing reciprocal recombination.

own. As a mechanism, it is incompatible with what we know about the normal semiconservative replication of DNA molecules. That is, this mechanism of DNA synthesis involves the production of an entirely new DNA.

At first semiconservative replication may not appear to be incompatible with copy choice. It would seem possible to copy one strand of one DNA molecule and then switch to the corresponding strand of the second DNA molecule, thereby making a recombinant DNA by a semiconservative process. However, as soon as this step is done, it becomes evident there is no way for the fully produced molecule to be released from the two "parent molecules" unless parts of the original (parental) molecules are broken and discarded (or recombined). The only difference between the traditional view of breakage and exchange and the semiconservative replicative method of copying is the order of the events. Recombination that results from a switch in the reading during semiconservative replication merely gives you exchange first, breakage second.

The copy choice mechanism has two features that are distinctly different from the breakage and exchange hypothesis. In the first place, recombination by copy choice can occur only through a process of DNA replication. Secondly, the recombinant DNA molecule so produced is a newly synthesized DNA containing none of the atoms found in the parental molecules. Fortunately, both parts of the hypothesis can be tested experimentally. The first clear test came as late as 1961. The rather small bacteriophage λ was used in the experiment. One strain was wild-type, and the other carried the genetic markers **c** and **mi**, mutants that affect the plaque "phenotype" and that are readily distinguishable from one another in the four possible genotypes: **+ +**, **c mi⁺**, **c⁺ mi**, and **c mi**. (The outside markers used to prove that recombination had occurred will be ignored in this description.) The wild-type phage was grown in ^{13}C and ^{15}N such that the DNA produced was all heavy. The mutant phages (**c mi**) were totally unlabeled. The phages were then placed on cells growing in a medium that contained the normal isotopes of carbon and nitrogen. In this way, any newly synthesized DNA would be unlabeled.

Recombinants (**c^+ mi** and **c mi^+**) were recovered and the DNA examined. If the mechanism of recombination is copy choice, all recombinants should have newly synthesized DNA, and no parental material should be carried over from the wild-type, heavy DNA. But such was not the case. The DNA was of intermediate density, indicating that both parental and nonparental material are present in the recombinant. Most importantly, the amount of heavy parental material from the wild-type DNA varied, as would be expected with the relative placement of the site of recombination in the chromosome. Since the **mi** gene is farther "down" the chromosome than the **c** gene, recombinants that were **c mi^+** had much less heavy parental (wild-type) DNA than did the progeny that were **c^+ mi** (compare Figure 23-10). These findings are quite inconsistent with copy choice, but they are consistent with breakage and exchange. Furthermore, some recombinant progeny were found that had more than 50 % heavy DNA. This finding suggests that the process of recombination does not *require* the replication of the DNA molecule. Were DNA replication required, the essential prediction of semiconservative replication is that at most only one chain of the original heavy DNA could be present in one molecule. However, should breakage and exchange occur in the absence of replication, it would be possible for a wild-type chromosome that was heavily labeled to receive only a small part of the unlabeled mutant chromosome (presumably that part carrying the **mi** gene), and thus the great bulk of the chromosome would be heavy. The finding of such recombinant progeny made it seem highly probable that most of the assumptions of copy choice are incorrect, that breakage and exchange is an event that can occur any time in the cell cycle, and that breakage and exchange occurs in prokaryotes as well as in the eukaryotes.

Confirmation of the breakage and exchange mode of reproduction in bacteriophage was provided in 1964. The experiments were done with two strains of bacteriophage T_4. One strain was labeled with ^{32}P and the other strain with

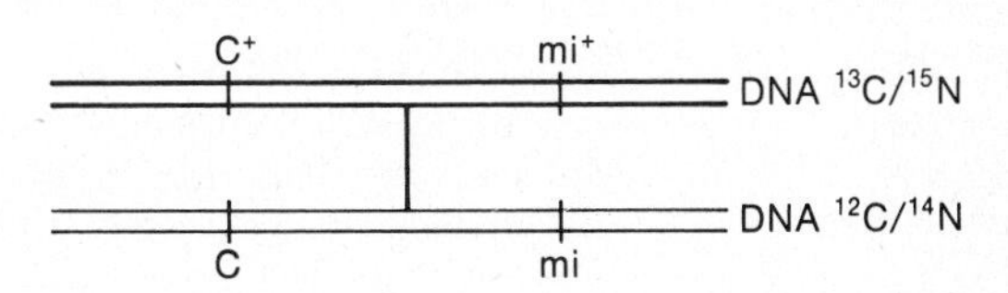

Figure 23-10. A test of copy choice. Any recombination between **c** and **mi** involves the exchange of heavy and light DNA. This contradicts the prediction of copy choice (compare to Figure 23-9).

5-bromouracil. The bromouracil DNA is very heavy and can be detected in that way. The ^{32}P-labeled DNA is radioactive. As in the preceding experiment, phages were plated out on cells and allowed to recombine. However, in this case, the cells were grown in the presence of cyanide. Under these conditions protein synthesis and replication of DNA do not occur. After allowing sufficient time for any recombinational events to take place, the cells were harvested and DNA was isolated. Surprisingly, recombinant DNA—that is, DNA that had both kinds of label present—was found. It thus appears that recombination can occur in the total absence of DNA replication, a feature that no form of copy choice could ever explain.

Even more suggestive was the discovery that heating the hybrid DNA molecules caused them to separate into two distinct fractions: one a heavy fraction composed exclusively of 5-BU DNA and the other exclusively ^{32}P-labeled DNA. Apparently the hybrid DNA molecules are formed by the joining together of two double helix fragments from the parental DNA molecules (Figure 23-11). The two parts of the molecule can be separated by merely heating the DNA to its melting temperature. Consequently, the fragments are joined only by hydrogen bonds. As the figure suggests, the DNA strands are broken at different places, and the overlapping parts serve to connect the two fragments. This suggestion is extremely interesting, for it makes the general process of recombination more reasonable. The chances of the two DNA molecules breaking at exactly the same place within the cistron seem very low. If, however, two molecules of DNA were to break in such a way that at least two of the chains were overlapping, it would be quite possible for them to form hybrid associations of the type found in these experiments. All that would be necessary is an enzyme (a ligase) that could make the two phosphate linkages to put the two DNA chains together into a single double helix.

The preceding experiments were sufficient to convince almost everyone that copy choice was not the mechanism of recombination in the prokaryotes. Since 1965 several experiments in eukaryotes have shown that copy choice is excluded in this case as well. In eukaryotes, experiments cannot be conducted by the extraction of DNA, for here recombination involves the exchange of whole parts of chromosomes

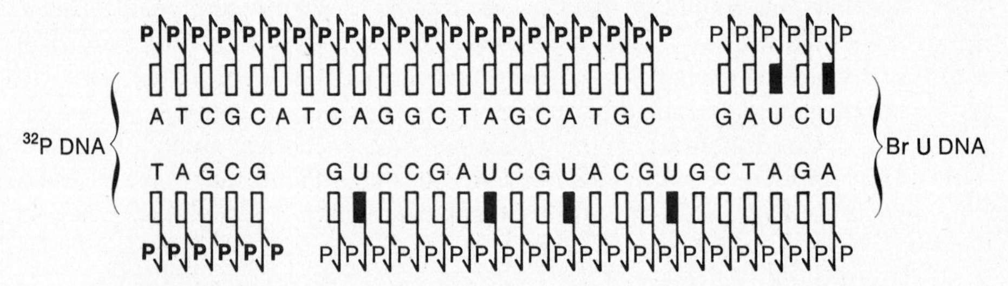

Figure 23-11. Hybrid DNA. A molecule is shown that is the result of breakage and exchange between 2 different DNAs. One previously had been labeled with ^{32}P and the other with bromouracil. Presumably such hybrid molecules are the first step in recombination. Later the gaps will be filled by enzymes similar to those in the DNA-repair complex.

that are more complicated than the single DNA molecules of the bacteriophage chromosomes. Furthermore, in eukaryotes, it is desirable to try to correlate the exchange parts of chromosomes with the stage in meiosis at which chiasmata can be seen to have formed (pachytene). Such studies can only be made by using radioactively labeled chromosomes. The technique is basically the same as the one used originally to show that the chromosome replicates semiconservatively (see Chapter 18). Premeiotic gonial cells are exposed to tritiated thymidine (^{3}HT) just prior to replication of the chromosomes. When such cells divide, they will have chromosomes that are labeled uniformly. (Since the DNA replicates semiconservatively, one-half the DNA molecule is labeled. A DNA molecule represents a chromatid.) In the next interphase (the one that immediately precedes meiotic prophase), these chromosomes will begin to replicate; but since the ^{3}HT was delivered as a short pulse, the cells are now growing in the absence of label. The unlabeled strand of the DNA molecule will produce an unlabeled chromatid, and the labeled strand of the DNA molecule will produce a labeled chromatid. Consequently, each chromosome entering meiotic prophase is half labeled and half unlabeled (Figure 23-12). Should exchange of parts occur between homologous chromosomes during meiosis, it would be possible to detect it by the trading of labeled and unlabeled parts between chromosomes. As the figure indicates, there

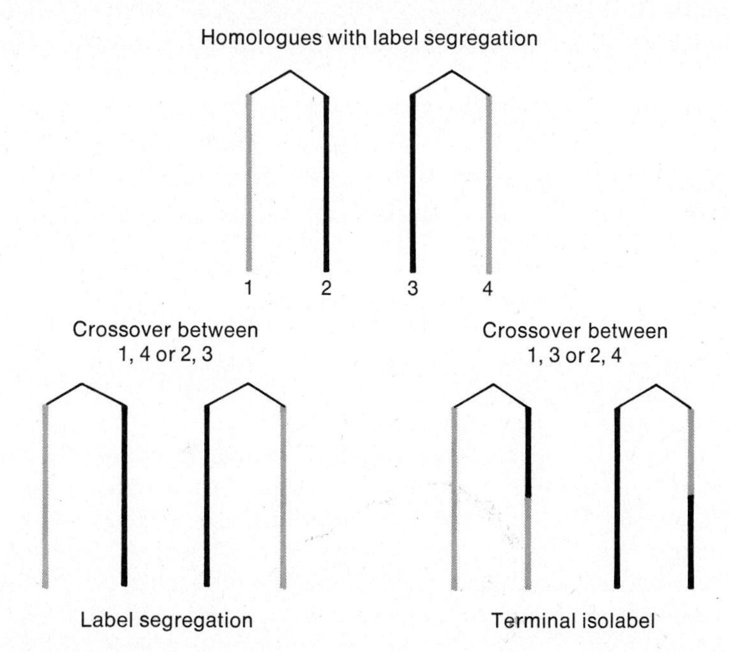

Figure 23-12. Exchange of labeled chromatids in crossing over. The diagram shows the basis of an experiment demonstrating breakage and exchange in eukaryotes. Each chromosome of a homologous pair has one labeled and one unlabeled chromatid. Certain crossovers will give a detectable exchange of labeled parts. (After Peacock, W. J. 1970. *Genetics* **65**:593.)

are a number of possibilities. Breakage and exchange between two unlabeled chromatids or between two labeled chromatids will be undetectable. However, exchange of parts between an unlabeled and a labeled chromatid will be readily detectable at the first meiotic metaphase by autoradiographic techniques (see Chapter 18). Such exchanges are clearly visible in properly prepared material, and this visible exchange of parts can now be used as a measure of the amount of exchange taking place within the cells during meiosis.

The first important discovery made with this technique was that the number of exchanges found in metaphase corresponds exactly to what we would expect on the basis of the number of chiasmata that are counted in pachytene. (Obviously we must correct the figures for the expected number of exchanges between two labeled or two unlabeled chromatids; however, making the simple assumption that breakage and exchange is a random phenomenon, the number of exchanges found fits nicely with the number expected.) This finding is interesting, for it confirms one of the earliest interpretations of crossing over—that the chiasmata seen at meiotic prophase are a visible reflection of exchange of parts of chromosomes.

In one case, using the orthopteran insect *Goniaea australasiae*, it has been possible to do an experiment that tests the copy choice hypothesis directly. In this insect, a temperature shock of 37°C delivered during meiosis greatly reduces the number of chiasmata observed. The temperature-effective period is limited; only those cells given the temperature shock during early pachytene appear to be affected. Prior to, or immediately after, that stage there is little effect. The question is: Will such a temperature shock have any effect on the amount of chromosomal exchange during meiosis? The answer, once again, is yes. As in the normal case, the number of exchanges is correlated directly with the number of observed chiasmata. When few chiasmata are seen in pachytene, there are few exchanges of labeled parts of the chromosomes. Consequently, it seems clear that the breakage and exchange of chromosomal material takes place during early pachytene. If that exchange is prevented from taking place, no chiasmata can be observed. The most significant fact, however, is that the temperature shock is applied well after the replication of DNA, which occurs in premeiotic interphase. Clearly no copy choice mechanism would explain these results. Rather, it appears that only the breakage and exchange mechanism is satisfactory.

It would be unfair to end this discussion without mentioning the many experiments that give contradictory results. Certain experiments seem to indicate that the effective period for agents that increase crossing over is the time of DNA synthesis. However, the bulk of the data now seem to be against this interpretation. How can we explain such contradictory results? Of course, we can always invoke experimental error or the Law of General Cussedness: under the most carefully controlled conditions, organisms will behave as they damned well please! But that is both unfair and unsatisfactory. It is far more likely that the mechanisms that control the exchange of genetic material during meiotic prophase are themselves under the control of other agencies of the cell. Interference with the metabolism of the cells can undoubtedly throw the recombination mechanism "out of kilter" and thus affect the total amount of recovered genetic recombination (either increase or decrease it). Under these circumstances the fact that certain investigators can find a wide variety of agents having a wide variety of effective periods should not be surprising. Taking that evidence as a whole, it seems probable that, in all organisms, some form of breakage and exchange is the mechanism by which genetic recombination is accomplished. Until we understand the breakage and exchange mechanism itself, however, we will be unable to provide critical tests to establish whether copy choice ever occurs in organisms.

The Mechanism of Recombination

As yet we are uncertain how recombination actually occurs, but some interesting suggestions are available. The best leads have come from the study of mutant strains of *E. coli* that are incapable of carrying out recombination of linked genes. These strains are called *recombinationless* or **rec**$^-$. The **rec**$^-$ mutants totally lack recombination, as well as the ability to be transduced. Apparently all the other systems are satisfactory. The phage are able to attach to the bacterial cells, and conjugation of the two bacteria is satisfactory. In both cases, DNA enters the recipient cells either from the bacteriophage or from the donor bacteria. All processes are satisfactory, but there is no recombination. It thus appears that these mutants lack the recombination mechanism.

An interesting feature of the **rec**$^-$ mutants of *E. coli* is that they are also UV sensitive. In other words, they appear to be defective in the DNA repair system. This situation initially suggested that the DNA repair and recombination mechanisms were the same. However, careful analysis has shown that this view is an oversimplification. As would be expected on the basis of the number of enzymes involved (see the preceding section on DNA repair), there are three distinctly different mutational sites in the UV-sensitivity system. Each controls one of the three enzymes needed for the dark repair of DNA. Similarly, there are at least two, and possibly more, mutational sites in the **rec**$^-$ system. But the **rec**$^-$ genes and the UV-sensitivity genes do not map at the same location. At least some parts of the two systems appear to be distinctly different. On the other hand, the fact that **rec**$^-$ mutants are UV sensitive suggests that at least some enzyme (or enzymes) is common to the two systems. The view is further supported by the well-established fact that most mutagenic agents, agents known to damage DNA structure, can alter the frequency of genetic recombination. This fact certainly suggests that the DNA repair system, or some system similar to it, is involved in the exchange of materials between chromosomes. Unfortunately, a definite answer to this question is impossible until all the enzymes of the two systems are completely characterized.

A number of hypotheses have been advanced to explain how breakage and exchange between homologous DNA molecules can occur. Two are given in Figure 23-13. The first is based largely on the results of the phage recombination experiment. It will be remembered that in this case the two different parts of the hybrid molecule were held together by hydrogen bonding in areas in which two chains overlapped. As envisioned in Figure 23-13(a), the two broken molecules pair

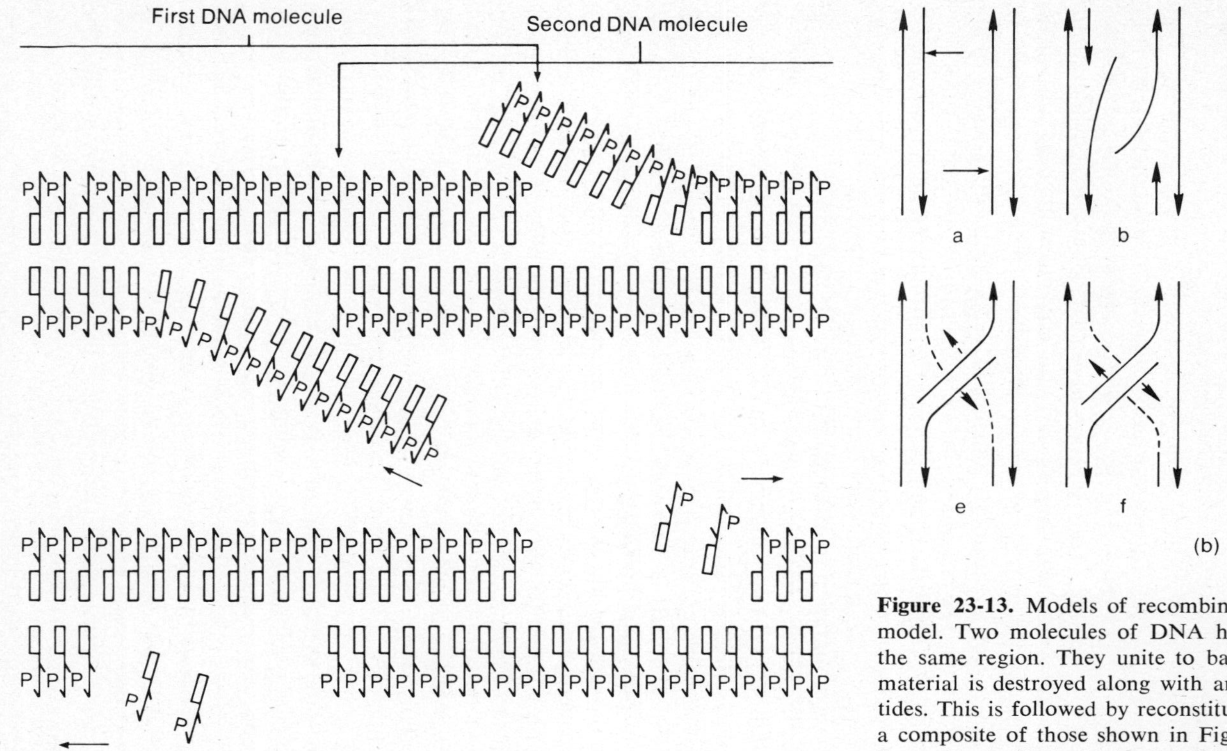

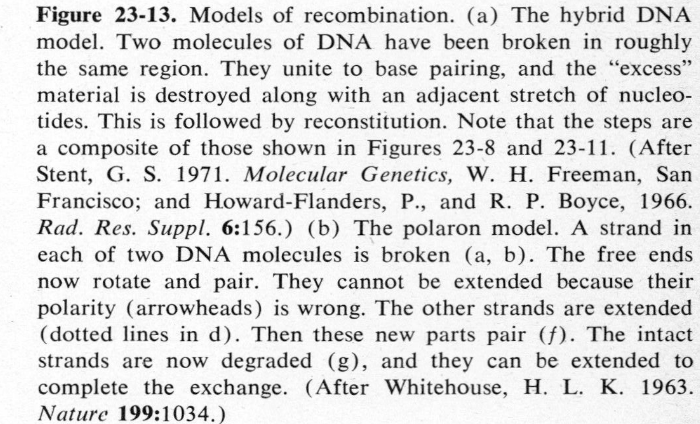

Figure 23-13. Models of recombination. (a) The hybrid DNA model. Two molecules of DNA have been broken in roughly the same region. They unite to base pairing, and the "excess" material is destroyed along with an adjacent stretch of nucleotides. This is followed by reconstitution. Note that the steps are a composite of those shown in Figures 23-8 and 23-11. (After Stent, G. S. 1971. *Molecular Genetics,* W. H. Freeman, San Francisco; and Howard-Flanders, P., and R. P. Boyce, 1966. *Rad. Res. Suppl.* **6:**156.) (b) The polaron model. A strand in each of two DNA molecules is broken (a, b). The free ends now rotate and pair. They cannot be extended because their polarity (arrowheads) is wrong. The other strands are extended (dotted lines in d). Then these new parts pair (*f*). The intact strands are now degraded (g), and they can be extended to complete the exchange. (After Whitehouse, H. L. K. 1963. *Nature* **199:**1034.)

in such a way that they are joined together, held so that the two molecules are aligned in the region of exact homology. The displaced chains are then removed, presumably by the same mechanism as is used for DNA repair (or at least one very similar to it), and the molecules are joined together. We can assume (with very little evidence) that such a process involves at least four different enzymes—the three enzymes of the DNA repair system (an exonuclease, to remove the excess material, a polymerase to replace part of the stretch removed, and a ligase to join the two chains covalently) and presumably another enzyme that breaks open the DNA molecule providing the initial "nicks" that permit the process to begin. The existence of such a "*nickase*" has been demonstrated, and one of the enzymes controlled by the rec$^-$ system of *E. coli* appears to have nickase properties. However, the general model is very much a postulate. Its most interesting features are that the initial breaks can occur almost anywhere in the DNA chain and that the two chains need not be broken in the same place. Furthermore, there is no reason why the process must be reciprocal. The other two parts of the chromosome could easily be degraded, in which case nonreciprocal events would be observed. Of course, the model could also account for reciprocal recombination by merely invoking the high probability of the recombination of the other parts of the two chromosomes. What is important is not that it is a nonreciprocal mechanism but that it can be either reciprocal or nonreciprocal, depending on the conditions within the cell (as yet unspecified).

The second model envisions a mechanism that is completely reciprocal. As indicated in Figure 23-13(b), the initial event is assumed to be a break in adjacent areas of two strands of paired chromosomes. The arrowheads indicate the polarity, and hence the direction of chain extension by the polymerase enzyme. It can be seen that two of the chains are capable of being extended. The other two are not (since the direction of extension is incompatible with their organization). As the two broken chains are extended, they, too, are free to pair, and thus a cross-point is generated between the closely paired DNA molecules. At this point it is necessary to invoke a new enzyme that now degrades the two *intact* chains in the region of the cross-point. This degradation makes possible the extension of these chains to complete the "cross over," and thus a totally reciprocal event has occurred. Careful study of this figure indicates that the kinds of enzymatic activity required are very similar to those given in the first example. It has the advantage of making the events exactly reciprocal, which certainly seems to be the rule in those eukaryotes that can be carefully studied. Furthermore, the initial breaks can occur anywhere, and it is only necessary to have an enzyme that can recognize distortion of DNA chains in the region where the initial events take place. Since this is characteristic of repair enzymes, it seems, in general, to be a reasonable process. Of course, the repair enzymes could equally well remove the area of DNA that is involved in the initial cross and rebuild the molecules to their original, noncrossover condition. Presumably such events must happen. If they do, of course, no recombination will be detected. Recombination in this model would result only from those cases in which the "intact" strands of the DNA are removed and repair proceeds via the crossed-strand mechanism.

On the basis of available evidence, it appears that there may be a difference between prokaryotes and eukaryotes in the frequency of nonreciprocal events. Most of the evidence indicates that genetic recombination in eukaryotes is primarily reciprocal (although there is good evidence that nonreciprocal events do occur rarely). In the prokaryotes, on the other hand, nonreciprocal recombination appears to be the rule. Although initially thought to indicate a difference in the mechanism of genetic recombination, it now appears that any such difference lies elsewhere. It is probable that the recombination enzymes are basically the same in all forms, and in all cases some kind of breakage and exchange must occur. However, in the eukaryotes, the rather complex synapsis of homologous chromosomes undoubtedly places a restriction on the recombination process that is not present in the prokaryotes. In the end, this restriction may be the difference between the two, and the high level of reciprocal events in eukaryotes may well be an indirect consequence of the synaptic mechanism; or it may be the protection of broken DNA strands from degradation by exonucleases, by the chromosomal proteins that encase the DNA. Recombination is important to organisms, for it increases the genetic variability. Whether the mechanism is reciprocal, however, is not as important. In either case, the variability of the total population is increased, and it is on that total population that the evolutionary processes must act. From the point of view of the geneticist, relatively little is likely to be added to our knowledge by the unraveling of the exact mechanism of recombination. The essential hypothesis, based on breakage and exchange, will remain valid and the biological consequences will remain the same regardless of which mechanism is used to effect the breakage and exchange process that leads to recombinational events.

REFERENCES

AUERBACH, C., J. ROBSON, and J. R. CARR. 1948. The chemical production of mutations, *Science* **105**:243–247. (This is a nice review of the early work with chemical mutagens by the people who did most of it.)

HANAWALT, P. C., and R. H. HAYNES. 1967. The repair of DNA. *Sci. Amer.* **216**:36–43. (An excellent, simple description of the repair process.)

HAYES, W. 1968. *The Genetics of Bacteria and Their Viruses.* 2nd ed. John Wiley & Sons, New York. (Chapter 13 presents an excellent summary of mutation, and Chapter 15 covers recombination at the molecular level.)

HAYNES, R. H., S. WOLFF, and J. TILL (Eds.). 1966. Structural defects in DNA and their repair in microorganisms. *Rad. Research Suppl. 6.* (A collection of original papers and reviews. For the advanced student mostly, but the summary by Setlow and the discussion conducted by Delbruck are worthwhile for any student.)

LOVELESS, A. 1966. *Genetic and Allied Effects of Alkylating Agents.* Pennsylvania State University Press, University Park, Pa. (A well-written, comprehensive review of an important part of chemical mutagenesis.)

Original Papers and Advanced Reviews

FREESE, E. 1959. On the molecular explanation of spontaneous and induced mutations. *Brookhaven Symp. Biol.* **12**:63–75. (One of the important papers on the molecular basis for mutation.)

MESELSON, M., and J. J. WEIGLE. 1961. Chromosome breakage accompanying genetic recombination in bacteriophage. *Proc. Natl. Acad. Sci. U.S.* **47**:857–868. (The report of the work that disproved copy choice for microorganisms.)

PEACOCK, W. J. 1970. Replication, recombination, and chiasmata in *Goniaea australasiae. Genetics* **65**:593–617. (A modern demonstration that breakage and exchange is the basis for crossing over in eukaryotes.)

TOMIZAWA, J., and N. ANRAKU. 1964. Molecular mechanisms of recombination in bacteriophage. II. Joining of parental DNA molecules in phage T_4. *J. Mol. Biol.* **8**:516–540. (One of a series of papers by these authors demonstrating the formation of hybrid DNA molecules as a basis for recombination by breakage and exchange.)

WHITEHOUSE, H. L. K. 1963. A theory of crossing over by means of hybrid deoxynucleic acid. *Nature* **199**:1034–1040. (An important paper that presents a molecular model to explain crossing over.)

QUESTIONS AND PROBLEMS

23-1. Define the following terms:

mutation frequency	back mutation
conditional lethals	polar mutations
transitions	transversions
fractional mutation	copy choice

23-2. What is the basis for the fact that different genes may have very different mutation frequencies?

23-3. Why does back mutation always occur at a much lower frequency than forward mutation?

23-4. How do spontaneous mutations occur?

23-5. What are temperature-sensitive mutations? (Compare your answer to the one you gave to Question 17-13.)

23-6. In what way are temperature-sensitive mutants particularly useful in the study of genetics?

23-7. In what ways is it possible to talk about "specific" mutagens? Could such agents provide a way to obtain directed mutation?

23-8. What is the significance of our failure to find fractional mutations in the progeny of treated Drosophila?

23-9. Since only one strand of the DNA molecule is transcribed into RNA, what is the advantage of a doubly stranded molecule?

23-10. Why do we feel certain that recombination occurs by breakage and exchange?

23-11. In what way are the processes of genetic recombination and DNA repair similar?

23-12. For a long time one of the mysteries of the breakage and exchange mechanism of recombination was how two homologous DNA molecules could break in *exactly* the same place and then exchange. (If that were not the case, the nucleotide sequence would change, and each recombination would give rise to a mutation.) How does the hybrid-DNA model remove this difficulty?

24/The control of gene action in prokaryotes

One of the most striking features of living systems is their ability to grow and differentiate in a highly ordered way. It suggests that the entire organism is programmed by its genetic material, so that at each stage of development certain goals are set by the genes. And when these goals have been achieved, the organism in some way "recognizes" that they have been and either stops or alters its pattern of development in some new direction. In the development from egg to adult, there may be a multiplicity of stages (larvae, pupae, adult) or a continuous development with an intervening infantile period, but in either case it is clear that certain genes express themselves at certain times and not at others. The changes in secondary sex characteristics at puberty are genetically determined (Chapters 27 and 28). Yet there is a long period of time in development when many of these changes are not evident, which suggests that at one period (puberty) certain genes are "turned on" and produce proteins that are essential for the control of particular developmental processes.

This control of genetic function is also evident at the level of the single cell. It is clear that there are definite timing mechanisms in the cell. For example, the replication of DNA occurs during a very limited period of the cell cycle (generally designated S for synthesis), whereas the transcription of the DNA to *m*RNA occurs largely in the period before S (designated G_1 for growth) or immediately after S (designated G_2). The DNA must be replicated before the cell divides; otherwise it would be impossible to have a truly mitotic division. Similarly, it is essential that the spindle proteins be formed at the time of division; otherwise chromosomal separation could not be effected. The cell must have a regulatory mechanism to ensure that each of the essential events in the cell cycle occurs at precisely the right time. We cannot think of the cell as a simple, unorganized set of biochemical reactions. It is the integration and control of these reactions that ultimately give the cell its direction, the historical aspect of which we call evolution.

A third line of evidence supporting the idea of differential gene activity comes from the fact that the enzymes of the cell may be present in different concentrations. In general, the amount of genetic material is uniform for all enzymes, one dose in haploids, two in diploids, and so forth. Thus some other method of control of the *amount* of synthesis must be present to ensure that certain enzymes are found in larger quantities than others. Consequently, we can address ourselves to the problem of the control of the *rate* of protein formation in cells, since it is the rate of formation that ultimately determines the steady-state level of enzymes within the cell (the level at which the amount lost is equal to the amount produced).

Since the structure of cells is determined to a great extent by the kinds of structural proteins available, and since the control of biochemical processes can be viewed as a regulation of enzymatic activity, it seems reasonable to propose that cellular control is to a great extent equivalent to the control of the synthesis of proteins. We know relatively little about the synthesis of structural proteins. Far more work has gone into the investigation of the synthesis of enzymes, for the obvious reason that they are easier to isolate and characterize on the basis of their enzymatic activity. However, what evidence we have so far (largely from studies of virus synthesis) indicates that there is no basic difference between the synthesis of structural proteins and enzymatic units. Consequently, it is safe to assume that studies of the control of particular enzymatic systems demonstrate principles that are quite general. In this chapter we will discuss control mechanisms in prokaryotes. The next chapter will treat the more complicated mechanisms of higher organisms.

Induction and Repression

Enzyme Induction

The most important advance in the study of the control of enzyme production was made through investigation of the *induction* of the synthesis of enzymes by substrates. Induction is a rather curious phenomenon. Most cells can utilize certain specific materials for growth (substrates) and cannot use other materials. However, in certain instances, when a cell is deprived of its normal substrate and grown on material that it does not normally utilize, the cell will "adapt" to the new substrate and gradually obtain the ability to use it in place of the normal one. The initial discovery of this adaptation was made with yeast cells, which normally use glucose as a carbon and energy source. However, they can be adapted to grow on galactose (another 6-carbon sugar), if they are incubated with this sugar in the absence of glucose. The most significant feature of this adaptation is that there is a lag period of from one to several hours before the cells begin to utilize the galactose, but after that lag period, the rate of growth is essentially the same on galactose as it was previously on glucose (Figure 24-1).

In the mid-1940s it was demonstrated that the lag period is, in fact, a period of synthesis for those enzymes that are required for the movement of galactose across the cell membrane into the cell and for the conversion of galactose to glucose (which is then used by the normal metabolic pathways). The utilization of galactose requires the synthesis of several enzymes, one of which is galactokinase, an enzyme that adds a phosphate group to the galactose as the first step in its conversion to glucose. The initial studies of the process focused on this enzyme, for it is easily characterized. Since no galactokinase could be detected in the cells prior to the exposure of the cells to galactose, it had to be assumed that the production of the enzyme by the cells is a completely adaptive response to an environmental stimulus. The substrate molecule is apparently capable of altering protein synthesis in such a way that a new enzyme will be produced, one that enables the cell to use the substrate.

The enzymes necessary for the utilization of glucose are called *constitutive enzymes.* They are made by the organism in sufficient quantity for normal metabolic activity *whether or not* the substrate is present. If yeast cells that have been growing on galactose for a period of time are inoculated into a medium containing glucose

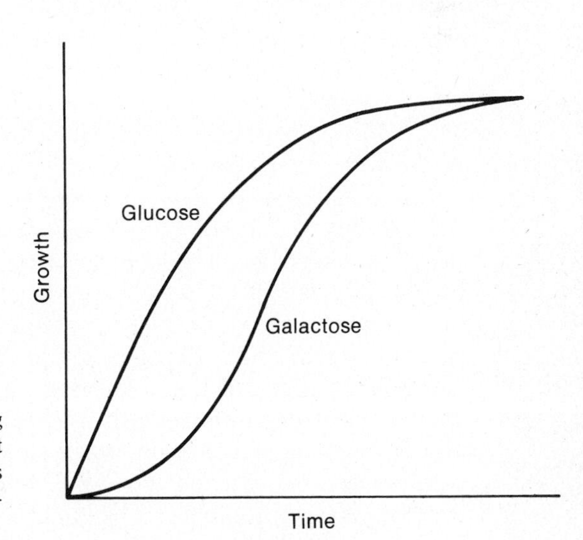

Figure 24-1. Enzyme induction. The cells show a lag period in their growth on galactose that is not present when they are grown on glucose. The enzymes needed for galactose utilization are synthesized during this period.

rather than galactose, growth continues normally with no lag period. This finding indicates that the enzymes necessary for glucose utilization are always present, even in cells that have been adapted to grow on galactose. On the other hand, if an organism that has been adapted to grow on galactose is transferred to a glucose medium and allowed to grow on it for a period of time, when the organism is reinoculated into the galactose medium, the lag period associated with enzymatic induction will reoccur. The cell is incapable of "remembering" that at one time it was adapted to galactose utilization. This situation suggests that the formation of the inducible enzyme is a protein-synthesis response that can be maintained only in the presence of the substrate. As the cells grow, the adaptively formed enzymes are diluted until they are of no further consequence. Unless the enzymes are synthesized constantly, the ability of the organism to grow on the substrate will be lost. From such studies it is evident that the major difference between constitutive enzymes and inducible enzymes is that the *constitutive* enzymes are produced in sufficient quantity to maintain normal metabolic activity whether or not the substrate is present, whereas the *inducible* enzymes are produced in sufficient quantity only in the presence of the substrate.

After the discovery of inducible enzyme systems, a wide variety of organisms was studied in an effort to determine the extent of the process. It now appears that induced enzyme formation is a relatively common phenomenon and that most cells have the ability to respond to at least a few compounds with the formation of enzymes that will use these compounds as substrates. In fact, this response is crucial in the maintenance of the proper concentration of certain important compounds within multicellular organisms. For example, in plants, the growth hormone auxin is produced at the tips (apices) and diffuses throughout the plant body. The presence of auxin molecules within various plant cells causes the formation of an oxidase that destroys auxin as an adaptive response. But all cells are not equally capable of producing this enzyme. Thus differences in auxin level can be maintained in various tissues by the differences in the induced enzyme response of which they are capable. The result, in turn, contributes to differences in the rates of growth of these various tissues.

It soon became evident that the substrate molecule acting as an inducer did not combine directly with the induced enzyme. Mutant forms of inducible cells were obtained that could produce only inactive protein. Nevertheless, they produced this protein only in the presence of the inducer. It is difficult to envision a complex between an inactive protein and its "substrate." Furthermore, several enzymes are produced at the same time, in response to a single type of inducer molecule. At least three enzymes are required for the utilization of galactose, and all appear at the same rate in response to the added galactose. There could be no usual enzyme-substrate complex in this case, since the substrate for the last enzyme in the reaction sequence cannot be present until the initial substrate has been converted into it. Finally, the inducer is not destroyed during the period of induction, as it should be if it combined with enzyme. All three facts suggested that the inducer forms a complex with some template that controls enzyme synthesis rather than with the enzyme itself. The fact that induction occurs by *de novo* synthesis of protein, rather than by "reshaping" existing protein, is in agreement with this postulate.

As investigation of the induced enzyme response continued, it was discovered that the inducer enters the cell through an active transport mechanism, which itself is induced by the substrate molecule. Unless the inducer is concentrated within the cell by such a mechanism, there is little adaptive response, since the substrate concentration is not high enough to give either maximum induction or maximum enzyme activity. The terminology of inducible enzymes was carried over to this active transport mechanism, and it was given the name *permease*, suggesting that the substrate induces an enzyme that is responsible for the transport of the molecule across the membrane. The nature of this permease is uncertain; however, it has the kinetic properties of an enzyme. What was most important in studies of this type was the demonstration that the lag period of growth was associated with the induction of the permease more than with the induction of the other enzymes. If cells are incubated for a brief period with the substrate and then removed, it is possible to demonstrate that a permease has been formed. When such cells are put in the presence of the inducer (substrate), there is no lag period at all in the formation of induced enzyme (and consequently the growth of the organism). Apparently the permease is more stable than the other enzymes, and once induced it remains active for a relatively long time. Thus the original suggestion that the inducer combines with some template that controls protein synthesis was reinforced by the finding that the response to the substrate, once inside the cell, is extremely rapid.

The problem raised by the preceding findings is the identity of the template with which the inducer combines. Since the amount of enzyme formed is proportional to the concentration of inducer (within limits), and since the response is rapid (once the inducer enters the cell), it seems reasonable that the inducer is releasing inhibition of that part of the protein-forming mechanism responsible for the production of a particular enzyme. (We can assume that the inducer reacts with something preformed in the cell, since the response is rapid.) There are three possibilities: the *m*RNA template, the gene, or some other molecule that controls enzyme synthesis. Since relatively few genes are associated with the production of a single enzyme (only one in haploid organisms), little substrate should be required to release all inhibition if the inducer acts on the gene. However, the yield of enzyme increases with inducer concentration over relatively wide ranges. Therefore it is unlikely that the inducer reacts directly with the gene.

Induction is a highly specific process. Galactose induces enzymes associated with galactose utilization and no others. Thus the responding mechanism in the cell must be able to "recognize" the inducer. It must be able to tell galactose from glucose, fructose, or other sugars. The response has a selectivity reminiscent of enzyme-substrate complexes. As with enzymes (Chapter 17), it is possible to trick the mechanism by supplying derivatives of the normal substrates. (In fact, such derivatives are frequently used to avoid the problems of gradual loss of inducer by the action of newly synthesized enzymes.) However, the process is still highly selective, with the various "pseudosubstrates" always bearing a strong resemblance to the natural inducer. Consequently, whatever the properties of the template, it is essential that it have binding possibilities for a stereospecific union with the inducer molecules. Obviously *m*RNA is an unlikely candidate for this type of reaction, as is the DNA of the gene. We are left, then, with the hypothesis that the inducer binds to an entity that controls protein synthesis without itself being directly a part of the synthetic machinery. We must ask whether there is a reality to go with this remarkably metaphysical construction.

Genetic Studies of Induction

The major advances in the study of inducible enzyme formation were made by studying the genetics of the process. A great deal of the work has been done with the lactose system of *E. coli*, but the results appear to be applicable to most inducible systems. (Lactose is a disaccharide of glucose and galactose, called a β-galactoside.) Three enzymes are required for lactose utilization—β-galactosidase, a permease, and a galactoside transacetylase. The subsequent utilization of the galactose produced would require the induction of the three enzymes of the galactose system discussed above. In 1961 Jacob and Monod published a review of the extensive literature on this subject and proposed a brilliant hypothesis to explain the results. What follows is largely based on their interpretation.

It early became apparent that the formation of inducible enzymes is under genetic control. Cells that are capable of making inducible enzyme (inducible cells) can be mutated to the constitutive condition, and they then produce the enzymes whether or not the substrate is present. For any particular enzyme system, this mutational event occurs in a single gene, and this gene maps at a place distinct from the genes carrying the information for the production of the enzymatic proteins. Since the function of the latter set of genes is the production of enzymes, they are given the name *structural genes* (they specify the structure of the proteins). The others are called *regulator genes*, since they merely control whether the structural genes will be inducible or constitutive.

The function of the regulator gene can be deduced from crosses of different strains of *E. coli*. The normal condition of the cell is that it can be induced to form β-galactosidase by the addition of substrate. Thus it has a regulator gene that makes the cell inducible ($\mathbf{i^+}$) and a structural gene responsible for the production of the enzyme in an active form ($\mathbf{z^+}$). Other mutant forms can be obtained: those that are constitutive and, therefore, have a mutation at the regulator locus ($\mathbf{i^-}$) and those that are incapable of forming an active enzyme because of a mutation in the structural locus ($\mathbf{z^-}$). An organism of the genetic constitution $\mathbf{i^-z^-}$ produces protein in the absence of inducer, but the protein is an inactive form of the enzyme.

As discussed in Chapter 12, the mating procedure in *E. coli* involves the "injection" of a chromosome from the donor strain (Hfr) into a recipient cell, with little exchange of cytoplasm. In this process the chromosome is transferred in a linear fashion with one end always entering the recipient cell first, followed by a gradual progression of the rest of the chromosome into the cell. Conjugating cells can be separated by mechanical agitation at any point during the process, and in this way it is possible to control the amount of genetic material transferred. By doing studies of the physiology of various "injected" cells, separated at different time intervals after mating begins, it is possible to detect the time at which various genetic units are transferred during the mating process (Figure 24-2).

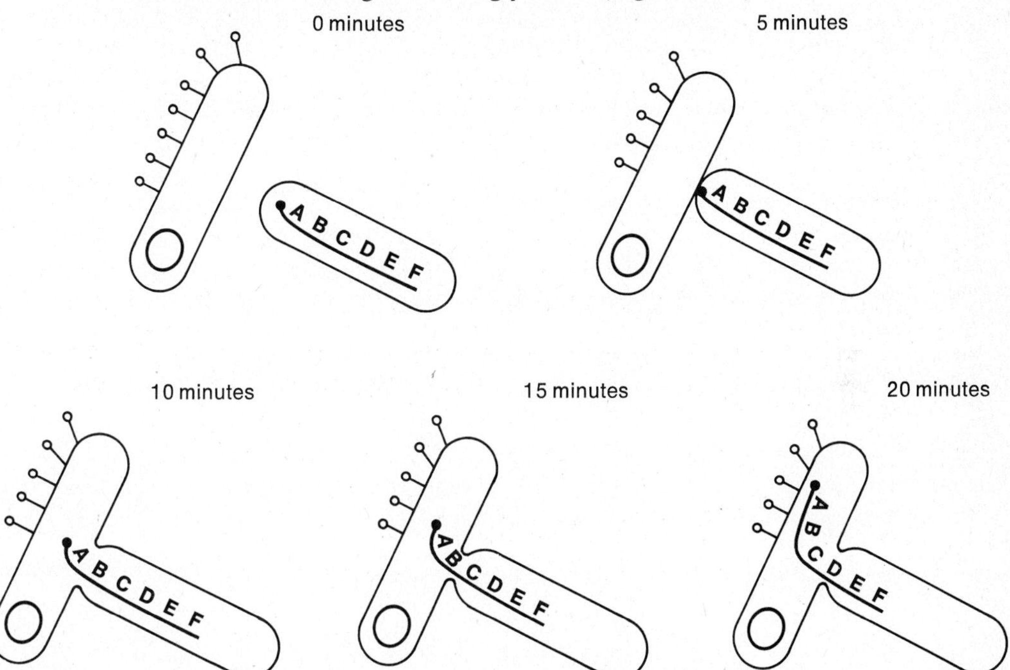

Figure 24-2. Mating in *E. coli.* The recipient cell is marked with bacteriophage. The donor cell injects its chromosome in a linear manner into the recipient cell. At certain times after mating has begun only certain portions of the chromosomes have been transferred. Separation of the cells at a particular time insures that only certain genes will have been transferred. (After Jacob, F., and E. Wollman. 1961. *Sexuality and the Genetics of Bacteria.* Academic Press, New York, N.Y.)

The mating can be performed in such a way that the genes are transferred from a bacterium that is both inducible and capable of making active enzyme ($\mathbf{i^+z^+}$) to a cell that is constitutive but incapable of making active enzyme ($\mathbf{i^-z^-}$). Such a cross produces the heterozygote $\mathbf{i^+z^+/i^-z^-}$. In this mating, it is observed that both the $\mathbf{i^+}$ and $\mathbf{z^+}$ genes enter at approximately 18 minutes. Therefore they must be closely linked. The recipient cell is constitutive and is constantly making an inactive protein that is the product of the $\mathbf{z^-}$ gene. However, as soon as the $\mathbf{i^+z^+}$ genes have entered the cell, active enzyme is produced as well. No inducer is required for enzyme formation, even though the $\mathbf{i^+}$ gene is transferred with the $\mathbf{z^+}$ gene. In other words, in a system in which an $\mathbf{i^-}$ gene has been active, the cytoplasmic constitution is such that the $\mathbf{z^+}$ gene can act in the absence of an inducer even though it is linked to the $\mathbf{i^+}$ gene.

If the converse experiment is done, and i^-z^- is injected into an i^+z^+ bacterium, no formation of constitutive enzyme occurs, even though the same heterozygote (i^+z^+/i^-z^-) is formed. Thus it cannot be the function of the i^- allele that permits constitutive production in the first case. (In other words, i^- is not dominant to i^+.) Rather, it is the nature of the cytoplasm in which the gene finds itself that determines whether the production of enzyme will be constitutive or inductive. A z^+ gene placed in cytoplasm that has been produced by an i^- cell functions constitutively whether or not an i^+ gene is also present; a z^+ gene placed in cytoplasm that has been produced in an i^+ cell functions inductively, whether or not an i^- gene is also present.

The interpretation that the **i** gene "conditions" the cytoplasm in some way is further supported by considering what happens in the first case as time passes. Gradually these cells (i^+z^+ donor/i^-z^- recipient) become inducible. After about one hour, these cells that had been producing β-galactosidase as a constitutive enzyme cease their production and will only begin production again when inducer is added.

The simplest interpretation of the preceding results is that the i^+ gene is synthesizing a *repressor* molecule that normally prevents the synthesis of the enzyme. This repressor must be made in excess and must be found in the cytoplasm of cells in which the i^+ gene has been active. In such cytoplasm, all alleles of the **z** gene will be incapable of producing protein as a result of the presence of repressor. However, should an inducer substrate be added, this repressor is inactivated (Figure 24-4). On the other hand, the i^- allele is apparently incapable of forming active repressor, and, consequently, cells that have developed in the presence of the i^- allele lack repressor altogether. Any **z** allele introduced to the system will immediately form the protein for which it carries the information. Once the i^+ allele has had a chance to synthesize enough active repressor, however, the i^+/i^- diploid becomes inducible, and the constitutive synthesis of β-galactosidase is shut down.

It is presumed that the substrate molecule that acts as an inducer can react with the repressor molecule and inactivate it. It is safe to assume that this reaction is rather specific, since a particular inducer will turn on the synthesis of only those enzymes responsible for its own utilization. For example, a sugar will induce sugar-utilizing enzymes but will not induce enzymes for the utilization of amino acids. This situation suggests a "recognition" of the inducer, usually a small molecule, by the repressor, a reaction similar to that between an enzyme and its substrate. Our hypothesis must be that the resulting inducer-repressor complex can no longer function to repress (prevent) the synthesis of the enzyme. A conformational change in the repressor molecule that causes its active site to be deformed (Figure 24-3) is presumably responsible.

One fact that remains to be incorporated into the model is that the inducer causes the production of several enzymes of a biosynthetic pathway (all directly, or indirectly, related to its own utilization). In both the lactose and the galactose systems, three separate enzymes are induced; and in certain cases, the number of enzymes produced is quite large. In all cases, the members of the enzymatic pathway appear simultaneously, in response to the initial inducer (rather than being sequen-

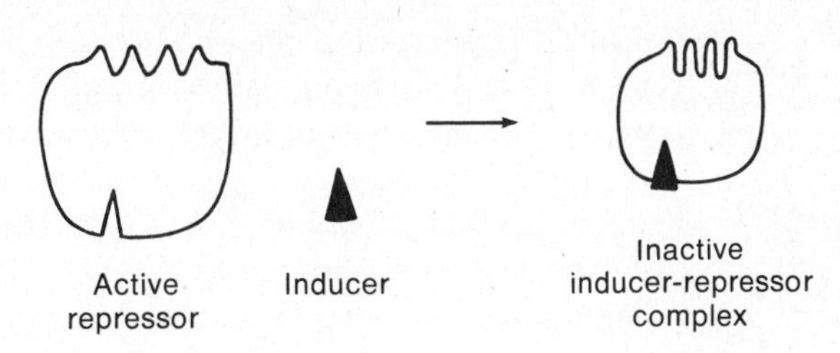

Figure 24-3. Repressor-inducer interaction. The normal repressor molecule has an open active site. However, if an inducer molecule reacts with it, the shape of the repressor changes causing the active site to be closed.

tially induced as the reaction proceeds). Thus the system is capable of responding at the level of several different structural genes, all of which are apparently under the control of a single regulator gene. We must conclude that either the repressor produced by the regulator gene can react with several different loci or that it can react with some other part of the genome, which, in turn, controls the expression of the structural genes. In *E. coli*, the latter alternative is apparently the case. Studies of the lactose system revealed that certain kinds of constitutive mutants remove the structural genes from the control of the regulator gene. Such mutants are constitutive under all circumstances, even when put into the heterozygous condition with known $\mathbf{i}^+$ genes present. When these mutants were mapped, it was determined that the mutational event had occurred at a place close to, but not within, the structural genes and at a place distinct from the regulator gene. Thus it was necessary to propose a third type of genetic locus, called the *operator*, which determines whether the structural genes will be repressed by the product of the regulator gene.

To clarify the preceding point, consider two different kinds of heterozygotes of the lactose system. (In this case, **i** and **z** have the meaning given to them previously; **o** is the designation of the operator locus, and **y** is a designation for the gene controlling the production of permease.) First, let us take the diploid $\mathbf{i}^+\mathbf{o}^+\mathbf{z}^-\mathbf{y}^-/\mathbf{i}^+\mathbf{o}^+\mathbf{z}^+\mathbf{y}^+$. Such a cell is completely inducible; and when inducer is added, it will make both active and inactive forms of β-galactosidase and permease. The $\mathbf{i}^+$ genes make active repressor and the $\mathbf{o}^+$ gene interacts with that repressor so that no synthesis of protein under the control of the structural genes **z** and **y** can occur.

Consider a second diploid, which we can designate $\mathbf{i}^+\mathbf{o}^+\mathbf{z}^-\mathbf{y}^-/\mathbf{i}^+\mathbf{o}^c\mathbf{z}^+\mathbf{y}^+$. Such a cell is constitutive for the production of active β-galactosidase and active permease. If inducer is added to this cell, inactive forms of the two enzymes are produced ($\mathbf{z}^-$ and $\mathbf{y}^-$ become active). Apparently the operator mutation can confer a constitutive nature on only those structural genes that are directly linked to it. The presence of $\mathbf{o}^c$ does not make the $\mathbf{z}^-\mathbf{y}^-$ genes (on the other chromosome) constitutive. They are still subject to control by the repressor produced by the $\mathbf{i}^+$ gene, since the operator linked to them is $\mathbf{o}^+$.

The foregoing experiments suggest that the operator gene can be mutated in such a way that it renders the structural genes associated with it insensitive to the repressor produced by the regulator gene. In microorganisms, particularly in *E. coli*, all the structural genes responsible for the production of the enzymes for a particular biosynthetic pathway are closely linked. In this case, the genes for the lactose system are directly adjacent to one another (as are those for galactose utilization). For this reason, it is simple enough to propose that the operator locus can control the transcription of all genes simultaneously and that under normal conditions the repressor molecule binds only to the operator locus rather than to the structural genes individually. As long as the repressor is bound to the operator locus, the RNA polymerase cannot move past it on the DNA, and the message contained in the structural genes cannot be transcribed into *m*RNA. On the other hand, should the operator locus be changed in such a way that it can no longer combine with the repressor molecule, the genes are read normally and the cell is constitutive. A diagram of the system developed to explain enzyme induction is shown in Figure 24-4.

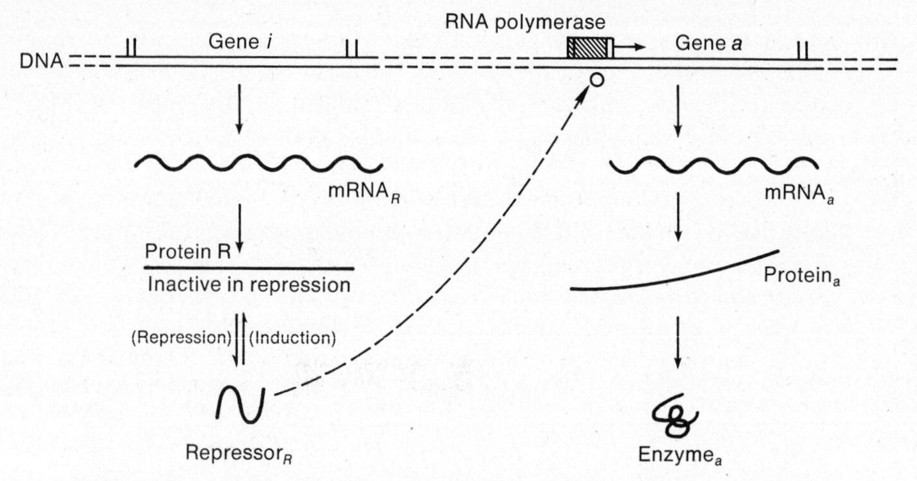

Figure 24-4. A general model for induction and repression of protein biosynthesis. The regulator gene, *i*, is responsible for the synthesis of a protein that can exist in two conformations. In one form the protein is inactive as a repressor. In the other (bottom), the protein has affinity for a chromosomal location, *o*, the operator. When the repressor protein is at *o*, RNA polymerase is unable to synthesize *m*RNA along gene *a*. When repressor protein is absent from *o*, RNA polymerase initiates synthesis of *m*RNA. The amount of the repressor available for inhibition of *m*RNA synthesis depends on the rate of synthesis of the repressor protein, the repressor's affinity for *o*, and the presence of molecules of small molecular weight that influence the interconversion of the two forms of the repressor. In induction, the substrate (etc.) converts the repressor molecule to a conformation that is inactive as a repressor. In repression, the end product (etc.) enhances the conversion of repressor molecules from the inactive form to a conformation that is active as a repressor. (After Hartman, P. E., and S. R. Suskind. 1969. *Gene Action*, 2nd ed. Reprinted by permission of Prentice-Hall, Inc., Englewood Cliffs, New Jersey.)

There is additional evidence to support the model proposed in Figure 24-4. Another complex locus involving an operator and structural genes is the one controlling the synthesis of purines. This region lies near the lactose and galactose

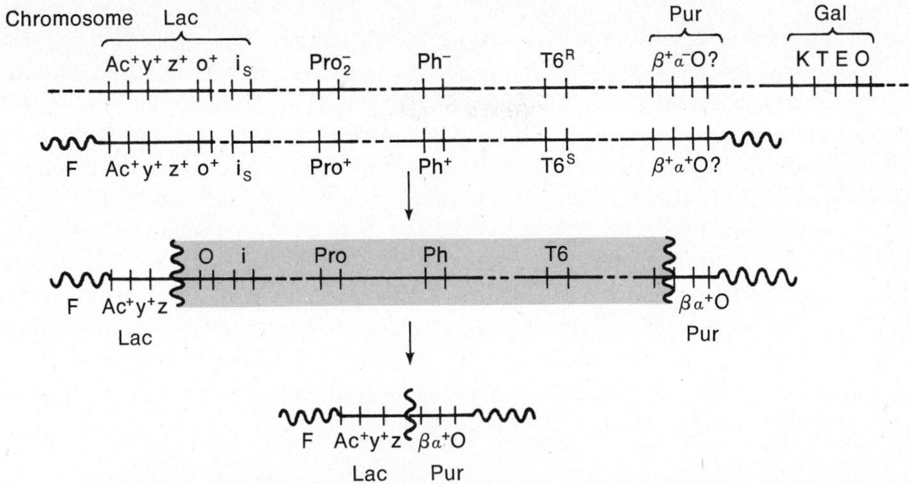

Figure 24-5. Deletion in *E. coli* joining the lactose and purine operators. The upper line presents a short region of the chromosome showing three well-defined operons, galactose (**gal**), purine (**pur**), and lactose (**lac**). The dotted lines indicate omitted regions of the chromosome. The line immediately below carrying the F genome is a small fragment of the chromosome (known as an episome) carried along in the cell. The episome underwent a deletion (shaded region) joining the **lac** and **pur** operons. Under appropriate conditions, such episomes can be tested for genetic function, and in this case the union of the two operons placed the **lac** region under the complete control of the **pur** operator. The structural genes are apparently completely at the mercy of whatever operator they happen to be attached to. In the normal chromosome, the intervention of the other genes (particularly its own operator locus) separates the **lac** operon from the **pur** operator. (After Jacob, F., A. Ullman, and J. Monod. 1965. *J. Mol. Biol.* **13**:704.)

operons discussed above (Figure 24-5). Certain mutants have been obtained that, on testing by both biochemical and genetic means, are apparently deletions of the region intervening between the purine and lactose operons. The result of one such deletion is to remove both the regulator and the operator locus of the lactose region and a small part of the β-galactosidase gene, as well as a small part of one of the structural genes of the purine region. In such mutants, the lactose structural genes now come under the specific control of the purine operator. They behave as if they were structural genes for purine biosynthesis. Whatever affects the purine operator also affects the lactose genes. Clearly such a demonstration indicates that the transcription of the genetic message is under the control of an operator locus and that the structural genes can only be transcribed in the absence of a repressor or when they are linked to an operator that can no longer combine with the repressor. Since the structural genes under the control of a particular operator locus act as a unit in transcription, they have been given the collective name of *operon.*

The model of induction involves two sorts of reversible interactions—a reversible interaction between the inducer and repressor molecule and a reversible interaction between the repressor molecule and the operator locus. If neither interaction is reversible, the system would remain turned on or off indefinitely, and yet the response is relatively rapid and concentration dependent. Consequently, we must assume that, at both levels, the binding is reversible and that even in $\mathbf{i^+o^+}$ cells a low level of enzyme should be present at all times. (There would be *some* occasions when the repressor is not attached to the $\mathbf{o^+}$ locus.) Such has proved the case. For example, it can be shown that approximately 1 to 5 molecules per cell of β-galactosidase are produced in the absence of inducer. When inducer is added, the concentration rises to approximately 5000 molecules per cell (a normal level for most enzymes). Consequently, the induction of enzyme synthesis cannot be viewed as an all-or-none phenomenon. There is a slow "leaking" of the enzyme to the cytoplasm, and the induction merely controls the *rate* of the process, increasing it to maximum levels at times when sufficient inducer is present to indicate a "need" for the production of more enzyme.

Isolation of the Repressor

The model presented in Figure 24-4 is entirely consistent with the observed genetic and biochemical facts. However, it remains metaphysical in that we have postulated a repressor molecule (a demon) responsible for the observed effects. Before the model can be considered satisfactory, it is necessary to demonstrate that such repressor molecules exist. This was demonstrated for the lactose system in 1967. Repressor molecules for other systems have been isolated; however, since the development of the model was based on the genetics of the lactose region, that work is discussed here.

The first problem is what to look for when searching for a repressor. Since the molecule is postulated to be the product of a gene ($\mathbf{i^+}$), it must be either RNA or protein (or both). However, the stereospecificity that the repressor shows for combination with inducers is so strongly reminiscent of enzymatic function that a protein seems to be the most likely candidate. Since the amount of any particular repressor protein present in any cell is likely to be small, it is not easy to extract total cellular protein and look for repressor molecules. However, the repressor molecules do have one property that should distinguish them from the bulk of other protein. They bind inducer molecules specifically. Fortunately, a mutant strain was discovered that produces a repressor having a strong binding affinity for the inducer molecule. The mutant is at the regulator locus and is designated $\mathbf{i^t}$ for "tight" binding. Since the most reasonable test for the presence of the repressor molecule is its ability to bind a specific inducer, such a strain is extremely useful in that it requires less of the repressor to give demonstrable binding. Furthermore, the strain can be made diploid for the **lac** region. With two copies of the $\mathbf{i^t}$ gene, the amount of repressor made is doubled. Such cultures of *E. coli* were grown, and the proteins were extracted and separated in a number of different fractions, on the basis of size, solubility, and similar features. The various concentrated protein fractions were then tested for their ability to bind a specific inducer. Once a fraction of protein had been found that gave evidence for specific binding of the inducer, it was possible to purify it to the point that only that one type of protein was present.

It was presumed that the purified protein isolated by the preceding method was repressor. A number of tests support this interpretation. The protein can be tested in vivo to demonstrate that, in fact, it can repress transcription of the **lac** operon. Furthermore, the protein can be extracted from $\mathbf{i}^-$, $\mathbf{i}^+$, and $\mathbf{i}^t$ cells, following exactly parallel extraction procedures. As would be expected, the $\mathbf{i}^-$ strain yields no molecules with the ability to bind the inducer, whereas both the $\mathbf{i}^+$ and $\mathbf{i}^t$ strains do. Similarly, as would be expected, the inducer is bound more tightly to the repressor isolated from $\mathbf{i}^t$ than it is to the repressor isolated from $\mathbf{i}^+$.

A study of the properties of the isolated repressor indicates that the binding of inducer is not affected by treatment with either RNase or DNase, but it is removed by treatment with proteolytic enzymes. Similarly, binding of inducer is removed by temperatures above 50°C. These properties are consistent with the proposition that the repressor molecule is a protein.

Having isolated the repressor molecule, the remaining question was whether it could be demonstrated that the isolated molecule would bind to the operator region of the *E. coli* chromosome. To do so, it was necessary to make heavily labeled repressor molecules in bacteria that had been made triploid for the $\mathbf{i}^+$ locus. (In addition to the region on its own chromosome, another **lac** region was carried on a lysogenic phage and a third carried on an F^+ episome.) One asset of such a system is that a relatively high concentration of repressor molecules will be produced. A second is that it is possible to separate the phage DNA from the others. Since the phage used is relatively small, the **lac** region represents a large percentage of the total phage DNA; thus the phage DNA is a good test system for the binding of the repressor to the operator locus.

Once the phage **lac** region DNA and the labeled repressor have been isolated, they can be mixed and then subjected to density gradient centrifugation. If the repressor binds to the DNA, the band of phage DNA isolated will carry the majority of label from the repressor. This proved to be the case. Therefore the repressor binds to DNA. If the repressor is first mixed with inducer and then added to the DNA, only a small amount of binding of the repressor to the DNA is observed, which is in agreement with the proposition that when the inducer binds to the repressor, the conformation of the protein molecule changes and the binding to the operator is removed.

The binding of the repressor is specific for the **lac** region. Tests indicate that there is no binding to DNAs that do not have the **lac** genes. But does the repressor bind to an operator locus? This question can be answered by the use of **lac** regions containing different kinds of operator mutations. (There are a variety of such mutations in which the amount of enzyme produced constitutively varies.) The repressor can be demonstrated to bind far more strongly to an $\mathbf{o}^+$ system than it does to a 1% $\mathbf{o}^c$ (that is, one that produces 1% of the normal amount of enzyme constitutively), and far more strongly to a 1% $\mathbf{o}^c$ than to a 20% $\mathbf{o}^c$. Consequently, it is established that the repressor made by the $\mathbf{i}^+$ gene binds specifically with an operator locus in the genome, since all other parts of the DNA are identical in these tests.

It is interesting that melting of the DNA before union with the repressor completely removes the binding. Apparently the repressor can only bind with doubly stranded DNA molecules. This finding is consistent with the postulate that the repressor acts to prevent transcription of the DNA message, a process that occurs only with doubly stranded DNA (Chapter 19).

Based on the nature of the repressor molecule, an estimate of the size of the operator locus is possible. The repressor must be able to recognize a stretch of approximately 12 bases in order to select a unique site on the *E. coli* chromosome. A 12-base sequence will specify one out of 1.6×10^7 possible locations (4^{12}), and there are only about 3×10^6 base pairs in the *E. coli* chromosome. Such a recognition distance (12 bases) would cover approximately 35 Å, about one turn of the DNA helix. The size of the repressor molecule isolated is sufficient to account for a recognition site of this magnitude. Consequently, we can assume that the operator locus is much smaller than the structural gene and that the amount of space on the chromosome devoted to such loci is rather small. When all the foregoing information is taken together, the model in Figure 24-4 seems extremely attractive. The only question remaining is whether this model, which is certainly well documented for the **lac** region, can be applied to all other cases of induction.

A reservation is necessary with regard to the operon hypothesis in the discussion of the genetics of higher organisms. It is certain that induction occurs in the eukaryotes, and the process seems to be the same in most respects. Consequently, there is no reason why the model of induction developed for *E. coli* should not be applied, in principle, to other organisms. However, there is good evidence that the structural genes controlling a particular biosynthetic pathway are seldom as highly organized in eukaryotes as they are in the microorganisms. In many cases, the structural genes are scattered among different chromosomes; thus the concept of an operon per se is probably of little use. If the model is to be applied, we must presume that each structural gene has its own site for the binding of repressor molecules. In certain cases, however, blocks of genes seem to operate together (see below), which would indicate that the binding site (operator) has the same structure for all members of one synthetic pathway. Until the eukaryotes have been studied in more detail, it will be impossible to decide exactly how far the model developed for *E. coli* can be extended.

The Operator and Polarity

The operon hypothesis requires a polarity of transcription of the DNA, which begins at the operator locus and terminates at the end of the block of structural genes associated with the particular metabolic pathway (presumably terminating at the position adjacent to the operator for the next block of structural genes). Such a mechanism implies the production of a single, long *m*RNA containing the message for the synthesis of several different enzymes. In agreement with this postulate is the fact that polycistronic messengers, messengers carrying information for several polypeptide chains, are common in microorganisms. The premise that several structural genes of the lactose system are transcribed into one message, under the control of a single operator, seems perfectly reasonable.

In both transcription of the *m*RNA from DNA and translation of that *m*RNA into a polypeptide chain, there are certain distinct start and stop signals. In transcription, the DNA-dependent RNA polymerase recognizes start and stop signals in the DNA. Similarly, in the translation process, the ribosomes attach to one end of the *m*RNA containing a start signal, and when they reach a stop signal, they detach. However, these signals can hardly be the same in both cases. The stop signal for the ribosomes is a nonsense triplet; but since these triplets appear in the *m*RNA, either as true stops or as mutational events, they must be transcribed from the corresponding code in the DNA into the *m*RNA. Otherwise they would not be there for the ribosomes to use. Clearly the RNA polymerase that transcribes the DNA into *m*RNA does not recognize these nonsense triplets as stop signals. In fact, considerable evidence now exists that the stop and start signals for this enzyme are far more complex than those for the regulation of ribosome activity.

DNA is a very large molecule that encodes many cistrons, not all of which are to be transcribed simultaneously (which is what control mechanisms are all about). At any particular time, some will be transcribed; some will not. The operon hypothesis requires a rather complicated start signal, the operator locus, that can bind with the repressor and be blocked so that the RNA polymerase cannot proceed past that point and the following DNA cannot be transcribed. A simple triplet mechanism would not be sufficient to provide enough variety to control the rather large number of different genes present in even a rather simple organism such as *E. coli.* Furthermore, a complex start signal will prevent mutational events in one operon from affecting other operons. Presumably the transcription process is reordered by the attachment of RNA polymerase at each new start in the molecule. The nature of the stop signals that separate cistrons within an operon remains an open question.

Once the RNA message has been generated, it is there to be read, and the translation process requires no complex starts or stops. Triplets can suffice. This situation implies that most of the control of gene action occurs at the level of the transcription of *m*RNA, and this seems to be the case. Certain controls of the translation process will be discussed at the end of this chapter. These translational controls, however, do not appear to be primary control mechanisms.

Enzyme Repression

In the preceding sections we developed a model to explain enzyme induction by substrate molecules, a mechanism that, within limits, permits the cell to adapt to changing environmental conditions. The converse of this process is *enzyme repression*, in which the accumulation of the end product of a biosynthetic pathway represses the synthesis of the enzymes used in its own production. Obviously this process is a complement to induction and, in the day-to-day life of any cell, extremely important. Without feedback repression of some type, the cell might overproduce certain end products in concentrations sufficiently high to be toxic. Feedback repression is a means by which the cell "knows" when it has enough of a substance. There are two ways to achieve this end. One is a method whereby an end product inhibits the activity of existing enzymes (such a mechanism actually does exist in cells and is frequently called feedback inhibition). This method is not as satisfactory as repression of synthesis of the enzyme. If there is no need for the end product, there is no need for the cell to expend energy to synthesize the polypeptides from which the enzymatic units are made. As long as there is a sufficiently high concentration of end product to act as a repressor at this synthetic level, the cell has a way of preventing formation of protein that is not needed in the overall economy of the cell.

Although the preceding section was developed only in terms of enzyme induction, many studies that led to the model presented in Figure 24-3 actually were done with the feedback repression system, and it was probably for this reason that the term "repressor" was first used to describe the product of the $\mathbf{i}^+$ gene. It was early observed that arginine represses the synthesis of the enzyme necessary for the first step in its own synthesis and that phosphate represses the synthesis of alkaline phosphatase, an enzyme that removes the terminal phosphate group from any of a number of substrates. Since these initial observations a number of enzymatic pathways have been investigated, and it now appears that feedback repression is a widespread process. The most interesting feature of this system is its close parallel to induction. In particular, the synthesis of an entire block of enzymes of a biosynthetic pathway is repressed by a single end product, and the repression is concentration dependent.

Whether a system is repressible depends on genetic control. Consider the case of alkaline phosphatase. The structural gene responsible for the production of this enzyme has a variety of mutant forms, each of which produces a somewhat altered alkaline phosphatase molecule. The synthesis of all molecular forms is repressed by phosphate. However, it is possible to obtain a mutation at an entirely different locus that makes the alkaline phosphatase nonrepressible. Such strains are capable of making the enzyme in the presence of large quantities of phosphate, and the form of the enzyme produced depends completely on the structural gene present in the system. The mutation that makes the system nonrepressible has no effect on the structural gene that carries the information for protein synthesis; instead the mutation is in a gene that is concerned with regulating the activity of the structural gene. Such a parallel with enzymatic induction is too obvious to be ignored.

To make the parallel between repression and induction even more evident, consider the situation shown in Figure 24-5. A deletion joined the **lac** structural genes to the purine operator. As might be expected, since purines are a vital component of all cells, the enzymes for the production of purine are constitutive. Consequently, the deletion causes the **lac** genes to produce enzymes constitutively. The test of the fact that the **lac** structural genes are under the control of the purine operator is, in fact, the repression of the production of the **lac** enzymes by purines. The purine concentration in the cell rises and causes repression at the **pur** operator locus. This process, in turn, shuts down both the **pur** and the **lac** structural genes. Thus the experiment not only indicates that the operator has control of the transcription of the genes in the microorganisms, but it also shows that an operator can function in either an inducible or a repressible system. This case is not unique, and it is generally assumed that repression and induction follow the same patterns, being merely obverse conditions of the same process.

The simplest way to account for repression is to assume that the repressible systems are ones in which an inactive conformation of the repressor molecule is produced by a regulator gene. Consequently, although the operator is capable of binding the repressor **(o^+)**, no active repressor is made to prevent the transcription of the structural genes. However, the inactive form of the repressor can bind with the end product of a reaction series, which then acts as a corepressor, causing a conformational change in the protein, which makes it an active repressor. (This is the reverse of the induction case.) When sufficient quantities of end product have accumulated, enough active repressor is generated by the formation of repressor-end product complexes to bind with the operator locus and prevent transcription of the structural genes.

The postulate that the repressor molecule is able to exist in two states (active and inactive)—states that can be altered by the presence of a relatively small substrate or end-product molecule (inducer or corepressor)—is is not simply a construct. We have already seen one demonstration of the reality of this model in the interaction of the inducer and repressor molecules. Extending it to the case of a feedback repression seems quite reasonable. Furthermore, a number of enzymes exhibit the same property. They can exist either in active or inactive conformations, and each state can be stabilized, or even promoted, by the presence of small molecules that may or may not have any relationship to the normal substrate of the enzyme. Studies of such enzymes have indicated that the activating (or inhibiting) molecule binds at a different site from the one at which the substrate binds, and the only apparent role of this activating molecule is to "lock" the enzyme into the proper conformation. Since these enzymes can have two different conformations, they have been given the name *allosteric* enzymes. Their presence in the cell is of great importance in regulating the overall metabolism. A detailed analysis of this kind of protein is outside the scope of a genetics book, and the interested reader may wish to refer to a biochemistry text. From our point of view, it is only important to recognize that allosteric proteins are relatively common in cells and that their properties are quite similar to those shown by the repressor molecules that have been isolated to date. On this basis, then, we can propose that regulator genes produce repressor proteins that have available to them at least two different states. In inducible systems, the repressor produced has, as its more favored condition, the ability to bind to a specific segment of DNA. In the repressible system, the repressor molecule is synthesized with an alternate configuration, and so it cannot bind readily with DNA. The function of the small molecule (inducer, corepressor) is to shift the equilibrium such that one state or the other becomes the more favored.

Structural genes are presumably of three types: inducible, repressible, or totally constitutive. It is unlikely that the structural genes exist in combined form (that is, both inducible and repressible), since such a control mechanism is needlessly complex. If we are to extrapolate from the findings with some of the allosteric enzymes that have many activators and inhibitors, it is possible for the feedback repression system (or the induction system) to become very complex indeed. Obviously the use of the model for enzyme induction to explain enzyme repression is still somewhat of a construct. It will probably be a more difficult problem to find inactive repressor molecules that are used in a repression system than the active ones that have been found for inducible systems. However, the problem is clearly soluble by much the same technique used for the inducible enzyme system, and it is presumed that future work in this area should provide the essential information to permit an evaluation of the model in its entirety.

Rate Control: The Promoter

The preceding discussion has centered around the control of transcription of the genetic material into *m*RNA. It has indicated that, in a general way, the rate of transcription depends on the concentration of active repressor. If the operator locus is blocked, no transcription can occur. When it is not blocked, transcription can occur. However, it is not a completely satisfactory explanation for the control of the rate of protein synthesis. It tells us which genes are likely to be turned on at a particular time and which turned off, but it does not indicate why there exist different rates of transcription of various genes that are turned on at the same time, and yet this is the case. Not all genes are transcribed at the same rate. The most obvious case is that of the regulator genes of the various operons. Clearly the high binding efficiency of the repressor molecules suggests that relatively little repressor must be made by these genes in order to maintain control of transcription. Estimates of repressor synthesis suggest that, in fact, this assumption is correct—that the various regulator genes (**i** genes) do make their repressors relatively slowly. It is also clear from the study of other kinds of systems that not all proteins are made at the same rate; those that are required in great quantity tend to be transcribed rapidly, and those that need to be present only in small quantities may be transcribed very slowly indeed. Another mechanism is needed to explain this differential rate of transcription.

A very interesting explanation of how the rate of transcription is controlled came originally from the studies of the **lac** operon. Certain mutants were obtained that had a greatly reduced production of enzyme, both in $\mathbf{i}^-$ cells that lacked active repressor and in $\mathbf{o}^c$ cells, which should be constitutive. Obviously something was interfering with the rate of transcription in these new mutants. Careful analysis has shown that such mutational events occur in a region immediately preceding the operator gene, and this special region has been given the name *promoter*.

Apparently the promoter is a region of attachment of the DNA-dependent RNA polymerase that transcribes the genetic message into *m*RNA (Figure 24-6). Mutations in the promoter region change the binding efficiency of the enzyme. Most mutations reduce the efficiency of binding, but in some cases the binding efficiency is increased. If the polymerase can attach only with great difficulty to the promoter, then successful binding of the enzyme to the DNA will be rare and very little *m*RNA will be produced. (This is presumably the case with the regulator genes, which produce their protein very slowly.) If the affinity between the

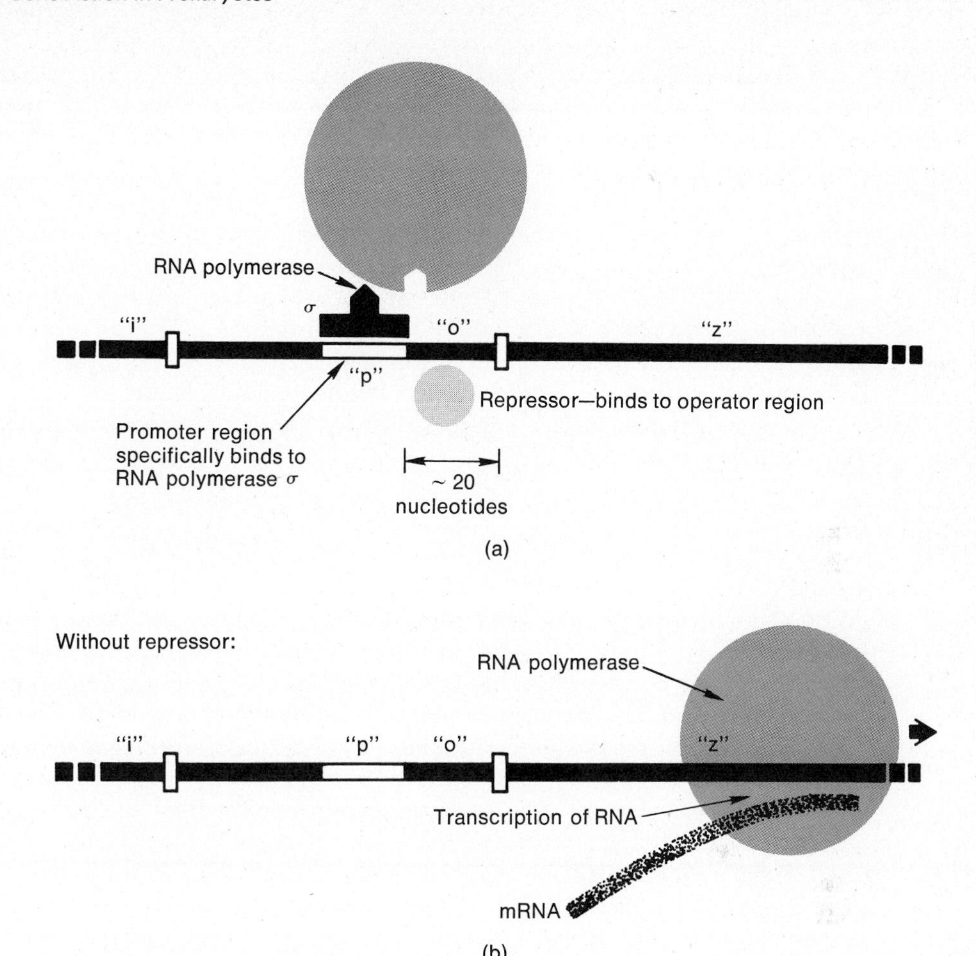

Figure 24-6. The promoter. A small region lying adjacent to the operator locus serves as the attachment point for the DNA-dependent RNA polymerase that transcribes RNA from DNA. The binding of the enzyme to the promoter is dependent on the presence of a specific sigma (σ) factor that "recognizes" the particular promoter region. If the operator locus is blocked by a repressor molecule, the polymerase may attach but not proceed along the cistron (a). If there is no repressor present (b), the polymerase attaches at the promoter locus and transcribes the message. Mutations of the promoter locus can either increase or decrease the binding efficiency of the polymerase σ and thus alter the rate of transcription. (After Watson, J. D. 1970. *Molecular Biology of the Gene*, 2nd ed. © 1970 by J. D. Watson. W. A. Benjamin, Inc., Menlo Park, California.)

polymerase and the promoter locus is great, there will be good efficiency of binding and the messages will be transcribed rapidly. Besides this control, there seems to be little that governs rate. Once the polymerase has been attached, the rate at which it proceeds along the DNA is the same in all cases. (The longer the cistron, the longer it takes to transcribe the message, of course.)

So far promoter regions have been identified only in *E. coli* and certain bacteriophages. We cannot be certain that this mechanism of the rate of control of *m*RNA transcription is a general one, but the fact that differential rates of protein synthesis exist in many cells is highly suggestive.

DNA-Dependent RNA Polymerase

The preceding discussion must focus our attention on the enzyme that transcribes DNA into RNA, which, for ease of reference, we will refer to simply as the polymerase in this section. Before proceeding with the discussion of control mechanisms, some of which directly involve the action of this enzyme, we will review what is known about its structure, for it is a rather unique functional unit.

The polymerase has a complex structure. The core enzyme consists of four different polypeptide chains

2α	39,000 MW	$1\beta'$	165,000 MW
1β	155,000 MW	1ω	10,000 MW

giving a minimum molecular weight of 370,000. In addition, the enzyme requires two detachable subunits: σ (sigma), a protein of 90,000 MW, which is required for attachment of the polymerase to the DNA molecule; and ρ (rho), a protein of about 60,000 MW, which is apparently required for termination of reading. The entire complex, then, involves six different kinds of polypeptides working together as an aggregate of almost 500,000 MW. This is a remarkable enzyme indeed. In fact, it is so large that it can be visualized in the electron microscope, and pictures of it attached to the DNA that it is transcribing are presented in Figure 24-7.

From presently available evidence, it appears that the σ factor recognizes the specific promoter region of 10 to 20 nucleotides. If the σ factor is not present, attachment cannot take place; if the promoter region has been mutated so that the σ factor no longer "recognizes" it, attachment cannot take place; or, finally, even if the σ factor is present and the promoter region is normal, transcription will not take place if the operator locus is bound by repressor. Consequently, there is a hierarchy of controls for the rate of *m*RNA transcription, made possible by the very complicated structure of the polymerase enzyme. Just how complicated this system can be will be made clear when we discuss positive feedback in the next section.

Positive Control

The Making of a Bacteriophage

Some of the most interesting work on the control of genetic activity in the 1970s has come from the unraveling of the processes involved when a bacteriophage replicates inside a bacterium. To understand this work, it is necessary to digress briefly to review the process by which phages are generated. Recall from earlier

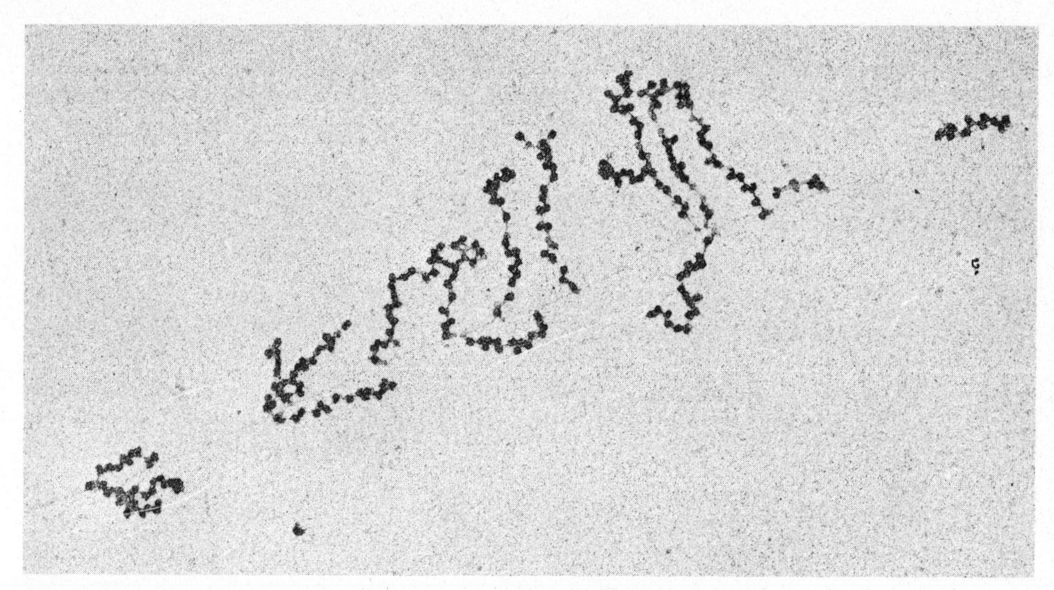

Figure 24-7. DNA-dependent RNA polymerase. In this remarkable electron micrograph, the enzyme can be seen attached to a very fine thread of DNA isolated from *E. coli.* The large size of this complex enzyme is made apparent here. (Photograph courtesy of O. O. Miller, Jr., Oak Ridge National Laboratory, and Barbara A. Hamkalo and C. A. Thomas, Jr., Department of Biochemistry, Harvard Medical School.)

discussions (Chapters 12 and 18) that only the DNA of the phage is "injected" into the cytoplasm of the host bacterial cell, and under its direction all the components necessary for phage production are synthesized. The normal time course for the development of T_2 phage can be used to illustrate the general events. When bacteria are examined by electron microscopy 4 minutes after infection, the bacterial chromosome can be seen to be disintegrating, but there is no evidence of new phage material. At 10 minutes after infection, the bacterial DNA is completely destroyed, but rather large amounts of new DNA are present. Upon isolation, this material can be shown to be identical to phage DNA. By 14 minutes after infection, the DNA has begun to condense and discrete phagelike particles can be seen. Disruption of the cells at this time results in a small yield of mature (infectious) phage. By 30 minutes, the bacteria contain a large number of mature phage, and shortly afterward the cells lyse.

The sequence of events is rather interesting. For example, the bacterial DNA is destroyed before replication of the phage DNA, thus ensuring that only phage DNA will be made. Similarly, the condensation of the DNA into a compact mass *precedes* the assembly of the head-coat protein, so that the DNA will be in the proper condition to be surrounded by the head coat prior to that protein being available. And, finally, the tail proteins are made after the head proteins, thus ensuring that the assembled phage head will be ready to receive the tail proteins once they are produced. The assembly itself can be shown, by a variety of mechanisms, to be nonenzymatic (purely random), a form of self-aggregation.

As the preceding summary suggests, it is possible to divide the life cycle of the bacteriophage into a number of periods. In the even-numbered T phages, the bacterial chromosome is destroyed early in the cycle. This process can only occur if the phage produces an enzyme that will destroy the bacterial DNA without damaging its own. (The protection of the phage DNA results from the fact that T-even DNA contains 5-hydroxylmethylcytosine in place of cytosine and that it carries glucose residues attached to it in a number of places. The DNase that will destroy bacterial DNA cannot attack the rather strange phage DNA.) This DNase is one of the first proteins made by the infecting phage. Others are a phage DNA polymerase, the several enzymes responsible for converting bacterial cytosine to hydroxylmethylcytosine, and the enzyme that adds the glucose residues to the finished DNA molecules. The latter are "early functions" of the phage genome, and without them no replication of the phage DNA could occur. At the other end of the cycle lies the production of the enzyme lysozyme that destroys the bacterial cell wall (lysis) and permits the release of the mature phage. Premature production of this enzyme would destroy the cell before the phages are assembled and would prevent effective reproduction. This is a "late function" of the genome. Between these early and late functions, head and tail proteins must be produced in the proper order. The question is: How is the proper sequence of events maintained?

Initially it was presumed that the control of early and late functions arose from the spatial separation of the various genes in the chromosome and that the time of transcription converted this separation into an effective temporal separation. If the early genes are at one end of the chromosome and the late genes at the other, the late functions would have to appear after the early ones. Although this description of the events is partially correct, the late functions appear too late (10 minutes after infection) to be explained on this basis alone. This fact leads to the natural hypothesis that the late genes are repressed; however, a thorough search for the repressor has failed, and it now seems more reasonable to propose that the late genes are under positive rather than negative control.

The molecular basis for positive control appears to be the sigma (σ) factor. Immediately after infection by T_4 phage, a small group of genes ("preearly") are transcribed by the bacterial RNA polymerase, using bacterial σ. One of the genes so transcribed produces a specific viral σ, without which it is impossible to read the next block of genes ("early"), which begins to function approximately 3 minutes after infection. The viral σ permits the attachment of the RNA polymerase to the promoter region immediately preceding the block of "early" genes. At the time the T_4 σ begins to appear, the bacterial σ disappears. (The mechanism of the disappearance is uncertain, but it probably involves an "antisigma" gene in the preearly block.) Consequently, there can be no further transcription of either the genes of the bacterial host or of the preearly phage genes. Concurrently with the appearance of phage σ there is a modification of the polymerase core enzyme. The change is relatively minor, involving a conformational (allosteric) change of the α polypeptides. This change is under the control of a preearly gene. The shift in the polymerase structure prevents the binding of the bacterial host sigma factor and makes it specific for the T_4 σ that recognizes the next (early) promoter region.

Transcription of the early genes produces phage proteins, one of which is yet another σ factor. This latter σ controls the attachment of the polymerase to the block of genes known as late genes. As in the transition from preearly to early function, the shift to late gene transcription also involves an alteration of the core enzyme. In this case, the change occurs in one of the β chains, which apparently alters the specificity of the enzyme for sigma activation. As before, the appearance of late σ largely displaces the early σ (although the reason may be because the early genes can no longer be read by the modified core enzyme), so that late functions are transcribed almost to the exclusion of early functions. Taken all together, the T_4 system is one in which positive control by means of specific σ synthesis is used as an essential part of the timing mechanism.

In the odd-numbered phage T_7, the majority of the DNA is transcribed by a completely new polymerase that is coded in the first gene on the chromosome. In this phage, only four genes are transcribed by the host polymerase (including gene 1), and the remaining 25 to 30 are read by the T_7-specific enzyme. Apparently it takes about 5 minutes for transcription to be detectable, and at that time all reading of the early genes stops. In this case, as suggested for T_4, one of the first genes transcribed by the phage polymerase produces an "antisigma" factor, so that the host enzyme can no longer operate.

How complicated the control of phage transcription may be is best illustrated by the life cycle of phage λ. This small, lysogenic phage can exist within the host in the prophage state (in which condition it is incorporated into the *E. coli* chromosome and replicated with it) or in the lytic state, detached from the chromosome and producing virulent phage. The interesting fact is that although the λ phage is part of the bacterial chromosome, almost none of its genes are transcribed into *m*RNA. The transcription is blocked by a repressor made at the $\mathbf{C_1}$ locus of the λ chromosome. This repressor attaches to the two operator loci immediately adjacent to and on either side of the $\mathbf{C_1}$ locus so as to prevent the polymerase from moving in either direction along the chromosome (Figure 24-8). Consequently, only one λ gene is transcribed, and it produces the repressor that holds the phage in the prophage (inactive) state.

Induction of the lytic condition may occur by any of a number of means; but, in all cases, there is a failure of the repressor to bind, and the transcription of the viral genes begins in both directions from the $\mathbf{C_1}$ locus. This genetic organization means that there are two early operons, one on each side of $\mathbf{C_1}$. One produces the proteins necessary for the recombinational events that will detach the λ DNA from the *E. coli* chromosome. The other operon contains a gene whose product specifically repres-

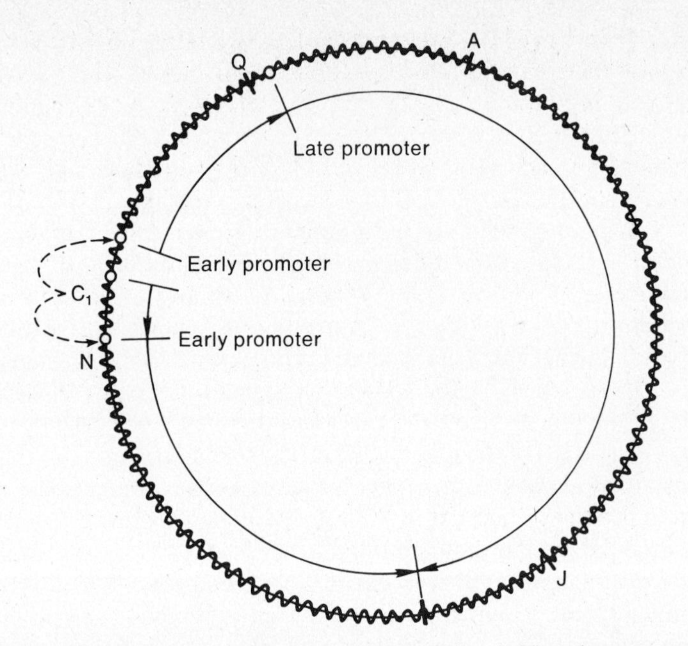

Figure 24-8. The λ-phage control mechanisms. The circular chromosome is read in both directions (solid arrows) from the promoters on either side of gene $\mathbf{C_1}$. In the lysogenic condition, $\mathbf{C_1}$ produces a repressor that blocks both early promoter loci (dashed arrows). In the lytic condition, transcription of the first few genes of the early region produces a repressor of $\mathbf{C_1}$ activity, and gene **N** produces a positive control protein for continued reading of the early operons. Gene **Q,** located just before the late promoter locus, produces a positive control factor that permits reading of the late operon. Gene **A** controls head protein, and gene **J** controls tail protein synthesis.

ses the $\mathbf{C_1}$ locus. Thus no further synthesis of λ repressor can occur, and the genes of the now detached phage DNA can be transcribed to produce mature phage particles.

The preceding paragraph describes a mechanism for the conversion of lysogenic to lytic phage. It is an example of how interacting repressors can control the timing of events within a cell, but it is far from a complete description of the growth of the λ phage particles. The complete transcription of the early operons depends on the action of gene **N**. In the absence of a functional **N** protein, no further transcription of the phage DNA can occur. Apparently the function of this particular protein is to neutralize the "stop" signal, so that the *E. coli* polymerase ρ factor cannot stop the reading and the transcription of λ DNA can continue.

Furthermore, the late functions, which comprise almost half the λ genome, will not be transcribed in the absence of a factor produced by one of the last genes in one of the two early operons. This gene **(Q)** produces a protein that is apparently a new sigma factor necessary for attachment to the late promoter locus. Thus the λ phage uses a positive control mechanism similar to that found in T_4, but it is used only for the late functions. This entire sequence of events is summarized in Figure 24-8.

The λ phage offers a remarkable example of the interaction of several different transcriptional controls to maintain the proper sequence of events in the life cycle of an organism. It is probable that the diversity of mechanisms is more a function of our extensive knowledge of the λ life cycle than a unique property of λ itself. In any case, it should be emphasized that temporal control in this phage also depends on the spacing of the genes in an operon. For example, the late genes code primarily for the head and tail proteins. The main head proteins are specified by the first genes of the operon, and the tail fiber protein (the final component of the tail) is one of the last genes to be transcribed. This means that the synthesis of the main head protein occurs at least 5 minutes before that of the tail fiber protein. In a life cycle as short as that of a bacteriophage, such delay is considerable. The same is true of the operon containing the **Q** gene. This gene is almost 4 minutes of transcription time from its promoter locus. Consequently, the first sigma protein is not present in detectable quantities until about 5 minutes after infection, and almost 10 minutes have elapsed before the σ concentration is sufficiently high to permit detectable levels of late *m*RNA. Clearly the separation of genes in the chromosome and the requirement for the accumulation of one gene product before the reading of the next operon commences can introduce a considerable delay in the appearance of the products of widely separated genes.

Another aspect of spatial separation in the genome can be best understood by calculations of the transcription time for two relatively simple systems. The polymerase of *E. coli* moves along the DNA at a rate of about 30 to 40 nucleotides per second. At this rate the complete transcription of all the λ DNA would require 25 minutes, and the transcription of the *E. coli* chromosome would take 35 hours! Clearly this is absurd, since the life cycle of the bacterium is only 20 minutes. Therefore we must assume that a number of operons are transcribed simultaneously. (This point has been noted above for the case of the two early operons of λ.) The organism, then, is faced with a "choice" between a few long operons or many short ones. The selection of the best type of organization depends on the results desired. The long, single operon of the late functions of λ permits the proper timing of the coat proteins for the phage. The head proteins are ready before the final tail fibers are made. For an organism with a more complicated life cycle, however, long operons could mean a wasteful delay in the production of important enzymes. Consequently, it is reasonable to propose that organisms that produce a large number of different enzymes, all of which are needed essentially simultaneously, would operate with the short-operon mechanism, with many, possibly the vast majority, of them being transcribed simultaneously. For other functions—the timing of cell division for example—the eukaryotic cell may well make use of the long operon. With this consideration in mind, it seems reasonable that the eukaryotes have divided their genome into many different chromosomes. When one has many small operons, possibly sharing the same kind of promoter and operator regions, it makes little difference functionally how they are organized; but structurally (particularly in terms of replication) the chromosomal mechanism may be preferable.

Finally, we should realize that the vast quantity of DNA in the cells of many higher organisms is far too great to be transcribed even in the relatively extended life cycles typical of their cells. It is quite probable that much of the DNA is never transcribed in any particular cell at any one stage of its development. In the multicellular organisms, each cell has much more information than it needs; and one of the key questions in differentiation is how a particular cell "knows" which part of its genome to use. Obviously the problem is a complicated one. The mechanisms outlined in this chapter have largely been derived from studies of prokaryotic cells, and there is no a priori reason why they must be presumed to operate as well in the cells of organisms with a discrete nucleus and a complex chromosomal structure. There are at best some suggestive hints that the mechanism is the same in all organisms, but we are a long way from describing what happens in the eukaryotes. In the next chapter we will consider some of the evidence from developmental biology that suggests, at least in the broadest outlines, that the control mechanisms discussed here are applicable to the higher organisms.

Control of Net RNA Synthesis

The foregoing discussion has centered on the control of transcription of *m*RNA from DNA. However, *r*RNA, and *t*RNA synthesis also appear to be regulated. In *E. coli*, there is a specific regulator gene, **RC**, whose product is sensitive to the level of free amino acids in the cell. There are two alleles, $\mathbf{RC}^{str}$ (stringent) and $\mathbf{RC}^{rel}$ (relaxed). When cells that are $\mathbf{RC}^{str}$ are deprived of amino acids, the production of both *r*RNA and *t*RNA falls by a factor of ten. In the case of $\mathbf{RC}^{rel}$ cells, the rate of synthesis drops, but there is still enough to give a net increase in both forms of RNA. Obviously this type of regulation is important to the cell, since there is no need to synthesize either RNA in the absence of amino acids. Presumably the system is the usual induction type of control, but the mechanism is not yet certain.

A net increase in *m*RNA concentration would require either that the rate of transcription increase or that the rate of destruction decrease. Since there seems to be a steady-state level of *m*RNA in most cells, it is interesting to determine how the messenger is destroyed. Surprisingly, we do not know for certain. Most of the evidence indicates that the destruction occurs in a polar manner, from 5′ to 3′, presumably following the last ribosome (as indicated in Figure 19-4c). However, no known enzymes have this specificity of attack. Furthermore, the kinetics for the destruction of **lac** *m*RNA are of a type that would be expected for a random event, thus suggesting a nonenzymatic destruction. For the present, we must leave it as an unsettled problem.

Translational Control

The majority of control mechanisms discussed were concerned with the control of transcription. From the point of view of the geneticist, such control is of the greatest importance and, in fact, it is the best-known mechanism. However, it is important to realize that there are instances in which translational control can be demonstrated as a means of controlling the rate of protein synthesis. For example, in the **lac** operon, three enzymes are coded for on a single polycistronic messenger, but they are not produced in equal quantity. In fact, they are produced in a ratio of 5 : 2.5 : 1. Some factor must be governing the rate of translation of the message into polypeptide chains.

In studies of the translation process it appears that the rate at which attached ribosomes travel along a message is invariant. Therefore the governing of translation rates through differences in the message would be negligible (except, of course, that relatively long messages would take longer to translate). On the other hand, the rate of translation could be controlled by the rate of attachment of the ribosomes to the message. If the affinity of the ribosomes for a particular message was low, they would attach rarely and the rate of translation would be reduced. In a polycistronic message, another alternative is available. The detachment at the stop signal might be more frequent in one case than another. Thus all later messages, following translation of the first (or more) message, would be lower. In either case, the mechanism involves the interaction between the ribosomes and the *m*RNA.

There is some evidence for the ribosomal-attachment hypothesis, and it comes from studies of the production of an RNA phage. In this case, the coat protein is made in much greater quantity than other proteins, even though the messengers for other proteins are present simultaneously and equally. It can be demonstrated that the attachment of the ribosomes to the *m*RNA at the unique ribosome binding site is different in two cases. The coat protein is translated from a region containing a highly efficient binding site. Consequently, many more ribosomes attach to this messenger per unit time than is the case for the other regions, which have low-efficiency binding sites. In addition, there is some suggestive evidence that the coat protein may interfere with the binding of the ribosomes to one of the other messenger regions. In this case, the production of one protein would inhibit the production of another protein at the translational level. This is an important suggestion, for it gives us one indication of control at levels other than transcriptional. It is to be presumed that such modes of control are significant in the cells of higher organisms where differential production of protein is one of the bases of differentiation, a subject to which we will turn next.

REFERENCES

BECKWITH, J. R. 1967. Regulation of the Lac operon. *Science* **156**: 597–604. (A good review of the control of transcription in the lac operon.)

BRENNER, S. 1965. Theories of gene regulation. *Brit. Med. Bull.* **21**:244–250. (A brief but perceptive review of the general problem.)

Cold Spring Harbor Symp. Quant. Biol. **26**. 1961. Cellular regulatory mechanisms. Cold Spring Harbor Laboratory, L.I., New York. (A mixture of review articles and original papers on the subject. For the advanced student.)

Cold Spring Harbor Symp. Quant. Biol. **35**. 1971. Transcription of genetic material. Cold Spring Harbor Laboratory, L.I., New York. (Several sections of this symposium bear directly on subjects discussed here. For the advanced student.)

HAYES, W. 1968. *The Genetics of Bacteria and Their Viruses.* 2nd ed. John Wiley & Sons, New York. (Chapter 23 is particularly relevant to the material discussed here.)

PTASHNE, M., and W. GILBERT. 1970. Genetic repressors. *Sci. Amer.* **222** (No. 6):36–44. (An excellent, simple presentation of the action of repressors and their isolation.)

Original Papers

GILBERT, W., and B. MULLER-HILL. Isolation of the lac repressor. *Proc. Natl. Acad. Sci. U.S.* **56**:1891–1898. (The paper whose methods and results are discussed in this chapter.)

JACOB, F., and J. MONOD. 1961. Genetic regulatory mechanisms in the synthesis of proteins. *J. Mol. Biol.* **3**:318–356. (This is a classic synthetic review of induction and repression. It contains the original formulation of the operon hypothesis. Well worth the effort to read. An example of good scientific prose.)

PARDEE, A. B., F. JACOB, and J. MONOD. 1959. The genetic control and cytoplasmic expression of inducibility in the synthesis of β-galactosidase by *E. coli. J. Mol. Biol.* **1**:165–177. (One of the important papers leading to the operon hypothesis.)

QUESTIONS AND PROBLEMS

24-1. Define the following terms:

enzyme induction	enzyme repression
constitutive enzyme	regulator gene
structural gene	repressor
inducer	operon
operator	promoter

24-2. What is the function of enzyme induction? Of enzyme repression?

24-3. How did Jacob and Monod explain enzyme induction?

24-4. What do you consider the best evidence for the validity of the Jacob-Monod model?

24-5. Why do we believe that enzyme repression is basically a modification of the enzyme induction mechanism?

24-6. What properties do proteins have that make them ideally suited to be repressors?

24-7. Why are the operator and promoter loci relatively short stretches of DNA?

24-8. How do promoter loci "isolate" the structural genes to which they are attached from mutational events in other regions of the same DNA molecule?

24-9. What is the role of the sigma (σ) factor in the control of gene action?

24-10. From what we know about λ phage, how can control mechanisms be used to maintain the prophage (lysogenic) condition?

24-11. What is the evidence that the synthesis of all forms of RNA is under genetic control?

24-12. Is the control of protein synthesis effected solely at the level of transcription? Explain your answer.

24-13. Refer to your answer to Question 22-8. How would you answer that question now?

25/Differential gene action in the development of eukaryotes

When developmental genetics was in its formative period, a vast amount of research was done to determine the inheritance of developmental abnormalities. This was the most obvious approach, and it was important work in that it showed clearly that genes control developmental processes as well as others. However, from the point of view of the modern scholar, that statement is almost self-evident, and so we will not review that very extensive literature here. Rather, we wish to focus on the problem of differential gene activity and how it can explain the complicated developmental patterns of higher organisms.

The egg develops into an adult by a process that is essentially equivalent to mitosis. How is it possible that the genes in the cells that determine the toenail are exactly the same as the genes in the cells that make the cornea of the eye? The answer is simple enough: certain blocks of genes are expressing themselves in one cell and not in the other. But the simple answer is not enough. We would like to understand how this difference in gene activity is achieved. Is it, for example, by the loss of certain parts of the genome during development, such that toe-forming cells and eye-forming cells are *not* equivalent in their genome?

One of the earliest and most influential suggestions that the genome of different parts of a multicellular organism may not be the same came from studies of the parasitic worm *Ascaris megalocephala.* The first cleavage division produces two cells of approximately equal size. One cell, labeled AB, gives rise to the dorsal part of the organism, and the other cell, labeled P_1, gives rise to the ventral part of the organism (Figure 25-1). At the next division the spindles of the two cells orient at right angles to one another, and at this time the two large chromosomes of AB break up into many small chromosomes, not all of which attach to the spindle. Thus when AB divides to give rise to two cells, A and B, some of the chromatin material is left behind in the cytoplasm and is lost. The division of P_1 does not involve this loss of chromatin. The two chromosomes are oriented normally on the spindle and are divided equally between two daughter cells, labeled P_2 and C in the diagram. At the next division there is further chromatin diminution in both A and B and in C as it divides. However, there is no loss in P_2 at division. This process continues with every cell except the most ventral product of P_2 losing chromatin at subsequent divisions. The lack of diminution in the ventral cell derived from P_2 is of great interest because this cell ultimately gives rise to the germ line (sperms and eggs) in the organism. If there is to be genetic continuity, it is essential that there be no loss of chromosomal material in the cells giving rise to the gametes, and no such loss is observed in that one cell line. The other cells will give rise to somatic parts of the organism, parts that do not function in reproduction. For these cells to lose a certain amount of genetic material would be acceptable, provided only that each one retains the material that is essential to its specific function later in development.

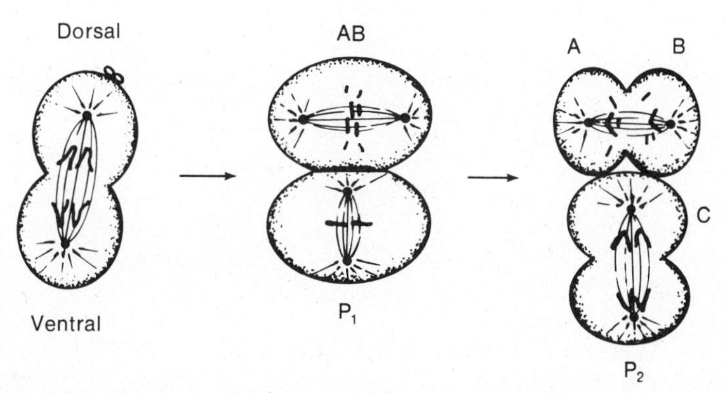

Figure 25-1. Chromatin diminution. At the left the Ascaris egg is seen at the first cleavage division. The two very large chromosomes are evident. As a result of the first division, two cells are produced, AB and P_1. When these divide at the second division, the chromosomes in cell AB break up and fragments of them are lost to the cytoplasm. Cell P_1 does not show this loss of material.

Ascaris, like many other invertebrates, has a mosaic development; that is, each cell of the early embryo gives rise to a specific part of the organism. If the cells are separated at the four-cell stage, each will give rise to certain parts of the organism and no others. We say that the fate of these cells has been fixed by the events that occurred during the first two cleavage divisions. Their potential has been restricted in that no one cell is capable of producing a whole organism. The mosaic pattern of development reinforces the concept that there may be a definite loss of chromosomal material in each cell, so that the fate of the cell is fixed irreversibly. It can never recapture the potential that it had as an egg to generate an entire organism except, of course, in the germ line. Such a view of development, which was common in the early part of this century, suggests that cleavage is a sorting out of determinants that can ultimately interact with the nucleus so that specific changes are progressively induced in the nuclei of the various cells.

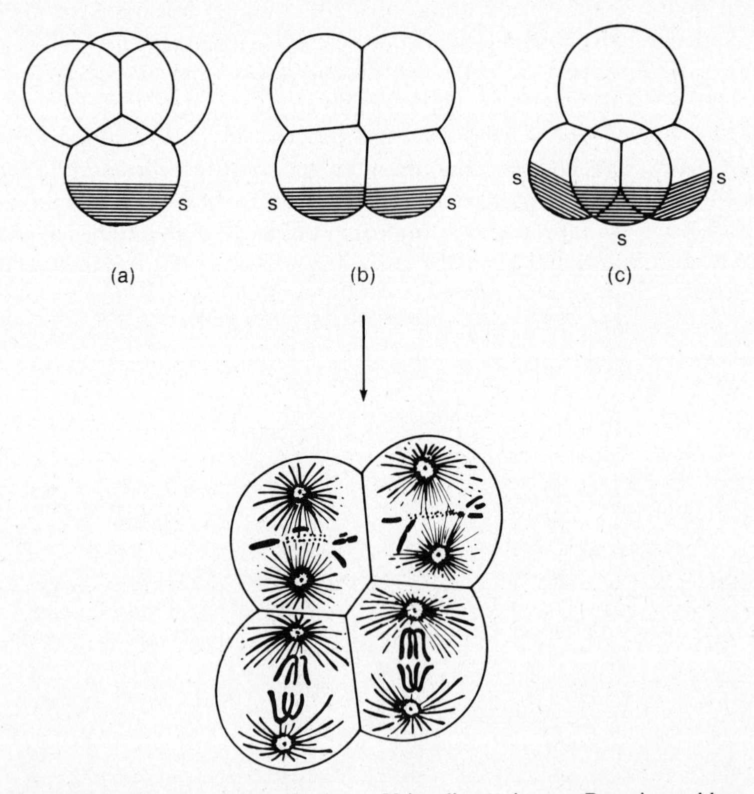

Figure 25-2. The role of the pole plasm. Using dispermic eggs, Boveri was able to get several different cleavage patterns in Ascaris eggs. In the three diagrammed,(a) has all of the ventral cytoplasm (pole plasm) in one cell;(b) has it divided in two cells; and (c) has it divided in three. Any cells receiving pole plasm do not show chromatin diminution. The lower figure shows the division in the case of (b) (After Wilson, E. B. 1924. *The Cell in Development and Heredity. 3rd ed.* © 1925 by Macmillan Publishing Co., Inc., New York. Renewed 1953 by Anna H. K. Wilson.)

To demonstrate the cytoplasmic location of the determinants, an intriguing experiment was done with dispermic Ascaris eggs. As you will remember from Chapter 8, dispermy produces abnormal spindle formation such that four spindles may develop in a single cell and produce four cells all at one time. Because of the differences in the orientation of the spindles, a variety of different kinds of dispermic Ascaris can be obtained, some in which one large cell contains the ventral cytoplasm, some in which it is divided into two, and some in which it is divided into three. In every case, there is no chromosome loss in the ventral cells (Figure 25-2). It appears that the ventral cytoplasm, or "pole plasm," determines that there will be no chromatin diminution. Any cells not receiving part of the pole plasm show loss of chromatin at successive divisions. With such experimental evidence, it is reasonable to conclude that the cytoplasm interacts with the nucleus and determines whether chromosomal material will remain complete—to form germ cells—or become incomplete—to further the differentiation of the somatic tissues.

The foregoing experiments, as well as numerous similar ones, did show that there is a sorting out in development. The egg cannot be considered a uniform blob of cytoplasm. For want of a better word, we can use the old concept of determinants and say that, at some time in the development of the egg, different kinds of material are put into the cytoplasm in different places. Which of these materials arrive finally in which cells is totally dependent on the way in which the egg cell divides—that is, is totally dependent on the orientation of the spindle within the cytoplasm at each division. In fact, it is for this reason that we refer to the early divisions of embryos as being cleavage divisions rather than mitotic divisions, since mitotic implies equational, and cleavage implies a sorting out of the cytoplasmic determinants, which can produce unequal results.

Unfortunately, the work with Ascaris is not conclusive. Very few organisms show chromatin diminution. Nor can we even be sure that the chromatin diminution that can be seen in these few cases represents the loss of important genetic material. After all, there is far more DNA in the cells of multicellular organisms than is necessary to code for the known proteins. Some of this DNA may be useless to the developing organism. (This subject was discussed in Chapter 22.) It is possible that it is some redundant or useless DNA that is lost in the case of Ascaris. Still, if the diminution process had no function, one wonders why it has been maintained. Since the genetics of this organism are not known, no unequivocal tests of the role of chromatin diminution have been forthcoming.

Although it is true that most of the invertebrates show a mosaic type of development similar to Ascaris, with the fate of the cells being fixed early on, the vertebrates do not. Presumably this difference is related to the very different developmental processes in the two groups. The vertebrates (and certain related invertebrates) undergo cleavage of the fertilized egg until a solid ball of cells, the morula, is formed. After further division a space opens in the interior, thus permitting the cells to form a hollow ball, the blastula (Figure 25-3). Once the blastula has formed, invagination of one wall of the hollow sphere occurs. This process brings a layer of cells inside, so that a multilayered structure, the gastrula, is

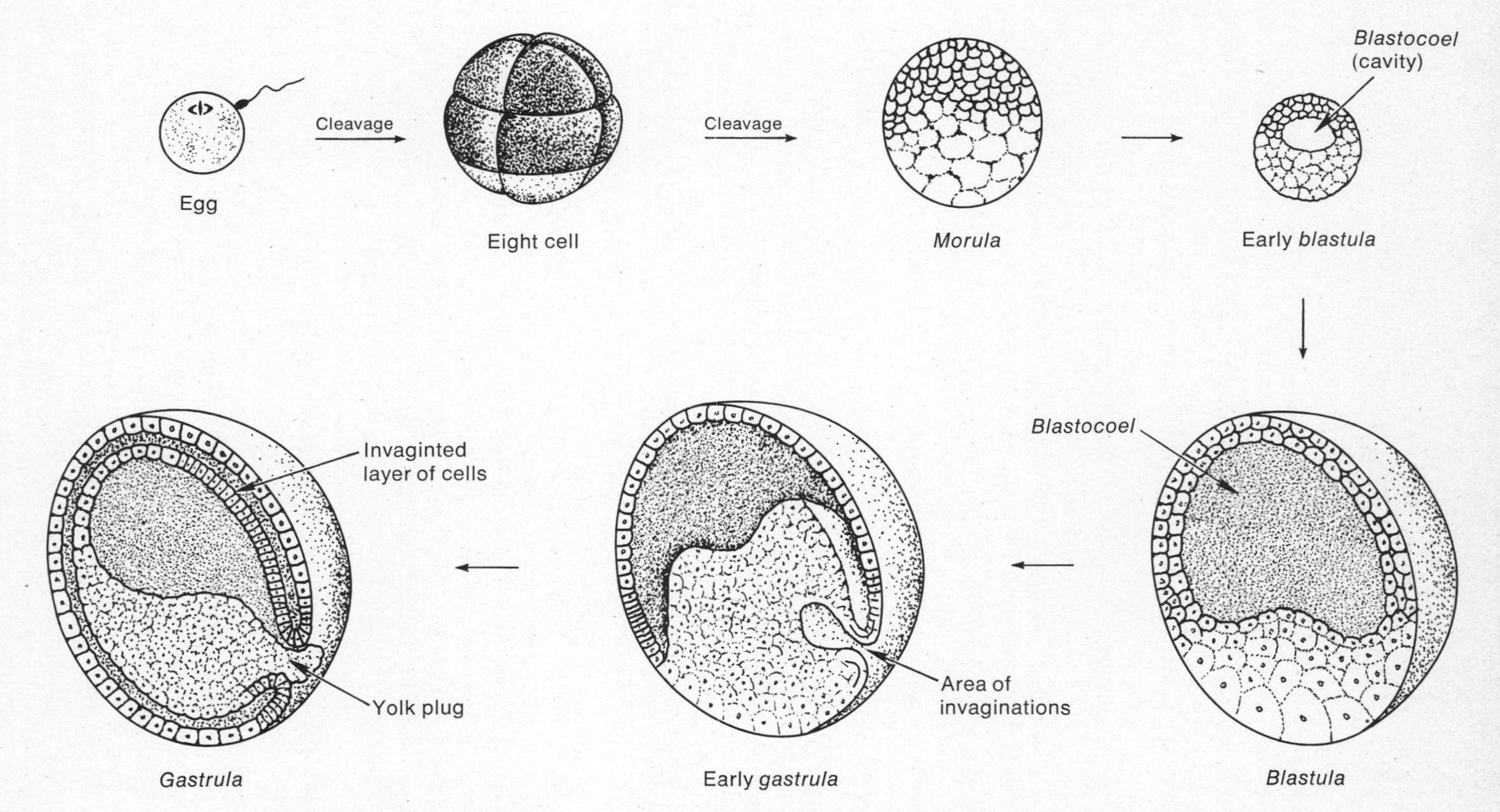

Figure 25-3. Early development of the frog. The fertilized egg divides to form a solid ball of cells, the morula. A cavity develops within the solid mass to make the blastula. Cells from the outer surface of the blastula invaginate, moving into the cavity, the blastocoel. Finally, the cells have moved into the cavity until it is filled completely. During this final stage, the gastrula, the fate of the cells is determined.

formed. The process of invagination, called *gastrulation,* converts the two-layered (external, internal) blastula to a three-layered (external, middle, internal) gastrula. In so doing, it provides the three "germ layers" (ectoderm, mesoderm, and endoderm) from which the various organs and tissue systems will develop. The most important feature of the process is that the interaction of the various layers as they come into contact with one another causes changes in the properties of the cells. This process, called *induction,* appears to be the point at which the fate of most cells is determined. Prior to the gastrula stage, the cells are quite flexible in their ability to form different parts of the organism.

The difference between the invertebrate and vertebrate modes of development raises the question of whether loss of chromosomal material occurs in the vertebrates. In order to answer this question in the late 1920s and early 1930s, a series of experiments was carried out to determine whether cytoplasmic determinants were affecting the character of the nuclei of amphibian embryos. In one experiment a loop of hair was used to tie off a small part of an egg (Figure 25-4). The

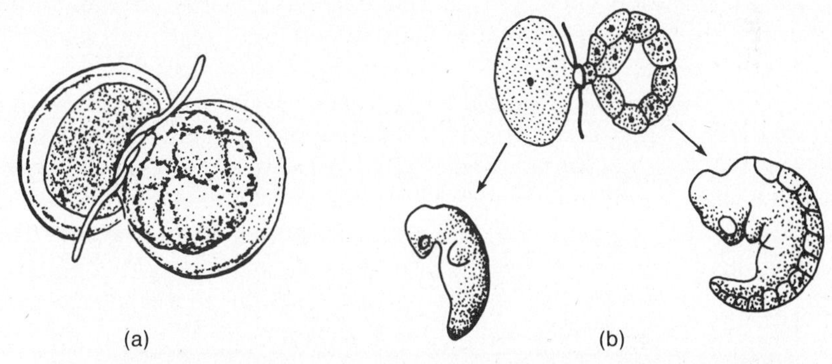

Figure 25-4. Ligation experiment. The egg was ligated with a hair loop just before the first division, so that one cell contains the nucleus and the other does not. The cytoplasmic bridge connecting the two is too small for the nucleus to pass. Cleavage occurs in the nucleated cell but not in the other (a). At a later stage the loop can be loosened slightly to permit one nucleus to enter the cytoplasm of the non-nucleated part. Constriction of the loop separates the two parts and development of each occurs normally. The non-nucleated part (on the left) develops more slowly than the part originally nucleated only because it starts later (b).

constriction made by the hair loop was too small to permit nuclei to pass out into the little piece of isolated cytoplasm or to permit any involvement of the isolated cytoplasm with the various spindles that formed at successive divisions. The rest of the egg cleaved normally with the small isolated piece of cytoplasm attached and remaining essentially dormant. The hair loop can be opened at any particular time in division—for example, at the 32-cell stage—and the nucleus from the cell immediately adjacent to the constriction will pass into the cytoplasm of the isolated part. Then the loop can be closed completely and a small cell cut off. This new cell will have the isolated cytoplasm of the egg and the nucleus admitted to it at whatever stage is chosen. Interestingly, such a cell, with a nucleus from a 32-cell embryo, developed to give rise to a perfectly normal embryo, although one reduced in size. This experiment showed clearly that, in the case of amphibia, there is no loss of genetic potential during the early cleavage divisions of the egg. Nothing of importance has been lost.

The only difficulty with the preceding very elegant experiment is that it cannot be carried through until later stages. Beyond a certain point, the hair loop interferes with normal development, and so it is impossible to test nuclei from the later, more obviously differentiated stages by this method. The necessary test is to transplant a nucleus from a late stage in development into the cytoplasm of an egg cell and see whether that nucleus can support the complete development of the organism. In the 1950s techniques were finally developed to make this test possible (Figure 25-5). The nucleus of the egg cell can be removed with a microneedle, in which case the cell heals and is quite viable for some time; or the egg nucleus can be inactivated with ultraviolet radiation so that it cannot divide or participate in any way in subsequent events. Once again, the egg cytoplasm is not damaged. The nucleus from any other cell can be transplanted into such enucleated eggs and the developmental potential of the nucleus can be assessed.

When nuclear transplant experiments were first performed in the American frog Rana, it appeared that all the nuclei of the late blastula stage were equal in their potency and could form a whole embryo. However, when nuclei were transplanted from gastrula, the results were rather uneven. Some nuclei seemed able to support complete development, whereas nuclei from other cells could not. Thus it appeared that there might be some progressive loss of genetic potential in the organism.

Shortly after these original experiments, another set of experiments was done with the South African clawed toad, Xenopus. In this case, the loss of potential seems much less. That is, the nuclei of a number of later stage cells can support complete development (at least to the larval stage). In fact, in certain cases it has been possible to take nuclei from an adult frog, transplant them into egg cells, and obtain a complete adult. Although the process cannot be done successfully with nuclei from any cell, it is perfectly clear that at least some adult cells contain their complete genetic potential. Thus if there is any loss of *potential,* it is highly selective; and in many cells of the somatic tissue no such loss occurs.

It would be nice to be able to say that the results of these experiments are clear-cut. They are not. The results with Rana and Xenopus are contradictory in several cases, and the techniques used by the two sets of experimenters are different. It is therefore impossible to state unequivocally that there is no loss of genetic potential during development. On the other hand, it should be clear that most of the arguments for the loss of genetic potential are not themselves convincing with regard to the loss of genetic material per se. In organisms that can regenerate lost parts, cells that have one particular fate can revert to a less-differentiated condition and then redifferentiate in a new direction. For example, cells of the upper arm may give rise to digits, and so forth. However, we do not mean that such cells have reverted to the embryonic condition during the process of regeneration but that

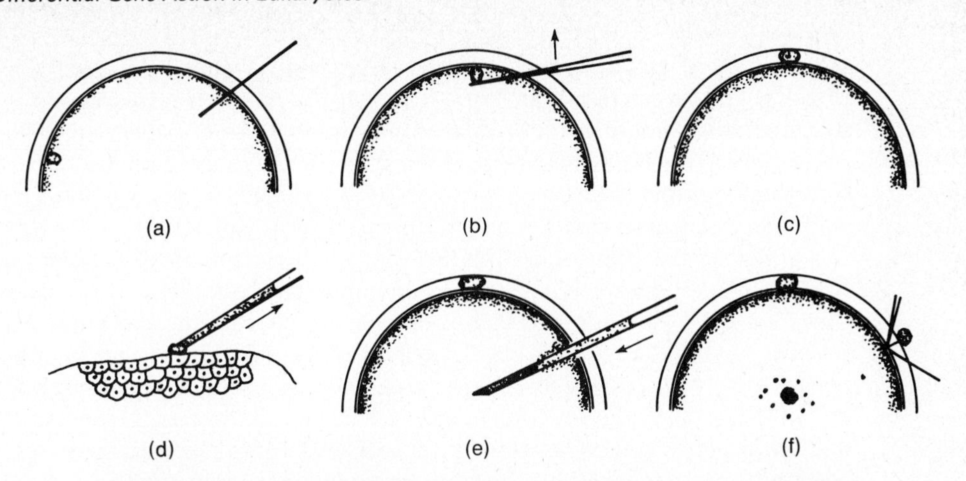

Figure 25-5. Nuclear transplantation. The cell is pierced with a very fine glass needle and the nucleus is pushed out of the cell into the space between the cell membrane and the outer protective coats of the egg (a, b, c). A cell from a blastula stage embryo is drawn up into a micropipette, which breaks the cell membrane and frees the nucleus (d). The micropipette is inserted into the enucleated egg and the material it contains is injected into the cell (e). The egg cell has been activated by the piercing of the membrane and now proceeds to divide (f). (After Briggs, Robert, and Thomas King. "Specificity of Nuclear Function in Embryonic Development" in *Biological Specificity and Growth.* Elmer G. Butler, ed. 1955. Princeton University Press. p. 126. Reprinted by permission of Princeton University Press, New Jersey.)

they are now able to express a potential that apparently had been lost in the course of development. In plants, the process is even more dramatic. There are now a number of cases in which the somatic cells of an adult plant can be isolated, grown in culture, and made to repeat the entire developmental process so as to produce a new, entirely fertile plant. No loss of genetic material is seen..

Finally, we must point out that the loss of genetic potential is not equivalent to the loss of genetic material. Genetic potential can be lost merely by repressing that function in some relatively permanent way. One way of losing genetic potential without losing the genes themselves would be by the formation of repressor molecules that, once attached to the DNA, will not dissociate (except with extreme difficulty). Under these conditions only extraordinary circumstances would permit the loss of repression and thus the induction of genetic function. Experiments designed to test the genetic potential of such cells might not provide the necessary stimulus to unlock the repression, and thus the potential cannot be demonstrated. In this context, one set of experiments with Xenopus is extremely interesting. Normally the cells of the brain do not divide. The nucleus has lost its potential to undergo mitosis, and it carries out its highly specific function of permitting the nerve cell to transmit signals. However, if the nuclei of such cells are removed and transplanted into enucleated eggs, they regain the potential to divide. Interestingly, this process takes more than one transfer. It is as if some material in the egg cell that is responsible for the maintenance of cell division can be used to remove the repression of the division function, but it may take several "doses" before the repression can be effectively removed. Transfer through the egg provides the necessary environmental conditions for unlocking a relatively rigid repressing mechanism. This is an interesting observation and one that fits nicely with the suggestion that loss of genetic potential, where it does occur, is probably the result of some sort of repression-induction mechanism rather than the physical loss of genes.

Yet another approach to this problem has been developed. It is the technique of cell fusion. It has been known for some time that cells of widely unrelated types may fuse when growing together in the same culture. The mechanism remains obscure, but it has been shown that infection of the cultures with a type of influenza virus, Sendai virus, causes a marked increase in fusion. Thus it has been possible to study the interaction of the two nuclei in the fused cell. In one of the most carefully studied systems, mammalian Hela* cells are fused with chicken erythrocytes. These latter cells contain nuclei but they are nonfunctional. In the process, the red cells hemolyze, and, in effect, only the chicken nucleus is transferred to the fusion cell. The heterokaryon so produced has a fully functional Hela nucleus and a repressed erythrocyte nucleus. After a time the erythrocyte nucleus swells and begins both DNA and RNA synthesis, and, still later, chicken antigens appear on the cell surface. We interpret this to mean that the nonfunctional nucleus of the erythrocyte can be restored to normal activity by the mammalian cell. If the chicken nucleus were making a repressor, the mammalian nucleus should be shut down. Consequently, it seems more likely that the mammalian nucleus makes a positive control factor that can "turn on" the chicken nucleus. In any case, the experiments suggest that the differentiation of the erythrocyte is a reversible process.

Differential Gene Activity

When one considers the bulk of the evidence, it seems most probable that differentiation is a process involving the differential activity of genes, with different parts of the genome acting at different times and at different places. As a result of the model that has been developed for induction and repression (Chapter 24), it is reasonable to propose such a mechanism, at least in general terms. Fortunately, there is evidence to support this proposition, the most impressive of which comes from a study of the chromosomal puffs of the Diptera.

The Dipterans have a rather unique chromosomal structure. The cells of certain tissues (particularly the salivary gland, hind gut, and Malpighian tubules) undergo a process of chromosomal replication without either separation of the components of the chromosome or cell division. As a consequence, giant polytene (many-stranded) chromosomes are produced. They are interphase chromosomes (in an extremely extended and physiologically active condition) that can be seen in

*Hela is the designation given to a line of human cells that has been carried in culture for many years. It was originally derived from a cervical carcinoma in a patient whose pseudonym was *Helen Lane*.

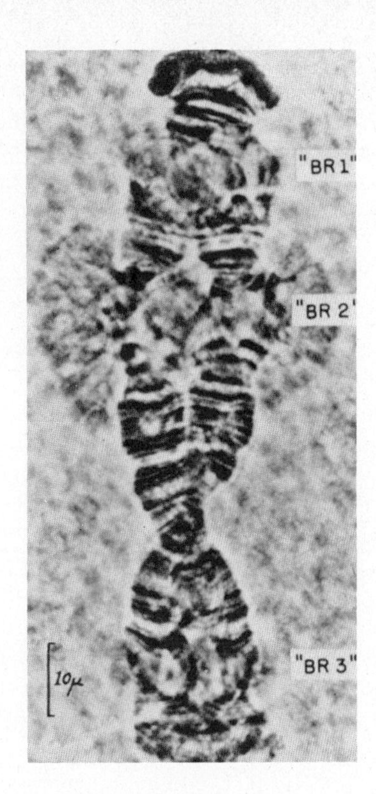

Figure 25-6. Chromosome puffs. The fourth chromosome of the Dipteran, Chironomus, is shown under phase contrast. The puffs are labeled as BR1, BR2, and BR3 (for Balbiani rings). From Beerman, W. 1959. "Chromosomal Differentiation in Insects" in *Developmental Cytology*. Dorothea Rudnick, ed. 1959. The Ronald Press Co., New York.

the light microscope because of their multistranded nature. The general structure of the salivary gland chromosomes was discussed in Chapter 13. In the study of such chromosomes it was observed that certain regions are "puffed" (Figure 25-6); they appear as greatly enlarged regions that do not stain well with stains specific for DNA. The giant chromosomes can be isolated from living cells, and the puffs are found to be a real feature of the chromosome rather than an artifact of fixation or staining. Investigation of the puffed regions indicates that they are areas of rapid RNA synthesis (Figure 25-7), and so it was suggested that they might be physiologically active regions of the chromosome. To confirm this postulate, particular chromosomes were removed from cells (by use of a micromanipulator) and their puffs were isolated. Extracted components of the puffed region were analyzed (by use of microelectrophoresis and a microspectrophotometer). The analysis revealed that the base composition of the isolated RNA is unlike either *r*RNA or *t*RNA. Furthermore, each puff produces RNA with a base ratio quite distinct from that of other puffs. These findings are consistent with the interpretation that *m*RNA was being produced in the puffs.

The most interesting feature of the puffed regions is that the pattern varies from organ to organ and from time to time. At a certain stage in development, a particular chromosome may have many puffs, and at another stage it may have very few. Similarly, the number of puffs present in a particular chromosome is not the

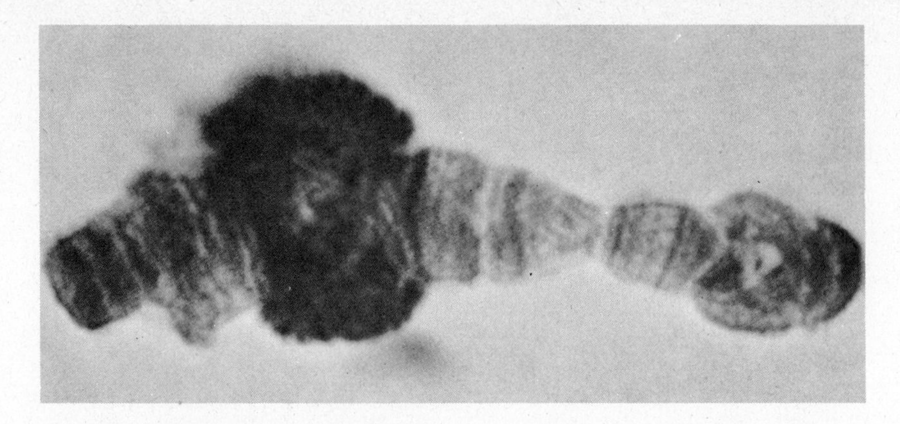

Figure 25-7. RNA in a chromosomal puff. The chromosome of Chironomus is stained with a dye specific for RNA and one specific for DNA. The DNA staining is intense in the bands but little is seen in the puffs. The puffs stain intensely for RNA. (Photograph courtesy of Dr. Claus Pelling.)

same in the salivary glands as in the Malpighian tubules at any particular time in development. Consequently, the suggestion of differential gene activity is supported by the observation that regions of the chromosome appear to undergo changes in the pattern of *m*RNA production.

Rather convincing evidence of the role of the puffs has been obtained in the midge Chironomus. In this organism the larval salivary glands produce a mucous secretory substance. Two different species of the organism have different kinds of mucus, one of which is granular in appearance. The presence of the granules is related to a particular protein. Significantly, in the species that has the granules, one particular chromosome (the smallest one) has four puffs, whereas in the species lacking granules, only three puffs appear. Therefore it appears that the production of a particular protein is under the control of a particular region of the small chromosome. When these species are crossed (which is possible under laboratory conditions), a hybrid is produced that has granules. However, one of the chromosomes of the hybrid has four puffs and the other has three. The chromosome from the granule-bearing species functions to produce the granules; the other chromosome is unchanged. Apparently the genetic character of a particular chromosome determines whether it can respond in the proper way at the proper time. Whatever the factors that elicit the response in the salivary gland from the gene making the protein granules, the chromosome of one species is incapable of responding to it.

It is worth noting that this situation is a visual demonstration of dominance in action. Granule-making is dominant in this cross, since granules are produced, and the dominance results in the activity of one gene while another gene is inactive. If we are to presume that the puffing is characteristic to the transcription of *m*RNA in this form, a fact that seems rather well substantiated by a variety of tests, then we must presume that, in this case, there is no transcription of the inactive gene. Thus the kind of mutation that we are dealing with here is not one in which the structural gene is changed and an inactive protein is made but rather one in which some defect in the transcription process (perhaps a promoter or an operator mutation) appears such that transcription cannot occur. Such control mutations can be of great benefit to eukaryotes, which, as already pointed out (Chapter 22), carry an extraordinarily large amount of DNA. Obviously these cells have developed a mechanism to prevent transcription of unnecessary DNA.

Further evidence to support the idea of a correlation between the presence of a puff and the production of a particular protein has come from studies of mutants of the granule-bearing species. In such mutants, the granules are absent and so is the puff. The relative position of the puff and the genetic map position of the mutant correspond. Similarly, modifier mutants have been discovered that map at other positions. These mutants suppress the production of granules, and there is no puffing of the chromosome. So it seems reasonable to propose that a chromosomal puff is a visible expression of the activity of a gene (or a block of genes).

Most significant are studies that have been made on the development of the larvae of Chironomus and Drosophila. During metamorphosis there are marked changes in the patterns of puffing of the salivary chromosomes. The process is under hormonal control. One of the important hormones is ecdysone, a substance that causes pupation. Larvae can be injected with the hormone and the pattern of development greatly altered so that larvae that normally would not pupate now do so. In such experimental cases, it can be demonstrated that the changes in the chromosomal puffing patterns induced by hormone treatment are exactly the same as those that occur in the normal process of development. Thus the hormonal stimulus is one that alters the pattern of genetic activity, and concomitant with this alteration is a change in the developmental processes of the organism.

The pattern of puffing associated with the changes induced by the injection of hormone is most suggestive. In Chironomus, the first response is a very large puff in chromosome 1, which appears approximately 30 minutes after injection. This puff is followed in approximately 30 minutes by another large puff in chromosome 4. The size of these puffs depends on the dose of hormone given. Apparently the hormone gives a direct stimulus to the chromosome. After 2 or 3 days, secondary puffs appear in the other chromosomes, and as they begin to appear, the original large puffs regress. The size of the secondary puffs is not related to the hormonal dose. In *Drosophila hydei*, a more complex pattern has been observed after ecdysone treatment. Twenty minutes after injection, three new puffs appear and, simultaneously, 18 existing puffs enlarge. As these latter puffs enlarge, 12 other preexisting puffs regress. Six hours after injection, five more puffs appear. Interestingly, of all these puff changes, only four showed the same magnitude of change and the same rate of swelling in response to hormone concentration. It appears that each puff site has its own unique response threshold.

Electron microscopy has shown that the puffed region of the giant chromosomes is apparently a place in which the DNA has lost the highly compact structure that is characteristic of the banded regions. It seems that the various individual strands of the DNA expand and take on the extended condition that might be expected to be necessary for transcription of *m*RNA. As far as can be determined, the threads are continuous from one end of the puff to the other, which would be expected if the chromosome is made up of rather long DNA molecules. The darkly staining band is merely a region in which these long molecules have become compacted. Unfortunately, exactly what controls the extension is unknown. In Drosophila, the swelling begins before RNA synthesis is initiated, and concurrent with the swelling there is a rapid accumulation of acidic proteins, which do not appear to be synthesized in the puffing region itself. The accumulation of acidic proteins in the puffed region causes a drop in the relative concentration of histones, but this may only be a dilution phenomenon. There is no clear evidence that the histone proteins actually detach from the puffed region. Present evidence suggests instead that the acidic proteins are necessary for the extension of the chromosome and thus, hypothetically, they may be considered positive control factors. The source of the acidic proteins and the control of their specific attachment to the appropriate region of the chromosome that must be "turned on" are completely obscure.

Studies with the Dipterans suggest the intriguing possibility that a hormone can act as an inducer to activate a block of genes. (It certainly appears to be the case that ecdysone activates those genes required for the pupation process.) In addition, there is suggestive evidence from mammals that hormones can act in this way. It has been known for some time that estrogen causes a remarkable increase in protein synthesis of the uterus, and, as might be expected, there is a concomitant increase in RNA synthesis. To a certain extent, this increase involves new *m*RNA synthesis, but much of it is a net increase in *r*RNA. This finding is interesting in that it suggests another possible control mechanism in the protein-synthesis system. Obviously the more ribosomes available, the more frequent will be their attachment to the *m*RNA. The result could be an increase in the number of copies made from a single message and hence an increase in net protein synthesis. What is even more interesting is that it has also been demonstrated that in some cases the ribosomes attach to the *m*RNA before it leaves the nucleus. (In microorganisms, one can see that the ribosomes are attached to the "free" end of the *m*RNA before transcription of the message from the DNA is complete.) Evidence suggests that unless such attachment has taken place, the *m*RNA cannot leave the nucleus. If true, an increase in the ribosomal material in the nucleus would act to accelerate the transport of the *m*RNA into the cytoplasm. Consequently, it can be seen that the stimulation of *r*RNA by estrogen could well account for the observed increase in protein synthesis.

On the other hand, the effect of estrogen on the liver is to stimulate the production of two specific proteins, neither of which has any role in the liver. Both are storage proteins required for egg yolk that are formed in the liver and accumulated in the ovarian tissue. It is interesting to note that injection of estrogens into roosters will cause their livers to produce these products, which are normally produced only by the female. This highly specific response must be mediated primarily by the production of two specific *m*RNAs. In this case, the general features of the process are so similar to enzyme induction that it is hard to avoid the conclusion that the hormone is acting to de-repress specific genetic loci.

Dosage Compensation

Another line of evidence for differential gene activity comes from the well-established phenomenon of dosage compensation. Such cases occur when there is a difference in the chromosomal constitution of the sexes (in man, XY males and XX females). In many organisms the Y chromosome carries few genes, and generally they are different ones from those on the X. This raises the question of whether, in the case of X-linked genes, protein production is equal in the male and the female. In most cases, it appears that XY is equivalent to XX, which raises the question of how different doses of a gene produce the same amount of product. It has been suggested that dosage compensation can be explained by the inactivation of one X chromosome in the homogametic sex (thus making XX equivalent to XY). The fact that the interphase nuclei of mammalian XX cells have a darkly staining, heterochromatic body (the Barr body, Chapter 28) was the original basis for such a suggestion, since it has long been argued that heterochromatic regions are genetically inactive. The number of such heterochromatic bodies can be directly related to the number of X chromosomes present (XX = 1, XXX = 2, and so on), and so it has been suggested that all but one of the X chromosomes become inactive.

It was possible to test this proposition by studies of the glucose-6-phosphate dehydrogenase activity in mammals. This enzyme is under the control of a gene located on the X chromosome, and it was easy to demonstrate that the enzymatic activity in peripheral blood cells is the same in both males and females. Apparently protein production in XY is equivalent to that in XX individuals. Furthermore, there is a mutant form in which there is a loss of the dehydrogenase. What kind of activity will be found in a heterozygous female? Such a female (X^+X^-) was found to have one-half the normal activity found in a normal female (X^+X^+). However, after careful investigation it turned out that one-half the cells had normal dehydrogenase activity and the other half had none. Therefore the apparent reduction by 50% is not caused by the action of the two genes in concert; rather, it would appear that one or the other of the X chromosomes is inactivated but that the inactivation is random. Consequently, a heterozygous cell would either have the normal gene active or the mutant *but not both.* If the mutant form expresses itself, there will be no dehydrogenase activity. If the wild-type gene is active, the cell will be perfectly normal. Such studies indicate that, at least in the case of the X chromosome, entire blocks of genes may be inactivated, which has led to the speculation that the chromosome regions of higher organisms that are known to be heterochromatic (and generally associated with a lack of detectable genetic content) may be genetic regions that are totally repressed.

The explanation of dosage compensation raises the interesting question of how the control mechanism operates. We have only preliminary and rather indirect evidence on this point. As noted in Chapter 22, heterochromatic regions of the chromosome are those in which the DNA appears to be supercoiled or highly compressed. Several lines of evidence suggest that under these conditions the RNA polymerase cannot transcribe DNA into *m*RNA. This situation is certainly true in bacteriophages. Once the DNA has replicated, it compresses prior to being enveloped by the head-coat protein. In that compressed stage, there is no longer any transcription of *m*RNA from viral DNA. Similarly, it can be shown in the eukaryotes that labeled RNA precursors are incorporated only in the regions of the chromosome where the chromatin is diffuse. In highly compacted regions there is little or no incorporation of RNA precursors. It has been possible to isolate the chromatin material from calf thymus and separate it into two fractions, diffuse and condensed, in vitro. Analysis of these fractions indicates that although 80% of the total DNA is present in the bulk condensed-chromatin phase, only 14% of the newly synthesized RNA is found associated with it. Interestingly, the RNA polymerase was found in roughly equal concentrations in the two fractions. The fact that the enzyme required for transcription is present suggests that the condensation of the DNA prevents the enzyme from attaching or at least transcribing the message. All this evidence clearly suggests that there is a mechanism in eukaryotes by which large blocks of DNA can be turned off, and this mechanism involves a condensation of the DNA such that the polymerase is inactive. According to this hypothesis, the highly compacted, heterochromatic regions of the chromosome are transcriptionally inert, but there is no reason to assume that a region that is heterochromatic at one stage of development is heterochromatic at another stage.

The Role of Chromosomal Proteins

The Histones

As pointed out in Chapter 22, the major fraction of chromosomal proteins consist of the histones. In the 1960s it was suggested that these proteins might act as specific repressors of genetic function. This hypothesis was based on the finding that the DNA mixed with histones in vitro could not be transcribed by the DNA-dependent RNA polymerase. Since then evidence has shown that the hypothesis is no longer tenable. First, there is too little variation in the histones; the total number of different molecules does not exceed ten. How could this small number of different molecules recognize the wide number of very different genetic loci that must be specifically repressed in eukaryotes? The problem is further complicated by the facts that completely unrelated species of organisms contain the same histones, even though their DNA must be different, and that the histone content is the same in cells that are quite different in function. Furthermore, the repressor molecules isolated from bacteria and viruses are much larger than histones and are in no way similar to them in amino acid composition. Finally, the addition of large quantities of histones to transplanted nuclei does not interfere with their ability to function in any way. In this case, it can be shown that chromosomes actually bind the histones without loss of function. Consequently, we must say of the role of the histones that, if they act as repressors, they must be rather nonspecific.

A nonspecific repression by histone fits the data well. Since the histones are basic proteins, they are rich in lysine and arginine. Consequently, they have a net positive charge. In fact, the charged groups seem to be clustered in the molecules, and these positively charged clusters could react with the negative charge of the phosphate backbone of the DNA. Such an interaction would greatly reduce the charge repulsion of the phosphate groups and facilitate supercoiling of the DNA. One consequence of the supercoiled condition is the inability of the DNA-dependent RNA polymerase to move along the genetic message and transcribe it. The initial finding that the addition of histones to extracted DNA prevents transcription in vitro is in agreement with the nonspecific repression hypothesis. There is other evidence as well. Perhaps the most highly suggestive bits of evidence are the ability of the lysine-rich histones to cause contraction of the loops of lampbrush chromosomes and the localization of one class of histones in heterochromatic areas. When all the evidence is considered, it seems likely that the histones act to prevent transcription in general by promoting the supercoiling of the DNA. However, since the histones are present at all times in all cells of eukaryotes, we must ask: How is the necessary genetic material derepressed so that it can be transcribed? While this question has no certain answer, there is some evidence that the nonhistone chromosomal proteins play an important role.

Nonhistone Chromosomal Proteins

The *nonhistone chromosomal proteins* (NCP) comprise about 20% of chromatin. Some of these proteins are enzymes that normally are found associated with the chromosome; among these enzymes are DNA polymerases, DNA-dependent RNA polymerase, and repair enzymes. However, the largest class is made up of acidic proteins of rather variant size. Their acid quality is the result of a high content of glutamic and aspartic acids. It is these latter, acidic, proteins that interest us as possible control factors, and for the rest of this section they are the only ones that are implied by the designation NCP.

In 1968, it was shown that the inhibition of RNA transcription caused by histones can be reversed by the addition of nonhistone proteins. This observation stimulated a large number of experiments that have shown that the NCP have the properties that would be expected in specific regulatory proteins—for example:

1. The NCP are far more varied than the histones.
2. The NCP pattern (the amounts of different types of NCP) varies with the stage of the cell cycle.
3. The NCP pattern varies with developmental stage.
4. The NCP pattern differs from species to species.
5. The rates of NCP synthesis and turnover are correlated with changes in the rate of RNA transcription.
6. There is a localization of NCP in the active regions of chromosomes.

7. NCP preferentially bind to the DNA of the organism from which they were isolated.
8. Some of the NCP bind specifically to repetitive sequences in the DNA and some bind to unique sequences.

The most important property of a specific regulator of gene action is its ability to control transcription of specific genetic loci. The hypothesis that the NCP act in this way has been tested in different ways. In one instance, chromatin was isolated from rabbit thymus and bone marrow. Each chromatin preparation was dissociated into DNA, histones, and NCP. The DNA and histones were reassociated to give preparations of NCP-free chromatin. These preparations were transcriptionally inert. The histones had completely repressed RNA synthesis. But when bone marrow NCP was added to these preparations, RNA characteristic of bone marrow cells was transcribed; and when thymus NCP was added, RNA characteristic of thymus cells was transcribed. At the very least, this showed that the NCP are cell-type specific. In an even more specific test, chromatin was prepared from mouse brain and fetal mouse liver. The fetal liver cells make the protein globin; brain cells do not. When a fraction of NCP from fetal liver was added to the chromatin from mouse brain, globin messenger was formed. Here, a specific gene was derepressed by NCP. Clearly, the NCP can turn on specific genetic regions.

From these experiments, we have learned that the histones act as relatively nonspecific repressors of genetic function and that the control of genetic activity is effected by derepression of specific loci by the nonhistone proteins. This is not to deny that there may be some degree of specificity in the action of histones. Some may bind preferentially to heterochromatin and some may have preferences for the different types of euchromatin. Nevertheless, the histone action is the same in all cases. Histones prevent transcription of RNA from DNA.

When specific NCP are present, specific genes or blocks of genes are derepressed. In this manner, the differential activity of genes is controlled by the differential synthesis of NCP. Although the factors that control the differential synthesis of NCP are unknown, it would be possible to set up in the transcription process an initial asymmetry that would be self-sustaining. Once one set of NCP was made, only certain genes would function. The products of these genes could be either a rather single-minded set of structural proteins and enzymes or they could include other nonhistone proteins that could derepress new loci at a later time. The presence of a single type of NCP could set this machinery in motion, and we can hypothesize that the "cytoplasmic determinants" of the egg cell could be different NCP that have been localized in the other parts of the egg. A nucleus finding itself in one compartment of the egg cytoplasm would have one set of genes derepressed. A nucleus in another part would have another set expressed. Consequently, two nuclei derived by mitosis from the same parent nucleus could have very different developmental fates. Of course, this whole discussion is highly speculative, but it does point out that this new positive control mechanism—derepression by the NCP—has important implications for the study of development.

The most reasonable hypothesis for the mode of action of the NCP relates the behavior of the NCP to their interaction with the histones. The histones bind to the DNA by virtue of their positive charge. The negatively charged NCP can compete with the DNA for histones and can thus free the DNA of these repressors. Of course, it is necessary to postulate that the NCP act in a similar manner to the repressor proteins that are known to control the operons of prokaryotes. That is, it is assumed that initially the NCP attach to specific regions of the DNA. In doing so they displace the histones in that region and the DNA is relaxed so that it can be transcribed. The importance of the negative charge in the displacement of histones by NCP is indicated by the finding that these proteins are frequently phosphorylated. The added phosphate groups increase the net negative charge of the NCP and make them more efficient in binding the histones. The fact that the phosphorylation of the NCP occurs during the time of maximum transcription of RNA is in agreement with the hypothesis. Furthermore, certain of the NCP have the ability to phosphorylate the histones. The phosphates reduce their net positive charge and further relax their binding to DNA. A cycle of phosphorylation and dephosphorylation that fits well with the cycle of RNA transcription in the cell has also been reported. All things considered, it seems most reasonable to assume that the NCP act to uncoil the DNA by loosening the bonds previously made between the histones and the DNA.

Stable m*RNA and Translational Control*

One of the differences between prokaryotes and eukaryotes is the relatively short life of *m*RNA in the former. This factor has prompted the idea that all higher organisms have stable *m*RNA. The idea is not strictly correct. It is true that the *m*RNA lasts for hours rather than minutes, but the life cycle of the two types of cells is very different as well. If we use the life cycle as a basis for comparison, it turns out that the *m*RNAs of the two groups of organisms have similar stabilities. This fact suggests that the longevity of the messenger is related to the length of time that the cell spends in "interphase"—a not too surprising conclusion, since the breakdown of the messenger is one of the factors in the control (timing) of cellular processes, On the other hand, in highly differentiated cells that are not dividing, there may be *m*RNA that is very stable. One of the most frequently cited examples is the immature red blood cell, which apparently produces all its hemoglobin by use of a stable *m*RNA. Since these cells are destined to produce about 90% of their protein in the form of hemoglobin, the fact that a stable messenger has been adopted is reasonable. This mechanism provides another, rather simple means of gene amplification, but it is presumably one that can only be used by cells that devote their lives to a single function. How the stability is built into the *m*RNA remains a mystery. It does suggest that the mere translation of *m*RNA to give protein is not the event that causes its normal breakdown.

In the early 1970s it was shown that there is a fraction of RNA that never seems to leave the nucleus. Generally these molecules are very large, but they are clearly not ribosomal. The relative stability of some of the *m*RNA of the eukaryotes suggests the possibility that this fraction is stored messenger that cannot leave the nucleus until some other event occurs, such as the attachment of ribosomes. This interpretation is supported by the observation that, in sea urchin eggs, there is unimpeded protein synthesis in the presence of actinomycin, an inhibitor that normally blocks the production of *m*RNA. It thus appears that the *m*RNA is already made and "waiting" to make protein. This stored *m*RNA is further demonstrated by the fact that the egg leaves its dormant state immediately after fertilization and begins rapid protein synthesis. The onset of this synthesis is much too rapid to be dependent on the production of new messenger, and it must be assumed that the *m*RNA for the initial stages of development is already produced and stored in the egg. Apparently the activation of protein synthesis caused by the sperm is effected by the joining of the ribosomes to the stored messenger, a process that may be necessary for the release of the *m*RNA from the nucleus. Unfortunately, we have no evidence as to what the translational control mechanism is, but this observation may provide an explanation for the large amounts of RNA that are found in the nuclei of all eukaryotes.

The work with cell fusion experiments supports the suggestion of stored messenger under translational control. As noted, human cells can be fused with those of the chicken. When the chicken nucleus swells initially, it can be shown that *m*RNA synthesis is taking place. Nevertheless, there is a lag in the release of this RNA from the nucleus, and no protein synthesis occurs until the nucleolus has developed. That the release of RNA is under the control of the nucleolus can be shown by irradiating it with a microbeam of ultraviolet. When this is done, the nucleolus is suppressed, and no release of RNA and no protein synthesis result, even though active transcription continues and nuclear RNA accumulates. Since the nucleolar organizer is responsible for the production of ribosomal RNA, the simplest interpretation of this finding is that normally the *m*RNA leaves the nucleus already attached to ribosomal elements (possibly as polysomes) and that, in the absence of ribosome production, no transport is possible. This same interpretation is used to explain the increased rate of protein synthesis observed in uterus that has been treated with estrogen (see above). It is still too early to be sure of the exact interpretation of these observations, but they suggest that translational control may be of great importance in eukaryotes. Such a translational control mechanism could explain the puzzling observation that protein synthesis is shut down at metaphase. Why is the existing *m*RNA (3-hour lifetime) not translated? It is possible that, with the disappearance of the nucleolus, the attachment of the ribosomes to the *m*RNA is inhibited.

The Control of Differentiation

Before concluding, let us return to the place where we began. How is it possible for cells having the same genetic material to be so different in their final form and function? The preceding discussion should have made clear that we do not lack mechanisms for the control of genetic function. Such mechanisms make possible the turning on of specific loci at specific times, to permit the change in function that is characteristic of cells in development. The kinds of control mechanisms known to operate in bacteria can be envisioned to function in eukaryotes. Only the relatively permanent contraction of some parts of the chromosome and the extensive amplification of genetic units in other parts can be considered special properties of eukaryotes. Neither condition, however, represents an absolutely irreversible change in the genetic material. The chromosomal proteins, which are absent in the prokaryotes, add a new level of control to the eukaryotic chromosome. Since the pattern of heterochromatic regions varies with the stage of development and since there is evidence that certain molecules can derepress nonfunctional regions, we may hypothesize that even the extreme cases seen in the eukaryotes are under positive control.

Although the control mechanisms now known to exist in various cells are adequate to explain both growth and differentiation in principle, we are a long way from having a detailed knowledge of how the various mechanisms interact. In particular, we lack any clear indication of how initial differentiation occurs. Once there are several different cell types, cellular interaction can be invoked to explain the progressive differentiation of the organism. (The problems of hormonal regulation and embryonic induction concern cellular interaction.) But we begin with a zygote that is a single cell, and somehow differences arise in the progeny cells even though they share the nuclear materials equally. The best evidence indicates that these differences occur very early in development, and that the various nuclei begin to assume independent functions in some way. In certain marine eggs, this period may be as early as the two-cell stage. Consequently, we must ask how such asymmetry can arise in the developmental system.

Two features, perhaps, are of overriding importance to the changes in the zygote that occur during cleavage. The first is that the cytoplasm of the egg cell is not uniform. Hundreds of experiments have shown that removal of certain portions of the egg cytoplasm is sufficient to prevent normal development, frequently with the loss of very specific parts of the embryo. Thus we can be sure that in many cases, probably in all, the egg cytoplasm has within it certain determinants of development. It is ironic that at this level of analysis we have advanced little from the ideas of Boveri. What are these determinants and how are they localized? Only rather vague answers are available, and even they must be postponed until we consider extranuclear inheritance (Chapter 26).

The second important feature of development is that there is relatively little genetic function during the time of cleavage. Much of what the embryo needs has been made prior to fertilization. Considering the tremendous quantities of DNA available in the egg, very little of it is being used. This situation suggests that in the early stages of development the majority of the structural genes are not of any great importance and that regulatory genes may give rise to the initial asymmetries that result in differences in genetic function during the later stages of development. Let us presume that a significant portion of the eukaryotic DNA is, in fact, "programming" DNA: genes whose specific function is to control the activity of other genes, probably through the production of both repressors and positive control factors. We can imagine the egg as being a cell whose conditions of development are such that, at maturity, mainly the programming genes are functioning. We could propose that sets of these genes exist, each of which controls a rather large but highly specific program of genetic function. (The master gene might turn on a set of regulatory genes that, in turn, control the activity of several sets of structural genes necessary for a particular functional adaptation.) When an appropriate stimulus is given (some molecule or set of molecules in the region of the egg where the nucleus finds itself), a complete block of genes, possibly widely scattered in the genome, would be turned on; and possibly another complete set of genes would be turned off by the functioning of the integrator units. The fact that such a hierarchy of control genes does exist is indicated by the response of the Dipteran salivary chromosomes to the hormone ecdysone.

We are still far from answering the problem of differentiation in multicellular organisms, but the process does not seem quite so mysterious as it did only a decade ago. The variety of control mechanisms that are now being discovered seem to hold the answer. It is presumed that the next "great era" of biology will be the unraveling of the detailed mechanisms whereby the egg cell finally expresses itself as an adult, multicellular organism.

REFERENCES

AMBROSE, E. J., and D. M. EASTY. 1970. *Cell Biology.* Addison-Wesley, Reading, Mass. (Chapters 12 and 13 give a good overview of the development of higher animals. They are far more detailed than this chapter but somewhat telegraphic.)

BEERMAN, W., and U. CLEVER. 1964. Chromosome puffs. *Sci. Amer.* **210** (No. 4):50–58. (A simple presentation of the work with Chironomus.)

EPHRUSSI, B., and M. WEISS. 1969. Hybrid somatic cells. *Sci. Amer.* **220** (No. 4):26–35. (A good introduction to the cell fusion experiments.)

HARRIS, H., J. F. WATKINS, C. E. FORD, and G. I. SCHOEFLI. 1966. Artificial heterokaryons of animal cells from different species. *J. Cell. Sci.* **1**:1–30. (One of the early reports of cell fusion studies.)

LYON, MARY F. 1972. X-chromosome inactivation and developmental patterns in mammals. *Biol. Rev.* **47**:1–35. (A good, advanced review of dosage compensation and other aspects of X-chromosome inactivation during development.)

WATSON, J. D. 1970. *Molecular Biology of the Gene.* 2nd ed. W. A. Benjamin, Inc., New York. (Chapter 16 presents an interesting view of development from the molecular point of view.)

Original Papers and Advanced Reviews

BELL, E. (Ed.). 1965. *Molecular and Cellular Aspects of Development.* Harper and Row, New York. (This volume is a collection of articles and reviews that cover all aspects of the problem. For the advanced student.)

BROWN, S. W. 1966. Heterochromatin. *Science* **151**:417–425. (A very good review of the subject.)

GURDON, J. B., and H. R. WOODLAND. 1968. The cytoplasmic control of nuclear activity in animal development. *Biol. Rev.* **43**:233–267. (A comprehensive review of nuclear transplantation experiments.)

QUESTIONS AND PROBLEMS

25-1. Define the following terms:

chromatin diminution
chromosome puff
gastrulation
dosage compensation
cell fusion

25-2. Explain how chromatin diminution could lead to differentiation during development. Are cytoplasmic determinants required for this process?

25-3. Explain how chromatin diminution could occur with no loss of genetic potential and thus be irrelevant in differentiation.

25-4. Explain the technique of nuclear transplantation. What has been shown about gene action using this technique?

25-5. Define the concept of differential gene activity. How can this concept be used to explain differentiation?

25-6. How could a small molecule, such as a hormone, regulate gene action?

25-7. Under proper experimental conditions some adult, highly differentiated cells can be made to return to a less-differentiated state (dedifferentiation) and then redifferentiate into cells that are distinctly different from the cells of their origin. How can you explain this process?

25-8. Compare the heterochromatization observed in dosage compensation to chromatin diminution as a mechanism for differentiation.

25-9. In many molluscs, separation of the two cells produced by the first cleavage division of the zygote will result in the production of two different and abnormal multicellular masses. In contrast, separation of the two cells of an amphibian zygote produces two perfectly normal, if somewhat smaller, larvae. Explain this difference.

25-10. Certain genes act early in development; others act quite late. Which of these two classes would you expect to show the more drastic effect of mutation? Explain your answer.

25-11. Mutation of certain genes can affect a large number of different parts of the organism. Such genes are called *pleiotrophic*. Nevertheless, our general premise is that one cistron produces one polypeptide. How do you reconcile these two statements? (You should be able to give at least two different explanations.)

26/Extranuclear inheritance: the role of cytoplasmic determinants

In the previous chapter the fact that the cytoplasm has a role in determining gene action was discussed. It is generally believed that the sorting out of "determinants" in the cytoplasm of the egg by cleavage division causes the asymmetry that results in differential gene activity and hence differentiation. It is necessary to ask what the evidence is for the existence of such determinants and for their "hereditary" role. So far our emphasis has been on heritable systems that behave in a Mendelian fashion. However, there are some instances of inheritance that are non-Mendelian in their behavior and for which the determinants lie in the cytoplasm and show a great deal of autonomy from the nucleus. What follows is far from an exhaustive discussion of the known examples. Instead we have chosen to present a few examples of the several very different systems showing extranuclear inheritance in the hope of demonstrating what we mean by cytoplasmic determinants. Armed with the principles illustrated here, it should be possible for the interested reader to read the more advanced reviews listed at the end of this chapter.

Maternal Influence

In the meal moth Ephestia, the skin and eye pigment of the larvae are under control of a simple allelic pair. The dominant allele **(A)** promotes pigment production. In the homozygous recessive condition **(aa)**, no pigment is produced. The working out of the mechanism for gene control of pigment formation in Ephestia was one of the earliest studies on which the one gene–one enzyme hypothesis was based (Chapter 15). Gene **A** controls the production of kynurenin, which is required as an intermediate in pigment synthesis.

When a heterozygous female makes eggs, half will carry the **A** allele, and the other half will carry the **a** allele. When such eggs are fertilized by sperm from homozygous recessive males **(aa)**, one-half the progeny should be heterozygous **(Aa)**, and the other half should be homozygous recessive **(aa)**. Consequently, one-half the progeny should lack pigment; however, initially, all the larvae carry some pigment. In time the pigment fades in half the larvae (those that are **aa**), but it is there initially. By the second generation, there is no pigment production at all in the homozygous recessive group.

Obviously, in this case, the mother **(Aa)** has transmitted to all the offspring, regardless of genotype, some of the kynurenin for pigment formation. The cytoplasm of the egg with the **a** genotype still has kynurenin from the action of the **A** gene prior to meiosis. Since there can be no continued production of kynurenin in the **aa** individuals, it is eventually used up and the color fades. Heterozygous individuals continue the production of kynurenin, and thus one-half the progeny **(Aa)** are pigmented, and the other half **(aa)** fade out gradually to the nonpigmented condition.

In this case, it is easy to see what is being transmitted. A simple chemical molecule is produced in sufficient excess to permit its utilization in the next generation. But it is a single-generation characteristic, and, for this reason, it is generally referred to as a *maternal influence* rather than a maternal inheritance. The influence is present in the immediate progeny but cannot be transmitted to subsequent generations.

A somewhat more complicated case comes from the snail *Limnaea peregra.* These snails can exist in either of two forms (polymorphic). In the first, the coiling of the shell is such that if you hold the shell in your hand, the opening will appear on the right. Such snails are said to exhibit *dextral* (right-handed) coiling. In the other case, the coiling is in the opposite direction (the opening would appear on the left), and these snails are called *sinistral* (left-handed). Referring to Figure 26-1, it can be seen that when a dextral snail is used as the female parent in a cross between a dextral and a sinistral snail, the offspring are all dextral. Such dextral offspring can be self-fertilized (the snails are hermaphroditic), and once again all the progeny are dextral. This is a rather strange result. It seems reasonable, on the basis of the first cross, that dextral is dominant to sinistral. Consequently, we would expect snails in the ratio of three dextrals to one sinistral, as the genotypes listed underneath the figures of the snails indicate. Yet no sinistral progeny appear. If, however, each of the snails from the F_2 is self-fertilized, in the following generation one-quarter of the

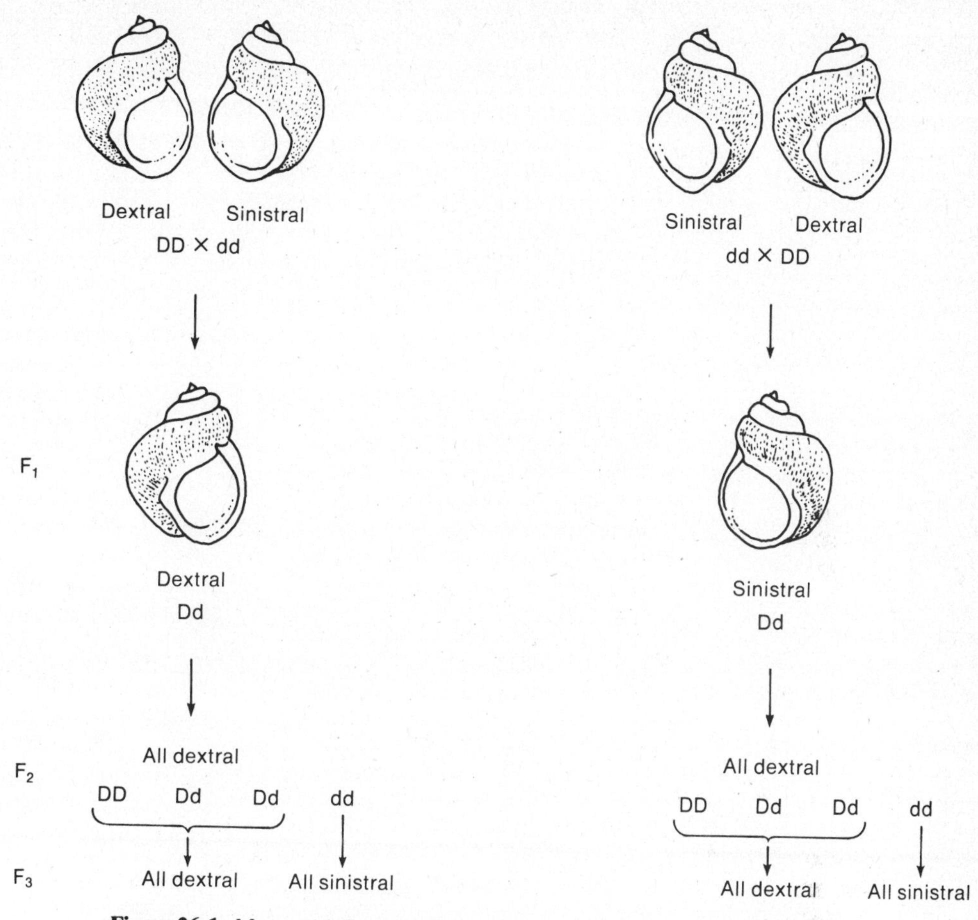

Figure 26-1. Maternal influence. The pattern of coiling of snail shells in Limnea shows the delay of gene expression by one generation in some cases. The pattern of coiling (dextral or sinistral) is that of the mother's genotype, not that of her phenotype. The cytoplasm of the egg is conditioned by the maternal genes to set the cleavage pattern prior to fertilization.

snails produce only sinistral snails. It thus appears that the phenotypic expression of the homozygous recessive sinistral condition is delayed by one generation.

Now examine the reciprocal cross, in which the dextral snail is used as the male parent. At the first generation, all are sinistral. On the other hand, when these phenotypically sinistral F_1 snails are self-fertilized, the progeny they produce are all dextral; and when the latter, in turn, are self-fertilized, it can be seen that three-quarters of them give rise to dextral snails and only one-quarter to sinistral
snails produce only sinistral snails. It thus appears that the phenotypic expression
the F_2 progeny consist of three-quarters dextral and one-quarter sinistral. But the phenotype of the F_2 generation is not consistent with this interpretation.

In examining both crosses carefully, the simplest interpretation is that the snail shell has the configuration that is characteristic of the *genotype* of the snail that is producing the eggs. In Cross I, the egg is initially produced by a homozygous dextral snail. All eggs, therefore, are dextral, and all the resulting snails are dextral. Similarly, when the F_1 is self-fertilized, its genotype is heterozygous, but it carries the dominant dextral gene. Therefore all its progeny are dextral. In the third generation, however, one-quarter of the F_2 snails are homozygous sinistral genotypically. The eggs they produce are only sinistral. Exactly the same line of reasoning can be used to explain the second cross. Initially the egg comes from a homozygous

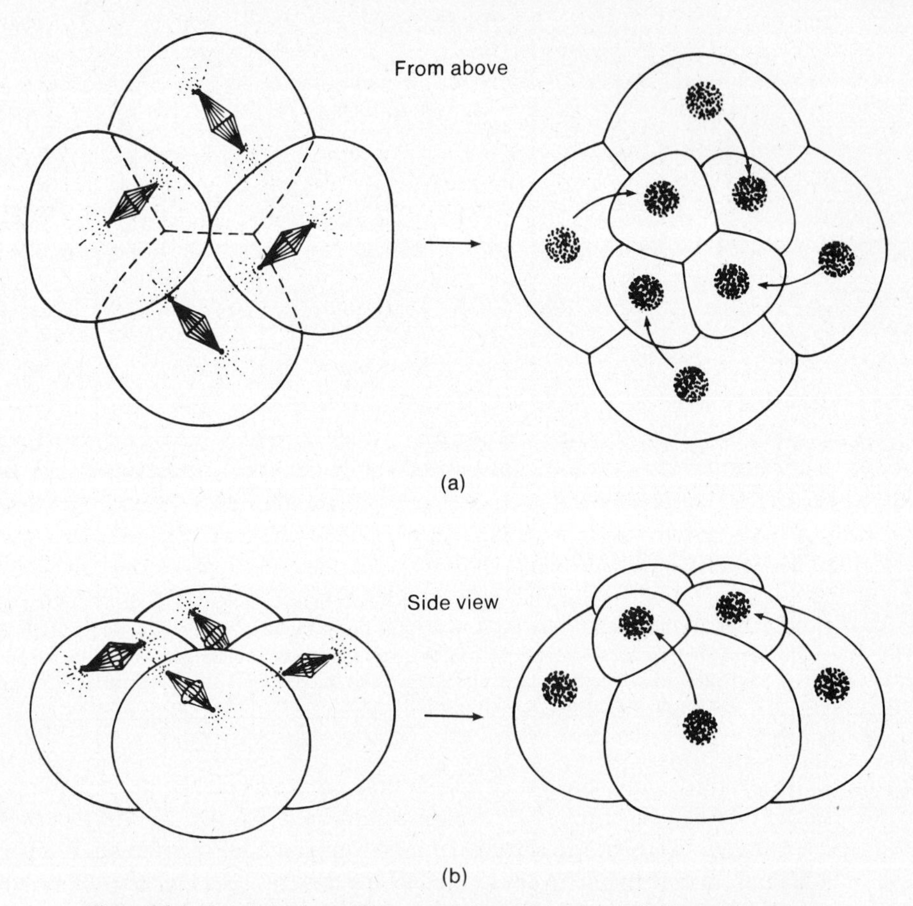

Figure 26-2. Spiral cleavage. This type of cleavage (early cell division of the zygote) occurs in a number of invertebrates. It is common in the mollusks, where it accounts for the spiral form of snails, etc. The spindles are set at an oblique angle in the cytoplasm (a and b). When division occurs, it produces four small cells and four large ones. The small, upper cells are pushed out of line with the large, lower cells. This establishes the basic spiral pattern for the rest of development. (After Bodemer, C. W. 1968. *Modern Embryology.* Holt, Rinehart and Winston, Inc., New York.)

sinistral snail. Therefore all the progeny of the F_1 are sinistral, but the genotype of the F_1 is the same as in the reciprocal cross. It is heterozygous and carries the dominant dextral gene. Thus all its progeny are, in fact, dextral. As before, the F_3 is consistent with this interpretation, since one-quarter of the progeny of the F_2 give rise only to sinistral snails.

The coiling pattern in snails is developed during the early stages after the fertilization of the egg. The orientation of the spindle at each of the successive cleavages is such that a spiral pattern develops (Figure 26-2) and the coiled structure of the snail is established. In other words, whether a snail is dextral or sinistral depends almost exclusively on the positioning of the spindle within the cytoplasm of the egg. Apparently one gene controls the final positioning. In eggs that are developing in the presence of the dextral gene, positioning of the spindle is fixed prior to fertilization, and afterward it will continue as a dextral snail. This is true even though only sinistral genes are carried in the zygote after meiosis. The cytoplasmic content of the egg is determined by the female parent. Her genotype has left its imprint on the cytoplasm, an imprint that endures during one generation.

Clearly this case is similar to that of Ephestia. Something is made that determines the positioning of the spindle and that can reside in the cytoplasm and act independently of the genotype of the developing organism. Unfortunately, we do not know what is made, but whatever it is, it cannot be perpetuated indefinitely. At the next generation, the new egg cells receive determinants under the influence of the maternal genes, and a snail that is phenotypically dextral but genotypically sinistral will give rise only to sinistral progeny.

The results in these two cases are reminiscent of the situation seen in the reciprocal crosses of *E. coli* in the inducible lac system (Chapter 24). When the $\mathbf{i}^+$ gene is present in the recipient cell, the cytoplasm contains a repressor, and thus the genes that are received from the donor cannot be transcribed. In all these cases, gene products are accumulating in the cytoplasm and are capable of exercising an effect that is only temporary. All can be considered cases of maternal influence, in which the phenotype of the organism may be inconsistent with the genotype; but in all cases the control ultimately resides in the nucleus and, in the long-term view, can hardly be called an inheritance in the strict sense.

Cytoplasmic States

Paramecium offers a variant on the systems best described as maternal influence. Since the cytoplasmic properties in this case are more long lasting and since the use of words like maternal and paternal in Paramecium is at best metaphoric, we have chosen to give it as a separate case. Before we can explain it satisfactorily, however, it is necessary to digress slightly and review the life cycle of this ciliated protozoan.

Paramecium has two kinds of nuclei—a small *micronucleus* and a very large *macronucleus*. The macronucleus appears to be highly polyploid and is irregular in its division habits. The micronucleus is regular in its behavior and is apparently the seat of Mendelian heredity. The best evidence suggests that there may be a division of function in which the macronucleus acts as a seat of gene transcription, a rather elaborate kind of gene-amplification system, with the micronucleus acting as a "germ line" for the organism. Consequently, from our point of view, it is possible to ignore the macronucleus and discuss only what happens to the micronucleus in the various kinds of reproductive behavior that the organism exhibits.

Paramecium can reproduce in three ways. Essentially, it can divide mitotically, a process called *binary fission.* A special name is given to this division because of the complicated sorting out of the various differentiated structures (rows of cilia, gullet, contractile vacuole), but, genetically, it is a mitotic division. The second method of reproduction is called *conjugation.* Here the two ciliates are joined together and the nuclei of both undergo meiosis, each producing four micronuclei. Three of the micronuclei die and one remains in each case. Each surviving nucleus then divides mitotically to produce two identical micronuclei. (It should be noted that this process is similar to the production of gametes in the higher plants.) The two cells are joined by a small cytoplasmic bridge through which each donates one nucleus to the other cell (Figure 26-4, top). The cells then separate, the two micronuclei fuse, and the exconjugants have exchanged genetic material in a truly sexual manner. Note that in this case, unlike the bacteria, both are donors and recipients.

The third method of reproduction is called *autogamy.* Here an unpaired, single individual undergoes meiosis, following exactly the same steps as those in conjugation. At the end of the process, two identical haploid nuclei are present in the cell. However, there is no partner, and hence no exchange of nuclei, so the two nuclei fuse and produce a new diploid organism. At first this process would appear to accomplish nothing, but, in fact, it accomplishes something rather unique. The organism that is produced is now exactly homozygous, since its genes are all derived from the replication of a single chromosomal set.

Fortunately, the different reproductive processes of Paramecium are under the control of the investigator. Consequently, when desired, a cross can be made through conjugation, and the exconjugants can be reproduced in large numbers by binary fission. Whenever a test of the genetic constitution of these organisms is desired, autogamy can be used to produce homozygous offspring. Paramecium is a remarkably good genetic tool.

The discovery of "cytoplasmic states" in Paramecium came from work with the antigenic proteins of the cell surface. Cells of all organisms are covered with large molecules (proteins, polysaccharides) that can act as antigens. That is, when these molecules, or the whole cells, are injected into the bloodstream of a mammal, specific antibodies are made against them. The antibodies can detect very small differences in the composition of molecules. This fact makes it possible for the investigator to detect differences in strains of Paramecium. Some strains have one set of antigens on their cell surface and other strains have another set. Since we have already demonstrated that the control of protein synthesis resides in the genes, it seems reasonable to propose that the differences in antigenic properties of the different strains are a reflection of differences in their genetic makeup, but on investigation this proposal did not appear to be completely true. For example, it is possible to take a strain of Paramecium that has been derived from a single homozygous animal by use of autogamy. With care, several lines, each characterized by different antigenic properties, can be derived from that homozygous strain. It is true that within each line individuals show a single antigenic type, but the several lines are distinctly different. Obviously we must ask how different lines of cells can be derived from an homozygous individual in a space of time that is far too short to permit the derivation of these lines by mutational events.

The first step in unraveling this puzzle was the discovery that when two different antigenic lines are crossed, the antigenic properties of the exconjugants are unchanged. Consider, for example, two antigenic lines, line A and line B. Two individuals, one from each of these lines, undergo conjugation. They exchange nuclei, and the exconjugants are genetically identical. Nevertheless, the exconjugant derived from line A continues to show line A antigenic properties, and the exconjugant derived from line B continues to show line B antigenic properties. Clearly, this inheritance appears to be cytoplasmically determined. The trading of nuclei does nothing, and little or no cytoplasm is exchanged in the process.

The situation is somewhat more complicated than just described. In fact, it turns out that the antigenic properties of the Paramecium depend on the temperature. For example, at 25°C a line may exhibit one antigenic property and at 29°C another. Thus apparently some environmental influence, as well as a cytoplasmic determinant, exists. Let us consider what happens with the genetically identical exconjugants derived from lines A and B under these circumstances. At 25°C, line A produces a pattern of antigens that is characteristic of line A. However, when the temperature is raised to 29°, the cells derived from line A show a pattern that is typical of line B. In other words, raising the temperature shifts the antigenic properties of line A to a line B type. This situation would be extremely difficult to understand were it not for the fact that the two lines are genetically identical. Apparently, in line A at 25°, one set of gene products is being produced. When the temperature is raised, the other set of gene products is favored, and the antigenic property of the system is switched. This experiment (and the other elegant experiments) indicates that the antigens are ultimately under genetic control. Only those antigens for which specific alleles are present can be produced. In some way, however, the cytoplasm is capable of determining which alleles will be expressed; and until such time as an accident of nature or an experimental intervention occurs, the cytoplasmic system is completely stable. Line A will produce line A antigens indefinitely unless something happens to upset the metabolism of the organism, in which case it may switch to a line B antigenic character.

It is not easy to explain the perpetuation of an unstable cytoplasmic state through time. Obviously we cannot be talking about the usual self-replicating, long-lived, "genetic" factor. It is hardly to be expected that the transition of 4°C is sufficient to change the form of such a determinant. Furthermore, the usual kind of repressor system invoked for the control of gene transcription does not seem to apply in this case. Since the genotypes of the two individuals are identical, both should be synthesizing the same kind of repressor. It is hard to see how one set of genes in line A synthesizes a repressor that blocks line B genes, while at the same time the line B genes are synthesizing a repressor that blocks line A genes in another cell. Even though it would be possible for a 4° temperature shift to change the conformation of a repressor molecule (the tertiary configuration of proteins is quite temperature sensitive in certain cases), the genetics of the situation and the stability of the cytoplasmic state do not favor a repressor interpretation.

At present, we are not certain what it is that determines the cytoplasmic state. However, a rather interesting proposal has been advanced that at least provides a model for understanding this kind of system. Simply stated, it proposes that each antigen is produced by a reaction sequence involving several steps (Figure 26-3). Intermediates produced in either reaction scheme can act as inhibitors in the

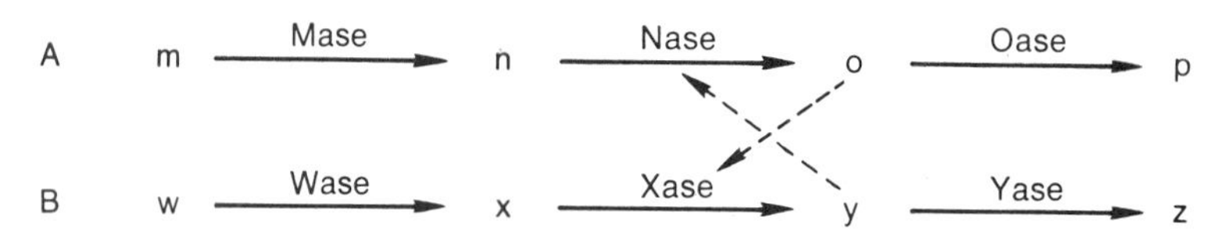

Figure 26-3. Cytoplasmic steady states. The diagram presents an explanation for the two different antigenic strains of Paramecium. If the A set of antigens has been produced, the cytoplasm of the cell will have a quantity of intermediate o. This intermediate inhibits the enzyme Xase necessary for the conversion of x to y. Consequently such cells will not make the B set of antigens, even though the necessary genes are present.

other. In this case, the reaction that proceeds will be the one that has already built up a sufficient concentration of the intermediate, such that the other reaction series is inhibited. In the example used above, the exconjugant derived from a line A cell has already been producing line A antigens, and the intermediates of the line A type are already in relatively high concentration. Therefore the pathway leading to line B antigens is inhibited. The same argument holds for the exconjugant derived from a line B cell. In both conjugants, both pathways are present, but only one pathway is operating. When something happens (such as a temperature rise) to upset the metabolism, the balance can be shifted, and now the formerly inhibited reaction pathway is free to run. When it begins to run, it obviously builds up a concentration of intermediates sufficient to prevent the production of the other pathway, and the cell is now in a different "cytoplasmic state." The overall proposition, then, is that the cytoplasmic state is equivalent to a metabolic steady-state condition. As long as there are no metabolic disturbances, the steady state is maintained, and only one set of genes is finally expressed. The cytoplasm in which these reactions occur and in which the intermediate products accumulate becomes the determinant of genetic expression.

It should be obvious that this situation is very similar to the cases of maternal influence discussed earlier. Only its relatively persistent character through both sexual and asexual reproductive cycles makes us treat it separately. Unfortunately, as yet we are not certain about the exact mechanism, and, until we are, it is necessary to treat the cytoplasmic steady-state hypothesis with some skepticism. At the present time, however, it does represent the simplest model to explain the results obtained in these crosses.

Cytoplasmic Particles

The preceding cases have dealt with molecular "determinants" in the cytoplasm, determinants that cause the expression of certain genes to be delayed or modified. There are other cases of cytoplasmic modification of inheritance in which a definite body can be associated with the behavior. The classic case is the killer phenomenon in Paramecium. The killer strains of this protozoan make a compound, *paramecin,* that kills other strains, generally called sensitive strains, when grown in the same culture. Interestingly, the killer strain can be distinguished from the sensitive strain morphologically, since killers have a granular cytoplasm and the sensitives are clear.

Since killers normally produce only killers and sensitives only sensitives, it seems reasonable to propose that there is a genetic mechanism behind the difference. To test for it, it is necessary to make crosses between the two. At first, doing so poses a difficulty, for if the two strains are together, the sensitives die. Fortunately, however, it is possible to wash the killer strains by repeated centrifugations until they have been freed of paramecin. In this condition, they are no longer killers for a period of time, and crosses can be made between killer and sensitive individuals. Homozgyous individuals prepaired by autogamy are crossed, as indicated in Figure 26-4. The exconjugants are of two types, granular and clear, but presumably they have identical genotypes. If the F_1 granular exconjugant is crossed with another F_1 granular, three-quarters of the progeny of the F_2 are killer and one-quarter are sensitive. The conclusions are that the killer phenotype is dependent on possession of the dominant allele and that the sensitive condition is a homozygous recessive.

The analysis is simple and straightforward until we observe what happens when crossing the F_1 sensitives with one another. When this cross is made, the F_2 are all sensitive, even though some of the F_2 must be homozygous dominant for the killer gene. Since the killer strains have a granular cytoplasm and the sensitives all have clear cytoplasm, it was proposed that the killer phenotype depends on both the **K** gene and a factor in the cytoplasm called *kappa.* The hypothesis is that, in the presence of the **K** gene, the kappa factor can be maintained. [Therefore, in the F_2, the individuals that are homozygous recessive **(kk)** become sensitive, since the kappa factor cannot be maintained in the strain.] On the other hand, the presence of the **K** gene is not sufficient to initiate the synthesis of the kappa factor, and, therefore, no kappa is ever present in the cells of the sensitive exconjugant type and all the

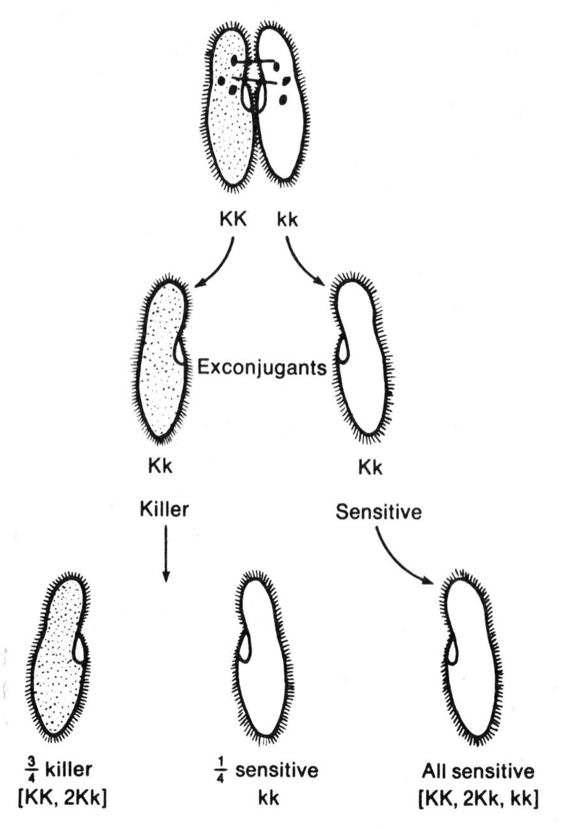

Figure 26.4. Inheritance of the killer phenotype in Paramecium. As the crosses on the left show, the killer allele is dominant to the sensitive. When the homozygous recessive condition occurs, the particles in the cytoplasm disappear and a sensitive phenotype is produced. The crosses on the right indicate that the killer phenotype cannot occur unless the cytoplasm of the cells contains *kappa* particles, regardless of the genotype of the cells. K maintains the particles but cannot initiate them.

cells are sensitive. This hypothesis certainly explains the data. Let us examine some tests of it.

The crosses described above were done rapidly. If, however, the cross between killer and sensitive is slightly prolonged, so that there is time for an exchange of cytoplasm as well as an exchange of micronuclei, then some of the kappa particles can move from the killer strain into the sensitive strain. When they do, all the F_1 individuals are killer in phenotype, and both the F_1 exconjugants produce F_2 generations that are three-quarters killer and one-quarter sensitive. This finding is in agreement with the initial hypothesis that the killer phenotype is dependent on the presence of kappa particles. Furthermore, it is possible to make Paramecium divide very rapidly by controlling the nutrient conditions. When done with a strain that is **KK** and that contains the kappa particles, occasional individuals arise that are sensitive. When these individuals are examined, it turns out that they have lost the kappa particle. Apparently the kappa particles cannot reproduce as rapidly as the cell can divide, and so they become greatly diluted until, by chance, certain cells are produced that lack the kappa particle altogether. Such cells are sensitives. This experiment, like the preceding one, is in agreement with the proposition that the **K** allele is incapable of initiating the existence of the kappa particle. The kappa particle itself, at least to a certain extent, is autonomous of the nucleus. But, of course, it is not completely autonomous; unless one **K** allele is present, the kappa particles cannot survive and are lost along with the killer phenotype.

Later investigations of the kappa particles indicate that they are viruslike particles, and thus their ability to reproduce, at least partially independently of the nucleus, is not surprising. They contain within themselves the necessary hereditary material to govern their own reproduction and to elicit the production of the substance paramecin. Here, in essence, is a genetic unit carried not on the chromosomes of the Paramecium but free-living in the cytoplasm, where, because of its intrinsic genetic properties, it is capable of a semiautonomous existence.

A similar case has been found in Drosophila. Fruit flies are normally resistant to the effects of high CO_2 concentrations. CO_2 is frequently used as an anaesthetic, and the fruit flies can remain in it for hours with no adverse effects. However, certain strains of Drosophila are very sensitive to CO_2; an exposure of only a few minutes leads to paralysis and death. The paralysis is the result of a neurological malfunction in the ganglia of the thorax, but the basis is unknown.

Sensitivity to CO_2 is inherited, but the inheritance does not follow Mendelian laws. There is no linkage to any of the chromosomes. Furthermore, the sensitivity can be transmitted by the injection of a cell-free filtrate of hemolymph into normal flies. This situation indicates that some nonchromosomal agent is responsible for the transmission of the sensitivity. The agent has been discovered to be a viruslike particle, called *sigma*, of about 40 millimicrons in diameter. Perhaps the most interesting feature of the inheritance of CO_2 sensitivity is that flies made sensitive by injection of hemolymph transmit the sensitivity to their offspring at very low frequencies, even though their hemolymph can be shown to contain high levels of the infective agent. Here transmission is always through the egg, never the sperm. On the other hand, in sensitive lines of flies derived from natural populations, transmission to the progeny is the rule; and the transmission can be effected by either eggs or sperm.

The explanation of the difference between sensitive strains and induced sensitive individuals is that, in the sexually transmitting strains from natural populations, the virus has become incorporated into the nucleus, whereas, in the induced sensitives, the virus particles are free in the cytoplasm. In the latter case, where viruses have entered the oogonia the particles will be transmitted through the egg cytoplasm. They will not be transferred by the sperm, since the sperm contributes little or no cytoplasm. In the "normal" transmission of sensitivity, the virus is contained within the nucleus (but not attached to a specific chromosome) and thus can be transmitted by either sperm or eggs in a regular but non-Mendelian manner.

Finally, we should point out that the inheritance of characteristics carried on episomes in bacteria is exactly analogous to the cases discussed here. These findings suggest that, in a great variety of organisms, a certain amount of cytoplasmic inheritance can be attributed to the existence of self-reproductive bodies that are at least analogous to viruses. We must distinguish these cases from the transmission of infectious diseases purely on the basis that they can pass through the sexual cycle of the organism. At this point the distinction becomes tenuous, and it will be some time before we can be sure how widespread the transmission of characteristics by cytoplasmic particles is.

Plastid and Mitochondrial Inheritance

One of the earliest examples of extranuclear inheritance came from the study of pigmentation in the four o'clock, Mirabilis, a study conducted in 1909 by Correns, one of the rediscoverers of Mendel. The strain of Mirabilis investigated showed variegated coloring; that is, part of the plant was green and part was white. This case of variegation was nonuniform—some parts of the plant had completely green branches, others completely white branches, and still others (the majority) variegated branches, with splotches of green and white mixed together.

Since reproduction is decentralized in most higher plants, each of the various branches of the plant produces flowers of its own. Thus flowers are produced on wholly white branches, some on wholly green branches, and some on variegated branches. When made to self-fertilize, the green branches produced only unvariegated green plants. The white branches produced only totally white progeny, which, lacking chlorophyll, died as seedlings. The variegated branches produced variegated progeny, green progeny, and white progeny. Furthermore, it could be demonstrated that the kind of progeny produced depended solely on the egg cell. If pollen from a white branch was placed on a flower from the green branch, only green plants would be produced; and, reciprocally, if pollen from a green branch was put on a white-branch flower, only white progeny were produced. The genotype seems to make little difference.

The explanation for this phenomenon is relatively simple. Color resides in plastids. If they make chlorophyll, they are green; if they have a defect such that they cannot make chlorophyll, they are white. To explain the results of the crosses, we need only assume that the plastids are autonomous bodies. In the egg cell of a variegated plant, both kinds of plastids will coexist. By chance alone, some cells will receive only green plastids at the time of division and others only white. The great majority, of course, will receive both green and white. Thus a variegated branch gives rise to three different kinds of progeny. A green branch contains only green plastids and therefore produces only green progeny; a white branch produces only white.

A somewhat similar case is found in variegated geraniums (Pelargonium). Here seeds derived from a white tissue give rise only to white progeny when self-fertilized or when crossed white by white. On the other hand, unlike Mirabilis, when the egg cell is derived from white plants but fertilized with pollen from green plants, progeny consist of green plants, white plants, and variegated plants. The distribution of progeny is apparently random, as would be expected on the basis of the interpretation already given for Mirabilis. Some plants receive only green plastids, some only white, and the majority a mixture. The difference in this case is the origin of the green plastid in the white egg. The solution to this problem came when it was found that, in the geranium, the male parent contributes some cytoplasm as well as nuclei. In this case, plastids can be transferred from the male parent. In Mirabilis, this process does not occur. Thus, once again, we are dealing with autonomous plastid inheritance over which the genes appear to have little or no control.

A sufficiently large number of well-established cases of plastid inheritance makes it seem reasonable to propose that the plastids are relatively autonomous bodies. On the other hand, certain studies of plastid inheritance show that this autonomy is not complete. Perhaps the best study is that of the *iojap* strain of maize. This strain shows green and white striping of the leaves, the name being derived from a combination of the original parent strain, *Iowa*, and the quite similar striped variety of maize, *Japonica*. Crosses between green and iojap strains indicate that the character is inherited in a simple Mendelian fashion with green dominant to striped (Figure 26-5, left). However, when the reciprocal cross was done (Figure 26-5, upper right), the F_1 exhibited three different phenotypes. Some of the plants were all green, some all white, and some striped. To explain this rather strange result, the explanation was put forward that the homozygous recessive iojap female has cytoplasm containing different types of plastids and, as projected before for Mirabilis, can give rise to three different kinds of progeny by chance assortment of the plastids alone.

When the iojap is used as a male parent, there is apparently no passing of cytoplasm and only green plastids are present. However, the green plastids of the F_1 are not completely autonomous. In the F_2, one-quarter of the progeny produced are homozygous recessive. These plants received only green plastids from the F_1 parent; nevertheless, the progeny are striped. It thus appears that the iojap gene in the homozygous recessive condition acts as a "mutator" for the plastids so that some green plastids are converted to colorless plastids.

It can be shown that the recessive allele does not merely modify the expression of the plastids. Heterozygous F_1 plants of the cross in which the iojap strain was used as the female parent were crossed to a homozygous green plant (Figure 26-5, lower right). From this cross only two kinds of progeny can be produced: **Ii** and **II**. All plants give rise to offspring that are both green, all white, and striped. Thus it would appear that once the variegated plastid condition has been established, it is autonomous of the genotype. The effect of the recessive homozygote, **ii**, appears to be solely that of initiating a disturbance of the plastids such that some plastids are converted to the white condition. As suggested initially, this finding indicates that there are interactions between nuclear genes and plastids, but it does not refute the proposition that the plastids are largely autonomous of the nucleus in their subsequent behavior.

The plastids are not the only particulates to show autonomous behavior. The mitochondria of certain organisms also show autonomy. The classic case of

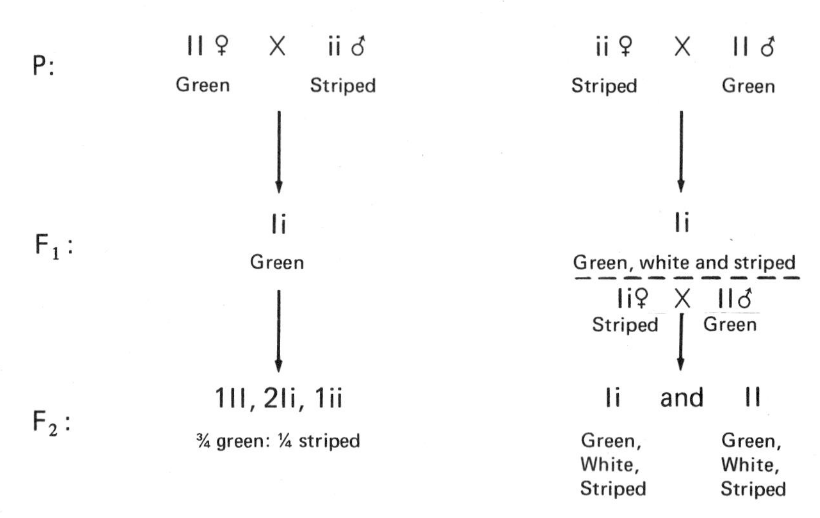

Figure 26-5. Plastid inheritance in maize. The iojap phenotype has striped leaves (green and white). The cross on the left indicates that iojap is recessive to the normal, green condition. The cross on the right shows a profound maternal effect. When an iojap female is used the F_1 plants show three types of phenotype: green, white, striped. The plastids in the egg cell of the iojap line are both green and white. The phenotype of F_1 plants depends upon which kind(s) of plastid(s) they receive. The lower right cross shows that the plastid phenotype is independent of the genotype once it has originated. The iojap gene is apparently a plastid "mutator."

mitochondrial inheritance is that of the *petite* phenotype in Saccharomyces (baker's yeast). Petite mutants grow only to one-half or one-third the diameter of the usual yeast colony. Their very slow growth is the result of a defect in the respiratory enzymes required for the aerobic utilization of energy supplies, enzymes that are located in the mitochondria exclusively. Such petite colonies breed true; however, petites give rise to nothing but petites. On the other hand, when crossed with normal yeast, no petite colonies appear even in subsequent generations. Thus although the condition appears, superficially, to be inherited as a recessive gene, in practice, the "gene" is lost as soon as it is bred with a normal yeast.

Subsequent tests showed that the petite colonies were caused by a defect in the mitochondria and that the defective mitochondria perpetuate themselves in the petite strain. When a cross is made between petite and normal, effected by the fusion of normal and petite cells, normal mitochondria are introduced. The normal mitochondria also perpetuate themselves in subsequent generations, and their presence is sufficient to give the yeast cells a normal phenotype, even though the defective mitochondria are also present. This case appears to be the same as plastid coloration; the mitochondria are independent of genomic control. Such mutants can also be found in Neurospora, so it is presumed that the autonomy of the mitochondria is not simply a unique feature of yeasts.

At first the autonomy of the mitochondria and plastids seems to be a dilemma. How can such cellular organelles be independent of nuclear control? The

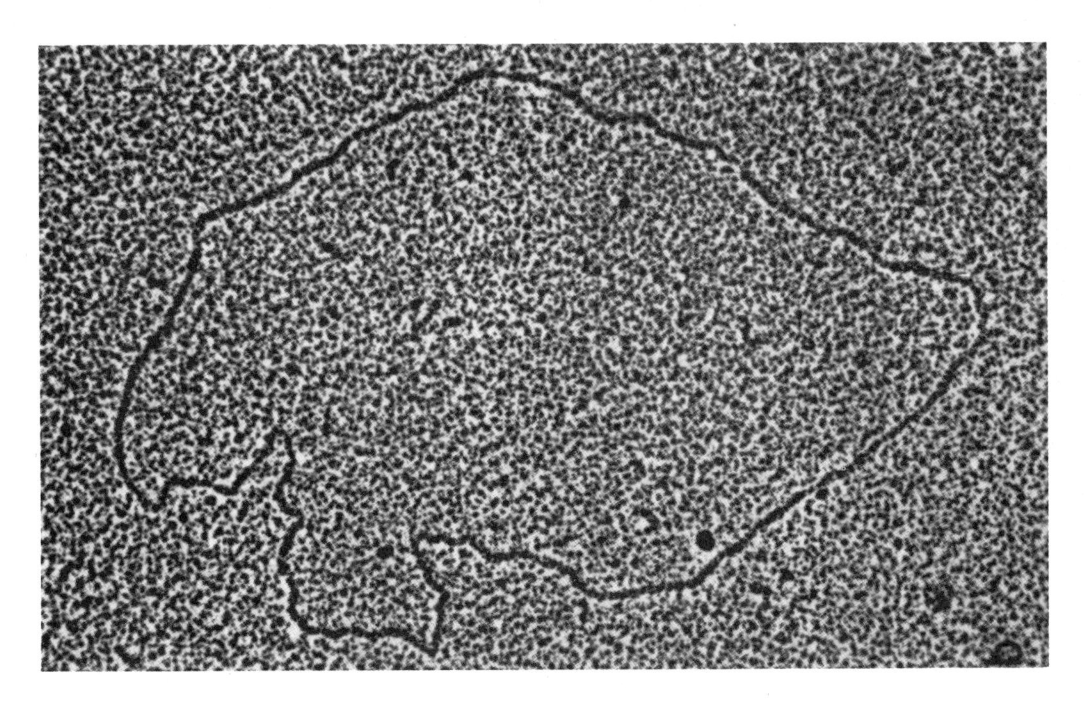

Figure 26-6. Mitochondrial DNA. A circular DNA molecule isolated from a mouse fibroblast cell. (From Nass, M. M. K. 1966. *Proc. Natl. Acad. Sci.* (*U.S.A.*) **56**:1215.)

question has now been answered by the discovery, in the mid-1960s, that both chloroplasts and mitochondria contain their own complement of DNA. An interesting feature of the DNA from particulates is that it appears to be present in a circular form of roughly the size found in bacteria (Figure 26-6). Furthermore, it has been shown that, like the prokaryotic cells, both the mitochondria and plastids contain their own protein-synthesizing machinery, which is more similar to prokaryotes than to the eukaryotic cells in which they reside. In fact, almost 20% of the DNA from the mitochondria of Neurospora apparently codes for the synthesis of mitochondrial ribosomes.

The mitochondrion appears to be composed of those proteins that are coded for by nuclear genes, probably the great majority of enzymes at least, and those coded for by mitochondrial genes, most probably largely structural proteins. Since the essential respiratory enzymes do not function well unless incorporated directly into the membranous structure of the mitochondria, it is possible that the mutants observed to give petite colonies are those in which the mitochondria are structurally deficient. On the other hand, certain of the enzymatic functions of mitochondria may well be coded by mitochondrial DNA. The chloroplast DNA is generally much larger than that of the mitochondria. This fact seems reasonable on the basis of the higher structural complexity of the chloroplast. Current studies indicate that the production of viable plastids results from the interaction of both nuclear and chloroplast genes.

The cellular particulates show their autonomy from the nucleus in another way. They divide out of synchrony with the rest of the cell. Division requires the replication of DNA, the assembly of the various structural proteins, and so forth. Obviously the lack of synchrony between the nucleus and the particulates is essential. When the nuclear DNA is being replicated, it is not being transcribed. Since a number of the enzymes and possibly other structural proteins of both particulates are made under the control of the nuclear genes, the assembly of the particulates can occur most readily at times when nuclear DNA is being transcribed. In this case, lack of *apparent* control of the nuclear genes over the replication process of the particulates is beneficial. In most cases, there are many mitochondria and plastids, and so a lack of synchronous division has no bad effects. Each daughter cell receives a complement of the particulates, which suffices for its continued existence. However, in some green algae, there is only a single chloroplast. In these organisms, it is essential that division of the plastid be synchronous with the division of the rest of the cell, and such synchronism occurs. Here a single mechanism is probably used to control both nuclear and particulate division.

The preceding discussion suggests that the most striking cases of extranuclear inheritance do not represent an abandonment of the normal genetic process but rather an amplification of that process through extrachromosomally located DNA. The reader should remember that in certain cases this method is used for the amplification of nuclear genes (Chapter 21). The difference in the two cases is that the extrachromosomal DNA generated during oocyte development is derived directly from the chromosomal DNA, whereas, in the case of the cellular particulate, the DNA remains away from the nuclear chromosomes at all times.

The "chromosome" of the particulates bears a remarkable resemblance to the chromosome of the prokaryotic cell, which suggests the possibility that the particulates represent prokaryotic symbionts living within the eukaryotic cell, for which they provide certain essential functions. Certainly the fact that the functions ascribed to mitochondrial DNA appear to be those for its own structural synthesis is strongly reminiscent of the bacterial viruses, whose DNA seems to carry primarily the information necessary to synthesize a new virus particle and little else. This hypothesis gains some substance from the fact that the structure and composition of some chloroplasts of higher plants are remarkably similar to certain present-day blue-green algae and that the highly involuted membranes of the mitochondria find their parallel in numerous bacterial cells. Like the bacteria, the particulates have DNA that is membrane bound and free of the histone proteins that are characteristic of the nuclear material of eukaryotes. Perhaps the most suggestive facts are that particulate protein synthesis is blocked by inhibitors that are most effective against bacterial cell systems and that the *t*RNA and *r*RNA of particulates are more closely related structurally to bacterial forms than to those normally found in the cytoplasm of eukaryotes. All this suggests some kind of common ancestry for different cells (or their parts), but the investigations are far from complete, and a discussion of the origins of cellular structure is outside the scope of this book. What is essential is the realization that the replication of the bacterium and a mitochondrion is not a fundamentally different process. Then it is not hard to understand the occasional autonomy of the particulates from nuclear control.

Structural Determinants

In the preceding cases, we have been able to discuss cytoplasmic inheritance in terms either of the products of nuclear genes that accumulate in the cytoplasm or of the existence of particles in the cytoplasm that have the properties of genes or chromosomes. There is one case, however, that does not exactly fit either description. It is the inheritance of the patterning of cilia in certain protozoa. The ciliated protozoa are highly differentiated single-cell organisms. The cells contain regions that are given over to quite specific functions: gullet, contractile vacuole, ciliary bands, and so on. The idea that growth is merely gene expression can be challenged by studies of these organisms. A few people with very good technical ability have been able to perform transplantation experiments with the ciliates. For example, a few rows of cilia can be taken off the cell and replaced in a reverse direction without the cell being killed! The organism now has all but a few rows of cilia beating in the normal direction, but the transplanted rows beat in the wrong direction. This process throws the cells off-course a little, but there seems to be no other problem. Consequently, we may conclude that the beat of the cilia is an autonomous function, an interpretation that is in agreement with what is known about the organization of cilia. What is much more surprising is that when these cells divide, they do so normally and produce organisms with the same pattern as the transplanted "parent." Furthermore, such characters persist even after the genetic constitution is changed by conjugation and autogamy. The inheritance of the structural pattern appears to be completely independent of the genes.

At first the observations made with transplanted ciliary rows would seem to indicate that the cilia are completely self-replicating structures. Of course, we know that, to some extent, their production is likely to be under genetic control, since there are species differences; but we cannot be certain how extensive that control is. Has the nucleus essentially lost control of the external coat of the ciliate? There is no good answer to this question at present. One of the critical questions, the origin of the ciliary proteins, remains unanswered. It is possible to suggest, however, that the loss of genetic control may only indicate that the assembly of the cilia is a process that occurs by self-aggregation of proteins around a discrete center (the basal body?). This center produces the specific pattern of organization in much the same way that a crystal in its mother-liquor can organize exact replicas of itself. Such an hypothesis would be consistent with the observation that the pattern is invariant. It also agrees with the fact that the timing of ciliary replication is properly correlated with nuclear division. Although these observations do not present us with a case of unequivocal loss of nuclear control, they do represent important evidence for the partial loss of such control to centers of organization within the differentiated cell. Part of the problem in obtaining the complete reversion of differentiated cells to the totipotent embryonic condition may reside in this feature of cellular organization.

A significant feature of this system is that the basal body from which the cilia are derived contains no self-replicating material. Viewed in overall terms, this case is not logically different from the earlier cases in which we proposed that the products of nuclear genes can accumulate in the cytoplasm and alter the pattern of heredity. In this case, we have merely gone one step further by proposing that once a pattern of structure (almost exclusively protein) has been established, that center can act as a source of organizational ability around which the products of the nuclear genes (individual proteins and polypeptides) aggregate. Consequently, they assume exactly the same orientation as the existing structures. In organisms such as the protozoa, in which there is no passage from differentiated to undifferentiated to differentiated, such centers of organization are maintained and become in and of themselves hereditary elements. Obviously, in organisms where there is a clear separation of the germ line from the somatic tissue, at each generation new centers of organization must be generated, and thus the genetic control of the process of differentiation is more complete. On the other hand, it is worth noting that in our discussion of differentiation (Chapter 25), it was necessary to propose centers of organization (determinants) within the egg, centers that are responsible for differentiation of function of nuclear elements. It is an interesting speculation to presume that, even in the case of the higher animals, such self-perpetuating organizational centers may be used for the control of differentiation.

REFERENCES

EPHRUSSI, B. 1953. *Nucleo-Cytoplasmic Relations in Microorganisms.* Oxford University Press, London. (A thin volume that deals particularly effectively with the petite mutants of yeast.)

L'HERITIER, P. L. 1958. The hereditary virus of *Drosophila. Adv. Virus Res.* **5**:195–245. (A review of the CO_2 sensitivity virus and others.)

RHOADES, M. M. 1955. Interaction of genic and nongenic hereditary units and the physiology of nongenic inheritance. In *Encyclopedia of Plant Physiol,* edited by W. Ruhland, **1**:2–57. Springer-Verlag, Berlin. (Probably the best review article on the subject of extranuclear heredity.)

SAGER, R. 1965. Genes outside the chromosomes. *Sci. Amer.* **212** (No. 1):70–79. (An excellent, simple presentation of part of the subject.)

SONNEBORN, T. M. 1959. Kappa and related particles. *Adv. Virus Res.* **6**:229–356. (A review of Paramecium killer phenotype and related materials.)

SONNEBORN, T. M. 1963. Does preformed cell structure play an essential role in cell heredity? In *The Nature of Biological Diversity*, edited by J. M. Allen. McGraw-Hill, New York. (A very interesting article that raises some important questions about the role of genes in development.)

Original Papers

LUCK, D. J. L. 1963. Genesis of mitochondria in *Neurospora crassa*. *Proc. Natl. Acad. Sci. U.S.* **49**:233–240. (A paper that discusses the role of mitochondrial DNA.)

STURTEVANT, A. H. 1923. Inheritance of the direction of coiling in Limnaea. *Science* **58**:261–270. (An original paper suggesting a maternal effect on gene action.)

QUESTIONS AND PROBLEMS

26-1. Define the following terms:

extranuclear inheritance	pigment variegation
maternal influence	petite mutants
cytoplasmic state	mitochondrion
kappa particle	plastid

26-2. What evidence is there that genes control the plane of cell division?

26-3. What are the developmental consequences of the proper placement of the spindle in the egg cytoplasm?

26-4. In one explanation of differentiation, it is postulated that the cytoplasm of the egg contains "determinants" that can influence gene activity. If these determinants were unequally assorted in the cytoplasm, the various cells resulting from the cleavage divisions would have different determinants and thus different gene functions. What evidence can you give in support of this idea?

26-5. How could genes ultimately control the postulated cytoplasmic determinants?

26-6. If a virus-borne disease is transmitted via the egg to all offspring, would you classify it as extranuclear inheritance? Explain your answer.

26-7. How would you answer the same question (26-6) if you were asked about the transmission of heroin addiction from mother to child?

26-8. Crown gall tumors of plants are caused by the passage of an agent from certain bacteria into the plant. The agent is probably DNA. The "infected" cells proliferate and form a tumor. In culture, these cells can be shown to have been "transformed" in certain important ways. What would be required for this change to be a case of extranuclear inheritance? What would be required for it to be nuclear inheritance?

26-9. How would you determine whether a particular trait was inherited via the nucleus or as a cytoplasmic particle?

26-10. The DNA isolated from mitochondria contains approximately 15,000 base pairs. How many polypeptides could this DNA code? Is this number enough to build a complete mitochondrion? Explain your answer.

26-11. In the snail Limnaea, right-handed shells are due to a maternal effect of the dominant gene D; left-handed shells are due to a maternal effect of the recessive allele d. Give the phenotypes and genotypes of the crosses below.
(a) DD female × dd male
(b) dd female × DD male
(c) F_1 female from either (a) or (b) × dd male
(d) F_1 male from either (a) or (b) × dd female
(e) Offspring of cross (c) individually self-fertilized

26-12. In Paramecium, the multiplication of the killer agent kappa depends on the presence of the dominant gene K. What types of offspring, killer or nonkiller, will originate in the following crosses? State separately the genotypes and phenotypes of the two exconjugants (1) and (2). Consider that several conjugating pairs are followed in each cross.
(a) (1) KK, not kappa × (2) KK, not kappa
(b) (1) Kk, not kappa × (2) KK, kappa
(c) (1) Kk, kappa × (2) Kk, not kappa.

26-13. You have two strains of Paramecium, each of which is heterozygous Kk, but strain I is a killer strain and strain II is a sensitive strain. What will be the genotypic and phenotypic results if
(a) many individuals of strain I undergo autogamy?
(b) many individuals of strain II undergo autogamy?
(c) individuals from strain I are conjugated with individuals from strain II, and the cross is made many times?

26-14. In yeast, a gene A in the presence of a self-duplicating factor + leads to the formation of large colonies. The allele a and/or the absence of the cytoplasmic factor (o) leads to the formation of small colonies. A normal strain, I, forms large colonies. Two different strains, II and III, form small colonies. When haploid cells of the three strains are crossed, the diploid zygote formed by the fusion produces four haploid spores after meiosis. The following types of colonies develop from these spores:

I × II	2 large: 2 small
I × III	4 large
II × III	2 large: 2 small

Further tests of the offspring show that the "large" colonies are indistinguishable from those of strain I, and the "small" colonies are indistinguishable from those of strain II.

(a) Give the genetic and cytoplasmic constitution of the three orginal haploid strains.

(b) What is the genetic and cytoplasmic constitution of the "large" and "small" colonies resulting from the above crosses?

(c) Why do the crosses shown above yield no "small" colonies like those of strain III?

26-15. A plant has green, white, and variegated branches due to its possessing both green and white plastids. Indicate the type(s) of offspring that you would expect from the crosses below.

(a) Green branch egg × green branch pollen
(b) Green branch egg × white branch pollen
(c) Green branch egg × variegated branch pollen
(d) White branch egg × green branch pollen
(e) Variegated branch egg × green branch pollen
(f) Variegated branch egg × white branch pollen
(g) Variegated branch egg × variegated branch pollen

27/Sex determination

The determination of sex is a developmental phenomenon, and questions about whether genetic mechanisms play a role in sex determination have been asked almost since the discovery of Mendel. As will be seen in this chapter, many things are known about the genetics of sex determination but enigmas remain.

Environmental Sex Determination

Determination of sex can be arbitrarily discussed under two major headings—environmental and genotypic. The first is called *environmental sex determination* because it would appear that the sex of the organism is determined primarily by external or nongenetic mechanisms. It seems as though the environment is playing a more important role than genetics in the final determination of the sex of the organism. It is possible, however, that the underlying genetic mechanisms of environmental sex determination have escaped analysis.

Hermaphrodites

Many organisms are true *hermaphrodites*—that is, a single organism that produces both functional male and female gametes. Many plants produce both male and female reproductive parts. Snails are another example of an hermaphroditic organism. In some snails, there is one sex gland that produces both sperm and ova. Certain cells develop into the male gametes, whereas other cells develop into the female gametes. This development takes place side by side in this sex gland. Finally, the familiar earthworm is a well-known example of an hermaphroditic organism, having both a functional ovary and a functional testis within the same organism.

Since, in these hermaphroditic organisms, all cells have developed mitotically from the original fertilized egg and are supposedly genetically alike, it is hard to imagine the presence of any genetic sex-determining mechanisms. Therefore such organisms are placed in the category of environmental sex determination. Doing so may simply be a mask to cover ignorance of exactly what mechanisms in the earthworm determine that the ovary develops in one place, the testis in another, and that development occurs properly.

Separate Sex Organisms

Environmental sex determination is not confined to hermaphroditic organisms, however. Several separate sex organisms, mostly marine worms of one kind or another, have bizarre sex-determining mechanisms that seem to have no underlying genetic explanation.

The first example that can be cited is a marine annelid, Ophryotrocha. This is a segmented worm that is a male if there are about 15 to 20 segments. On the other hand, if the segment number is greater than 15 to 20, it is a female. If some segments of a normal female, including those that contain the genital region, are cut off, a testis will be regenerated and the worm will develop into a male.

A second marine annelid, Dinophilus, provides another example of apparent environmental sex determination. Here the female produces eggs of two different sizes. The large eggs, when fertilized, develop into females; the small eggs develop into males. It is interesting to note that, in the adult worms, the females are much larger than the males. It might be imagined that the size difference in the eggs could be explained genetically by supposing that, during meiosis, a segregation of the chromosomes occurs such that one large egg and one small egg develop. However, this explanation fails, for it is known that the difference in size in the two eggs is present *before* meiosis occurs. Also, no sex chromosomes are known to exist, and so it would appear that sex determination in Dinophilus depends on the quantity of egg cytoplasm available for the development of the embryo. It is probable that, at some stage in the development of the ovary, a division occurs in which the cytoplasm is unequally divided.

A third example of environmental sex determination, and one of the most interesting, involves the organism Bonellia, another marine worm with which

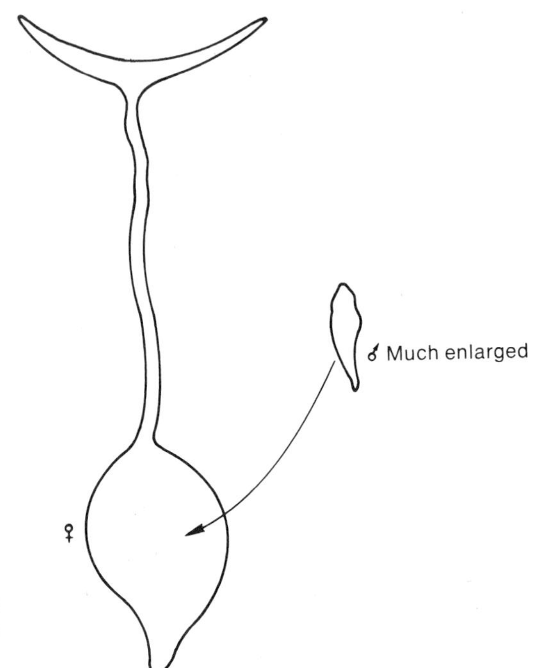

Figure 27-1. The marine worm, Bonellia. The male is much smaller than the female and lives in the female cloaca.

considerable experimental work has been done. Bonellia is a sexually dimorphic organism in which the female is large and the male comparatively small. (See Figure 27-1.) The male is parasitic and lives inside the cloaca of the female. All eggs produced by the female are alike and have the same sex potential. The egg, on fertilization, develops into a free-living ciliated larva that, if in sea water without adult females around, will attach to the substrate and develop into a female. However, if the ciliated larva finds itself in sea water with adult females around, the larva will settle on the female proboscis. The larva takes up material from the female and develops into a male, migrating down the proboscis to the cloaca of the female.

A number of interesting experiments have been performed with Bonellia. If larvae are raised in the laboratory in tanks that contain natural sea water and no Bonellia, the ciliated larvae settle down and develop into females as they do in nature. If the tanks contain artificial sea water—that is, water to which various salts in the approximate concentration found in nature have been added—the larvae again develop into females. But if the artificial sea water is agitated or stirred by some mechanical device, the larvae develop into males!

It is even possible to raise *intersexes*, in which the organism is neither male nor female, by allowing larvae to spend some time on the proboscis of the female and then removing them and forcing them to live away from the females. These and other interesting experiments have led to the tentative conclusion that a sex hormone is involved, but the true mechanism of sex determination in Bonellia remains unknown.

Genotypic Sex Determination

In the second category of sex determination—*genotypic*—genetic mechanisms are known to account, at least in part, for the determination of sex.

Gynandromorphs

The first example concerns sex mosaics in insects, particularly Drosophila. In a sex mosaic, part of the body is phenotypically male and part is phenotypically female. Such sexual mosaics are called *gyandromorphs*, or *gynanders*. (See Figure 27-2.) Gynanders, in a frequency

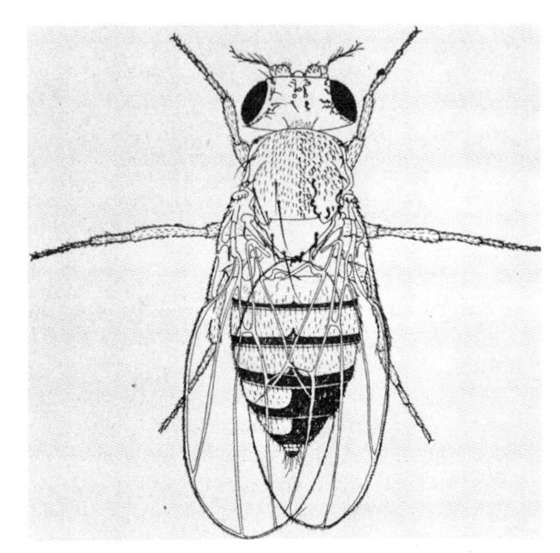

Figure 27-2. A gynandromorph of Drosophila. The female parts of the body are recognizable by the possession of straight bristles, the male parts by singed bristles. Note the male foreleg (right) with its "sex comb" and the dark coloration of the male tip of the abdomen (right). (Semidiagrammatic.) (After Stern, C. 1954. *Amer. Scientist* **42**.)

of about 1 in 2000, had been observed early in the Drosophila work at Columbia University. Morgan and Bridges explained gynandromorphs with the following assumptions. They knew that in Drosophila most cells develop phenotypically autonomously. The cells assume a phenotype dictated by their genotype, regardless of the cell's environment. They therefore assumed that the "sex phenotype" of each cell would develop autonomously. This assumption becomes more plausible with the knowledge that sex hormones are absent in Drosophila. Morgan and Bridges further assumed that gynandromorphs occur because of a chromosome loss during mitosis. They postulated that all gynanders begin genetically as females with two X chromosomes. Occasionally, during a mitotic cell division, segregation of the X chromosomes is abnormal and one pole receives only a single X chromosome. If such chromosome loss occurs early in development, say at the first mitotic division, in theory a gynandromorph would be expected that was roughly half male and half female. (See Figure 27-3.)

Gynandromorphs are found that are nearly 50/50, and the pattern may be either bilateral, where one side of the body is male and the other side female, or

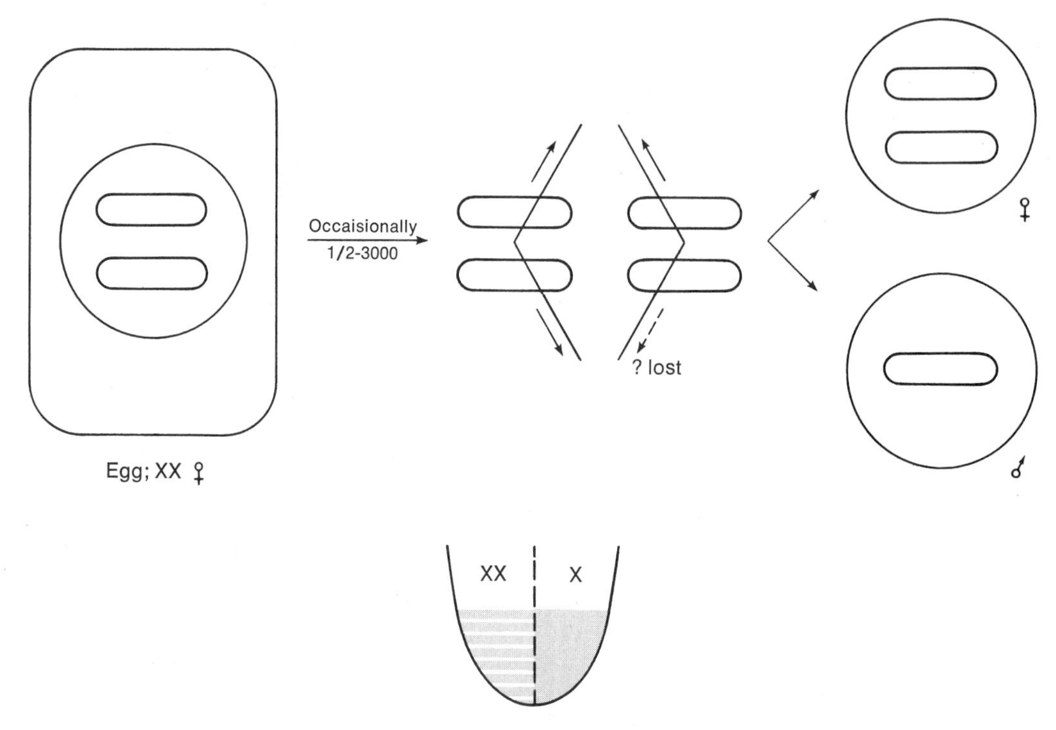

Figure 27-3. Morgan and Bridges' interpretation of the origin of gynandromorphs in Drosophila. During mitosis in an egg which is genetically XX and female, an X chromosome is lost. This results in one of the daughter cells (and all her descendants) being XX and female and the other daughter cell (and all her descendants) being XO and male. In some instances this will result in the abdomen of the fly showing one half with the typical female banding pattern and the other half with the typical male pigmentation.

anterio/posterior, where the front half of the body is one sex and the rear half the other sex. Such unusual gynandromorphs are detected phenotypically in Drosophila by noting the banding pattern of the abdomen, the presence or absence of the sex combs on the forelegs, and the phenotype of the external genitalia.

In Drosophila, truly 50/50 gynandromorphs are very rare, partly because of the peculiar type of early cleavage in these flies. During mitosis a number of nuclear divisions, without accompanying cytoplasmic divisions, occurs. These nuclei then migrate to the periphery of the egg where cytoplasmic divisions occur. Although the migration of nuclei to the periphery is not random and there is little mixing of the nuclei, true 50/50 gynandromorphs are not observed in high frequency because of the nuclear migration.

In order to test their hypothesis for the origin of gynandromorphs in

Drosophila, Morgan and Bridges designed the following experiment. A female that was homozygous for the sex-linked recessive gene yellow was mated with a male hemizygous for the sex-linked recessive gene singed. Yellow is a body-color mutant that also affects the color of the bristles, and singed is a mutant in which the bristle is reduced to a small gnarled shape. Because the bristles develop autonomously and are found all over the fly's body, each bristle can serve as an indication of whether any chromosome has been lost. From this mating, all females should be phenotypically wild type. If the hypothesis that gynandromorphs result from the loss of an X chromosome is correct and if chromosome loss is random with respect to whether the maternal or paternal X is lost, then two kinds of gynanders should be observed. If the maternal X is lost, a gynandromorph would be produced in which the bristles on the female part of the fly are wild type and those on the male part of the fly are singed. On the other hand, if the paternal X is lost, the female parts would again have wild-type bristles but the male parts would now have yellow bristles. In fact, Morgan and Bridges found these two types of gynandromorphs occurring in approximately equal frequency, substantiating their hypothesis that gynandromorphs arise in Drosophila by loss of an X chromosome. (See Figure 27-4.)

Double Fertilization

Some time after the Morgan-Bridges experiment, in a mating of a wild-type female Drosophila with a male that was heterozygous for the second chromosome autosomal dominant gene Curly and the third chromosome autosomal dominant gene Stubble, a gynandromorph was obtained in which the female part was Stubble but not Curly and the male part was Curly but not Stubble. Curly is a wing mutant and Stubble is a bristle characteristic. Each of these dominant mutants is lethal when homozygous. For several reasons, it is not easy to explain the origin of this gynandromorph by chromosome loss. To do so would require that the male part of the gynandromorph had lost two chromosomes, an X to make it male and the third chromosome containing Stubble to make it Curly but not Stubble. Aside from the unlikelihood of the loss of two chromosomes, it is known that the loss of an entire large autosome as the second or third chromosome in Drosophila is lethal. Another explanation for this gynandromorph must be sought. The most likely possibility is that the egg contained two nuclei, perhaps due to the suppression of polar body formation, and that each nucleus was fertilized by a separate sperm. (See Figure 27-5.)

Double fertilization of two "egg" nuclei is reasonably common in the Lepidopterans. In Bombyx, the silkworm, the female is the heterogametic sex (see Chapter 9). It was shown by Goldschmidt that, on occasion, double fertilization by two X-bearing sperm occurs. If this process happens to a binucleate egg in which one nucleus contains the X chromosome and the other the Y, the double fertilization can result in a gynandromorph. (See Figure 27-6.) Goldschmidt's hypothesis was corroborated by cytological investigation. In the honey bee, Boveri demonstrated

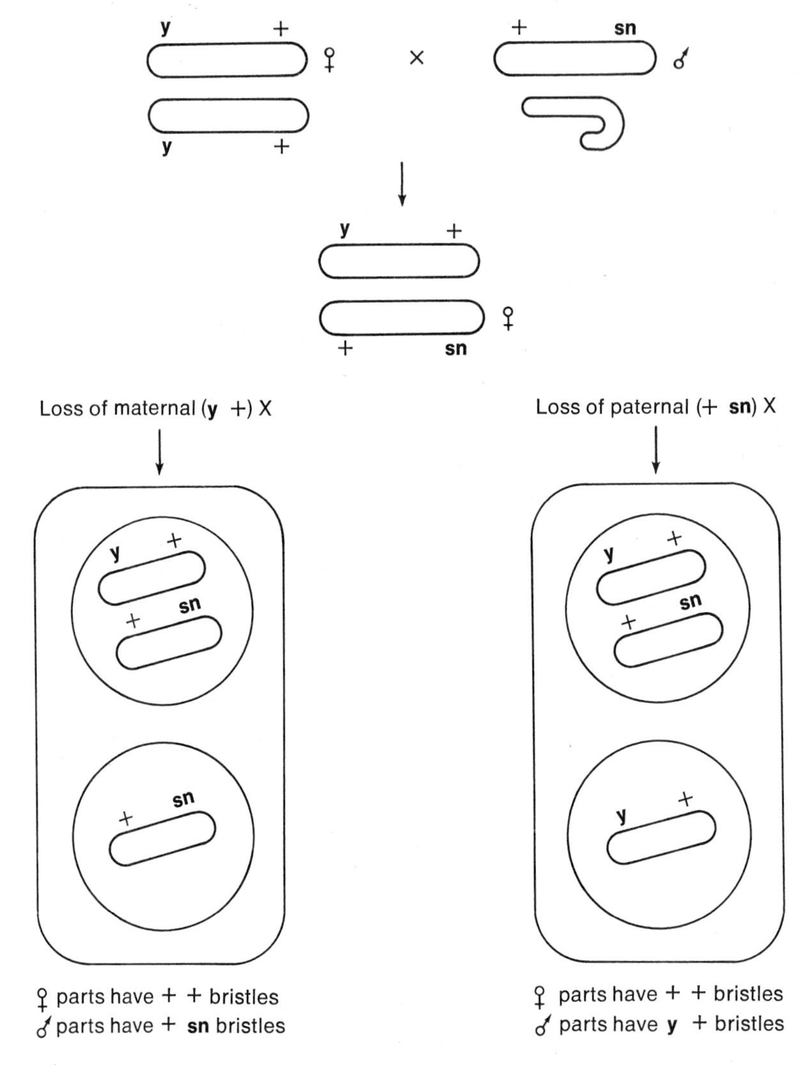

Figure 27-4. An experimental test of Morgan and Bridges' hypothesis concerning the origin of gynandromorphs. Both types of gynanders occurred with equal frequency, confirming that gynanders arise through the loss of an X chromosome and that either the maternal or paternal X may be lost. **y** = yellow body; **sn** = singed bristles.

that double fertilization can occur. He knew that in the honey bee the diploid female produces haploid eggs. If the haploid egg is fertilized, a $2n$ or diploid female results; but the haploid egg can also develop unfertilized, giving a $1n$ haploid drone. Boveri observed gynandromorphs in which the female part was biparental, as would be expected from the fact that females arise from a fertilized egg, whereas the male part

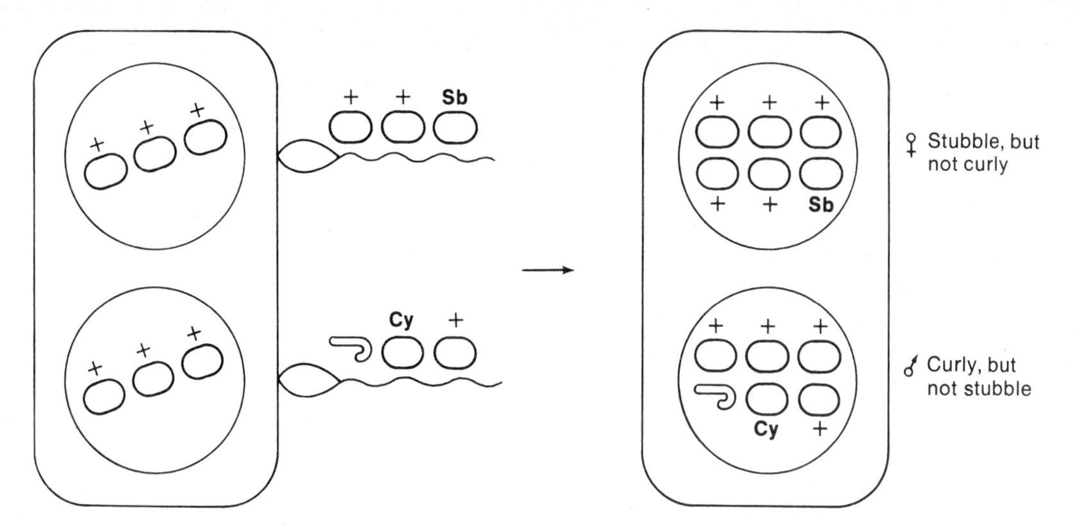

Figure 27-5. The origin of a gynandromorph in Drosophila through double fertilization of an egg nucleus and probably a polar body nucleus. The parental male was heterozygous for the dominant second chromosome mutant Curly (**Cy**) and the dominant third chromosome mutant Stubble (**Sb**).

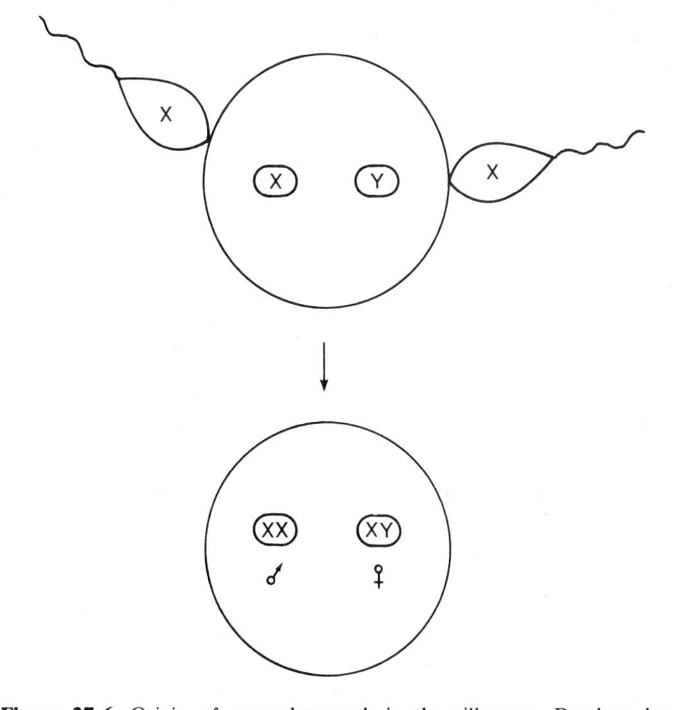

Figure 27-6. Origin of gynandromorph in the silkworm, Bombyx, by double fertilization. Since the female is the heterogametic sex, it is possible that one nucleus receives the X chromosome and the other the Y chromosome. If so, double fertilization can result in an XX nucleus which will be male-determining and an XY nucleus which will be female-determining.

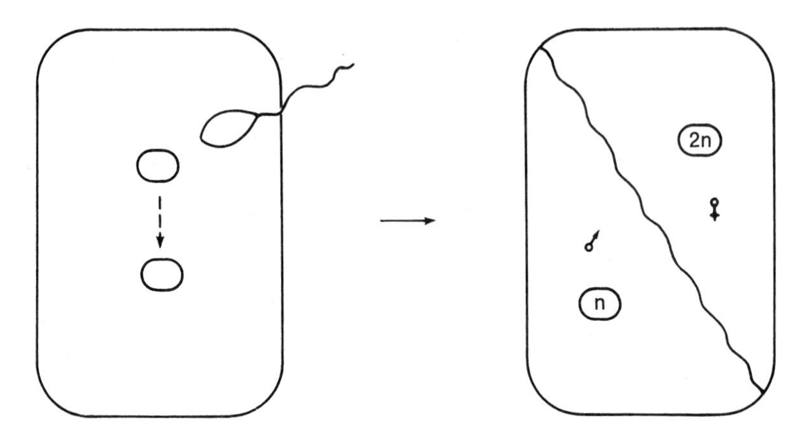

Figure 27-7. Boveri's interpretation of gynandromorphs in the honeybee. If the egg nucleus prematurely divides before fusing with the sperm, then fertilization may occur with one egg nucleus but not the other. The fertilized nucleus will be diploid ($2n$) and female-determining, showing characteristics of both parents. The unfertilized nucleus will develop parthenogenetically. It will be haploid ($1n$), male-determining and wholly maternal in phenotype.

of the gynandromorph was phenotypically completely maternal, as would be predicted if an unfertilized female egg nucleus developed with no paternal contribution. (See Figure 27-7.)

Balance Theory of Sex Determination

When Morgan's analysis of sex linkage in Chapter 9 was discussed, it was pointed out that sex determination in Drosophila was ostensibly under the control of sex chromosomes. Two X chromosomes result in a female and an XY gives a male. It is now necessary to present a refinement of that mechanism for sex determination proposed by Calvin Bridges in 1922.

Sex Determination in Drosophila

The analysis began with the discovery by Bridges of a triploid female. This fly, which has each chromosome present three times, probably arose from fertilization that involved an unreduced egg or sperm. A triploid female can by symbolized as having the chromosomal constitution 3X 3A, where A stands for a single *set* of autosomes.

In Chapter 9, while describing nondisjunction in Drosophila, the meiotic behavior of three synapsing homologs was discussed. It was asserted that when three homologs synapse, they do so two at a time. This assertion may mean either that throughout their length two chromosomes are intimately paired and the third is unpaired, or, more commonly, it means that at any given region of the chromosome two of them will be paired and one of them will not, and this situation may vary along the length of the chromosome. The net result, however, is the same. When there are three homologs, two of them generally end up at one pole and one at the other pole. This generalization should be true for all chromosomes, so that the triploid female observed by Bridges produces 16 different eggs with respect to the number and arrangement of chromosomes. Each of these eggs, of course, can be fertilized by one of two kinds of sperm, either an XA sperm or a YA sperm, thus resulting in 32 possible combinations. (See Figure 27-8.)

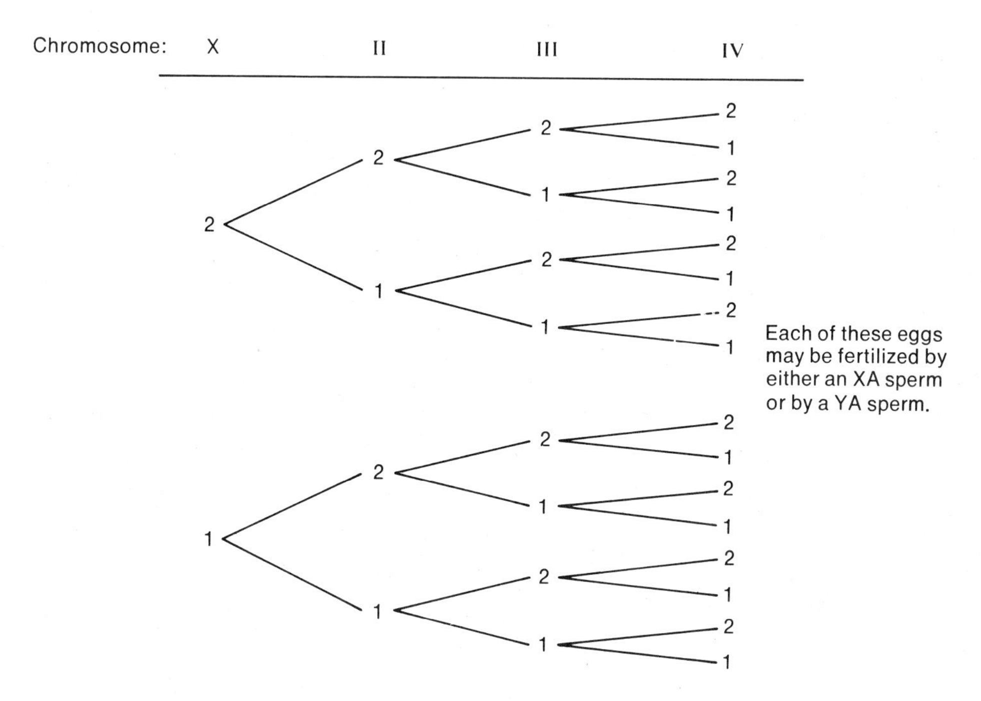

Figure 27-8. Theoretical distribution of chromosomes after meiosis in a triploid Drosophila. Sixteen different kinds of eggs can be produced and each may be fertilized by either of two kinds of sperm giving 32 possible zygotes. Many of these zygotes, however, are lethal. Also, the number of fourth chromosomes is unimportant.

All these combinations need not be considered because Bridges discovered that, in order to have a viable zygote, there must be either two second chromosomes *and* two third chromosomes or three second chromosomes *and* three third chromosomes. For survival, there must be the same number of large autosomes. Furthermore, it was discovered that the number of fourth chromosomes did not matter, probably because of their small size and hence small gene content. Therefore only four viable egg possibilities, each of which can be fertilized by two kinds of sperm, are relevant. (See Figure 27-9.)

EGGS	SPERM	ZYGOTE	PHENOTYPE
2X2A	XA	3X3A	Female
	YA	2XY3A	Intersex
2X1A	XA	3X2A	Super-female
	YA	2XY2A	Female
1X2A	XA	2X3A	Intersex
	YA	XY3A	Supermale
1X1A	XA	2X2A	Female
	YA	XY2A	Male

Figure 27-9. The zygotes and the phenotypes of the adults resulting from fertilization of eggs from a triploid female in Drosophila. Only those eggs that will give viable zygotes are listed. The symbolism 2A refers to 2 second and 2 third chromosomes. The fourth chromosome is relatively unimportant and is ignored.

An analysis of these results suggests several conclusions. It would appear that the Y chromosome has no effect on the determination of sex because a 2X 2A Drosophila is phenotypically identical to the 2XY 2A Drosophila. Similarly, the 2X 3A Drosophila is similar to the 2XY 3A Drosophila. This conclusion should not come as a surprise. Bridges' work on nondisjunction in Drosophila (Chapter 9) suggested that the Y chromosome played no role in sex determination. What is more revealing is the observation that the intersex 2X 3A differs from the normal female 2X 2A by an extra set of autosomes. Bridges concluded that the autosomes have a capacity for maleness, whereas the X chromosome has a capacity for femaleness. The Y chromosome has no effect at all.

The intersexes are sterile individuals, somewhat intermediate in phenotype between females and males. Superfemales and supermales are also sterile, and they show certain morphological differences from females and males, respectively. It has been pointed out that the terminology supermale and superfemale is an unfortunate one, and several suggestions have been offered as a substitute, such as *metafemale* and *metamale*, meaning beyond female and beyond male. The logic for such terminology will soon be apparent. It may be recalled again from the discussion of nondisjunction in Chapter 9 that the 3X 2A (superfemales) were almost always lethal.

From the preceding results and subsequent experiments, Bridges saw a pattern for sex determination in Drosophila that he was able to reduce to a simple quantitative relationship. (See Figure 27-10.) Bridges realized that it was the ratio of X chromosomes, which are female determining, to autosomes, which are male determining, which was responsible for the final sex of the organism. If the ratio is 1.0, a normal female results. If the ratio is 0.5, a normal male is produced. A ratio

	SEX	X	A	$\frac{X \to ♀}{A \to ♂}$
Sterile ⟵	Super-female	3	2	1.5
	Female	4	4	1.0
	Female	3	3	1.0
	Female	2	2	1.0
Sterile ⟵	Intersex	2	3	0.67
	Male	1	2	0.5
Sterile ⟵	Supermale	1	3	0.33

Figure 27-10. Bridges' balance theory of sex determination in Drosophila in which the sex of the fly is determined by the ratio of X chromosomes to autosomes.

between 0.5 and 1.0 gives an intersex. Finally, a ratio greater than 1.0 or less than 0.5 results in a superfemale and supermale, respectively, more properly called now metafemale and metamale. Figure 27-11 shows this relationship in a slightly different way.

Given the foregoing interpretation, another question can be asked. Does the X chromosome and/or the autosomes contain specific sex-determining genes at a particular locus or a few loci? Or are these genes spread throughout the genome as though it were a polygenic system? The answer was easily obtained for the X chromosome, for experiments involving the X chromosome are technically simpler to do. By genetic experiment it is possible in Drosophila to add to flies that are originally 2X 3A intersexes various fragments of the X chromosome. These fragments vary in size from relatively small pieces to an almost complete X chromosome. If one or, at the most, a few discrete loci on the X chromosome are responsible for femaleness, the addition to intersex flies of *particular* small fragments of the X should not result in a shift toward femaleness. On the other hand, if the female genes are polygenic and spread randomly throughout the X chromosome, then the addition of any fragment should push the fly toward femaleness, and there should be a rough correlation between the size of the X fragment added and the phenotypic effect. The experimental results showed that the latter assumption was the correct one. Any X fragment added pushed the intersex toward femaleness, and the larger the fragment, the greater the push.

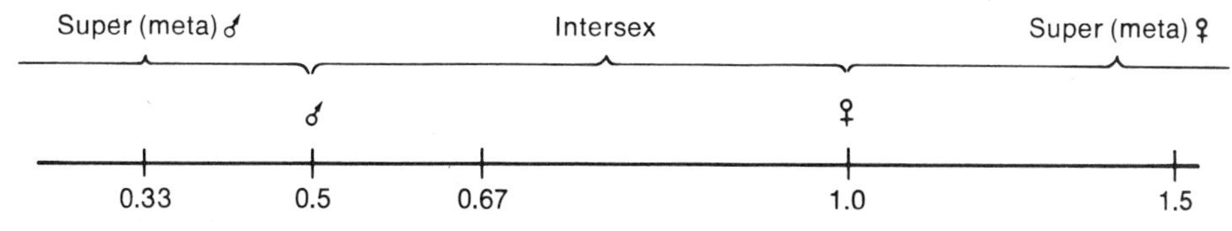

Figure 27-11. Another view of Bridges' balance theory of sex determination in Drosophila. The figures below the line represent the ratio of X chromosomes to autosomes.

With respect to the autosomes, although technically it is not as simple to do the experiment as comprehensively as with the X chromosomes, approximately the same experiment has been performed. From these experiments it was concluded that the third chromosome behaves like the X in that no single short autosomal region is responsible for the shift toward maleness. The same conclusion was reached for the second chromosome. It would appear that the autosomal genes are also polygenic in their effect on sex determination.

Sex Determination in Lymantria

With the elucidation of Bridges' analysis of sex determination in Drosophila, geneticists began to speak of the *balance theory of sex determination.* By it they simply meant that sex determination, at least in part, was due to a balance of various chromosomes and genes. In the Drosophila example, the X chromosome, on the one hand, and the autosomes as a set on the other provided the balance. Historically the balance theory of sex determination was probably first suggested by Richard Goldschmidt, who, starting in 1911, carried on an extensive series of experiments with the gypsy moth Lymantria. This moth has a wide geographical range, occurring in Europe, North Africa, and Asia. Goldschmidt made crosses between males and females from different geographical regions. For example, moths from Europe were crossed with moths from Japan. In these crosses he often obtained intersexes, either in the first or second generation; but if he crossed Lymantria from the same geographical region, intersexes were rarely produced.

From the results of many such experiments in Lymantria, Goldschmidt concluded that sex in the gypsy moth was determined by an interaction or balance of factors for maleness, carried on the X chromosome, and for femaleness, carried in the cytoplasm. He also added the concept of strength by proposing that in some races the strength of the factors for maleness or femaleness differed from the "strengths" in another geographical race. Within any one geographical race, however, natural selection had balanced the various strengths so that intersexes were rarely found. But when Lymantria from different geographic regions were crossed, the cross might involve a strong male determiner from one region and a weak female determiner from another, and, in certain instances, the result would be an intersex.

Although the evidence obtained by Goldschmidt on Lymantria was not as straightforward as in Drosophila, a careful reading of Goldschmidt's papers suggests that he was also proposing a balance theory of sex determination, and so he properly deserves recognition along with Bridges.

Sex Determination in Melandrium

An examination of sex-determining mechanisms in a few other organisms will show some variations on the theme of a balance between various factors during development. In the plant Melandrium, there are sex chromosomes with the female having two X chromosomes and the male having an X and a Y. In Figure 27-12 are

2X	2A	Females	XY	2A	Males
2X	3A		XY	3A	
3X	3A		2XY	3A	
4X	4A		3XY	3A	
			4XY	4A	

Figure 27-12. Sex determination in the plant, Melandrium. Unlike Drosophila, in Melandrium the Y chromosome plays an essential part in sex determination and is a male determiner. The X chromosome is a female determiner. The role of the autosomes is not clear but some experiments suggest that some autosomes have female determiners.

listed plants with various chromosomal arrangements and their accompanying phenotypes. Several conclusions can be drawn from these results. Unlike Drosophila, in Melandrium the Y chromosome clearly has a role as a male determiner, for whenever a plant has a Y chromosome, it is male; whenever it lacks a Y, it is female. The role of the autosomes is in dispute. One investigator found that they played no role. Another investigator showed in a different strain of Melandrium that some autosomes, at least, are female determiners. The X chromosome has been shown unequivocally to be female determining.

Sex Determination in Habrobracon

The final example of sex determination to be discussed results from many years of research by Whiting at the University of Pennsylvania. He worked with the parasitic wasp Habrobracon. Whiting knew that, in nature, the female was always diploid and the male haploid. He also knew that the female produces eggs that are haploid and that the male produces sperm that are haploid. If an egg is fertilized, a diploid female results; and if the egg is unfertilized, it may develop parthenogenetically into a haploid male.

In the laboratory, however, Whiting observed that, in addition to diploid females and haploid males, some diploid males occur.

A series of experiments led Whiting to the following interpretation for sex determination in Habrobracon. He assumed that there was a locus on the X chromosome that is responsible in some way for the determination of sex. Furthermore, he assumed that many alleles were present, symbolized $\mathbf{S}^1$, $\mathbf{S}^2$, $\mathbf{S}^3$, and so on. Unfertilized eggs develop into haploid males. If fertilization occurs and the diploid zygote contains two different sex alleles, Whiting hypothesized that such a heterozygote develops into a normal diploid female. However, if the zygote is homozygous for one of the sex alleles, it will develop into a diploid male. Because the number of different alleles is large, the probability of obtaining homozygotes in nature, by chance, is low. Moreover, it turns out that homozygous diploid males are not very viable and selection against them in nature is extreme, thus adding to the unlikelihood of ever discovering one. In the laboratory, because of the inbreeding experiments that a geneticist performs, the probability of obtaining a diploid that is homozygous for the sex allele is much higher. (See Figure 27-13.) Habrobracon would seem to be an example of a sex-determining mechanism that is genic rather than chromosomal.

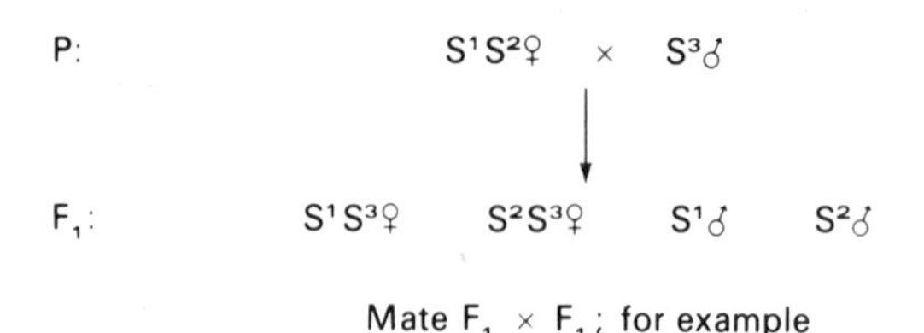

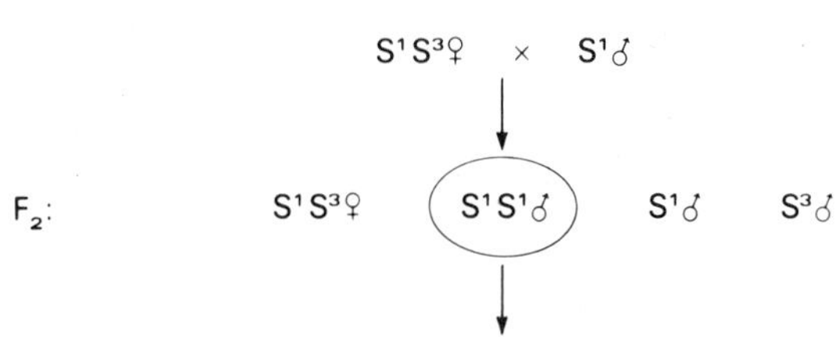

Figure 27-13. Sex determination in the parasitic wasp Habrobracon. Unfertilized eggs develop parthenogenetically into males. Fertilized eggs develop, usually, into females. Sex determination is controlled by a sex-linked multiple allelic series of genes, S. If the zygote is heterozygous for this gene it develops into a female. If the zygote is homozygous for this gene, rare in nature, it develops into a male.

REFERENCES

BRIDGES, C. B. 1925. Sex in relation to chromosomes and genes. *Am. Nat.* **59**:127–137. (Available in *Classic Papers in Genetics*, edited by J. A. Peters, Prentice-Hall, Englewood Cliffs, N.J., 1959.) (The analysis of sex determination in Drosophila showing that sex in this organism is due to a balance between the X chromosomes and the autosomes.)

GOLDSCHMIDT, R. B. 1955. *Theoretical Genetics.* 563 pp. University of California Press, Berkeley and Los Angeles. (In this fascinating book Goldschmidt devotes many pages to sex determination in general and to his contributions to sex determination in particular.)

MORGAN, T. H., and C. B. BRIDGES. 1919. *The Origin of Gynandromorphs.* Carnegie Institute, Washington, D.C., No. 278. (A description of gynanders in Drosophila and their cause.)

QUESTIONS AND PROBLEMS

27-1. Define the following terms:

environmental sex determination	intersex
genotypic sex determination	superfemale
gynandromorph	supermale

27-2. Explain briefly how sex is determined in:
(a) Bonellia
(b) Ophryotrocha
(c) Dinophilus.

27-3. Would you expect sex chromosomes to be discovered in any of the organisms mentioned in Problem 27-2? Why or why not?

27-4. In Drosophila, is it possible to obtain a gynandromorph that is more than 75% male? Explain why or why not.

27-5. The male part of a Drosophila gynander has a white eye, the female part a red eye. The mother of the gynander was homozygous red, the father hemizygous white. Explain the origin of the gynander.

27-6. A Drosophila male carries the recessive genes for yellow body and singed bristles on his X chromosome. A mating with a female produced the following progeny:

141 wild-type females
154 yellow females
136 wild-type males
147 yellow males
1 yellow gynander, partly singed and partly nonsinged

(a) What is the genotype of the parental female?
(b) What is the genotype of the zygote that resulted in the gynander immediately after fertilization?
(c) What are the genotypes of the female and male parts of the gynander?

27-7. What is meant by the "balance theory of sex determination"?

27-8. The line below can be used to represent a scale of sex increasing in femaleness from left to right.

Increasing maleness ⟷ Increasing femaleness

For the following sex types in Drosophila, state their name and enter by letter (a, b, c, etc.) the positions on the line.

(a) 2X 2A
(b) 3X 3A
(c) 2X 3A
(d) 3X 2A
(e) 4X 4A
(f) 1X 3A
(g) 2XY 3A
(h) 3X 4A

27-9. In Drosophila, an intersex is an individual that may be either XX 3A or XXY 3A. A triploid female that is homozygous for the recessive sex-linked gene for yellow body is mated to a wild-type male. She produces some intersexes that are yellow and others that are wild-type.

(a) Explain the origin of these two types by giving the genotypes of the gametes involved.

(b) Draw the anaphases of the meiotic divisions that gave rise to the female gametes. (For simplicity, assume that meiosis occurs in a single division.)

27-10. Briefly state a difference and a similarity in sex determination between Drosophila and Melandrium.

27-11. Describe sex determination in the parasitic wasp Habrobracon.

27-12. Goldschmidt said that sex determination in Habrobracon is nothing more than another manifestation of the balance theory of sex determination. What do you think he meant?

28/Sex determination in humans

With sex determination, as with other things scientific, man's principal curiosity is with man. It is interesting to know the mechanism for sex determination in Drosophila or Habrobracon but even more interesting is how sex is determined in man. It seems reasonable to suppose that this question was asked by primitive man, for it is not a profound observation to note that there are two sexes in humans, usually in equal frequency.

In the twentieth century, when the principles of genetics were being elucidated in plants, chickens, Drosophila, and other organisms, scientists wondered whether the same principles applied to human genetics. In many cases, it was possible to test the principles directly. For example, in Chapter 9 it was shown that sex-linked inheritance in man follows a pattern identical to sex-linked inheritance in Drosophila. Are the mechanisms for sex determination in man the same as in Drosophila? In the first third of the twentieth century no experimental evidence in

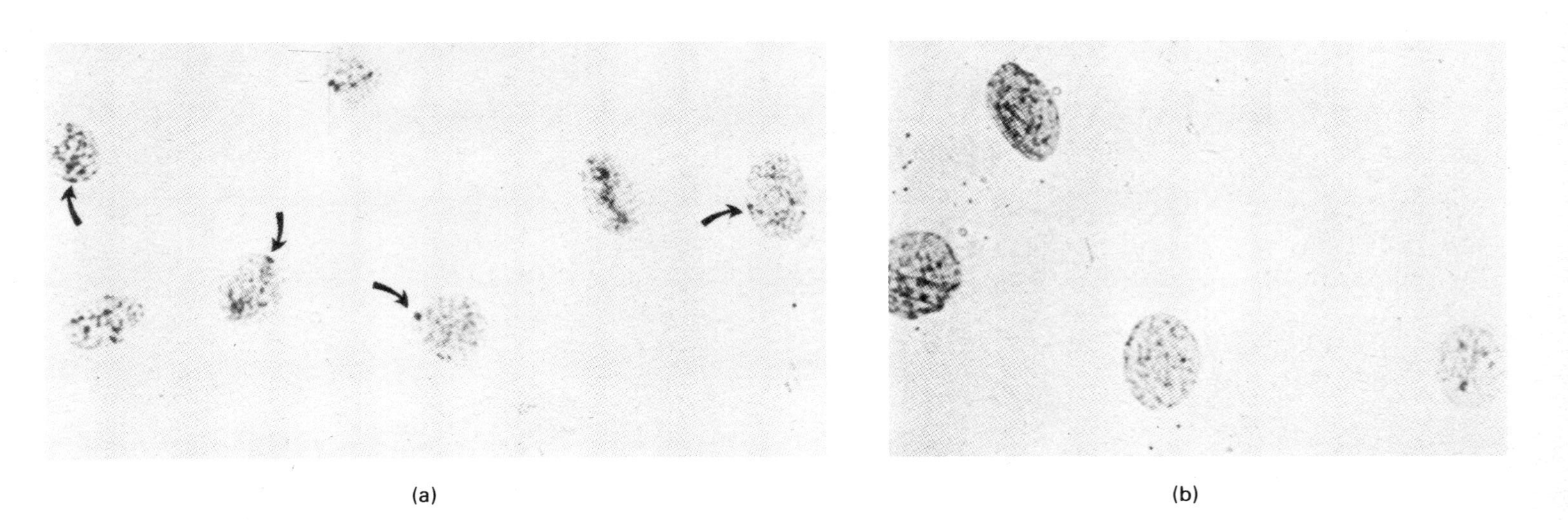

Figure 28-1. (a) Positive smear for sex chromatin (buccal smear). These cells, taken from a normal female, show Barr bodies (sex chromatin) in most cells. The presence of one Barr body in most of the buccal cells indicated that the patient has two X chromosomes. The arrows point to the Barr bodies. (b) Negative smear for sex chromatin (buccal smear). Photograph of cells from the buccal mucosa, stained for sex chromatin and magnified 1100 times. There are no small darkly stained areas on the edges of the cells. The small dark areas in the centers of the cells are not sex chromatin. These cells were taken from a normal male. (From Redding, A., and K. Hirschhorn, 1968. *Guide to Human Chromosome Defects.* Birth Defects Original Article Series 4. The National Foundation.)

humans was available, so geneticists in the 1920s and 1930s assumed that sex determination in man was similar to that in Drosophila. It was known that man had sex chromosomes and that the male was XY and the female XX. In the absence of evidence for or against, the assumption that sex determination in man was similar to that in the fruit fly seemed reasonable.

Sex Chromatin

In 1949 the Canadian scientist Barr, while looking at cells from the spinal ganglion of the cat, noticed that they had a dimorphic pattern. It seemed to him that he could tell by the appearance of the nucleus whether the cell came from a female or a male cat. He extended this research to other mammals, including man, with similar results. Present in the nucleus of cells from females was a structure, called the *sex chromatin*, that was absent from the nuclei of cells from the male. (See Figure 28-1.) It turns out that spinal ganglion cells are not the only source for nuclear sexing. In a relatively simple preparation, cells from the oral mucosa can be obtained and tested for the presence of sex chromatin.

Turner's Syndrome

The discovery of nuclear sexing caused some excitement among scientists. The possibility of telling the sex of an individual by examining only a few cells held great promise. However, as Barr's technique was extended, a few exceptions to the general rule were discovered. Females existed who did not have sex chromatin. It turns out that these females are not completely normal. They express a variety of clinical abnormalities, the most prominent of which is gonadal dysgenesis, or underdeveloped ovaries. In addition, there may be a webbing of the neck, usually a lower IQ, and a number of other physical abnormalities. Clinicians call a family of abnormal traits that tend to occur together a *syndrome*, and these women were said to have *Turner's syndrome*. They occur in a frequency of about 1 in 3500 female births, and their cells show no sex chromatin. (See Figure 28-2.)

Klinefelter's Syndrome

Also known were males whose cells showed sex chromatin. These males had gonadal dysgenesis and a number of other symptoms, including enlarged breasts, tallness, and a lower IQ. Such males are said to have *Klinefelter's syndrome*, and they occur in a frequency of about 1 in 500 male births. The cells of these males have sex chromatin. (See Figure 28-3.)

Geneticists proposed several possible explanations for these exceptions, none of which was supported by evidence. Turner's syndrome could be explained by

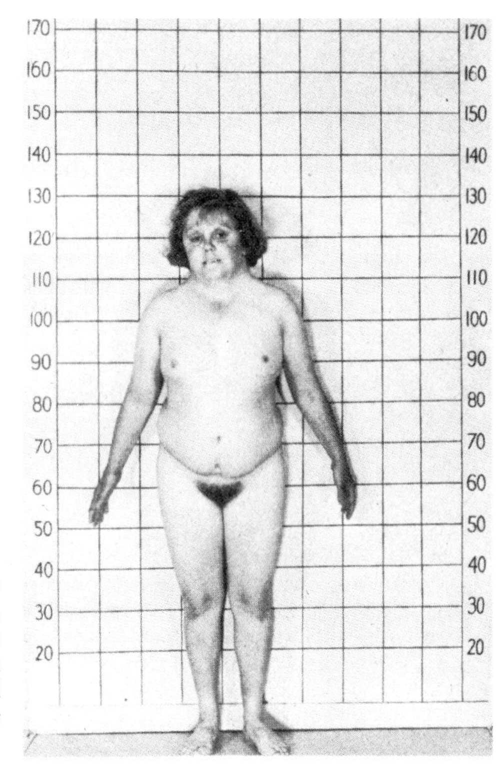

Figure 28-2. The Turner syndrome. The features are female external genitalia, short stature, webbed neck, broad shield-like chest with widely spaced nipples and undeveloped breasts. (From McKusick, V. A. 1969. *Human Genetics*, 2nd ed. Prentice-Hall, Inc., Englewood Cliffs, New Jersey.)

genetically XY individuals who had undergone sex reversal, so that the gonad, instead of developing into a testis, develops into an immature ovary. Sex reversal is well known in other organisms, particularly chickens, and this genetic hypothesis could explain Turner's syndrome and the absence of sex chromatin. Klinefelter's syndrome can be explained in a similar way by assuming that these individuals begin as females and that the gonad, instead of developing into an ovary, develops into an incomplete testis. However, many geneticists were not satisfied with these assumptions, and the clinical pathology of the gonads raises serious questions about sex reversal.

Down's Syndrome

Certain syndromes in man are unassociated with sex differentiation. One of the best known is *Down's syndrome*, which was originally called Mongolian idiocy or Mongolism. This syndrome is well known for a variety of reasons. First, it is a relatively common abnormality, occurring in a frequency of about 1 in 700 births. The physical symptoms presented by this syndrome also aroused interest among the

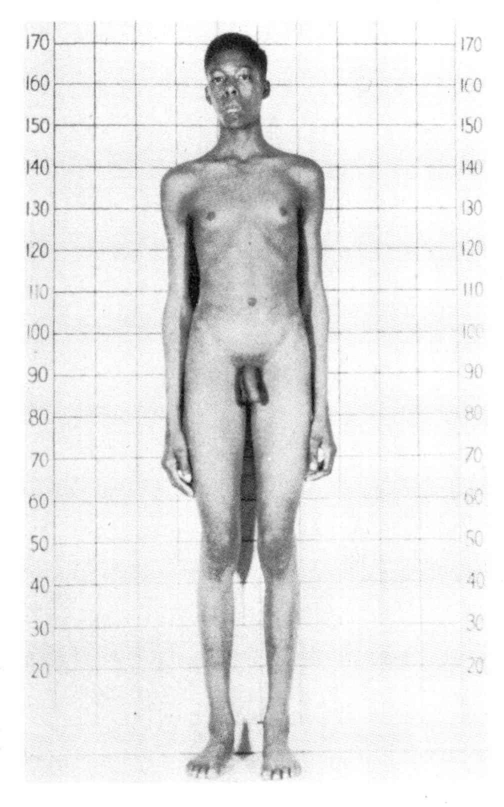

Figure 28-3. The Klinefelter syndrome. The external genitalia are male-type but testes are very small. Patients are long-legged and usually have female-like breast development. (From McKusick, V. A. 1969. *Human Genetics*, 2nd ed. Prentice-Hall, Inc., Englewood Cliffs, New Jersey.)

general public. The original name of Mongolism arose because an abnormality of the epicanthic fold of the eye caused such individuals to have a superficial resemblance to Orientals. The result, as one might imagine, led to old wives' tales that stated that Mongolism was caused by Oriental ancestry. This assertion is completely false. The other prominent symptom in Down's syndrome is a greatly lowered IQ, generally in the range of 70 or less. In addition, other clinical symptoms are observed, including abnormalities of the sole and palm pattern of the foot and hand. (See Figure 28-4.)

Relation to Maternal Age

Finally, another feature of Down's syndrome that attracted attention is the correlation of the incidence of Down's syndrome to the age of the mother. As shown in Figures 28-5 and 28-6, as the age of the mother increases, the probability of her having a Down's syndrome birth is greatly increased. It can be demonstrated that this finding correlates with the age of the mother and is not correlated to the age of the father nor to the rank in birth order. This rather interesting fact has serious social implications, for previously, and it may still be the case, certain groups in the population, particularly the professional groups—doctors, lawyers, teachers, and so on—tended to have children later in life than the general population; and it might be

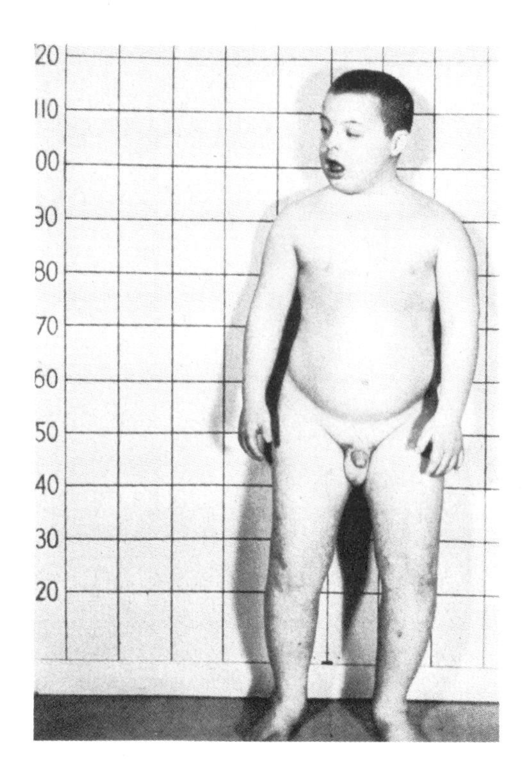

Figure 28-4. Down's syndrome. The clinical features are mental retardation, a peculiarity in the folds of the eyelids, short stature, peculiarity of the palm prints. (From McKusick, V. A. 1969. *Human Genetics*, 2nd ed. Prentice-Hall, Inc., Englewood Cliffs, New Jersey.)

expected that the frequency of Down's syndrome would be higher among professionals than in the general population. The data support this conjecture, and, ironically, those groups in the population who possess high intelligence have the highest incidence of Down's syndrome children, with an accompanying low IQ.

Many investigators, both geneticists and physicians, diligently sought the cause of Down's syndrome. The literature of the 1930s, 1940s, and 1950s suggests

Figure 28-5. Distribution of Down's syndrome with respect to maternal ages at their birth. The expected frequency is based on a control population. After the maternal age of 35 the incidence of Down's syndrome rises rapidly to a frequency of nearly 3%. (After Penrose, L. S. 1965. *The Biology of Mental Defect*, 3rd ed. Sidgwick and Jackson.)

MATERNAL AGE GROUP	RATIO OF OBSERVED TO EXPECTED	INCIDENCE PER 1000 BIRTHS
15–19	0.39	0.6
20–24	0.30	0.5
25–29	0.56	0.8
30–34	0.54	0.8
35–39	1.85	2.8
40–44	5.08	7.6
45–49	18.31	27.5

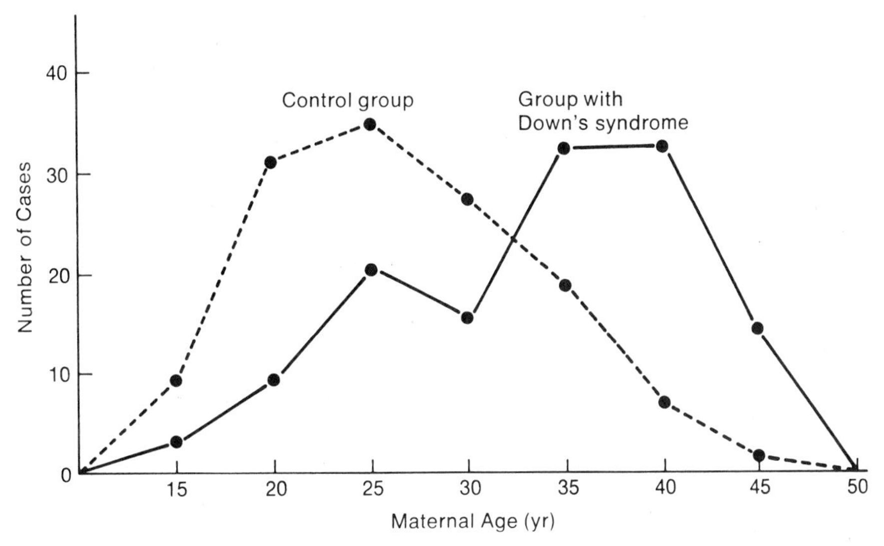

Figure 28-6. Frequency distribution of births at different maternal ages for Down's syndrome based on data from Figure 25-5. (After Penrose, L. S. 1965. *The Biology of Mental Defect*, 3rd ed. Sidgwick and Jackson.)

almost as many causes for the syndrome as there were people studying it. The geneticists' analysis presented evidence that strongly suggested an inherited component. The British human geneticist L. S. Penrose was particularly active in this field, and he provided evidence from twin studies and family pedigrees that pointed to a genetic cause, at least in part, for Down's syndrome. Unfortunately, however, either other explanations were not excluded or the samples studied were too small, and so it was difficult to draw unequivocal conclusions. In the early 1950s the cause of Down's syndrome was not known.

The Chromosome Number of Man

Before returning to sex determination in man, another, apparently unrelated question must be asked. What is the chromosome number of man? This seems a simple-minded question of no great import, and it should be easy to obtain the answer. After all, cytologists have been discovering chromosome numbers in various organisms for years, and it should be no great challenge to do the same for man. Interestingly, however, the problem is not as simple as it appears. Different investigators reported slightly different numbers, but the consensus in 1950 was that man had 48 chromosomes. Figure 28-7 shows a typical representation of the evidence for 48 chromosomes in man. It should be emphasized that this particular

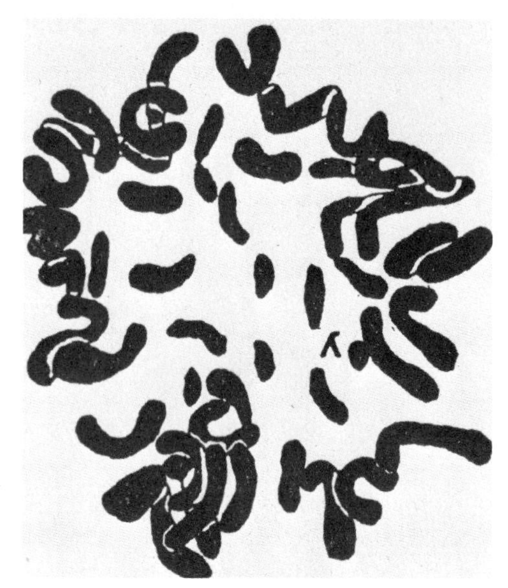

Figure 28-7. Spermatogonium of a white man showing 48 chromosomes including the small Y chromosome. Count the chromosomes and verify the accuracy of 48. (From Painter, T. S. 1923. *J. Exp. Zool.* **37**.)

figure of human chromosomes was especially selected because of its high quality. Such figures were very influential in persuading people that the chromosome number of man is 48. It should be apparent to the reader that an accurate determination of the chromosome number from this particular figure is not a simple matter.

Tjio and Levan's Technique

In 1956 two plant cytologists, Tjio and Levan, had developed and refined a cytological technique in plants that they applied to human cells. The principles of the technique are quite simple. They used a hypotonic solution to cause the chromosomes to swell, and they added a mitotic poison, such as colchicine, which arrests cell division during metaphase. The technique of slide preparation also caused the chromosomes to distribute themselves in a monolayer. Using this procedure, they hoped that the chromosomes of human cells could be observed at a stage when they are large and easy to study.

46 Chromosomes

Tjio and Levan obtained human embryonic fibroblast cells cultured from explants of legally aborted embryos. They counted only 46 chromosomes in their preparations and so published their paper in 1956 under the title of "The Chromosome Number of Man." Interestingly, when they searched the literature, they found several citations for 46 chromosomes, but the results were always equivocal. Tjio and Levan, although reasonably certain that their observations were correct, were concerned about the source of the cells and expressed the hope that verification would be quickly sought in spermatogonial

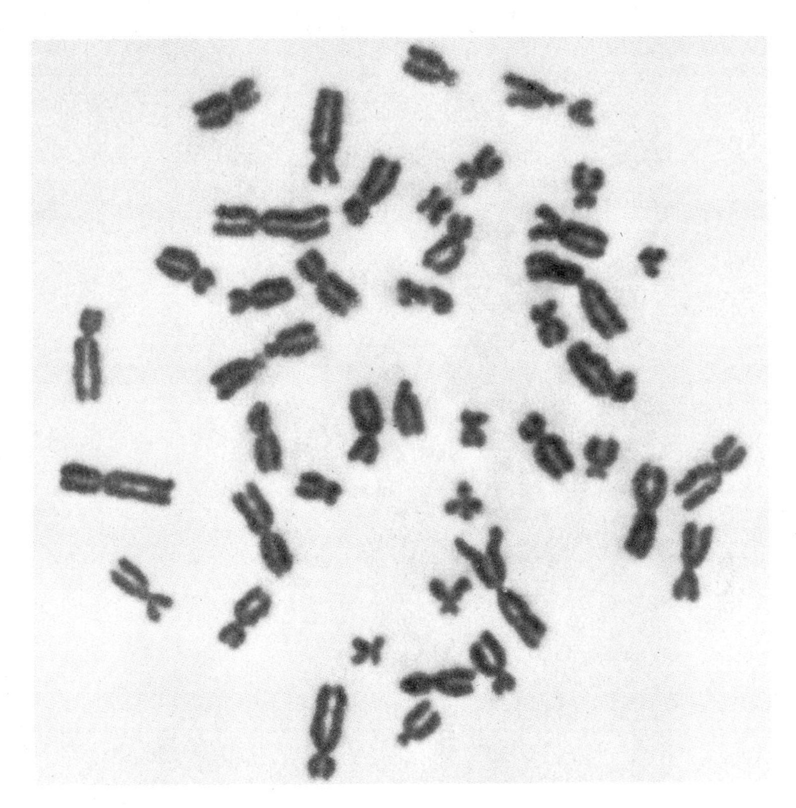

Figure 28-8. The human karyotype. Chromosomes of a single human white blood cell in the metaphase stage of mitosis.(From McKusick, V. A. 1969. *Human Genetics*, 2nd ed. Prentice-Hall, Inc., Englewood Cliffs, New Jersey.)

tissue. An interesting historical footnote is that in 1921 Painter, in an analysis of human chromosomes, indicates that in the clearest equatorial plates he observed 46 chromosomes.

The uncertainty was quickly resolved as other investigators, using the Tjio-Levan technique and studying human cells from a variety of tissues, found that the chromosome number is 46, not 48. (See Figure 28-8.) This story would not be so interesting if the only factor involved was the chromosome number of man. What is significant is the technique, for many investigators, on learning of the new procedure for culturing human chromosomes, immediately understood its potential significance. They asked: Are some of the abnormalities of man—perhaps some of the syndromes—chromosomal in origin? Previously it had been impossible to answer this question because of the difficulties of culturing human chromosomes, but with the new technique the question could be answered. It was not long before the first answer was available. In 1959, in France, Lejeune and his collaborators did a chromosomal analysis of individuals afflicted with Down's syndrome and discovered that these patients had 47 chromosomes! A more careful analysis indicated that the

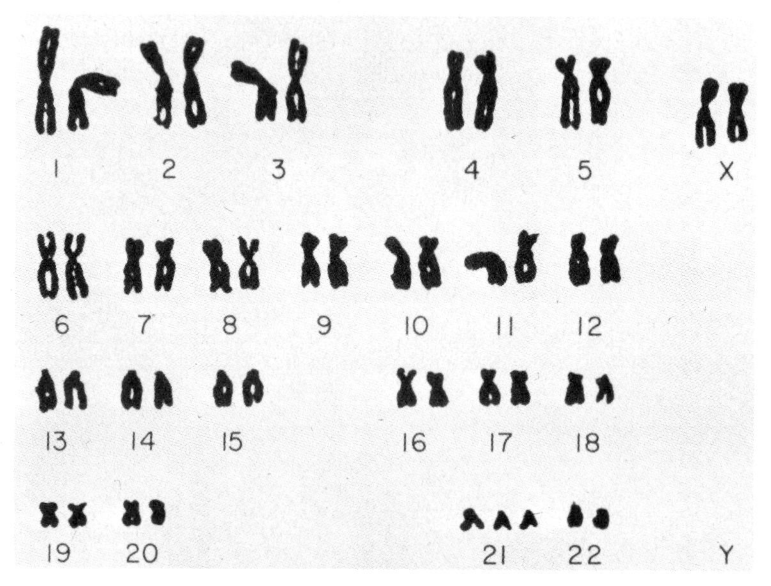

Figure 28-9. Karyotype of a male with Down's syndrome.(From Hamerton, J. L. ed. 1971. *Human Cytogenetics*, vol. II. Academic Press, New York.)

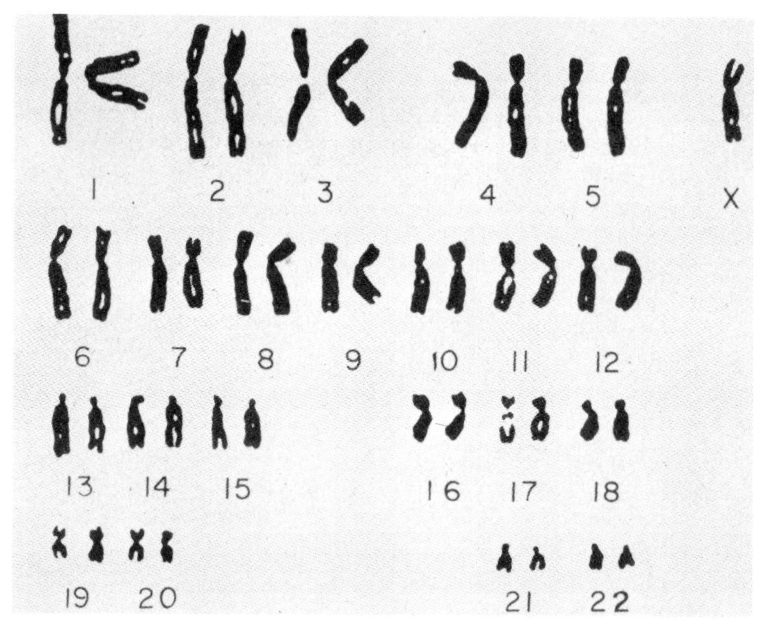

Figure 28-10. Karyotype of Turner's syndrome with an unpaired X chromosome and 22 pairs of autosomes. (From Hamerton, J. L. ed. 1971. *Human Genetics*, vol. II. Academic Press, New York.)

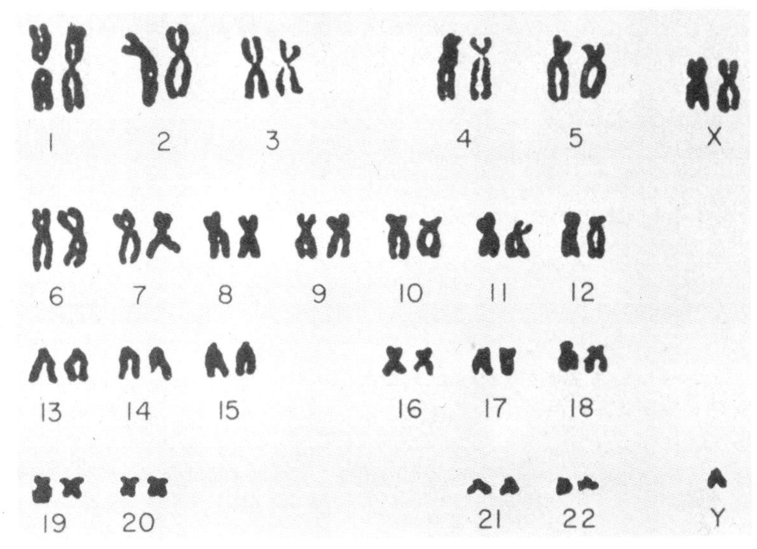

Figure 28-11. Karyotype of Klinefelter's syndrome showing two X chromosomes, one Y and 22 pairs of autosomes.(From Hamerton, J. L. ed. 1971. *Human Genetics*, vol. II. Academic Press, New York.)

extra chromosome was always the same—one of the smaller chromosomes, chromosome 21, according to the conventional system for numbering human chromosomes. (See Figure 28-9.) In the same year Ford, in England, showed that individuals with Turner's syndrome had only 45 chromosomes. The missing chromosome was a sex chromosome, so that such people were XO. Klinefelter's syndrome was characterized by 47 chromosomes, and the extra chromosome was also a sex chromosome, so that such people were XXY. (See Figures 28-10 and 28-11.)

In the following years many investigators studied the *karyotype*, which is the chromosomal complement, of various syndromes and abnormalities. From this human karyotyping, a number of chromosomal anomalies were discovered that involved the sex chromosomes in some cases and the autosomes in others. It is assumed that the majority, if not all, of these chromosomal abnormalities are caused by nondisjunctional events in either the paternal or maternal parent. In some of the sex-associated syndromes it has been possible, by the use of sex-linked markers, such as color blindness, to determine in which parent the nondisjunction has occurred.

Sex Determination in Man— The Y Chromosome

An examination of Turner's and Klinefelter's syndromes in greater depth reveals that they provide information about sex determination in man. In Chapter 9, in the description of nondisjunction in Drosophila, it was stated that an XO fly is a male and an XXY fly is a female. Consequently, the conclusion was drawn that the Y chromosome plays no role in sex determination in Drosophila. In Turner's and Klinefelter's syndromes, however, the situation is different. The XO individual is a female and the XXY individual is a male. In 1959 the British investigator C. E. Ford drew the conclusion that in man, the Y chromosome is a male determiner. A study of the other sex-associated chromosomal abnormalities reinforces this assertion. In every instance, the presence of a Y chromosome results in the individual being malelike, and the absence of the Y chromosome results in the individual being femalelike. There seems no doubt that, in man, the Y chromosome is a male determiner.

An interesting coincidence is that at almost the same time that Ford was publishing his paper about Turner's and Klinefelter's syndromes in man, two geneticists from Oak Ridge, Welshons and Russell, published an important paper in which they pointed out that sex determination in the mouse, also a mammal, was different from Drosophila in that the Y chromosome was a male determiner in the mouse.

X Chromosome Inactivation

It is now time to return to some of the enigmas raised earlier in this chapter. For example, why do individuals with Turner's syndrome show no sex chromatin, and why do individuals with Klinefelter's syndrome have sex chromatin? There seems to be a relationship with the number of X chromosomes that an individual has. Mary Lyon suggested that, in female mammals, one of the two X chromosomes becomes functionally inactive and heterochromatic early in development (Chapter 25). It is believed that the sex chromatin is the cytological manifestation of the inactivation of one of the X chromosomes. If so, it is expected that normal females with two X's show one sex chromatin, whereas normal males, with only one X, show none. The evidence from a number of different human chromosomal anomalies is persuasive that the amount of sex chromatin found in the nucleus is related to the number of X chromosomes. The relationship is that there is one sex chromatin for each X chromosome less one. Klinefelter's syndrome is expected to show sex chromatin because there are two X chromosomes. Individuals with three X chromosomes show two sex chromatins. (See Figure 28-12.) It is thought that inactivation of one of the X chromosomes plays a role in dosage compensation in mammals, such that males, which ordinarily have only one X chromosome, and females, which have two, will show the same phenotype for sex-linked recessive mutants.

Does the knowledge that Down's syndrome is due to an extra chromosome help in understanding the correlation of this abnormality with maternal age? The answer is yes and no. Yes, is the sense that it was known for many years that the frequency of nondisjunction in Drosophila increases with the age of the female. No, in the sense that the cause of this correlation with age is not known. Many fascinating

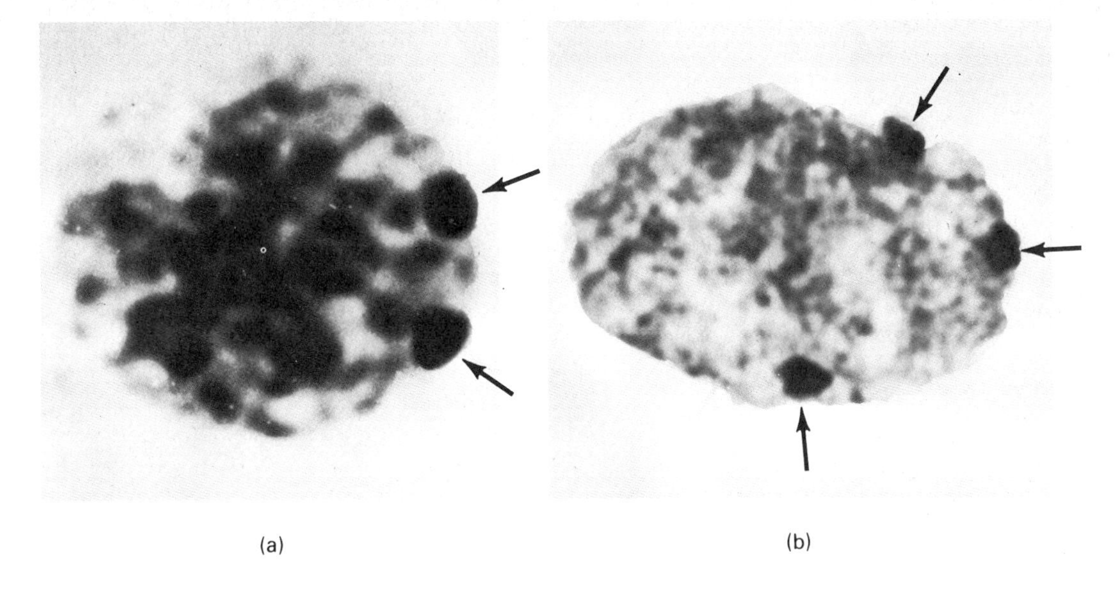

Figure 28-12. (a) Positive buccal smear with two Barr bodies. This cell, with two Barr bodies (arrows), was taken from a female patient with three X chromosomes rather than the usual two. (b) This cell with three Barr bodies (arrows) was taken from a male patient with four X chromosomes and one Y, rather than the usual one X and one Y chromosome. (From Redding, A., and K. Hirschhorn. 1968. *Guide to Human Chromosome Defects.* Birth Defects Original Article Series 4. The National Foundation.)

facts have been discovered about human cytogenetics, human disease, and behavior anomalies, but they are outside the scope of this book.

It seems appropriate to reiterate that the breakthrough in human cytogenetics, with the accompanying discovery of the cause of several human diseases, was due to a simple cytological procedure for culturing human chromosomes. Many important scientific breakthroughs follow from simple and apparently unglamorous experiments.

REFERENCES

BARR, M. L., and E. G. BERTRAM. 1949. A morphological distinction between neurones of the male and female, and the behavior of the nucleolar satellite during accelerated nucleoprotein synthesis. *Nature* **163**:676–677. (The discovery of sex chromatin in mammals.)

FORD, C. E., K. W. JONES, P. E. POLANI, J. C. DE ALMEIDA, and J. H. BRIGGS. 1959. A sex-chromosome anomaly in a case of gonadal dysgenesis (Turner's syndrome). *Lancet* **1**:711–713. (Available in *Papers on Genetics*, edited by Levine, C. V. Mosley Co., St. Louis, 1971.) (One of the first papers describing a human karyotype. Suggested also that the Y chromosome in man is a male determiner.)

PENROSE, L. L. 1965. *The Biology of Mental Defect.* 3rd ed., 374 pp. Sedgwick and Jackson, London. (A good source of information about the etiology and genetics of Down's syndrome by the person who contributed much of the information.)

STERN, C. 1973. *Principles of Human Genetics.* 3rd ed., 891 pp. W. H. Freeman and Co., San Francisco. (An excellent text to consult for additional information on human genetics.)

TJIO, J. H., and A. LEVAN. 1956. The chromosome number of man. *Hereditas* **42**:1–6. (Available in *Papers on Genetics*, edited by Levine, C. V. Mosley Co., St. Louis, 1971.) (The pioneer paper that made human cytogenetics possible.)

WELSHONS, W. J., and L. B. RUSSELL. 1959. The Y chromosome as the bearer of male-determining factors in the mouse. *Proc. Natl. Acad. Sci. U.S.* **45**:560–566. (Showed the role of the Y chromosome in sex determination in mammals. An excellent example of scientific method.)

QUESTIONS AND PROBLEMS

28-1. Define the following terms:
sex chromatin
syndrome
karyotype

28-2. What is the significance of Turner's and Klinefelter's syndromes with respect to an understanding of sex determination in humans?

28-3. What genotypes and phenotypes will result from the following fertilizations in humans?
Normal egg ×
(a) nondisjunctional sperm with no sex chromosomes
(b) nondisjunctional sperm with both sex chromosomes
Nondisjunctional egg with both sex chromosomes ×
(c) normal X sperm
(d) normal Y sperm
(e) nondisjunctional sperm with no sex chromosomes
(f) nondisjunctional sperm with both sex chromosomes
Nondisjunctional egg with no sex chromosomes ×
(g) normal X sperm
(h) normal Y sperm
(i) nondisjunctional sperm with no sex chromosomes
(j) nondisjunctional sperm with both sex chromosomes

28-4. For the following types in humans, indicate how many sex chromatins will be seen:

(a) XX
(b) XY
(c) XXX
(d) XXY

(e) XO (f) XXXY
(g) XYY (h) XXXXY

28-5. Color blindness in humans is due to a sex-linked recessive mutant. How might the inheritance of color blindness be utilized to determine where nondisjunction occurred among the parents of
(a) Turner's syndrome?
(b) Klinefelter's syndrome?

28-6. There is a syndrome in man, XYY, for which some investigators have claimed a phenotype involving increased aggressiveness. This premise is based on evidence that more XYY individuals are found in penitentiaries than would be expected due to chance. Other investigators, however, say that there may be other causes for the penal confinement besides the extra Y some. Regardless of which hypothesis is eventually shown to be correct, what is the origin of an XYY individual?

28-7. Why is it that most chromosomal anomalies in man, such as Down's syndrome, are characterized by more than phenotypic abnormality?

28-8. It might be thought that many human abnormalities would be shown to be due to trisomy, like Down's syndrome, but, in fact, very few other trisomies have been discovered. Why do you suppose that is so?

29/Genetics and evolution

PART I

When a new science, such as genetics, is developing, as it was in the early part of the twentieth century, it is to be expected that a number of the early investigators in the field would be invited to give public lectures about the new science. This situation was typical of genetics, and early in the twentieth century the British geneticist and colleague of Bateson, R. C. Punnett, was giving such a public lecture before an audience in London. Instead of using flower or cotyledon color as examples of early Mendelism, the lecturers often used examples from human genetics. Typically, at this time, the inheritance of human eye color was used as an example of simple Mendelism. The example states that brown eyes are dominant to blue eyes, and from a cross of a brown-eyed homozygous parent with a blue-eyed parent, a 3 : 1 ratio of brown eyes to blue is obtained in the F_2. It is now known that this example of inheritance of eye color in man is not 100 % correct, but it is a reasonably good approximation to a simple Mendelian monohybrid cross.

At the end of the talk, as is usual, the lecture was opened to questions, and one member of the audience asked Dr. Punnett why, if brown eyes were dominant to blue eyes, there are so many blue-eyed people. Dr. Punnett was evasive in his answer, saying something to the effect that heterozygotes are available to generate the blue-eyed individuals. However, he did not clearly perceive the answer to the question and went on to other questions from the audience. When he returned to Cambridge, he recounted the incident to his colleague and cricket-playing friend, the mathematician G. H. Hardy. Hardy immediately understood the question that was asked and saw the solution. His formulation is now recognized as fundamental for an understanding of genetics and evolution.

Hardy–Weinberg Law

The problem can be formulated in a slightly different way. Imagine that there are two populations of equal size. Suppose that one population is composed entirely of the genotype **AA** and that the second population consists only of the genotype **aa**. Further suppose that the two populations are brought together and that the individuals interbreed at random. Finally assume that there is no migration, no selection, and no mutation. What will be the genotypes of the resulting population in the next and in all following generations, and what will be the frequencies of the two genes **A** and **a**?

Hardy was able to show, with some simple algebra, that if the initial frequency of the **AA** genotype was p and if the initial frequency of **aa** was q, and if these were the only two alleles involved, so that $p+q=1$, then it would follow that the distribution of genotypes in the next and all subsequent generations will be p^2 **AA** : $2pq$ **Aa** : q^2 **aa**, given the assumptions stated. This formulation, worked out in 1908 by Hardy and also independently by the German physician Weinberg (although this fact was not discovered until some years later), has become known as the *Hardy–Weinberg Law*, or the Hardy–Weinberg Equilibrium, and it is the foundation of population genetics.

Geneticists today state the Hardy–Weinberg formulation in a slightly different way. Under the conditions postulated by Hardy, one population consisted only of the genotype **AA**. It follows also that it consists only of the gene **A**. A similar argument can be advanced for the second population, which consisted only of the genotype **aa**, and therefore only of the gene **a**. The values p and q can be assigned not to genotypes but to gene frequencies. That is, in the first population, the frequency of the gene **A** $= p$; and in the second population, the frequency of the allele **a** $= q$. If the frequency of gene **A** $= p$, then the frequency of **AA** should be $p \times p = p^2$. Similarly, frequency of **aa** should be $q \times q = q^2$, and **Aa** should be $2 \times p \times q$. The coefficient 2 is necessary because there are two ways that **Aa** can occur: **A** from one parent and **a** from the other, and the reverse.

Application of Hardy–Weinberg Formula

The Hardy–Weinberg formula can be applied quantitatively to populations to answer several questions. In humans there is a blood group, known as the MN group, which is composed of individuals belonging to one of three blood types—M, MN, or N. Two alleles are involved, $\mathbf{L^M}$ and $\mathbf{L^N}$, with no dominance, so that the three genotypes, $\mathbf{L^ML^M}$, $\mathbf{L^ML^N}$, and $\mathbf{L^NL^N}$, are responsible for the three phenotypes.

Assume a population composed of individuals as shown in Table 29-1. Is such a population in equilibrium with respect to the MN blood group? To determine this, p and q must be calculated and applied in the Hardy–Weinberg formula.

Let p = the frequency of $\mathbf{L^M}$ and q = the frequency of $\mathbf{L^N}$.

Table 29-1. Hypothetical Distribution of Individuals for the M-N blood group

BLOOD GROUP	NUMBER OF INDIVIDUALS
M	240
MN	460
N	300
	1000

In the population, there are 240 M individuals, each having two $\mathbf{L^M}$ alleles for a total of 480 $\mathbf{L^M}$ alleles. In addition, there are 460 MN individuals, each having one $\mathbf{L^M}$ allele. The total $\mathbf{L^M}$ alleles in the population are

$$480 + 460 = 940$$

In a similar fashion, the total $\mathbf{L^N}$ alleles in the population are

$$460 + (2 \times 300) = 1060$$

Since the 1000 individuals represent 2000 total alleles, the frequency of $\mathbf{L^M}$ (p) and of $\mathbf{L^N}$ (q) can be calculated as follows:

$$p = \frac{940}{2000} = 0.47 \qquad q = \frac{1060}{2000} = 0.53$$

If the population is in equilibrium, then, according to Hardy–Weinberg, there should be the following distribution of phenotypes:

$$p^2\,\text{M}, \qquad 2pq\,\text{MN}, \qquad q^2\,\text{N}$$

or

$$(0.47)^2\,\text{M}, \qquad 2(0.47 \times 0.53)\,\text{MN}, \qquad (0.53)^2\,\text{N}$$

Table 29-2. The Expected Distribution of Blood Groups when $p = 0.47$ and $q = 0.53$.[a]

	Number of Individuals	
BLOOD GROUPS	OBSERVED	EXPECTED
M	240	221
MN	460	498
N	300	281
	1000	1000

$\chi^2 = 5.82$; $p_{2 d.f.} > 0.05$

[a] A statistical analysis of the observed data indicates that the deviation is not significant and that the population is in equilibrium.

In Table 29-2 these values are calculated and the expected distribution of types is given, based on a population of 1000.
The goodness of fit can be tested with chi square; and in this example the *p* value is greater than 0.05, thus indicating that the population is in equilibrium. This finding suggests that, for this population, mating is random and there is no selection for any of the three phenotypes.

Another application of Hardy-Weinberg would be to calculate the frequency of heterozygotes for a trait determined by a single pair of genes with one allele dominant. For example, in humans, albinism is due to the recessive allele **a**; and normally pigmented people, therefore, are either **AA** or **Aa**. If it were determined that the frequency of albinism in a certain population is 1/20,000, what is the frequency of *carriers*—that is, heterozygotes who are phenotypically normal but carry the **a** allele?

If the assumption is made that mating is at random and there is no selection (assumptions probably not true for albinism), then the distribution of genotypes should be

$$p^2\ \mathbf{AA}, \qquad 2pq\ \mathbf{Aa}, \qquad q^2\ \mathbf{aa}$$

where $p =$ **A** and $q =$ **a**.

Since the only genotype unequivocally identified is **aa**, it is possible to calculate q, because **aa** $= q^2$. In the example, the frequency of albinos was given as 1/20,000. Therefore

$$q^2 = \frac{1}{20{,}000}$$

$$q = \sqrt{\frac{1}{20{,}000}} = \frac{1}{141}$$

The frequency of carriers

$$(\mathbf{Aa}) = 2pq = 2 \times \frac{140}{141} \times \frac{1}{141} = \frac{1}{70}.$$

In summary, given the assumptions of random mating and no selection and the fact that the frequency of albinism is 1/20,000, then it follows that one in 70 individuals is a carrier for the albino gene.

Processes of Evolution

There seems to be a paradox here because evolution implies change with respect to time, and the Hardy–Weinberg formulation states that the population remains constant. This seeming paradox can be resolved by examining the factors or forces that might upset the genetic equilibrium. It should not come as a surprise that some of these factors will contradict the very carefully outlined assumptions made about the conditions of the initial populations. The remainder of this chapter will be concerned with the geneticists' view of the major factors in evolution, indicating how they play a role in modifying the Hardy–Weinberg Equilibrium. The four processes in evolution are mutation, recombination, natural selection, and isolation.

Mutation

The first important factor of evolution is *mutation.* It was pointed out more than a hundred years ago by Charles Darwin that any attempt to understand the mechanism of evolution must start with an investigation of the sources of hereditary variation. Darwin was able to satisfy himself that hereditary variations are always present in wild as well as domesticated species. But the origin of the variation remained obscure to Darwin, who was not afraid to confess his ignorance on this point. He pointed out clearly that one of the important discoveries needed in science concerned the laws of heredity, and he implored all scientists to seek them. We know that they were not to be found during Darwin's lifetime, for Mendel's experiments were unknown and unappreciated until 1900—18 years after Darwin died.

When an evolutionist looks at a population of organisms and tries to reconstruct what the population has faced in its evolutionary history, he comes to the following simple first approximation. He assumes that any population of organisms in nature is adapting to the given environmental situation in which it finds itself. A given population or species of organisms occupies an *ecological niche,* which is defined as the sum total of all the environmental impacts on the organism.

Adaptation

It can be argued that as the population adapts to its ecological niche, its collective genes will determine the population's success in the evolutionary sense. As an oversimplification, assume that there is one particular genotype that best adapts the population to a particular environmental niche. If this assumption is true, it can be shown that the population will tend toward homozygosity for that single genotype that best adapts it to the ecological situation. That is, this genotype will tend to predominate in the population. The genetics of this situation will be discussed in more detail later. By this process of increasing adaptation, the population of organisms will become more and more homozygous, for in the long term the genotype that gives the greatest adaptation to the particular environment in which it is living will be selected. The tendency for a population to become homozygous can be spoken of as *adaptation.*

Adaptability

It can also be argued, however, that adaptation serves the well-being of the species only as long as the environment stays constant. Anyone who has studied the geological record of the earth knows that the environment does not remain constant. In most areas of the earth, organisms are subjected to daily cycles of environmental change involving heat and light and to seasonal cycles of change in addition to long-range geological disturbances. If it were true that populations tended to become homozygous for the one or few genotypes that best adapted them to a given environmental situation, then it should also be true that, if the environmental situation changes, the population might not have the genetic resources available to cope with the change. Adaptation works well as long as a constant ecological niche is assumed, but such constancy is not the rule in nature. Therefore it can be argued that the population that tends toward homozygosity is also likely to become extinct. For the geneticist, the argument could be made that it would best serve the interest of the population for it to maintain a high degree of heterozygosity. It needs sufficient genetic variability to allow adaptation to a different and changing environment. The tendency of a population to become heterozygous is defined as *adaptability.* These assumptions would seem to argue for both homozygosity and heterozygosity, a paradox. How the paradox is resolved may determine the evolutionary success of the population in the long run.

If the assumption of the necessity for a population to maintain heterozygosity is true, then what is the source of the variability? The geneticist would certainly answer, as a first approximation, that it must be mutation, both gene and chromosomal. In fact, do such mutations play a significant role in the evolution of natural populations?

Evidence for Mutation in Natural Populations

Individuals chosen randomly from any natural population usually are phenotypically similar in many, although not all, phenotypic characteristics. If Drosophila, for example, is sampled from a natural environment, the odds are high that every fly will have the same eye color, bristle shape, and so on. This is what is meant by the term *wild type.* The wild type is that phenotype observed in nature. What is true of flies is also true of other organisms. In mammals, for example, rabbits observed in nature will have the same wild-type appearance. The coat color will be gray or agouti. This fact is also true of most natural populations of squirrels.

The question of real interest is whether the organisms that appear wild type phenotypically are also wild-type genotypically. Experimentally, it is possible to analyze natural populations of Drosophila for genetic variability. Drosophila is chosen for study because it is easily manipulated genetically and has been thoroughly researched. To test for genetic variability on a given chromosome, say the second chromosome, a female is used that has the dominant wing mutant Curly on one

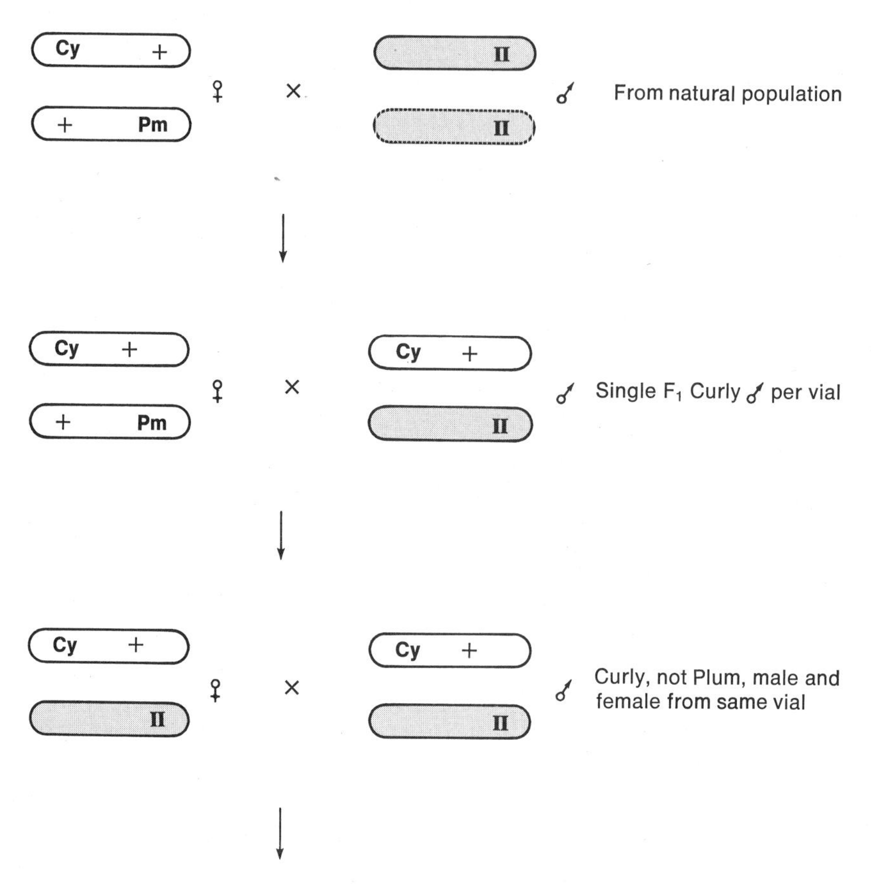

Figure 29-1. Scheme for analyzing genetic variability in chromosomes in natural populations of Drosophila. Through this breeding program, specific second chromosomes from nature are made homozygous. If they contain, for example, a recessive lethal gene only Curly flies will occur in the final generation. Associated with Curly is an inversion complex to prevent crossing over. **Cy** = Curly; **Pm** = Plum.

second chromosome and the dominant eye-color mutant Plum on the homologous chromosome. Both Curly and Plum are lethal when homozygous. Also associated with these dominant markers are several inversions, so that crossing over is suppressed in females containing these inversions.

Male Drosophila are collected from nature and mated to virgin females of the Curly-Plum stock just described. F_1 Curly males are mated singly to Curly-Plum females. The Curly, but not Plum, male and female progeny from each vial are mated, and since these Curly males and females came from a single F_1 male, each has the same second chromosome contributed by the parental wild-type male from the natural population. Recombination does not occur in male Drosophila and is suppressed in females by the inversion complex. If the second chromosome from nature has no mutant genes, the expected progeny from the cross of the Curly male by the Curly female would be 2 Curly : 1 wild-type fly, since Curly is lethal when homozygous. If, however, such a second chromosome contained a mutant—for example, an autosomal recessive lethal—then homozygosity for this chromosome would mean homozygosity for the recessive lethal and the non-Curly flies would die. Therefore if only Curly flies occur in the progeny, the conclusion is that the second chromosome from nature contained a recessive lethal. (See Figure 29-1.)

Using a breeding program similar to the one in Figure 29-1, Th. Dobzhansky, a prominent population geneticist, sampled wild chromosomes of *Drosophila pseudoobscura*, a different species from the familiar *melanogaster.* The karyotype of *Drosophila pseudoobscura* is different from *melanogaster* in that there are three pairs of rod chromosomes, a large V-chromosome pair, which is the sex chromosome, and a small dot chromosome pair.

Figure 29-2. The percentage of chromosomes II, III, and IV in *Drosophila pseudoobscura* that contain genetic variants. The figures are percentages of all (second, third, etc.) chromosomes tested. (Data from Dobzhansky, Holz, and Spassky. 1942. *Genetics* **27**.)

GENETIC VARIANTS	*Chromosomes Containing Genetic Variants* (%) II	III	IV
Lethals and semilethals	21	14	26
Viability modifier			
deleterious	21	31	41
favorable	1	0.4	0.5
Sterility genes	14	–	8
Visible mutants	4	3	2

In Figure 29-2 are data from natural populations of *D. pseudoobscura.* The data clearly indicate that most "wild" chromosomes contain a high percentage of genetic variability. It is interesting to note that many of the genes are deleterious. Lethals, semilethals, and sterility genes are not likely to be advantageous to the organism. Experiments with other species of Drosophila have verified Dobzhansky's results, and those few experiments that have been performed on other organisms also confirmed the *pseudoobscura* evidence. The conclusion drawn from these experiments is an important one. It is quite clear that although natural populations appear phenotypically homogeneous, the individual organisms constituting the population are not genetically the same. A great storehouse of variability *is* present in natural populations, and the term wild-type is a myth when considered at the genotypic level. If the results obtained from Drosophila are representative of natural populations of other organisms, there seems no question that the concept of adaptability—that is, the presence of heterozygosity—in nature is a real one. Moreover, an examination of the kinds of mutation isolated from natural populations of Drosophila shows them to be identical to the laboratory mutations of Drosophila that geneticists have become familiar with over the years. In other words, there is no reason to suppose that naturally occurring variability is in any way different from laboratory-occurring variability. This is another way of saying that mutation is a reality in nature and that it serves as the generator of the variability that is assumed to occur in all natural populations.

Why are Mutations Deleterious?

It might be asked: Why are most mutations discovered in nature deleterious? The answer is not certain, but the following reasonable arguments could be suggested. It must be realized that a natural population may have existed, in the evolutionary sense, for a very long time, certainly thousands of generations. It seems reasonable to suppose that during this time the population has become well adapted to the environment. Without even assuming absolute constancy of the niche, it can also be assumed that during this long evolutionary period most possible mutations have occurred at one time or another and have been selected against or rejected as not being the most highly adapted. In other words, the phenotype that is observed in a given population of organisms at any instant of time undoubtedly represents the result of many generations of evolution. The presently occurring genotype is the one that has been selected for. Thus any newly arisen mutations have already occurred in the evolutionary history of the organism and have been selected against. Consequently, they are, by definition, deleterious to the population.

In conclusion, it can be said that mutation provides the raw material for evolution. It supplies the genetic variation to a population that makes possible the new genotypes that may be necessary for the evolutionary success of the population. A final question that might be asked is: How important is mutation for evolution?

Recombination

By reason of the arguments already listed, mutations would seem to be very important, and in the long term, of course, they are. To reiterate, mutation is the ultimate source of variability. However, if the argument about the necessity for variability as a compensation for environmental change is correct, it is fair to ask how significant mutation will be at the time of environmental change. Remember, there is no insistence that the environmental change be instantaneous or cataclysmic. The geneticist's answer as to the importance of mutation is that mutation alone is insufficient because spontaneous mutation rates are very low, on the average of 10^{-6} per locus (Chapter 14). And such a low rate is not likely to be sufficient to provide the necessary variability at the time it is needed by the population. Mutation, however, does not act alone, and the second factor that is considered important for evolution, *recombination*, must now be considered. Given time, of which natural populations have enough, mutation will build up a large storehouse of variation. Recombination of this genetic variation provides the many possible genotypes that might mean the difference between survival and extinction for a given population. Figure 29-3 illustrates the significance of recombination. Notice that an organism with 23 heterozygous pairs of genes generates over 8 million different kinds of gametes. The number 23 was selected because it is the number of chromosomes in man. If all 23 pairs of chromosomes differed from each other, in one sex alone there would be over 8 million combinations.

NUMBER OF HETEROZYGOUS GENE PAIRS	NUMBER OF DIFFERENT KINDS OF GAMETES PRODUCED
1	2
2	4
3	8
23	8, 388, 608
⋮	⋮
n	2^n

Figure 29-3. The number of gametes produced by an organism heterozygous for a various number of genes. If the organism were heterozygous for only one pair of genes, **Aa**, two kinds of gametes would be produced. As the number of heterozygous gene pairs increases the number of different kinds of gametes increases exponentially.

Importance of Crossing Over

If the number of possible genotypes were limited by the number of chromosomes involved, then the variability would not be great, particularly in organisms like Drosophila that have small numbers of chromosomes. However, crossing over occurs and, in theory, generates as many combinations as there are different gene pairs. Since it is believed that the number of genes in most organisms is relatively large, the possible number of genotypes from recombination due to crossing over becomes almost infinite. In man, for example, the number of gene pairs is assumed to be on the order of 50,000 to 100,000. Assume the more conservative estimate of 50,000 pairs of genes, and also assume that 99% of these genes are homozygous and identical in all human beings, with only 1% of the total genotype varying from individual to individual. This leaves 500 possible heterozygous gene pairs. The number of combinations, therefore, in any given individual may be on the order of 2^{500}. With two individuals involved, the number of possible genotypes that might be produced is on the order of $2^{500} \times 2^{500}$! Is it any wonder, then, that no two human beings, with the exception of identical twins, have ever been or ever likely will be genetically identical?

It seems persuasive that recombination combined with mutation is a potent force for generating variability. Suppose that, by chance, a favorable combination is evolved that greatly increases the adaptive value of the population. What is to prevent, at the next generation, that favorable genotype from being recombined? One possible device to prevent indiscriminate recombination is inversion. Recombination is greatly reduced in inversion heterozygotes; and if such inversions occur in nature, it could be argued that they might be selected for as a means of preventing continual erosion of favorable genotypes through recombination. In fact, inversions are found in natural populations of Drosophila and grasshoppers, for example, and, if looked for, might well be discovered in natural populations of many different organisms.

Before concluding the discussion of recombination, a word of caution is necessary. It might seem from the arguments advanced that it is easy for a population to survive. Whenever an environmental catastrophe confronts the population, it seems as if recombination of all the variation that has accumulated with time provides that favorable genotype that now allows the population to survive in the new environment. It is as though the population reacts directly to the environmental exigencies. Of course, it does not. Mutation and recombination are random events, and it is only by chance that those genotypes that allow the population to survive are produced. Some evolutionists have calculated that about 95% of evolutionary trials in the past have led to extinction, and what is viewed in the living world today represents only a small fraction of evolutionary attempts. Most discussions of evolution tend to dwell on evolutionary success and, as a result, may appear misleading.

Natural Selection

When Darwin wrote his monumental book *On the Origin of Species* in 1859, the title also contained the words "by natural selection." Natural selection is the third factor in evolution to be considered, and the following pages will present it from the geneticist's point of view. Selection can be defined genetically by using an oversimplified example. Suppose that a population is composed of many organisms. Assume for simplicity, that each organism has a different genotype. Further assume that one of these genotypes is better than any of the others, that it has the highest *adaptive value*. How is such a conclusion about adaptive value reached? The geneticist does it by assigning the highest adaptive value to the genotype that contributes the most to the next generation. Look at it in a slightly different way. Each organism may contribute some progeny to the next generation. This process will be accomplished through gametes that contain genes. By definition, the genotype that constitutes the largest fraction of all genotypes of the next generation is said to have the highest adaptive value. The geneticist uses the concept of a *gene pool*, which is defined as the total of all genes in the gametes of a population. Each organism contributes to the gene pool, and from the gene pool arise the organisms of the next generation. *Natural selection* can, therefore, be defined as the difference in gene pools from one generation to the next.

Wingless Insects

How might selection work on the variability that mutation and recombination combine to generate in the population? An example that influenced Charles Darwin is instructive. When Darwin sailed on the voyage of the *Beagle*, his travels took him up and down the coast of South America, and he often examined the fauna of islands lying off the mainland. He noticed that, in many instances, the insects on these oceanic islands were identical in phenotype to those on the mainland, with one important exception. The insects of the island often had no wings or at best small vestigial wings.

This observation made a great impact on Darwin, and he assumed that the winged insects of the mainland have a higher adaptive value, since mobility will greatly increase the facility for mating, for gathering food, and for escaping danger. However, Darwin reasoned, on the oceanic islands where winds are known to blow at high velocity, perhaps the ability to fly might be selectively disadvantageous because an airborne insect could be blown out to sea. Thus Darwin began formulating the concept of natural selection, in which he visualized that different environments act differentially on given phenotypes.

There is no way of knowing unequivocally whether Darwin's interpretation about wingless insects is true, but there is evidence strongly supporting his view. Two French experimenters tested Darwin's assumptions. In a population cage, which they built to house a very large number of Drosophila (see Figure 29-4), they introduced in known frequencies some wild-type flies and some vestigial wing flies. The population was allowed to breed for many generations, supplied from time to time with fresh food. At the end of the experiment, almost no vestigial flies remained. Selection apparently was very strong against vestigial and in favor of wild type.

They decided to repeat the experiment, but this time they placed the population cage on the roof of their laboratory in Paris. The roof environment was quite different from that inside the laboratory. Outside the flies were subjected to winds, whereas inside they were sheltered. At the end of the experiment, which was carried out exactly as before, they observed that this time the vestigial flies were in the majority. Although this experiment is not absolute proof of Darwin's theory of natural selection, it seems reasonable to conclude that, in different environmental situations, the same genotype can have different adaptive values. Also, it seems reasonable that the example of wingless insects on oceanic islands suggests how evolution occurs. Mutations to wingless forms are produced randomly from time to time. On the mainland, homozygotes for these mutants are selected against. On the

Figure 29-4. A population cage. (From Dobzhansky, Th., 1947. *Evol.* **1**.)

oceanic islands, however, a wingless fly might have a strong selective advantage, and the result would be a rapid buildup of a population of wingless insects.

In the twentieth century, one of the principal criticisms of Darwin's theory by those unwilling to accept it was that there was no good evidence for natural selection. Although natural selection was an interesting idea, experimental proof was difficult to find. Some diehard skeptics hold to this view today, but the evidence for natural selection is compelling.

Industrial Melanism

A classical example of natural selection involves pigmentation in the Lepidoptera, moths and butterflies. If, early in the nineteenth century, one were a butterfly collector in England, the specimens usually captured were lightly pigmented yellows and whites. Occasionally, however, a rare capture of a darkly pigmented or *melanic* form was made. This capture was cause for great celebration among Lepidopteran collectors.

In this century it has been possible to test genetically the dark forms and the light forms with respect to their adaptive values. The experiments indicate that, under laboratory conditions, the melanic form is superior in viability to the light form. To return to the nineteenth century, in certain areas of Europe melanic forms were being captured in greater and greater frequency. There seemed to be a change in the population of Lepidoptera. In retrospect, it is seen that the change in frequency of dark forms was associated with the industrialization that occurred in Europe in the midnineteenth century. It is reasonable to suppose that while the dark form may have superior viability to the light form, in the lush green countryside that abounded prior to the Industrial Revolution, the black forms, because they were more conspicuous, were more vulnerable to predation by birds. This susceptibility to predation overcame whatever inherent physiological superiority might have existed. However, with industrialization, which in its early stages contributed a great amount of soot and pollution to the environment, the black forms now were less vulnerable to predation due to protective coloration, and the light forms became more conspicuous. As a result of the change in environmental background, the melanic forms greatly increased in frequency in the population.

A considerable body of experimental evidence testing the various parts of this hypothesis has been compiled, principally by the English geneticist H. B. Kettlewell. The conclusion he draws—that there has been a change in the genetic structure of the population due to industrialization—is persuasive. Today *industrial melanism* is considered a prime example of evolution by natural selection because the genetic constitution of a population has changed with time.

Many other examples of a similar nature, in which the genetic constitution of a population has changed with respect to time, can be cited. Many of these examples involve the use of insecticides. In California, for example, scale insects are a great pest in the citrus groves. For years they were kept under reasonable control with hydrocyanic gas. Early in the twentieth century, however, various communities reported to the state agriculture department that fumigation was ineffective. If the concentration of gas was increased, some control was restored, but scattered isolates in the state reported that even the increased concentration of HCN was insufficient to kill the red-scale insect.

Experiments have indicated that the resistant and nonresistant strains probably differ by a single pair of genes. The most reasonable hypothesis is that resistance to HCN arose in the population through random mutation and then was subsequently selected for in the extreme environment of hydrocyanic gas. Although it is quite clear that such mutation and selection have not occurred in all populations, those in which they have occurred provide compelling evidence for natural selection.

Adaptive Polymorphism

It was pointed out earlier that a superficial examination of natural populations in many species would show that the members were phenotypically similar except for a few scattered mutants. It was also shown that a genetic examination of the population indicates that it is unlikely that any two individuals share a genotype in common. There are populations, however, in which the heterogeneity is visible in one way or another. In Chapter 26 mention was made of snail populations that contain individuals with right-coiling and left-coiling shells. Populations that show two or more variants are called *polymorphic*. Phenotypically, polymorphic species are observed infrequently. Genotypically, all populations are polymorphic to some extent.

It was argued at the beginning of this chapter that polymorphism within a species increases adaptability, thereby providing for the possibility of genetic change. In defining selection, the oversimplification of a "single best genotype" was used, but, in fact, a single genotype can never serve a population confronted with changing environments as well as several genotypes can.

Inversions in Natural Populations of Drosophila

One of the best-known and thoroughly analyzed examples of polymorphism involves the presence of chromosomal inversions in wild populations of Drosophila. Chromosomal polymorphism due to inversions is known in natural populations of over 30 species of Drosophila. The experiments to be cited were conducted largely by Th. Dobzhansky and his associates and cover a span of 30 years, beginning in the 1930s.

Dobzhansky studied *Drosophila pseudoobscura* and observed, in collections of natural populations, several different gene arrangements for the third chromosome. The inversions were examined by studying the salivary gland chromosomes. (See Chapter 13.) The various inversions were usually named for the geographic locality at which they were collected in the Sierra Nevada Mountains in the western part of North America. The arrangements that concern our discussion are called Standard (ST), Chiricahua (CH), and Arrowhead (AR).

Dobzhansky's experiment was similar to the one with wingless Drosophila done by the two French investigators. Flies carrying two different inversions, ST and CH, were placed in a population cage and allowed to breed for about one year. The initial frequency of ST was 11% and that for CH 89%. Figure 29-5 shows the results of the experiment. In approximately 4 months the frequency of ST in the population quadrupled. Subsequently, the ST frequency increased more slowly, to about 70 %, after which the curve leveled off and no significant changes were observed.

An analysis of this experiment leads to several conclusions. First, it appears that carriers of the ST arrangement have an adaptive advantage, for the frequency of ST increased rapidly at the expense of CH in the early stages of the experiment. If this observation is taken to its logical conclusion, it can be shown that eventually only ST would remain in the population, because CH is being selected against. However, it is observed that such is not the case. An equilibrium is reached at about 70% ST and 30% CH. Dobzhansky concluded that the best explanation for the establishment of such an equilibrium is that the heterozygote, ST/CH, has the highest adaptive value of the three possible genotypes. The second highest adaptive value belongs to the homozygote ST/ST, with the homozygote CH/CH having the lowest adaptive value. No other explanation can so simply explain the persistence of both types in the population over a long period of time.

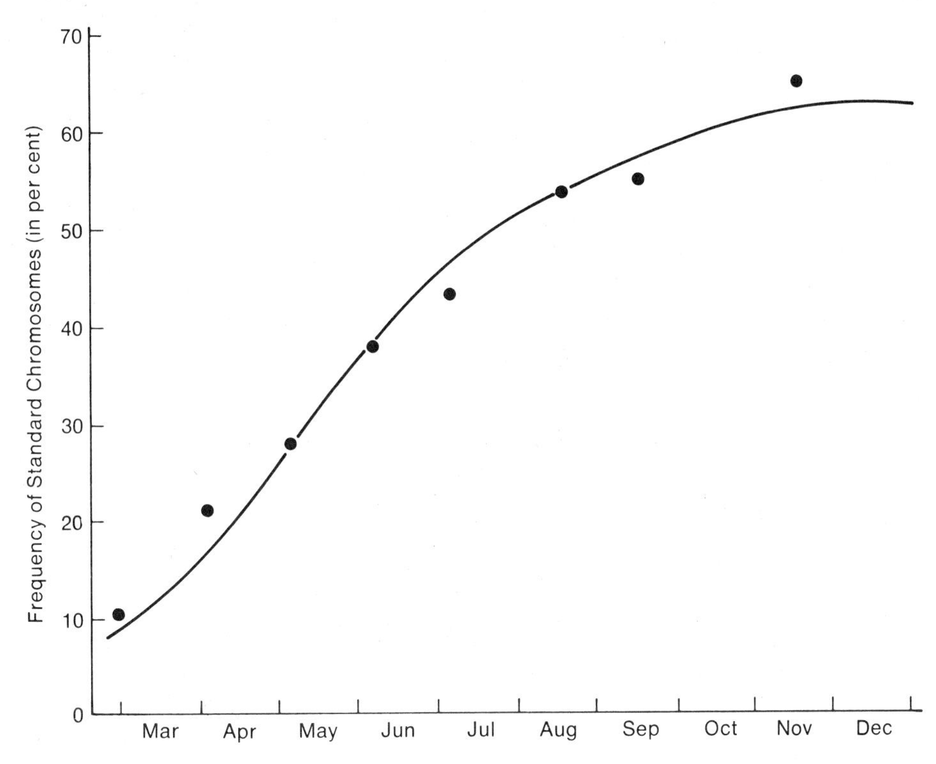

Figure 29-5. Frequency of Standard chromosomes (in per cent) in different months in the population cage No. 35 (From Dobzhansky, Th., 1947.)

Additional experiments involving other combinations of gene arrangements gave similar results. Although one gene arrangement initially appeared to have selective advantage, it did not replace the second gene arrangement. Instead an equilibrium was reached.

Another observation that Dobzhansky had made over a period of years involved the relative frequencies of the various gene arrangements in nature. He obtained his data by placing collection containers at various localities in the mountains in order to obtain samples from the wild populations. These samples were brought to the laboratory, and salivary gland chromosomes were examined to determine the relative frequency of the various arrangements. Figure 29-6 presents the results of 6 years' collecting. Notice that the different arrangements do not occur with the same frequency. During the cooler months ST is relatively frequent, whereas CH is relatively infrequent. During the hot months this relationship is reversed. The frequencies do not remain constant throughout the year but fluctuate, with ST reaching its low point in the hot part of the year, when CH attains its peak frequency. Dobzhansky also noted that these frequencies were cyclic. Within errors

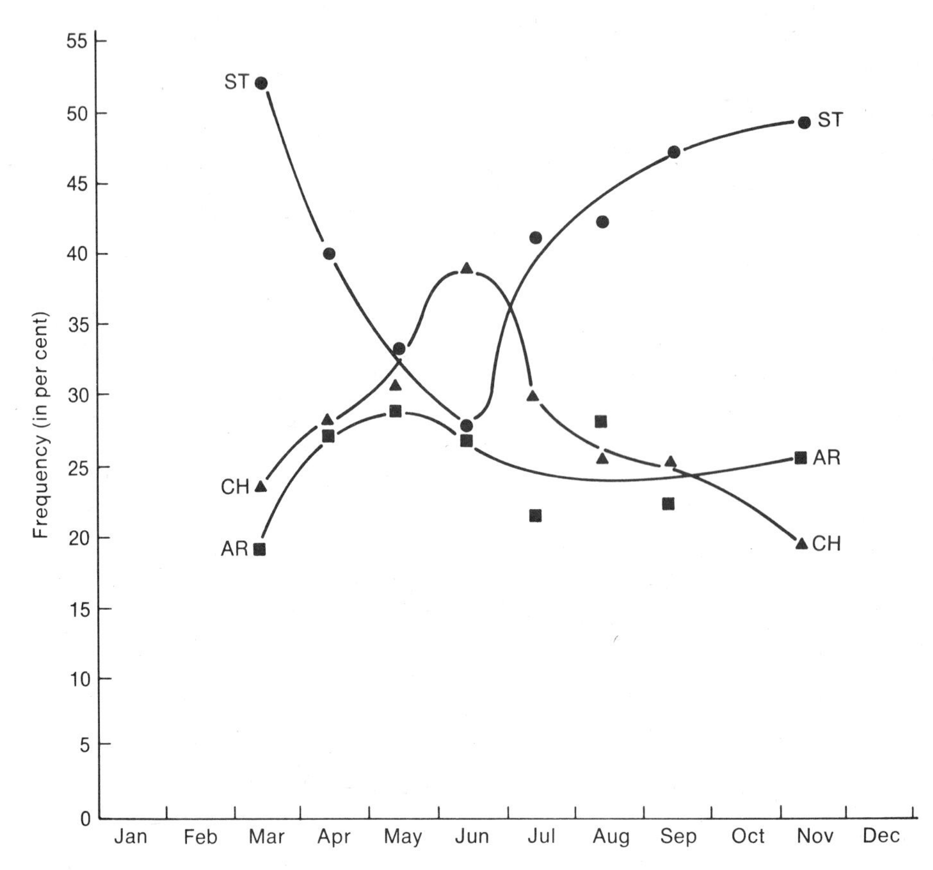

Figure 29-6. Changes in the frequencies of chromosomes with Standard (circles), Chiricahua (triangles), and Arrowhead (horizontal rectangles) gene arrangements in the population of Pinon Flats, California. Ordinate, frequencies in per cent; abscissa, months. Combined data for six years of observation. (From Dobzhansky, Th. 1947. *Evol.* **1**.)

of random sampling, the frequencies of the different arrangements for any one month were the same from one year to the next.

Dobzhansky hypothesized that selection for the heterozygote results in the polymorphism observed in the population becoming balanced. Natural selection, by favoring the heterozygote, will maintain both arrangements in the population, no matter how poorly adaptive they may be when homozygous. Dobzhansky also postulated that this *balanced* or *adaptive polymorphism* was itself the result of natural selection, and he looked for experiments to test this assumption.

The geographic range of *D. pseudoobscura* throughout the western part of North America is extensive, ranging from the Canadian Rockies down through the Sierra Nevadas in the United States to Arizona and into Mexico and Guatemala. Even though this range is extensive and nearly continuous, it is known that flies from one population, say in California, do not interbreed with flies of a second population—for example, in Mexico—because the range for flies from an individual population is not very extensive. Dobzhansky, therefore, decided to repeat his population cage experiments but with flies that came from populations geographically remote.

When ST and CH from the same geographic region (the experiment described earlier; Figure 29-5) are tested, the heterozygote is adaptively superior. If the same experiment is done by using two different arrangements, ST and AR, again from the same region, heterozygote superiority is observed in the cage experiment. Other combinations of inversions give the same result.

However, when Dobzhansky tested ST and CH that came from different geographical regions, only occasionally was heterozygote superiority observed. Usually one or the other homozygous combination had the highest adaptive value. Similarly, the test of ST and AR from different geographical locations only randomly resulted in heterozygote superiority. In a few instances, the heterozygote had the highest adaptive value, but generally one homozygote or the other was likely to have the highest adaptive value.

Dobzhansky reasoned that the probable interpretation of these facts involved selection. The chromosomes with different inversions carry different complexes of genes within them. In any given locality, these various combinations have been acted on over time by natural selection; and selection has resulted in balanced polymorphism. Dobzhansky argued that the two different arrangements had become mutually adjusted or *coadapted*, so that the inversion heterozygotes possess the highest adaptive values. But in different geographic regions, although the inversions are essentially the same, the genes within them are likely to be different; and so the test involving two arrangements from different locations results in no heterozygote superiority because natural selection has had no chance to act on the various combinations that might have been formed. The result, therefore, is random. Sometimes one genotype is favored, sometimes another. Dobzhansky concludes that heterozygote superiority is the outcome of a long process of adaptation to the environment, and adaptation that results in heterozygote superiority of the individuals in a population increases the relative advantage of the population as a whole.

The hypothesis is strengthened, maintains Dobzhansky, by the observation of the cyclic nature of inversion frequencies that are observed in nature (Figure 29-6). Even though the ST arrangement has the highest adaptive value under cold conditions, and even though occasionally there are extreme winters in the Sierra Nevadas that favor all carriers of such ST arrangements, no matter how severe nor how prolonged the winter might be, spring and warmer weather eventually arrive. If selection entirely favored ST, severe and prolonged winters might result in ST completely taking over in the population. When spring and warmer weather ultimately come, however, ST no longer has the highest adaptive value. If the population has eliminated the CH arrangement, then the fitness of the population as a whole is decreased. Thus Dobzhansky argues that selection for the heterozygote increases the versatility of the population, so that no matter how severe the selective pressures become at any one time, the higher adaptive fitness of the heterozygote will maintain all genetic posibilities in the population.

Protein Polymorphism

Dobzhansky's analysis of inversions from wild populations of *D. pseudoobscura* clearly established the presence of balanced polymorphism. Ideally, however, it would be important to determine how much genetic variation exists in any given population, particularly at the gene-product level. Experiments to test this question were done by Hubby and Lewontin, again utilizing *D. pseudoobscura*.

Their experiment proceeded from the hypothesis that a gene mutation will result in an amino acid change in the polypeptide controlled by the gene. In at least some instances, such an amino acid alteration should affect the electrostatic charge of the enzyme or any other protein of which the polypeptide is part (see Chapter 17). By utilizing *starch gel electrophoresis*, Hubby and Lewontin were able to discriminate between slightly different polypeptides.

Electrophoresis is a technique that separates similar but not identically charged macromolecules. Essentially the procedure consists of introducing relatively low current to the solution. When such a current is applied positively charged molecules will move toward the cathode (negative pole) and negatively charged molecules will migrate toward the anode. The technique is sophisticated enough to achieve complete separation of proteins having similar but different charges.

By exploiting genetic differences in electrophoretic mobility, Hubby and Lewontin surveyed allelic variation from five natural populations of *D. pseudoobscura*. Their results showed that approximately 30% of all proteins tested were polymorphic and about 12% of the loci of any individual were heterozygous.

A somewhat similar study of enzymes of human blood showed that 20 or 30% of the loci examined were polymorphic. Since the methods used for these studies are likely to result in underestimates, it seems safe to conclude that polymorphism, probably balanced polymorphism, is widespread in natural populations.

Isolation

It is interesting to recall that Darwin's classic work in 1859 was entitled *On the Origin of Species*. Up to this point nothing has been said about how species arise. Darwin rejected the concept of special creation and substituted for it his principle of natural selection. The twentieth-century geneticists who have reinterpreted Darwin in the light of an understanding of genetic principles have essentially supported his concepts. The geneticists interested in evolution are often referred to as *neo-Darwinians* and their assumptions about evolution include four processes: mutation, recombination, selection, and isolation. Three have been discussed, but before considering isolation, a species should be defined.

Species

The definition of a species has always been a contentious point among evolutionists. Some argue that it is not possible to define a species with any precision, and so such definitions are best left unattempted. Others insist that there is no reality to a species, and, therefore, any attempt at definition is futile. Most neo-Darwinians do not hold this view, however. It is possible to define a species, and the fact that, in some instances, examples can be shown that are not covered by the definition is of little importance. It should also be pointed out that different kinds of biologists may define species differently. The most prominent neo-Darwinian, Dobzhansky, gives the following definition:

> An ancestral species is transformed into two or more derived species when an array of interbreeding Mendelian populations becomes segregated into two or more reproductively isolated arrays. Species are, accordingly, systems of populations; the gene exchange between these systems is limited or prevented in nature by a reproductive isolating mechanism or perhaps by a combination of several such mechanisms. In short, a species is the most inclusive Mendelian population. [From Dobzhansky, 1970, p. 357.]

Dobzhansky emphasizes that species are dynamic and not static units. It should come as no surprise, therefore, that examples in nature can be found in which it is difficult to say exactly whether there is one species or another. Such a situation is analogous to walking through a forest. There must come a time in that walk when it would be hard to say whether you are still going into the forest or whether you are coming out. It is not worth quibbling over. Secondly, note that the definition of species introduces the concept of isolation and isolating mechanisms, the fourth process of evolution.

Isolating Mechanisms

Why are isolating mechanisms needed for evolving populations? The argument might go like this. It has already been shown that the process of mutation generates a great variety of gene and chromosomal variants and that recombination will act on this storehouse of variability to produce numerous gene combinations. Some of these combinations will be favored in a given ecological situation, and the result will be a population that at a given instant has a high adaptive value. However, it can be argued that if interbreeding were indiscriminate, if the population that is highly adapted can interbreed with any other migrating population, then such favorable combinations that have been built up would be broken up by interbreeding. At least, there would be no mechanism guaranteed to maintain these harmonious combinations. If species are to continue, the gene combinations that have been generated by mutation, recombination, and selection must be protected from disintegration through indiscriminate interbreeding, and it is proposed that isolating mechanisms serve this function. The importance of isolating mechanisms has been emphasized by Dobzhansky, and the following discussion is based largely on his arguments.

Dobzhansky divides isolating mechanisms into two major groups. The first is *geographic isolation*, in which the interbreeding and gene exchange between populations are limited or prevented because the two populations occupy separate geographical areas and do not come into contact with each other. Therefore there is no possibility of gene exchange. The second major group of isolating mechanisms is *reproductive isolation.* In this instance, the populations are not necessarily physically separated but are prevented from exchanging genes through one or more different isolating mechanisms. Table 29-3 lists a number of different isolating mechanisms. Other authors might add one or two other isolating mechanisms to this list.

Table 29-3. Reproductive Isolating Mechanisms

1. *Premating* or *prezygotic* mechanisms prevent the formation of hybrid zygotes.
   - A. *Ecological* or *habitat isolation.* The populations concerned occur in different habitats in the same general region.
   - B. *Seasonal* or *temporal isolation.* Mating or flowering times occur at different seasons.
   - C. *Sexual* or *ethological isolation.* Mutual attraction between the sexes of different species is weak or absent.
   - D. *Mechanical isolation.* Physical noncorrespondence of the genitalia or the flower parts prevents copulation or the transfer of pollen.
   - E. *Isolation by different pollinators.* In flowering plants, related species may be specialized to attract different insects as pollinators.
   - F. *Gametic isolation.* In organisms with external fertilization, female and male gametes may not be attracted to each other. In organisms with internal fertilization, the gametes or gametophytes of one species may be inviable in the sexual ducts or in the styles of other species.
2. *Postmating* or *zygotic* isolating mechanisms reduce the viability or fertility of hybrid zygotes.
   - G. *Hybrid inviability.* Hybrid zygotes have reduced viability or are inviable.
   - H. *Hybrid sterility.* The F_1 hybrids of one sex or of both sexes fail to produce functional gametes.
   - I. *Hybrid breakdown.* The F_2 or backcross hybrids have reduced viability or fertility.

FROM DOBZHANSKY, 1970.

An examination of Table 29-3 indicates that the reproductive isolating mechanisms found toward the bottom of the table involve situations where the populations are probably found together and may even interbreed, but for various reasons no fertile progeny are issued. At the top of the list are isolating mechanisms that tend to separate the two populations and prevent interbreeding, although the separation is not as absolute as we would find for geographic isolation.

In discussing the role of isolating mechanisms in evolution, Dobzhansky argues the following points. First, the buildup of isolating mechanisms takes a very long time. Secondly, it is argued by most evolutionists that *speciation*, which is used often as a synonym for evolution, begins with geographic isolation. A population, for example, might be split into two or more subpopulations by some sort of geographic barrier. If this situation occurs and persists for a long time, it is argued that each now separate population will go its own genetic way. The two populations are likely to assemble separate genetic endowments. If the two populations come together again at some later time, the genetic divergence that almost certainly took place probably results in isolating mechanisms of the zygotic type (bottom, Table

29-3). In other words, the populations may interbreed, but no fertile progeny are produced. If true, it is argued that selection will act very efficiently and rapidly on these two populations to build up prezygotic isolating mechanisms (top, Table 29-3), so that the two populations no longer engage in inefficient and energy-consuming interbreeding that results in no fertile progeny. Rapid selection for prezygotic isolating mechanisms prevents any interbreeding.

In summary, most evolutionists believe that speciation occurs with *allopatric* populations, populations that are separate from each other, in contrast to *sympatric* evolution where speciation occurs when two populations are found in the same area. The buildup of isolating mechanisms takes a long time, and it should be emphasized that most populations are probably isolated from each other not by one but by a combination of several isolating mechanisms.

REFERENCES

CAVALLI-SFORZA, L. L., and W. F. BODMER. 1971. *The Genetics of Human Populations.* 965 pp. W. H. Freeman and Co., San Francisco. (Although the emphasis is on human populations, this is an excellent detailed treatment of modern evolutionary techniques.)

DOBZHANSKY, TH. 1937. *Genetics and the Origin of Species.* 364 pp. Columbia University Press, New York. (A very important and influential book on evolution. It was the first book to present the neo-Darwinian point of view and to synthesize much of the population genetics research of the early 1930s.)

DOBZHANSKY, TH. 1947. Adaptive changes induced by natural selection in wild populations of *Drosophila*. *Evol.* **1**:1–16. (A short and well-written review of the early work on adaptive polymorphism inversions in *Drosophila pseudoobscura*.)

DOBZHANSKY, TH. 1970. *Genetics of the Evolutionary Process.* 505 pp. Columbia University Press, New York. (Dobzhansky's most recent book on genetics and the origin of species.)

DOBZHANSKY, TH., A. M. HOLZ, and B. SPASSKY. 1942. Genetics of natural populations, VIII. Concealed variability in the second and fourth chromosomes of *Drosophila pseudoobscura* and its bearing on the problem of heterosis. *Genetics* **27**:463–90. (A demonstration that natural populations contain much genetic variability.)

HARDY, G. H. 1908. Mendelian proportions in a mixed population. *Science* **28**:49–50. (Available in *Classic Papers in Genetics*, edited by J. A. Peters. Prentice-Hall, Englewood Cliffs, N.J., 1959.) (Hardy's classic contribution to populations genetics.)

KETTLEWELL, H. B. D. 1961. The phenomenon of industrial melanism in Lepidoptera. *Ann. Rev. Entomology* **6**:245–262. (An excellent review of the fact of industrial melanism and some of the experiments testing the hypothesis of natural selection.)

LEWONTIN, R. C. 1974. *The Genetic Basis of Evolutionary Change.* 346 pp. Columbia University Press, New York. (An excellent and up-to-date book on genetics and evolution, with a thorough discussion of the facts and theory of balanced polymorphism in natural populations.)

QUESTIONS AND PROBLEMS

29-1. Define the following terms:

ecological niche
adaptability
adaptive value
gene pool
industrial melanism
polymorphism
species
sympatric
allopatric
carrier

29-2. In qualitative terms, what is meant by the Hardy–Weinberg equilibrium?

29-3. If you analyzed a population for a trait known to be due to a single pair of genes and your analysis revealed that the population was not in equilibrium for that trait, what possible reasons could you offer for the lack of equilibrium?

29-4. Assume a trait controlled by a single gene pair (A, a) and in which there is intermediate inheritance such that all three phenotypes are distinguishable. Two different populations were sampled for this trait with the following results:

	Phenotype			
POPULATION	AA	Aa	aa	TOTAL
I	400	400	200	1000
II	984	832	184	2000

(a) Is Population I in equilibrium?
(b) Is Population II in equilibrium?

29-5. A number of populations were sampled for distribution of the MN blood group. State whether each of these populations is in equilibrium or not.

(a) 100% M
(b) 100% MN
(c) 100% N
(d) 25% M, 50% MN, 25% N

(e) 2% M, 96% MN, 2% N (f) 9% M, 42% MN, 49% N
(g) 20% M, 40% MN, 40% N

29-6. Samples of three large populations of Drosophila mating at random in population cages were found to have the following genotypes:

	Genotype		
POPULATION	AA	Aa	aa
I	57	169	29
II	92	199	77
III	21	183	50

Compare, statistically, each of these distributions with that expected on Hardy–Weinberg assumptions and account for any differences.

29-7. In humans, the Rh blood group is determined by two alleles R,r. Rh negatives are rr, whereas Rh positives may be either RR or Rr. What is the frequency of heterozygotes if the frequency of Rh negatives is
(a) 16%?
(b) 36%?
(c) 100%?

29-8. What is the evidence that populations are heterozygous for many recessive genes?

29-9. What is the significance of the fact that many populations are heterozygous for many recessive genes?

29-10. What is the significance of recombination for evolution?

29-11. What does the example of industrial melanism teach us about evolution?

29-12. There are some people who take antibiotics frequently, often for rather mild symptoms, such as a cold. Is there any danger in the indiscriminate use of antibiotics? Explain your answer.

29-13. What is meant by adaptive or balanced polymorphism?

29-14. Give an example of adaptive polymorphism and indicate its importance in evolution.

29-15. Discuss the significance of isolating mechanisms in evolution.

29-16. Many species of ducks, if brought together by man, are known to produce successful hybrids. Does this fact affect your answer about the significance of isolating mechanisms?

29-17. The Indian water buffalo and the African water buffalo, generally considered two separate species, have been known to hybridize successfully in zoos. Does knowledge of this fact change your understanding of what a species is?

29-18. Goldschmidt proposed a theory ("saltation" theory) in which he believed that evolutionary changes came about suddenly by large "macromutations." For example, he suggested that birds evolved from reptiles by the sudden formation of a "very birdlike reptile." What objections might a neo-Darwinian present to Goldschmidt's theory?

29-19. Assume that a large population of organisms has been evolving in a stable environment for one million generations. Assume that after this period of time *no more* mutations occur. The mutation rate is now zero. Finally, assume that the environment begins to change. What is likely to be the fate of the population and why?

29-20. Consider two islands, on each of which are located large populations of a single species of animals.

(a) On Island 1 the population has been evolving for a million years in a relatively stable environment. Suddenly the environment changes drastically. Assuming the population adapts to the new environment, what evolutionary processes would you say were responsible for the adaptation?

(b) On Island 2 the population suddenly becomes split into two parts by a mountain range. After very many generations the two parts of the original population come together again. What do you think the nature of this reformed population might be? Indicate the evolutionary processes involved in your answer.

30 / Genetics and evolution

PART II

In Chapter 29 the four main processes of evolution were discussed. Absent from the discussion was any mention of the importance of population size. This subject has occupied the interest of several investigators, but it is primarily the American geneticist Sewall Wright who developed the most elegant and complete treatment of the subject. Wright, in the early 1930s, in a largely mathematical paper, showed the theoretical importance of population size. His argument will be presented non-mathematically in a slightly different way.

Genetic Drift

Assume two islands that are exactly similar in environment. Assume that on each island is placed one self-fertilizing plant that is heterozygous for one pair of genes,

$\mathbf{A}^1\mathbf{A}^2$. Further assume that regardless of the number of F_1 plants that might be produced, on each island only one plant survives. What will be the genotype of the surviving plant on each of the islands, and if a proliferation of plants occurs in subsequent generations, what will be the genotypes of all the plants on each of the islands?

From this oversimplified example it can be shown that the genotype of the single F_1 plant will be determined largely by chance. If many F_1 progeny were produced, the Mendelian distribution of $1\mathbf{A}^1\mathbf{A}^1 : 2\mathbf{A}^1\mathbf{A}^2 : 1\mathbf{A}^2\mathbf{A}^2$ would be expected; but if only one plant survives, and it is assumed that there is no differential selection for genotypes on either island, then the surviving genotype on each island will be determined by random processes alone.

For example, suppose that, by chance, the genotype that survives on island 1 is $\mathbf{A}^1\mathbf{A}^1$, and further suppose that, by chance, the genotype that survives on island 2 is $\mathbf{A}^2\mathbf{A}^2$. Rather dramatically, the genotypes on the two islands have been fixed and now differ significantly from each other. Stated another way, if the population size is small, there will be random and rapid fluctuation of gene frequencies because of chance alone. This chance fixation of genotypes has been called *genetic drift.*

A few geneticists have used the concept of genetic drift to provide an explanation for what are called *nonadaptive* characters. These traits seem, on the surface, to be unaffected by natural selection. It is very difficult to prove that any trait is truly nonadaptive or neutral with respect to selection, but if there were, differences in the frequency of genes could be accounted for by invoking the concept of genetic drift.

What are the consequences of genetic drift? It would seem to suggest that an understanding of population size is important for the study of evolutionary processes. Figure 30-1 shows the relation of population size to selection and mutation. If the population is small, selection plays a relatively minor role in changing gene frequencies, and drift becomes the more important factor. Conversely, if the population is large, drift plays a minimal role, and selection will be the more important factor changing gene frequencies.

I. Comparison of selection and population size:
   If population is small: Drift (selection)
   If population is large: (drift) Selection

II. Comparison of mutation and population size:
   If population is small: few mutations; species will be constant with little or no evolution. Tends to lead to extinction.

   If population is large: many mutations. This produces a reservoir of genetic variability upon which selection can act.

Figure 30-1. A comparison of population size with selection and with mutation assuming, for simplicity, either a small or large population.

A comparison between mutation and population size results in the following conclusions. If the population is small, there will be few mutations. The species will tend to stay genetically constant, and there will be little or no evolution. Often the population will become extinct. On the other hand, if the population is large, there will be many gene loci available and hence many mutations. The result will be a reservoir of genetic variability on which selection can act.

An examination of the simplified examples given in Figure 30-1 reveals a paradox. On the one hand, there is rapid evolution due to genetic drift if the population is small; but since small populations may have little genetic variability, there is not much variation for drift to act on. On the other hand, if the population is large, there is considerable variability, but selective pressures act slowly compared to genetic drift. What, then, is the ideal population size for rapid evolution?

Population Structure

Wright, among others, has considered that question and proposed a model. Whether the model is correct or not is not known. But it is important to realize that population size plays a role in evolution and that geneticists have tried to analyze evolution by examining population structure.

Wright resolved the paradox by proposing a population structure as shown in Figure 30-2. He proposed a number of relatively small populations that are connected with each other through random but infrequent migrations. The totality of the population is large, so that the variability present in toto is considerable; but each subpopulation is relatively small, so that drift allows for rapid gene fluctuations.

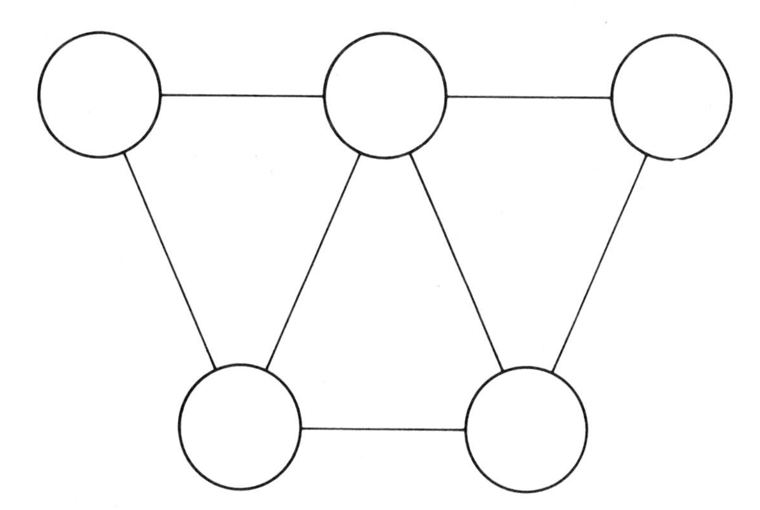

Figure 30-2. Sewall Wright's model of population structure for the most rapid evolution. The circles represent relatively small populations. The total population size including all the circles is large. The lines connecting the circles represent infrequent migrations from one isolate to another.

Should random fixation lead to a deleterious genotype in one of the isolates, then the subpopulation is likely to become extinct. But the niche will not be lost to the population because the few migrants from other isolates will move in and repopulate it. By a similar argument it can be demonstrated that if drift succeeds in providing one or more favorable genotypes, they will not be confined to the single niche but can be spread throughout the entire population by migration. The reader might try to construct other population models that would be efficient for rapid evolution.

It might be asked why genetic drift is discussed here rather than with the four major processes in Chapter 29. The answer is that it is hard to establish the fact that drift has played an important role in the evolution of living organisms. The difficulty lies in the fact that it is not easy to make accurate estimates of population sizes that occurred a long time ago. It has been shown experimentally that if population size is small, then drift plays a major role in the change of gene frequencies. Whether it *has* played a major role in the evolution of living forms is still somewhat conjectural.

Polyploidy

In 1928 the Russian cytogeneticist Karpechenko crossed a radish, Raphanus, by a cabbage, Brassica, in his laboratory. As might be expected from an intergeneric cross, the F_1 plants that he obtained were mostly sterile. It may come as a surprise that any hybrid is obtained, but for some time it has been known that, in the laboratory, interspecific or intergeneric crosses may yield a sterile hybrid. However, under laboratory conditions, Karpechenko kept trying for fertile seeds from the F_1 hybrid and in a few cases was able to obtain F_2 plants.

These F_2 plants were rather interesting. They showed no segregation, although segregation would be expected from a cross of F_1 by F_1. (Recall Mendel's dihybrid crosses.) Secondly, the F_2 were phenotypically similar to the F_1, again a rather unusual and unexpected result. And, finally, the F_2 were fertile. It was possible to cross them and obtain F_3 and later generations with no great difficulty. The F_3 obtained also showed no segregation, were similar phenotypically to the F_2, and were fertile. Karpechenko called the hybrid Raphanobrassica, but he was disappointed in his hope of combining the good qualities of the radish with the good qualities of the cabbage. Unfortunately, just the opposite happened, and Raphanobrassica proved to have little economic value.

His results, however, are of great interest. Because he was a cytogeneticist, Karpechenko was interested in chromosome numbers. He knew that the diploid number of Raphanus was 18 and that the diploid number of Brassica was also 18. We would expect the F_1 plant to have 18 chromosomes and it did. When Karpechenko made a cytogenetic analysis of the F_2 plants, he observed that there were 36 chromosomes, twice the expected number.

Karpechenko analyzed the situation in the following way. From meiosis in the parentals a Raphanus gamete is produced that has nine chromosomes, symbolized 9R, and a Brassica gamete that also has nine chromosomes, symbolized 9B. Because these forms have evolved separately for a long time, it is not surprising that they are genetically unlike; and it is postulated that, in meiosis of the F_1 plant, the nine Raphanus and the nine Brassica chromosomes are sufficiently dissimilar genetically that they do not pair with each other. The usual synapsis of homologous chromosomes expected in meiosis is not obtained in this intergeneric hybrid. Because of the failure of synapsis, the meiosis is disorderly and chaotic, and the chromosomes segregate randomly to the poles, almost always resulting in duplication-deficiency gametes that are nonfunctional. Occasionally, however, there is an abortive division in the production of gametes by the F_1 plant such that a few exceptional unreduced gametes may be formed that have 18 chromosomes, 9R and 9B. Should two such gametes result in fertilization, an F_2 plant will be formed that has 36 chromosomes, 9R + 9R + 9B + 9B. The F_2 plant is expected to show no segregation and to be phenotypically identical to the F_1 from which it came. The plant should be perfectly fertile because when *it* undergoes meiosis, the chromosomes each have a homolog to pair with.

Reproductive Isolation

Another interesting feature of Raphanobrassica is that when it is crossed back to either parent, Raphanus or Brassica, the cross fails to yield fertile progeny. If crossed to Raphanus, for example, a hybrid is obtained, but it produces no fertile seed. Since the hybrid has 9R chromosomes from the Raphanus parent, 9R chromosomes from Raphanobrassica, and 9B chromosomes from Raphanobrassica, and since during meiotic synapsis the 9B chromosomes will have no homologs, thus resulting in a disorderly meiosis, only nonfunctional gametes are produced.

It can be said, therefore, that Raphanobrassica is reproductively isolated from each of the two parental species. By definition, there has been produced in the laboratory, in the short time of a few generations, a new species. Some scientists have described this process as cataclysmic evolution! Raphanobrassica is called a *tetraploid*, for the new plant has four times (36) the basic number (9) of chromosomes.

Causes of Reproductive Isolation

Why is the F_2 hybrid reproductively isolated from the parents? One reason, as stated above, is that any hybrid formed cannot produce functional gametes because of meiotic difficulties. But a number of physiological reasons can prevent the formation of any hybrid. One of the first polyploids studied was Oenothera by De Vries early in the twentieth century. Oenothera has a diploid chromosome number of 14. De Vries discovered a tetraploid form in which the diploid number was 28. The 28-chromosome form was found to be reproductively isolated from the 14-chromosome form. De Vries recognized that, in the normal cross, a pollen grain that is haploid grows through a style that is diploid. In Oenothera, however, depending on how the cross is made, there is either a diploid pollen grain growing through a diploid style (if the maternal plant is the diploid and the paternal plant is the tetraploid) or, in the reciprocal cross, there is a haploid pollen grain growing through a tetraploid style.

De Vries recognized that there may also be a physiological relationship between the embryo and endosperm that could, in certain cases, lead to incompatibility. In the normal situation, a $2n$ embryo develops in conjunction with a $3n$ endosperm. In a cross of a diploid by a tetraploid, there is, again depending on how the cross is made, a $3n$ embryo growing with a $4n$ endosperm (if the maternal parent is the normal diploid) or a $3n$ embryo growing with a $5n$ endosperm (if the maternal parent is the tetraploid). In addition to these problems, there may also be physiological difficulties in seed formation.

Geneticists have recognized two kinds of tetraploids. The radish by cabbage cross in which the hybrid is derived from two different diploids is called an *allotetraploid.* Also possible, and known to occur in plants, is a tetraploid that arises from multiplication from within a single diploid so that all the homologs come from the same source. This tetraploid is called an *autotetraploid.*

Polyploidy in Nature

Is there any evidence in nature for speciation through polyploidy similar to what Karpechenko was able to achieve in the laboratory? If such examples are looked for, it is quickly realized that many are available. One of the earliest and best-described examples was provided by the Swedish geneticist Müntzing and involved the genus Galeopsis, the hemp nettle. Galeopsis is a well-known species of plant studied originally by Linnaeus, who described three species, *Galeopsis pubescens, Galeopsis speciosa,* and *Galeopsis tetrahit.* In the laboratory, Müntzing crossed *pubescens* by *speciosa.* He knew that each had a diploid chromosome number of 16. As might be expected from an interspecific cross, the F_1 were almost all sterile. But with the patience of the laboratory experimenter, he continued the cross and eventually was able to obtain an F_2 plant. The F_2 plant proved to be fertile, showed no segregation, looked like the F_1, and had 32 chromosomes. More interestingly, it looked like *Galeopsis tetrahit,* a naturally occurring species.

The laboratory tetraploid was crossed by *tetrahit* and was shown to hybridize very readily, yielding fertile progeny. Müntzing came to the inescapable conclusion that *tetrahit* in nature arose as an allopolyploid from *pubescens* by *speciosa.*

Prior to the experiment by Karpechenko, two tobacco geneticists, Goodspeed and Clausen, in 1925 discovered what is recognized as the first laboratory case of polyploidy. They were crossing *Nicotiana tabacum,* which has a haploid chromosome number of 24, by *N. glutinosa,* which has a haploid chromosome number of 12. They obtained an F_1 that had a chromosome number of 36 and that was sterile as expected. However, one fertile exception showed a phenotype typical of a plant that had doubled its chromosomes. Cytological examination verified 72 chromosomes, 24 *tabacum* pairs plus 12 *glutinosa* pairs. The hybrid was called *N. digluta,* and all the gametes formed by it had 36 chromosomes.

How did this F_1 plant arise so that it contained 72 chromosomes? The assumption is that, during mitosis, the spindle accidentally broke down. Afterward the nuclear membrane was reconstituted, but since the chromosomes had already replicated, there were now 72 chromosomes. Subsequently, normal cell divisions produced a growing point that, due to the somatic doubling, now has 72 chromosomes. If a branch develops from it, the reproductive parts will have 72 chromosomes, and meiosis will be orderly, producing fertile seeds.

Winge Hypothesis

In 1917 Winge, on theoretical grounds, postulated that polyploids may originate by partial failure of cell division. The "Winge hypothesis" received experimental verification from the experiment of Goodspeed and Clausen. Experimentally, it is now possible to induce chromosome doubling with substances such as colchicine, which act as mitotic poisons and interfere with normal spindle function. Figure 30-3 shows a few examples of polyploid series in plants.

PLANT	HAPLOID NUMBER	POLYPLOID SERIES
Triticum (wheat)	(7)	14, 28, 42
Chrysanthemum	(9)	18, 36, 54, 72, 90
Solanum (night shade, potato)	(12)	24, 36, 48, 60, 72, 96, 108, 120, 144
Papaver (poppies)	(7)	I: 14, 28, 42, 70
	(11)	II: 22, 44

Figure 30-3. Some polyploids in plants. The number in parentheses is the haploid or basic number of the series.

Polyploidy in Animals

If, as seems probable, polyploidy has played an important role in evolution, why was it not considered a major factor in evolution in Chapter 29? Speciation by polyploid is treated as another "special case" because the evidence suggests that it is confined principally to plants. If there were no polyploidy in nature, it would be expected that there are as many organisms with an odd number of chromosomes as with an even number. If, however, polyploidy has occurred, there should be more species with an even number of chromosomes. For example, assume two species, one with a haploid chromosome number of 3, the other with 4. If polyploidy occurs in each, the result will be two forms with a chromosome number of 12 and 16 (2×6, 2×8); and the haploid number for these two forms will be 6 and 8, respectively, *both* even. Figure 30-4 shows some haploid chromsome numbers in plants and animals. It is interesting that in both forms there are more species with an even number of chromosomes, but the situation is much more pronounced in plants.

Why is polyploidy rarer in animals? Although this question cannot be answered with certainty, several possibilities exist. Polyploidy is expected more

	Haploid Chromosome Number	
	EVEN	ODD
Plants	1646	767
Animals	644	426

Figure 30-4. A comparison between plants and animals of haploid chromosome number. If chromosome number is random there should be expected as many forms with an odd number as an even number. If polyploidy has played a role there should be more even numbers, for two times an odd or an even number is an even number. Note that there are proportionally many more plant species with an even number than animal species.

frequently in hermaphroditic forms than in species with separate sexes. The reason is that two unreduced gametes are needed in order to form a tetraploid. The probability of this happening separately in a male and female with subsequent fertilization is very rare. However, in a plant that produces both male and female gametes, only one event is necessary to form unreduced gametes (see the Nicotiana example of Goodspeed and Clausen). It is revealing, perhaps, that the majority of animal species in which polyploidy is suspected are hermaphroditic forms.

Another possible deterrent to speciation by polyploidy in animals might be the sterility of certain triploid forms. If fertilization results between a normal ($1n$) and an unreduced ($2n$) gamete, a triploid individual ($3n$) is formed. In Chapter 27 it was pointed out that in some animals, such as Drosophila, there is no such thing as a triploid male. The nearest to it is a 2XY 3A individual that is a sterile intersex. So various sex-determining mechanisms in animals may act as barriers to speciation by polyploidy.

REFERENCES

DOBZHANSKY, TH. 1937. *Genetics and the Origin of Species.* 364 pp. Columbia University Press, New York.

DOBZHANSKY, TH. 1970. *Genetics of the Evolutionary Process.* 505 pp. Columbia University Press, New York. (Both books, the second essentially a fourth edition of the first, give good descriptions of genetic drift and the importance of population size.)

KARPECHENKO, G. D. 1928. Polyploid hybrids of *Raphanus sativus* L. x *Brassica oleracea* L. Zeitschrift für induktive Abstammungs-und Veresbungs lehre **48**:1–85. (A description of the laboratory-produced species *Raphanobrassica.*)

MAYR, E. 1963. *Animal Species and Evolution.* 795 pp. Harvard University Press, Cambridge, Mass.

STEBBINS, G. L. 1950. *Variation and Evolution in Plants.* 643 pp. Columbia University Press, New York. (Both books are useful complements to Dobzhansky's books on evolution. Both present the neo-Darwinian point of view, one by an animal systematist and the other by a plant geneticist.)

QUESTIONS AND PROBLEMS

30-1. Define the following terms:

genetic drift
tetraploid
allotetraploid
autotetraploid

30-2. Does genetic drift play a more important role in a large or a small population? Explain.

30-3. Does selection play a more important role in a large or a small population? Explain.

30-4. Discuss briefly the influence of population size as an evolutionary force.

30-5. Assume that on an island (I) a *single, annual self-fertilizing* plant that is heterozygous for one locus, Aa, is introduced and that on another environmentally identical island (II) many such plants are introduced. Assume further that on Island II for many generations only one individual offspring survives but that, finally, multiplication occurs, whereas on Island II always many plants survive. After a thousand years, what genotypes do you expect

(a) on Island I?
(b) on Island II?

30-6. Assume that on each of a hundred environmentally identical islands a *single, annual* self-fertilizing plant heterozygous for two independent loci (AaBb) is introduced. Assume further that for very many generations only one individual offspring survives on each island but that, finally, multiplication occurs. What genotypes will be found on the different islands?

30-7. Assume that on an island a large population of a single species of animals has been evolving for a million years. Suddenly the population is reduced to a very small size. What might happen to this population? Indicate the evolutionary processes involved in your answer.

30-8. Agronomists often try to cross two different but related species of crop plants and attempt to obtain an allotetraploid from such diploid hybrids.

(a) Why are they interested in obtaining these allotetraploids?
(b) Why is the diploid hybrid not used?

30-9. Discuss briefly the influence of polyploidy as an evolutionary force.

30-10. The allopolyploid *Raphanobrassica* has 36 chromosomes in its somatic cells. In backcrosses to Raphanus ($2n=18$), it gives rise to a few hybrid plants. What do you expect the pairing of chromosomes to be in the meiotic divisions of these hybrids?

30-11. A new species of grass, *Spartina townsendi*, was discovered and it proved to be an allotetraploid derived from the native species *stricta* ($n = 28$) and a recently introduced species *alterniflora* ($n = 35$).

(a) What is the diploid chromosome number of *S. townsendi*?
(b) Outline its probable mode of origin.

30-12. An allotetraploid is formed from species A ($2n = 16$) $\times$ species B ($2n = 20$). The allotetraploid is crossed back to species A and a "sterile hybrid" is formed.

(a) How many chromosomes does the "sterile hybrid" have?
(b) How will these chromosomes pair during meiosis?
(c) Why is the hybrid sterile?

30-13. The mule, which is the offspring of a male donkey and a mare, is a vigorous animal, well suited for hard work. However, it is sterile. Why?

30-14. Chrysanthemums are known with diploid chromosome numbers of 18–36–54–72–90. Could such species have arisen by a process of autopolyploidy? Why or why not?

Index

A

D

E

F

G

N

T